Soils of Wisconsin

With contributions by
Marvin T. Beatty
Gerhard B. Lee

Illustrations by
R. D. Sale
G. W. Stanley
M. L. Czechanski
R. J. Olmstead
F. D. Hole

Published for the Geological and Natural History Survey,
University of Wisconsin-Extension

Soils of Wisconsin

Francis D. Hole

The University of Wisconsin Press

Bulletin 87
Soil Series 62

University of Wisconsin-Extension
Geological and Natural History Survey
Soil Survey Division

M. E. Ostrom, State Geologist
F. D. Hole, in charge, Soil Survey Division

In cooperation with
Department of Soil Science, College of Agricultural and Life Sciences
and
Soil Conservation Service and Forest Service, USDA

Published 1976
The University of Wisconsin Press
Box 1379, Madison, Wisconsin 53701
The University of Wisconsin Press, Ltd.
70 Great Russell Street, London

First printing

Printed in the United States of America

For LC CIP information see the colophon

ISBN 0-299-06830-7

Contents

APPENDICES

List of Illustrations

List of Tables

Preface

Observations of soils of Wisconsin have been recorded by a great number of specialists in soil genesis, classification, and mapping, starting with T. C. Chamberlin in the 1880s. This publication makes available to the public a comprehensive summary of information thus far gathered on the soil resources of the state. It includes definitions of terms and concepts relating to factors and processes of soil formation, soil properties, modern soil classification, and descriptions of major soil regions of Wisconsin and of the specific kinds of soils that characterize them. A map sequence, beginning with an index map of Wisconsin (Fig. 1-1) and continuing through a series of pedogenic factor maps (Plates 2 through 6) to the final soil map (Plates 1 and 7), presents geographic relationships of soils to the environment.

The importance of identifying and grouping soils in scientific and applied classification systems lies in the functions that the soils perform in the landscape. Soils redistribute precipitation, filter and decontaminate aqueous solutions and suspensions, make mineral nutrients available to plants and animals and people, and support structures such as roads and buildings.

The variety of soils in a given area is astonishing. The difficulties of characterizing, classifying, and mapping so many soils are great—"There are few subjects," T. C. Chamberlin wrote in 1883, "upon which it is more difficult to make an accurate and at the same time intelligible report than upon soils." To cope with these difficulties, procedures of soil analysis, classification, and cartography have been developed by the international fraternity of soil survey specialists, in which Wisconsin workers have long been active. The soil scientist excavates the object of his inquiry, but only at points carefully selected on the basis of long experience and spaced closely or widely depending on the high or low intensity of land use envisaged. Soil survey information is reported in this book in both general and technical terminology. To help the reader with the latter, the principal soil classification terms are arranged in a chart, Fig. 4-2, which appears in Chapter 4, and are explained in the Glossary, Appendix 1.

A basic understanding of the soils of Wisconsin helps the observer to "read" each landscape for practical purposes or simply for the pleasure of it. Every site has a particular combination of geologic materials that were shaped long ago into hills, plains, and valleys. A layer of soil several feet thick covers the hills like a fragile fabric, many features of which have been impressed on the land by the action of climate, plants, and animals in the course of many thousands of years. The present vegetative cover preserves it from erosion by wind and water. The soil itself serves as a reservoir of nutrients and moisture to support plant growth, and as a filter for water percolating to the water table. Sound conservation practices annually reduce soil erosion losses, even on cropland, to less than one thousandth of the plow layer.

The people of this state have the power to control the land-use pattern and to adapt it to the capabilities of the hundreds of different kinds of soils. We in Wisconsin have long been concerned with good stewardship of the land. The purpose of this publication is to provide a tool for the effective discharge of this responsibility.

Acknowledgments and Previous Reports

Many persons have contributed information and assistance as this report was prepared. The help and encouragement of George F. Hanson and Meredith E. Ostrom, State Geologists and Directors (1953-1972 and 1972——, respectively), Geological and Natural History Survey, and their critical review of the report have been invaluable in bringing this work to completion. Helpful suggestions from the following workers are also acknowledged with gratitude: Perry Olcott, Johannes Bouma, and Marvin T. Beatty of the Geological and Natural History Survey; Gerhard B. Lee of the College of Agricultural and Life Sciences; A. J. Klingelhoets, P. H. Carroll, and R. E. Fox of the U.S. Soil Conservation Service. The list of soil scientists who have worked in the field and laboratory during a span of three quarters of a century would be long indeed, as the bibliography indicates. If this book were dedicated to anyone, it would be to them, who, in the words of F. H. King (1911), walked through the fields and forests "to learn by seeing." Important observations have come, via the author's field notebooks, from colleagues at other land grant institutions and from soil correlators of the U.S. Soil Conservation Service, including the following: J. K. Ableiter, William DeYoung, R. B. Grossman, Lacy Harmon, C. E. Kellogg, I. J. Nygard, A. H. Paschall, R. W. Simonson, and G. D. Smith. Wisconsin county soil correlation reports of the National Cooperative Soil Survey have been useful.

The high quality of execution of illustrations and maps is the contribution of R. D. Sale, M. L. Czechanski, and coworkers of the University of Wisconsin Cartographic Laboratory at Madison.

Grateful acknowledgment is made of the advice on the manuscript given by Donald F. Kaiser, Director of Editorial and Library Services of the University of Wisconsin-Extension, and Thompson Webb, Director of the University of Wisconsin Press, and of the perceptive and meticulous editing done by editors Elizabeth Steinberg of the Press and Linda B. Halsey and Harriet M. Shetler of Extension.

In addition to the earlier reports on the soils of the state by T. C. Chamberlin (1883) and A. R. Whitson (1927), one by Wilde, Wilson, and White (1949) on silvicultural aspects of our soils has been helpful. The writer has consulted the work of Curtis (1959) on vegetation of Wisconsin, and Martin's (1932) *Physical Geography of Wisconsin,* to which readers are referred.

M. L. Jackson and coworkers have made important contributions to the understanding of the mineralogy of Wisconsin soils.

Part I

Introduction to Genesis and Classification of Wisconsin Soils

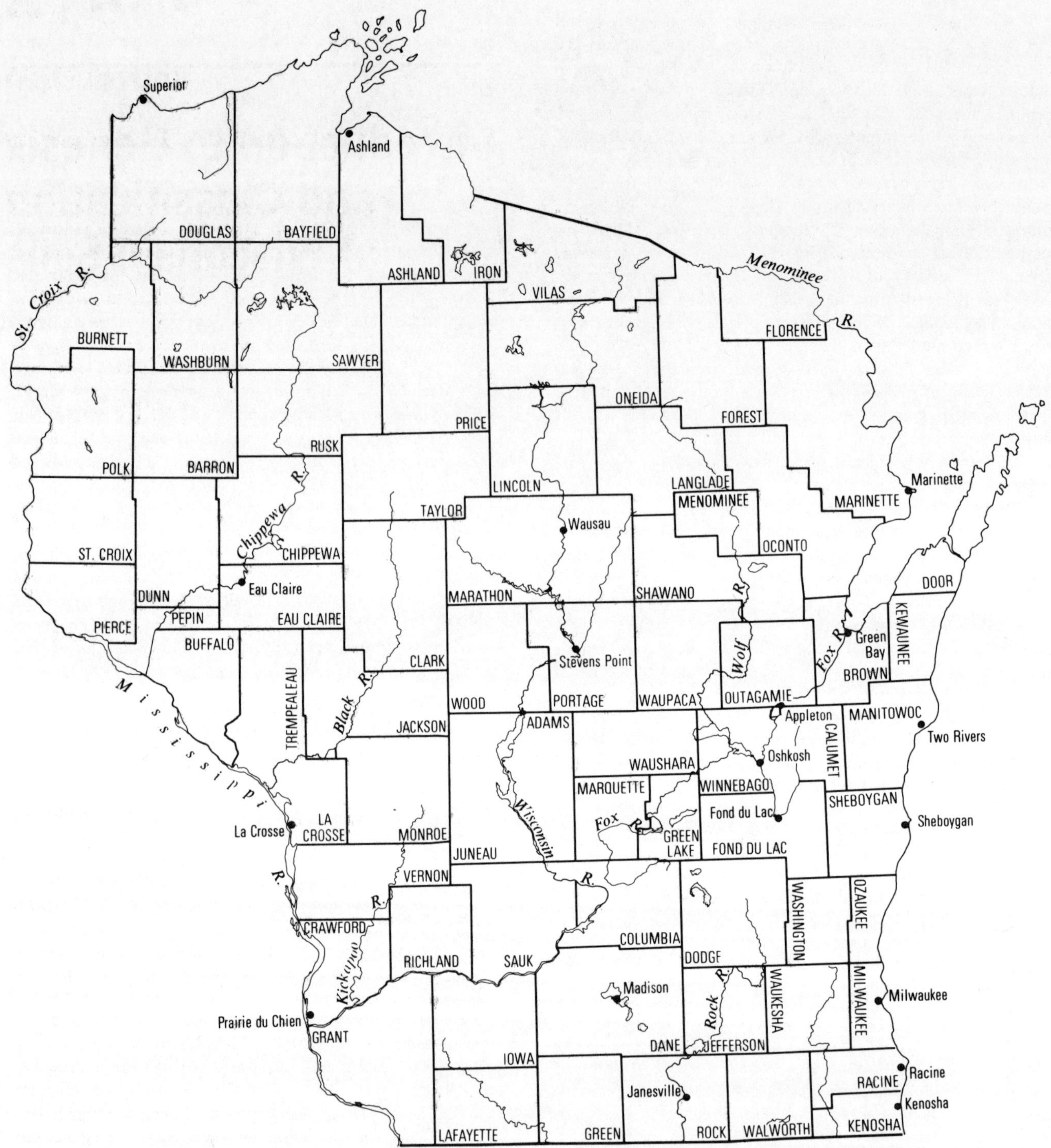

Figure 1-1. Index map of Wisconsin, showing main rivers, county boundaries, and principal cities.

CHAPTER 1

Introduction

THE SOILS OF WISCONSIN

The soils of Wisconsin constitute a layer as much as 4 or 5 feet thick that has a dry weight of nearly 300 billion tons and covers about 35 million acres. We are not used to thinking of our soil resource in terms of tonnages as we think of gravel deposits and ores. The fact is that soil is a prime organo-mineral resource of importance to the stability of buildings and highways, the economy of our cities, the productivity of our farms and forests, the quality of our nutrition, and the vitality of wildlife and wilderness areas. Eroded soil materials in the nearly one million acres of Wisconsin now under lakes and streams influence water quality by yielding or by adsorbing pollutants. The soils of this state have been the subject of extensive soil investigation for nearly a century, during which time a vast amount of information on specific areas and soils has been reported. Three colored wall maps of the soils of the state have been published (in 1882, 1927, and 1968). Our present knowledge of the soils of Wisconsin, of which this report is a summary, is a basis for sound land management and for future research in pedology and soil-water conservation.

Areal analysis of the soil map (Plate 1) indicates that Wisconsin subsoil materials are deep (more than 2 feet thick over consolidated bedrock in 95% of the state). A third of the soil area is derived from glacial outwash sand and gravel, a third from glacial till loams, a tenth from deposits in ancient glacial lakes, and a tenth from bedrock-derived residuum. Wetland soils (both organic and inorganic) are shown on the map to cover nearly 10% of the state, but bodies too small to show probably more than equal that. About 40% of the land is covered with a foot or more of weathered loess, from which some of the most productive soils have formed. Wisconsin is crossed by a southeast-trending climatic and ecological tension zone that separates cool-summer forest soils on the northeast from warm-summer prairie and prairie-forest transition soils on the southwest. Forest soils have formed on two thirds of the area, and prairie- and savanna-influenced soils have developed on the remaining third. Clayey soils cover about 10% of the state; silty soils, about 40%; loams and sandy loams, 25%; sands, 20%; and peats and mucks, about 5%.

The state has nine major soil regions, with interspersed wetlands constituting a tenth unit. (A) The southwestern region is a two-story landscape with productive ridge and valley soils separated by wooded steep shallow soils and rockland. The great variety of soils and topography contributes to a scenic landscape, but demands careful land and water management to handle problems of erosion and waste disposal. (B) The southeastern upland has an intricate soil pattern of wetlands and plains, hills, drumlins, and ridges, including the prominent Kettle Moraine. The contribution of dolomitic limestone to subsurface materials is associated with a high level of native subsoil fertility. (C) Soils of the central sandy uplands and plains inherit their coarse texture from Cambrian sandstone and deposits of sandy outwash from the melting of glaciers thousands of years ago. Buried silts and clays of ancient lake beds, rocky sandstone crags of scattered mounds and mesas, and intermixtures of dry and wet soils are features of the landscape of the central sandy region. (D) Soils of the western sandstone uplands, valley slopes, and plains include shallow silt loams and deep sandy loams, some steep stony land, and imperfectly drained flats. (E) Northern and eastern sandy and loamy reddish drift uplands and plains have a mixture of sands, reddish silts, and clays, interlayered in some areas. (F) The northern silty uplands and plains are characterized by shallow to moderately deep silty soils over compact acid loamy and sandy glacial drift. Soil drainage is a problem in many places. (G) The vast northern loamy uplands and plains consist of acid stony and somewhat sandy soils on glacial moraines, drumlins, eskers, and outwash flats. (H) The northern sandy uplands and plains include sands of extensive pine and oak barrens on pitted glacial drift and nearly level outwash plains. (I) Soils of the northern and eastern clayey and loamy reddish drift uplands and plains have high clay content and plant nutrient levels. Management problems on these soils include tillage, drainage, and disposal of liquid wastes. (J) Stream bottoms and wetlands of the state are occupied by mineral soils and peats and mucks with high water tables. Naturally well-drained alluvial soils are not extensive.

DEFINITION OF SOIL

Soil is considered here as a natural unit in a pedological (soil science) or ecological sense, rather than in an engineering sense. It is geologic material altered by plants and animals and climate. Soil is a body of mineral and organic matter occurring naturally on the land surface of the earth, interrupted by bodies of "not-soil," which include rock outcrops, lakes, and streams. The soil envelope of the earth, sometimes called the *pedosphere,* lies at the interface between the biosphere and the lithosphere. It is a blend of the biologic and geologic materials. The pedosphere may be thought of as a mosaic, made up of specific soil bodies. In plowed fields, dark bodies of soil commonly show up distinctly in a setting of higher lying, light-colored soil bodies. Thus, examples of the basic unit of soil classification, the soil body, may be easily seen in the landscape when the fields are newly plowed. Soil bodies in Wisconsin range from 1 to 7 feet deep, and are commonly 4 feet thick. The lower surface of a soil body is marked by the limit of leaching of naturally occurring lime (carbonates) in many soils and by the lower limit of common rooting of native perennial plants in all soils. Loose material below this level is not considered soil in the pedological sense, as indeed loose material on the moon is not.

Because soil bodies are hundreds or thousands of feet across,

Table 1-1. A brief explanation of common soil horizon designations

Major categories of horizons[a]		Soil horizons in grassland soils (Mollisols)	Soil horizons in deciduous forest soils (Alfisols)	Soil horizons in hemlock forest soils (Spodosols)
Organic		O1 Litter layer. Dead vegetation (grasses and forbs)	O1 Forest litter layer. Fallen leaves, twigs, seeds, branches, and other plant and animal materials	O1 Forest litter layer. Fallen needles, twigs, cones, branches, logs, and other plant and animal materials
		O2 Humus. Thin black residue from decomposition of O1 material	O2 Humus. Thin black residue from decomposition of O1 material	O2 Humus. Black residue from decomposition of O1 material
Mineral	Of solum (A and B)	A1[b] Dark surface soil. A mixture of mineral and humidified organic matter	A1[b] Dark surface soil. A mixture of mineral and humidified organic matter	A1[b] —— Usually absent ——
		A2 —— Usually absent ——	A2 Lighter colored subsurface soil out of which some fine material has been washed	A2 Light-colored subsurface soil that has been bleached by downward-percolating water
		A3 or AB Transitional horizon	A3 or AB Transitional horizon	A3 or AB Transitional horizon
		A&B Zone of interpenetration of A and B	A&B Zone of interpenetration of A and B	A&B Zone of interpenetration of A and B
		B1 Transitional horizon	B1 Transitional horizon	B1 Transitional horizon
		B2t[c] Clay-enriched subsoil layer; much of the clay washed down from the A horizon	B2t[c] Clay-enriched subsoil layer; much of the clay washed down from the A horizon	Bhir Subsoil layer enriched in organic matter (h) and iron oxide (ir)
		B3 Transitional horizon	B3 Transitional horizon	B3 Transitional horizon
Mineral	Of subsolum	C Underlying mineral horizon like that from which the A and B horizons formed	C Underlying mineral horizon like that from which the A and B horizons formed	C Underlying mineral horizon like that from which the A and B horizons formed
		IIC Underlying mineral horizon unlike that from which the A and B horizons formed	IIC Underlying mineral horizon unlike that from which the A and B horizons formed	IIC Underlying mineral horizon unlike that from which the A and B horizons formed
		R Bedrock	R Bedrock	R Bedrock

[a]See Soil Survey Staff, 1951.

[b]Where buried, this is called the A1b horizon. Where the A1 and/or A2 horizons are plowed, an Ap horizon results.

[c]This may be subdivided into B21t, B22t, B23t, from the top of the B2t horizon downward. Where strongly cemented, the designation is B2tm. Where weakly cemented, as in a fragipan, the designation B2tx is used. In the lower sequum of bisequal soils, prime symbols are added, viz. A′ and B′ (not shown in the table). In this volume the symbol (B) is used sometimes to indicate a cambic (weak) B horizon.

they are studied in depth at only a few selected points. Emphasis is placed on relatively small columns, called *pedons* (see Chapters 4 and 5), and their cross sections, called *soil profiles* (Fig. 4-2). The contrasting layers of soil are called *soil horizons.* The topsoil is designated as the A horizon, the subsoil is called the B horizon, and the underlying geologic material is the C horizon. There are several subdivisions of these horizons, as shown in Table 1-1 (see Fig. 4-2). An observer can usually break a soil horizon into smaller, fragile units. These are called *peds.* Blocky peds, about the size of the thumb, are common in many subsoils in Wisconsin.

Figs. 1-3 through 1-8 are sketches of some representative soil profiles, showing horizons and patterns of peds in the horizons. These features are discussed more fully in Chapters 4, 5, and 17.

SOILSCAPES

The soils of Wisconsin consist of all the soil bodies in the state. They occur as populations of soil bodies arranged on the land surface in characteristic patterns. The term *soilscape* is proposed (see Chapter 6 and the Glossary) to designate a particular natural association of soils in that complex of soils, vegetation, animals, and cultural features that we call landscape. The terms *soilscape* and *soil association* are used interchangeably in this book. A particular soilscape is characterized by a topographic sequence of soils downslope from the top of a hill or ridge to and including an adjacent depression[1] (Figs. 2-51, 2-52). Land operators have to take into account the nature

1. Although the toposequence is the most common kind of soilscape in Wisconsin, the lithosequence and the biosequence are also represented. See Appendix 1, Glossary.

of the soils and these groupings of them, whether the soils are used for purposes of wilderness, or agriculture, or urban and industrial development (Hole, 1973, 1974b). A statistical study of the properties of our soilscapes and a comprehensive numerical classification of the soils in them are still in the future.

WISCONSIN SOILS RELATED TO PRINCIPAL PHYSIOGRAPHIC REGIONS

A soil map relating Wisconsin soils to principal geomorphic features (Plate 2) shows that certain steep slopes and escarpments (Fig. 1-2) are followed by important soil boundaries, such as the boundary of Soil Region B trending north, just west of Madison, and the outer edge of the east-west glacial moraine in north-central Wisconsin (separating Soil Regions F and G). Each soil region consists of two or more separate bodies of soils. The ten major soil regions of Plates 2 and 7 are generalized from the large colored soil map (Plate 1).

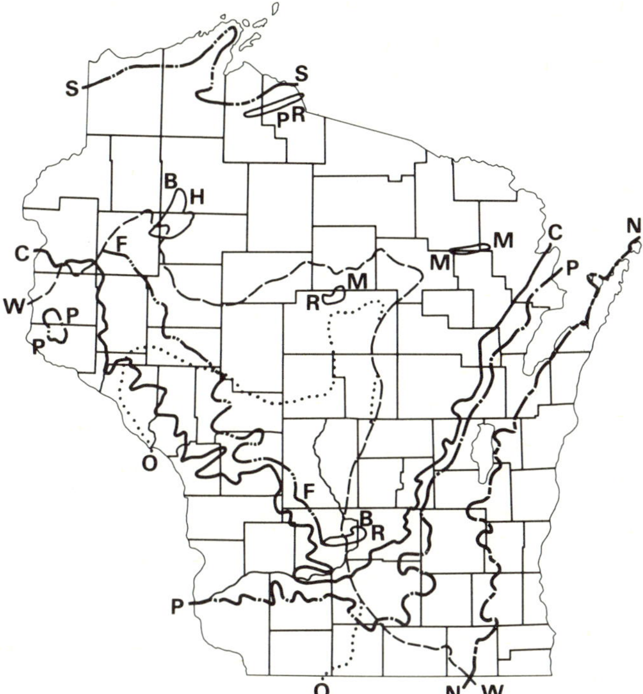

Major Escarpments

- **N N** Silurian ("Niagara")
- **P P** Sinnipee (Platteville-Galena)
- **C C** Prairie du Chien
- **F F** Franconia of the Upper Cambrian
- **S S** Superior (glacial drift over Precambrian)

Ranges

- **B H** Barron Hills (Blue Hills)
- **B R** Baraboo Range
- **R M** Rib Mountain
- **M M** McCaslin Mountain
- **P R** Penokee-Gogebic Range

Glacial Drift Boundaries

- **O O** Older glacial drift
- **WW** Woodfordian (Cary) glacial drift

Figure 1-2. Major escarpments, Precambrian ranges, and glacial boundaries in Wisconsin.

THE PLACE OF WISCONSIN SOILS IN THE WORLD SOIL CLASSIFICATION

Soils are classified on the basis of horizons and other features seen in cross sections (soil profiles) on the walls of excavations 4 or more feet deep, and on the basis of physical and chemical properties as determined in the field and laboratory. Ten major kinds of profiles are recognized by the world soil classification system of the United Sates Department of Agriculture (Chapter 5). Six of these are represented in Wisconsin (see Fig. 5-1 in Chapter 5 and at the end of the section of plates). Their relation to world soils is shown in the key, Table 1-2. It can be seen that very old, highly weathered soils (Ultisols and Oxisols) are lacking in this state, as are soils formed under climates with long dry seasons (Aridisols, Vertisols; see Buol, Hole, and McCracken, 1973). The orders characteristic of a humid temperature climatic region are represented by sketches in Figs. 1-3 through 1-8.

THE PLACE OF WISCONSIN SOILS ON THE WORLD SOIL MAP

Wisconsin soils are related not only taxonomically to world soils, as indicated above, but also geographically. The world soil map (Soil Conservation Service, 1967b) shows two major soil boundaries in the state. These are arbitrarily generalized as straight bands in Fig. 1-9. They delineate three soil regions that may be related to the Prairie, Alleghenian, and Boreal floristic provinces (Curtis, 1959). The first is in the southwestern part of the state and is an extension of grassland soils (Mollisols), both well drained (Udolls) and wet (Aquolls), in the humid end of the great prairie of the United States and Canada. Similar soils are present in Argentina, Uruguay, south-central USSR and adjacent portions of Europe, north-central China, and Australia. The second and central major soil region is an extension of woodland soils of the northeastern United States. These fertile soils are called Alfisols (Gray-Brown Podzolics) with associated acid Fragiochrepts (Acid Brown Forest soils), and wetlands containing Aquolls (Humic Gleys), Albaqualfs (Planosols), and included Entisols (Regosols such as those on sand dunes). Similar soils are present in England and adjacent portions of central Europe, north-central China, and northeastern Australia. The third general region is in the northern part of the state, where the soils are like those of southern Canada, Scotland, Scandinavia, northern Europe, and the eastern half of northern USSR. They are cool climate forest soils called Spodosols (Podzols) and Boralfs (Gray Wooded soils), intermixed with wetland soils called Histosols (Bog soils) and Inceptisols (Humic Gleys). These terms and others in Table 1-2 are explained in Chapter 5 and subsequent sections.

Table 1-2. Generalized soil key to the soils of the world (USDA classification)

(The six soil orders represented in Wisconsin are shown in boxes [see corresponding Figs. 1-3 through 1-8]. Some Mollisols have cambic B horizons.)

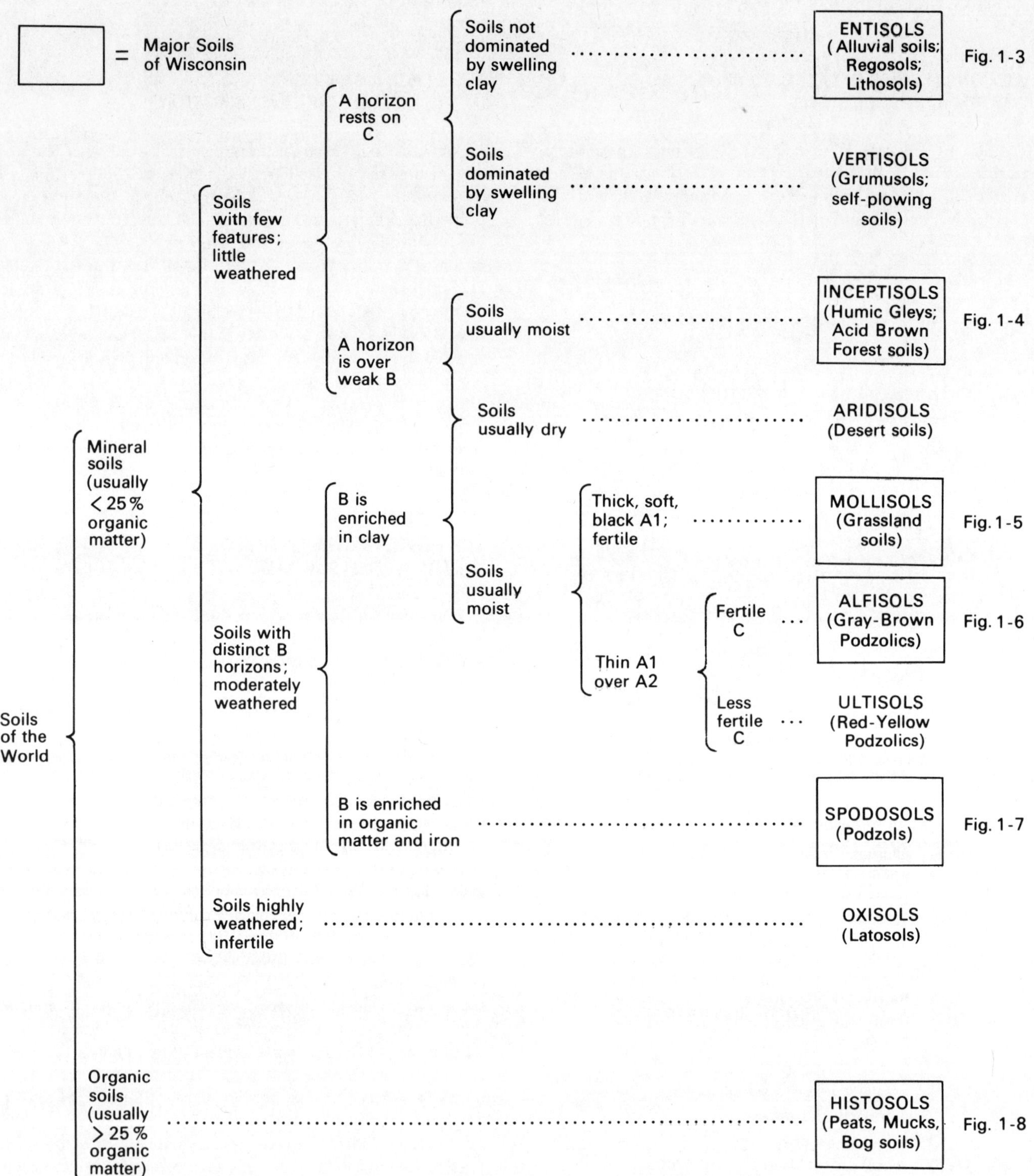

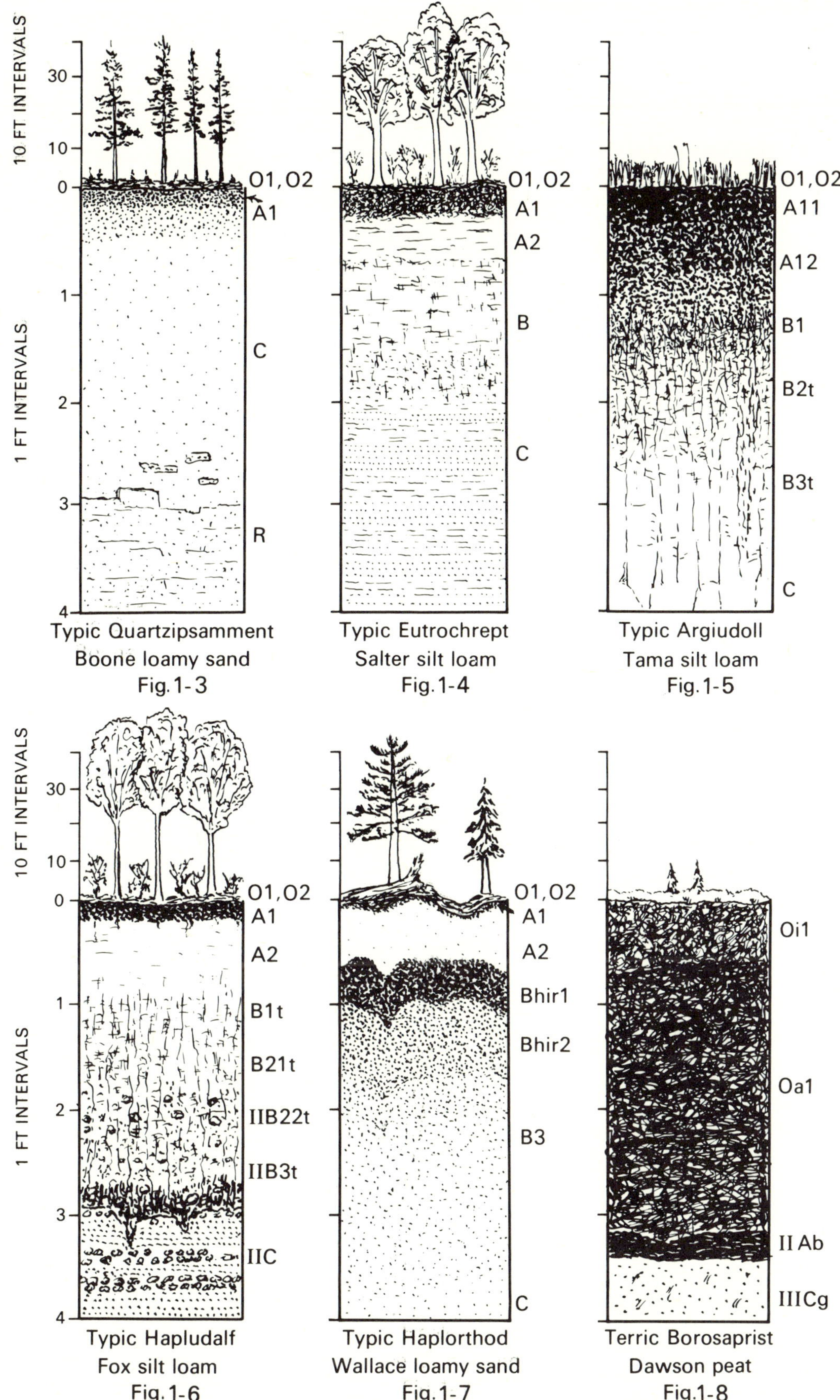

Figures 1-3—1-8. Profiles of soils representative of orders found in Wisconsin.

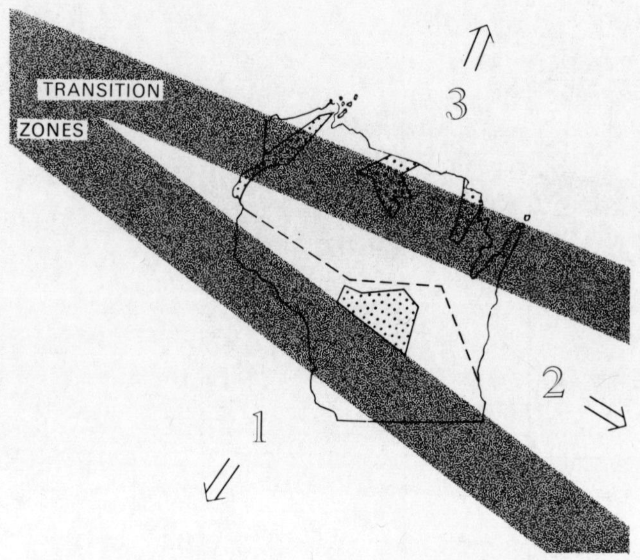

1 Well to poorly drained soils of prairie and associated woodlands (Udoll-Alfisol-Aquoll soil region that extends southwest to Nebraska)

2 Well to poorly drained forest soils with associated wet prairies and sandy Regosols (Alfisol-Aquoll-Entisol soil region that extends east to the Appalachian Mountains)

3 Well to poorly drained soils of mixed conifer-hardwood and swamp conifer forest lands (Spodosol-Histosol soil region that extends to Alaska and Nova Scotia)

Stippled areas represent sands and peats. (Psamment-Aquent-Fibrist soils)

The dashed line represents the ecological "Tension Zone" of Curtis (1959)

Figure 1-9. Three major soil regions of Wisconsin, generalized.

INTRODUCTION

CHAPTER 2

Factors of Formation of Wisconsin Soils

Early woodsmen, fishermen, hunters, and farmers of Wisconsin, and the American Indians before them, observed that different soils have different capacities to support life, to withstand the erosive action of wind and rainstorms, and to respond to management. Many of these people noted some important relationships between soil properties and environmental factors. They noted that soils were more fertile in limestone country than in quartz sandstone country; that many prairie soils were consistently more productive of crops than forest soils were (Jenny, 1961); that "Maple-Basswood land" was more productive of agricultural crops than "Pine-Hemlock land"; that sandy soils of cool sections of northern Wisconsin contained more organic matter than sandy forest soils of the warmer southern counties; that soils at the bottoms of hills were typically darker, deeper, and more moist than soils at the crests; and that yields of crops were outstanding on many artificially drained lowland soils. These relationships may still hold true today even though modern farming practices tend to equalize fertility levels between soils.

From these and similar observations we may conclude that soil formation is influenced by minerals inherited from rock, original vegetation, climate, and lay of the land (topography). Most of our soils are hundreds to thousands of years old, and have had time to change at least moderately in response to these controls. Students of soils and land systematically observe soil characteristics and relate them to features of the landscape and to conditions of a particular parcel or region. This method of studying landscapes was first practiced by Humboldt (Dolan, 1959), Dokuchaev (1879), and Hilgard (1892), and has been applied in Wisconsin by Chamberlin (1883), Whitson (1927), and numerous other workers. As this information is interpreted on the basis of accumulated information, soils may be rated[1] for purposes of land-use planning (see Table 2-1). A

1. Some of the principal soil factors mentioned in this chapter as especially influencing yields of crops have been incorporated by Berger, Hole, and Beardsley (1952) into a practical soil productivity scorecard for cropland, by means of which a piece of land can be rated and compared with other fields and farms. Other soil scorecards are available for rating for pastureland, woodland, wildlife land, and soil judging events (Aebischer et al., 1969).

Table 2-1. Description and ratings for five soils of Wisconsin

Soil name and region	Notable factors of soil formation	Special soil features observed	Inches of water available to plants in 3 ft. of soil	Common corn yields[a] (bu/acre)	Red pine yield[b] (BF/acre)	Support of roads
Tama silt loam (a Typic Argiudoll; prairie soil, in southwestern Wisconsin)	Prairie vegetation; loess deposit	Thick, dark surface soil; silty texture; nearly level	8	120	——[c]	Fair
Lapeer loam (a Typic Hapludalf, Gray-Brown Podzolic, in southeastern Wisconsin)	Deciduous forest cover; limy glacial till	Shallow, dark surface soil; brown clay loam subsoil; rolling landscape	4	80	500	Good
Plainfield loamy sand (a Typic Udipsamment, Regosol, in central Wisconsin)	Jack pine, Hill's oak cover; glacial outwash sand	Thin, dark surface soil on brown sand	2	40[d]	500	Excellent
Withee silt loam (an Aquic Glossoboralf, a northern Gray-Brown Podzolic, in north-central Wisconsin)	Deciduous forest cover; acid loess over acid glacial till	Mottled, yellowish tight subsoil; gently undulating	6	80	——[e]	Poor
Pence sandy loam (a Typic Haplorthod, a Podzol soil, in northern Wisconsin)	Coniferous forest; acid glacial outwash sand and gravel	Thin, pale surface soil on reddish-brown subsoil; level to rolling	3	60	575	Excellent

[a] 15% moisture corn. The climatic factor is involved. The number of frost-free days ranges as follows: Tama and Lapeer soils—160 days; Plainfield and Withee soils—140 days; Pence soil—120 days. Early frost frequently prevents maturation of the grain on the Pence soil.

[b] BF/acre = boardfeet per acre according to Scribner rule: one board foot is one foot square and one inch thick.

[c] This prairie soil is used almost entirely for agriculture. If pine trees are to be planted, the prairie soil is first inoculated with appropriate microorganisms called mycorrhizae.

[d] Proper irrigation and fertilization increase yields to more than 100 bushels per acre on Plainfield sand.

[e] This soil is somewhat poorly drained, and hence not suitable for red pine (*Pinus resinosa*).

number of charts that have been devised for this purpose are available from the U.S. Soil Conservation Service and the University of Wisconsin-Extension.

The purpose of this chapter is to describe the factors responsible for the formation of soils of Wisconsin, namely (1) initial or parent material, (2) original cover of plants and associated animals, (3) climate, both regional and local, (4) topography, including landscape position of each kind of soil and extent of each major slope gradient, and (5) age and degree of development of the soils.

INITIAL MATERIAL OF SOIL

Initial material[2] of soil is that material present when soil formation begins. For example, sandy soils of Wisconsin (Soil Regions C and H, Plates 1, 2, and 7) began with quartz sand mixed with grains of silicate minerals, including feldspar and mica. Initial material includes consolidated and unconsolidated mineral deposits—granite, sandstone, limestone, shale, glacial drift, loess, dune sand—and organic matter. Peat and muck soils, which may be as much as 40 feet deep in Wisconsin bogs, formed from plant remains rather than from rock material. In all soils, even sands, some organic material (forest litter, prairie grass roots) is quickly incorporated during the very first stages of soil formation. Acreage figures for major soils of Wisconsin (Chapter 21) reveal that nearly 90% of the area of the state is occupied by soils that are known to contain less than 25% (and in most cultivated areas less than 5%) of organic matter (by dry weight) in the first 6 inches of the A horizon. We speak of the common soils as mineral soils (Table 1-1) because even in the topsoil they consist largely of grains and fragments of minerals such as quartz and feldspar. The other soils are, by distinction, called organic soils.

2. This term is considered more precise than the more common and somewhat anthropomorphic synonym, "parent material."

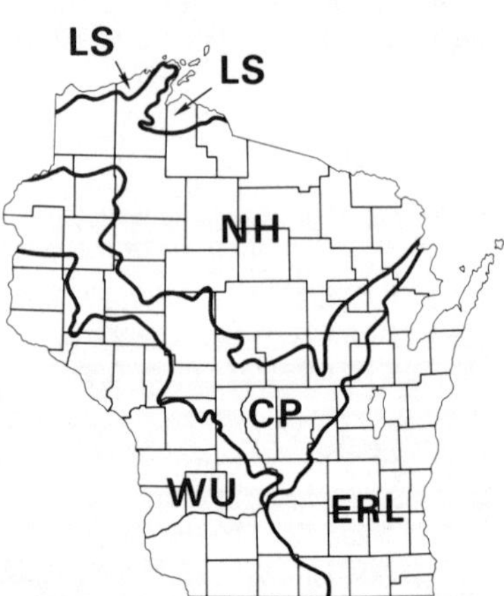

CP Central Plain (underlain by upper Cambrian sandstone)

ERL Eastern Ridges and Lowlands (underlain by Ordovician, Silurian and Devonian dolomites)

LS Lake Superior Lowland (underlain by Upper Keweenawan sandstone and conglomerate)

NH Northern Highland (underlain by Precambrian crystalline rocks)

WU Western Upland (underlain by Ordovician dolomites)

Figure 2-1. Major physiographic divisions of Wisconsin (after Dutton and Bradley, 1970).

Bedrock

Bedrock protrudes at the surface or lies at shallow depths in many places on steeplands of the "Driftless Area" of southwestern Wisconsin and elsewhere at scattered sites along cuestas and around quartzite masses (Figs. 2-1, 2-2, 2-3). As early as 1883, T. C. Chamberlin listed the four main kinds of rocks that have contributed materials to the soils: limestone, sandstone, shale, and crystalline rock. Precambrian crystalline rocks form the Northern Highland core of the state (Fig. 2-1). It is a southern extension of the ancient Canadian Shield. The rocks, older than 600 million years, include granite and gneiss, gabbro, diorite, basalt, quartzite, sandstone, shale, and associated metamorphic rocks (Plate 3). Lying against these ancient rocks, like shingles laid in reverse order on a roof, are interbedded Paleozoic sandstones, dolomites, and shales (Tables 2-2, 2-3). The bedrock formations lie buried under Pleistocene deposits (glacial drift) over about three fourths of the state (Tables 2-4, 2-5), but are represented in fragments in the drift.

Table 2-2
Estimated proportionate extent of bedrock units in Wisconsin

Unit (Plate 3)	Area[a] mi²	Area[a] %
Devonian System		
Devonian formations (chiefly dolomitic shale)	300	0.5
Silurian System		
Silurian dolomitic formations	5,000	9.0
Ordovician System		
Maquoketa Formation (shale and dolomite)	900	2.0
Sinnipee (Platteville-Galena) Group (dolomite with some limestone and shale)	5,000	9.0
Ancell Group (including St. Peter Formation) (sandstone, shale, and conglomerate)	1,800	3.0
Prairie du Chien Group (predominantly dolomite)	4,700	9.0
Cambrian System		
Upper Cambrian formations (sandstones with some dolomite and shale)	14,600	27.0
Precambrian		
Upper Keweenawan (Lake Superior) formations (sandstone with some shale and conglomerate)	2,400	4.5
Quartzite, slate, and iron formations	600	1.0
Gabbro and basalt	5,430	12.0
Granite and undifferentiated igneous and metamorphic rocks	13,670	23.0
	54,400[b]	100.0

[a]Areas determined by the cut-and-weigh method on the basis of the Geologic Map of Wisconsin, Plate 3.

[b]The figure for the total land area of the state is from Table 19-1, rounded off to the nearest hundred.

Table 2-3. Estimated proportionate extent of bedrock and residuum recognized in Wisconsin soil profile descriptions[a]

Bedrock unit	Area of bedrock outcrops mi²	%	Area of loess- and glacial drift-covered bedrock and residuum mi²	%	Total bedrock and residuum (with or without loess and drift cover) recognized in soil profiles mi²	%
Devonian System	0	0	0	0	0	0
Silurian System	163	0.3	0	0	163	0.3
Maquoketa Formation	54	0.1	270	0.5	324	0.6
Sinnipee Group	163	0.3	3,200	6.0	3,363	6.3
Ancell Group	54	0.1	210	0.4	264	0.5
Prairie du Chien Group	163	0.3	3,200	6.0	3,363	6.3
Upper Cambrian Series	2,176	4.0	1,358	2.5	3,534	6.5
Precambrian	326	0.6	0	0	326	0.6

[a]Areas are determined on the basis of total soil area of Wisconsin from sums of soil associations (Table 19-1) that include soils in the profile descriptions of which bedrock is mentioned. Baraboo silt loam (B3) does not contribute to the bottom line in the second column of this table because it is inextensive. This also applies to some other soils.

In the "Driftless Area" of southwestern Wisconsin, silty soils derived from loess (Table 2-6) over limestone are usually separated from it by a reddish-brown cherty clay one to several feet in thickness, much of which is considered to be residual from impurities and shaly layers in the carbonate bedrock (Tables 2-3, 2-7). Residuum from acid rock underlies soils in limited areas in north-central Wisconsin (Fig. 2-4).

It can be seen that bedrock geology is important in soil studies with respect to (1) the character and proportion of insoluble residues in the rocks, particularly limestones, (2) the extent to which structure of rocks is inherited by soil (as in soil associations F14, F15, F23; and A4 in Pierce County), (3) the fertility of the lithologic units affecting plant growth, (4) the extent of erosion and deposition of rock materials and their residues in the landscape, and (5) the geomorphic significance of the rocks (see the section on topography and soil-water relationships in this chapter).

The gross similarities that do exist between bedrock pattern and soil pattern are somewhat unexpected, considering that most of our soils have not formed directly on bedrock. Apparently the work of glacial ice, quiet and running meltwaters, and winds has been localized enough in shifting loose materials to blur but not to obliterate the broad geologic pattern.[3]

Glacial Drift

Glacial drift is all the rock material deposited by glacial ice and meltwaters derived from it. Wisconsin has such a remarkable display of these materials that a major stage of glacial activity has been called the Wisconsinan (see Table 2-10). Advances, stagnation, and melting of continental glaciers have left about three fourths of Wisconsin covered with a blanket of glacial drift (see Plate 4 and Table 2-4). These deposits are less than 100 feet thick over half of the state, and 100 to 300 feet thick in northern, east-central, and southeastern counties and in the Mississippi and Wisconsin River valleys. These materials consist of stratified or washed drift, i.e., outwash, inwash, ice-

3. Pebble counts show that about 15% of the stones in till of southeastern counties were imported from the Canadian Shield of northern Wisconsin and Canada; the rest are from more local bedrocks.

Table 2-4. Estimated proportionate extent of units of the Wisconsin glacial deposits map (Plate 4) (compared to Table 2-5)[a]

Unit: Form	Unit: Material	Area mi²	Area %		Estimates from Table 2-5(%)
End Moraine	Till	7,612	14	48	36
Ground Moraine	Till	18,515	34		
Plains, unpitted	Outwash	3,262	6	25	30
Plains and hills, pitted	Outwash	10,330	19		
Lake Basin Plains	Lacustrine deposits	7,612	14		9
"Driftless Area" Terrain with included valley bottoms and benches	Miscellaneous glacial deposits and residuum of surrounding uplands	7,069	13		25
		54,400	100		100

[a]Extent determined by the cut-and-weigh method from the Wisconsin glacial deposits map, Plate 4. The total land area of the state is derived from Table 19-1.

Figures 2-2—2-31. Maps showing soil association areas identified by selected properties.

Figure 2-2. Steep stony land and bedrock outcrops are prominent.

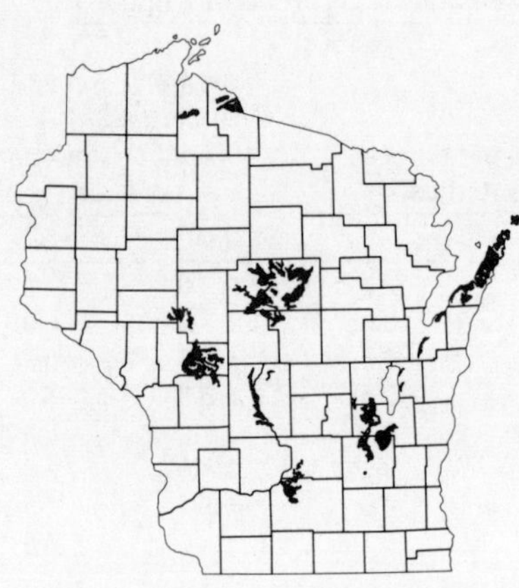

Figure 2-3. Landscape is not steep; bedrock lies near the surface in many places.

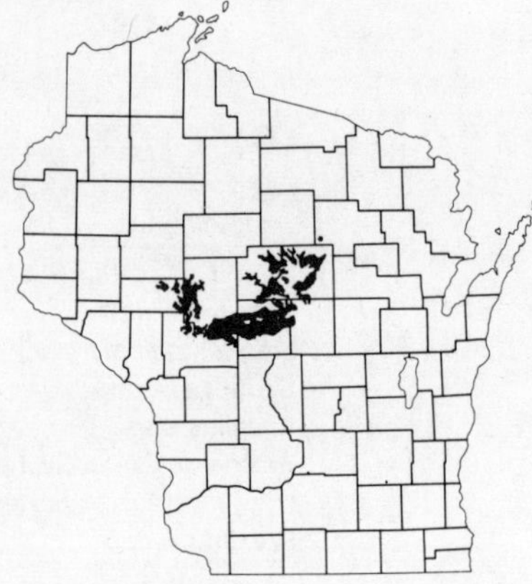

Figure 2-4. Substratum is commonly acid bedrock residuum (not derived from limestone).

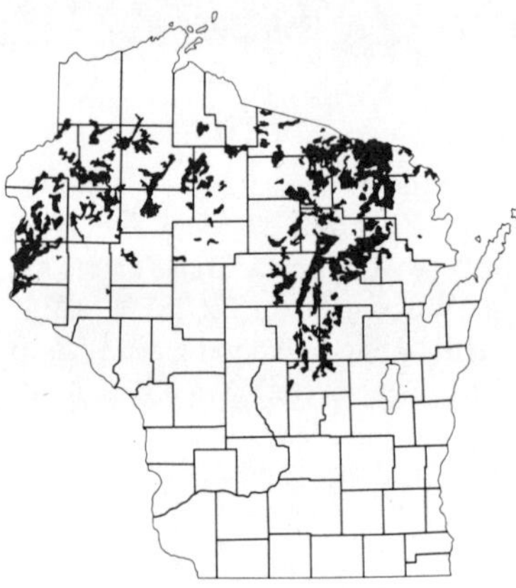

Figure 2-5. Silty and loamy soils are commonly underlain by a substratum of acid sand and gravel glacial outwash.

Figure 2-6. Substratum is commonly calcareous (within 6 feet) sand and gravel glacial outwash.

Figure 2-7. Substratum is commonly sand glacial outwash which is calcareous within 6 to 10 feet.

Figure 2-8. Substratum is commonly calcareous sandy loam to loamy sand glacial till.

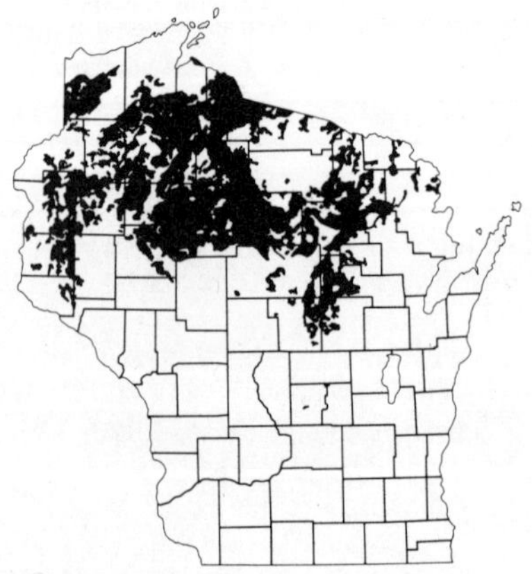

Figure 2-9. Substratum is commonly acid sandy loam to loamy sand glacial till.

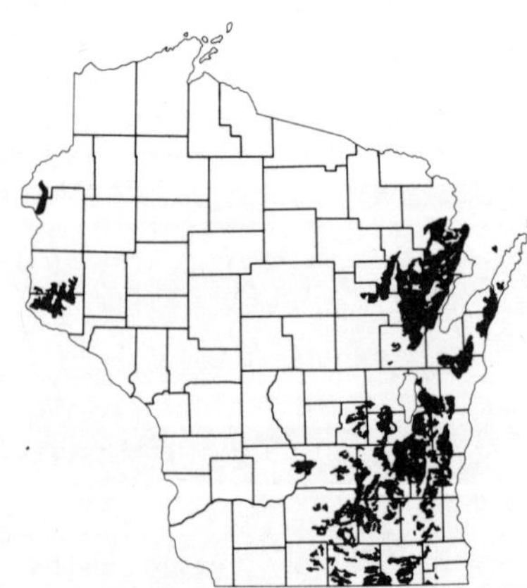

Figure 2-10. Substratum is commonly calcareous loam to sandy loam glacial till.

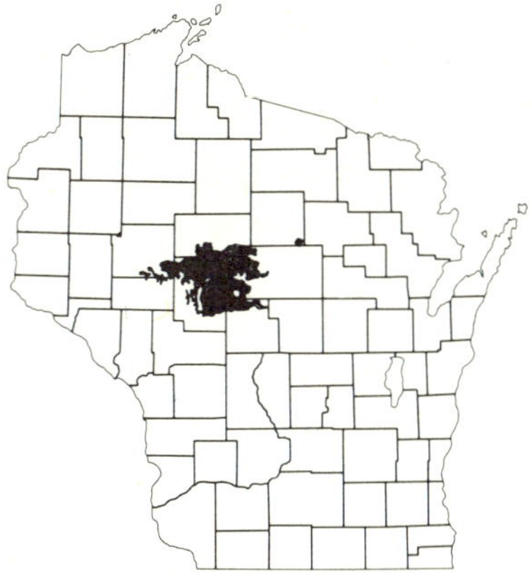

Figure 2-11. Substratum is commonly acid loam to sandy loam glacial till.

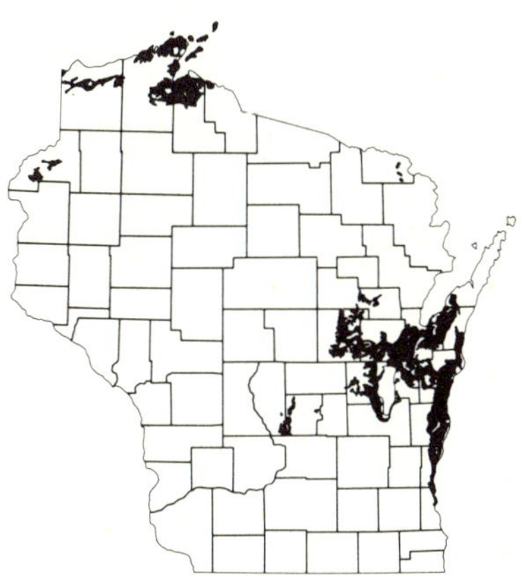

Figure 2-12. Substratum is commonly calcareous silty clay loam to clay glacial till.

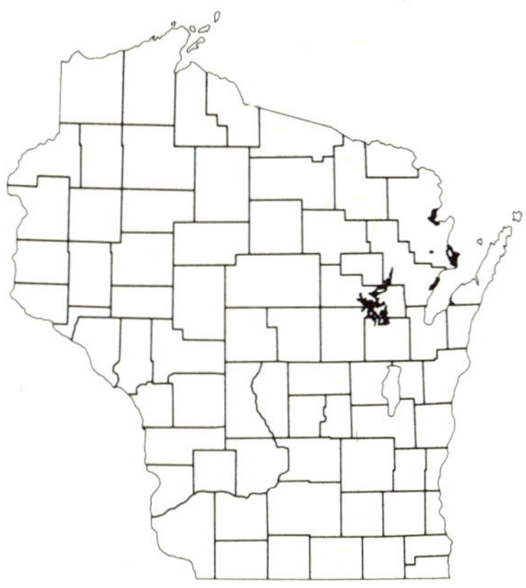

Figure 2-13. Substratum is commonly acid to neutral sandy glacio-lacustrine deposits.

Figure 2-14. Substratum is commonly acid medium- to fine-textured glacio-lacustrine deposits.

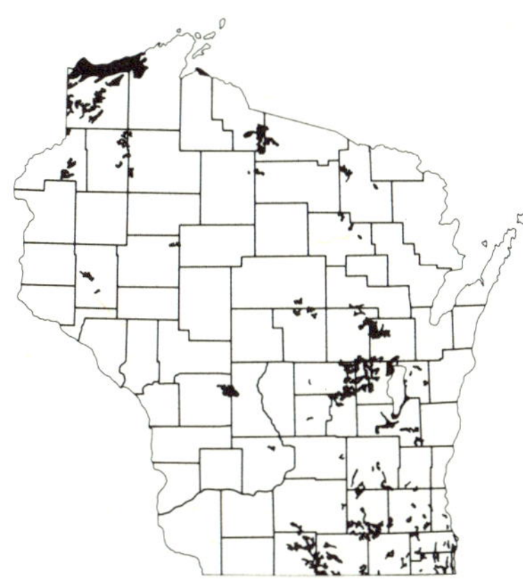

Figure 2-15. Substratum is commonly calcareous to neutral medium- to fine-textured glacio-lacustrine deposits.

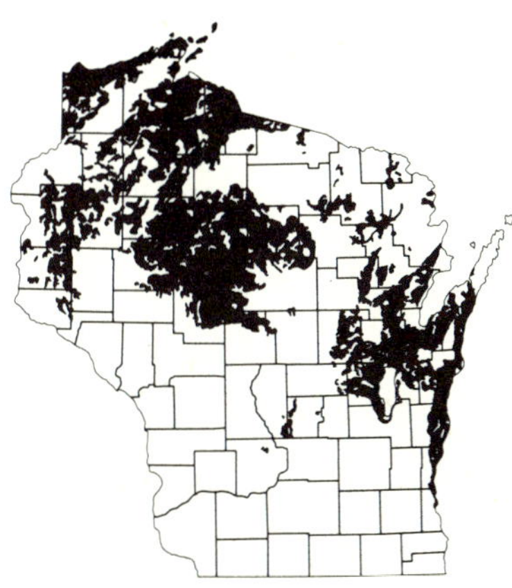

Figure 2-16. Substratum is commonly reddish-brown (5YR-2.5YR) glacial till.

Figure 2-17. Substratum is commonly old (pre-Woodfordian) glacial till.

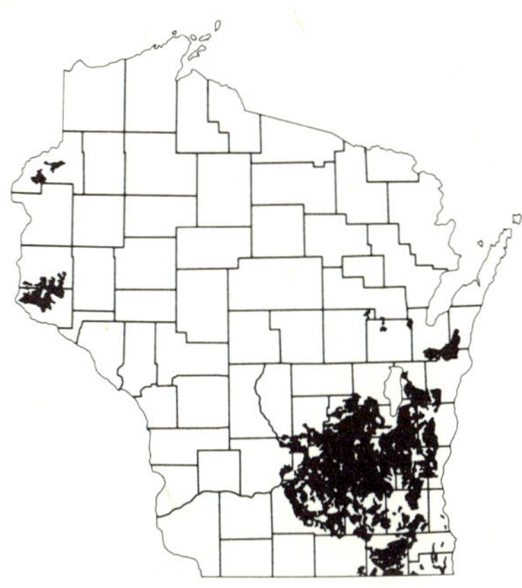

Figure 2-18. Substratum is commonly yellowish- to grayish-brown (10YR) glacial till.

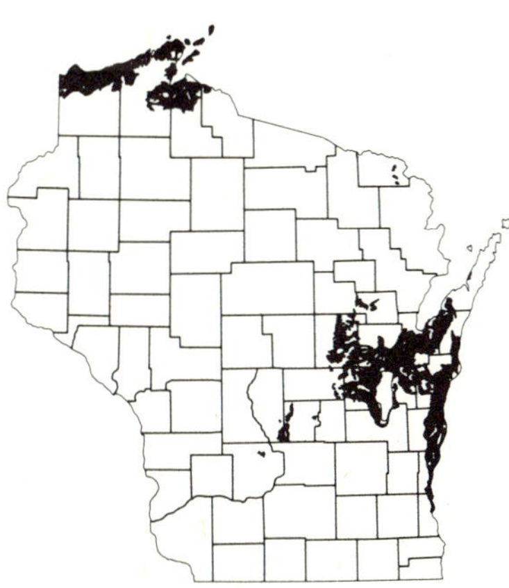

Figure 2-19. Substratum is commonly reddish-brown (5YR-2.5YR) clayey glacial till.

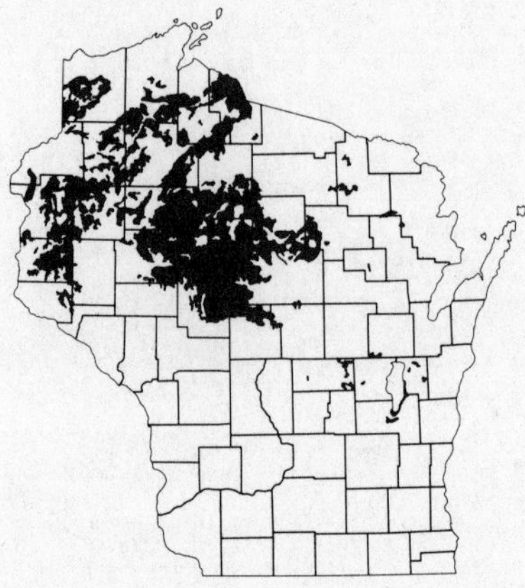

Figure 2-20. Substratum is commonly reddish-brown (5YR-2.5YR) loam to sandy loam glacial till.

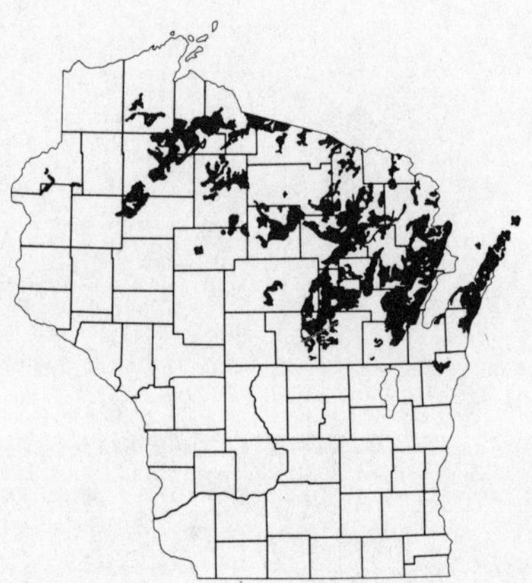

Figure 2-21. Substratum is commonly pink (7.5YR) sandy loam glacial till.

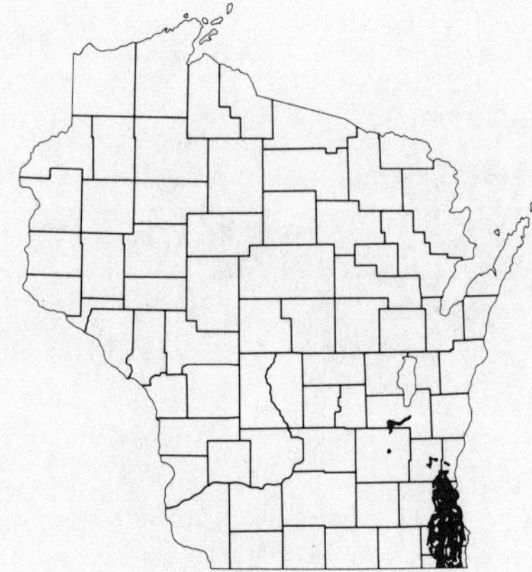

Figure 2-22. Substratum is gray to olive-brown (10YR-2.5Y) glacial till.

Figure 2-23. Sandy surficial soil horizons commonly total 18 to 36 inches in thickness over a finer textured substratum.

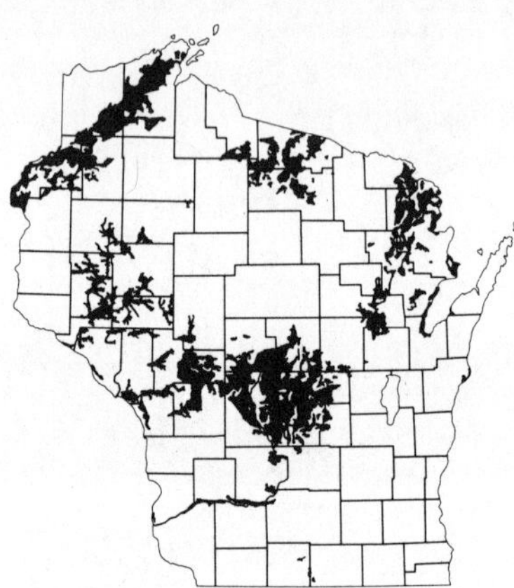

Figure 2-24. Sandy surficial soil horizons commonly total more than 36 inches in thickness and are underlain by acid sand glacial outwash.

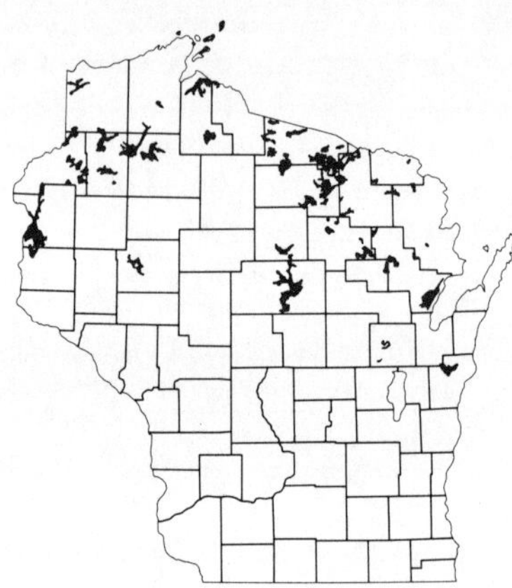

Figure 2-25. Loamy surficial soil horizons commonly total 10 to 24 inches in thickness over either coarser or finer textured materials.

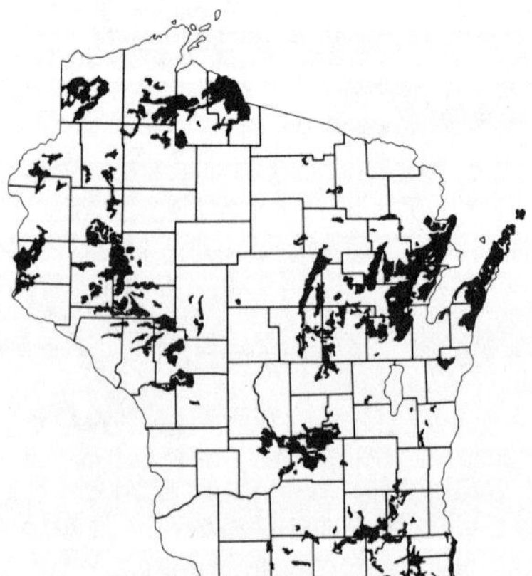

Figure 2-26. Loamy surficial horizons commonly total 24 to 40 inches in thickness over either coarser or finer textured materials.

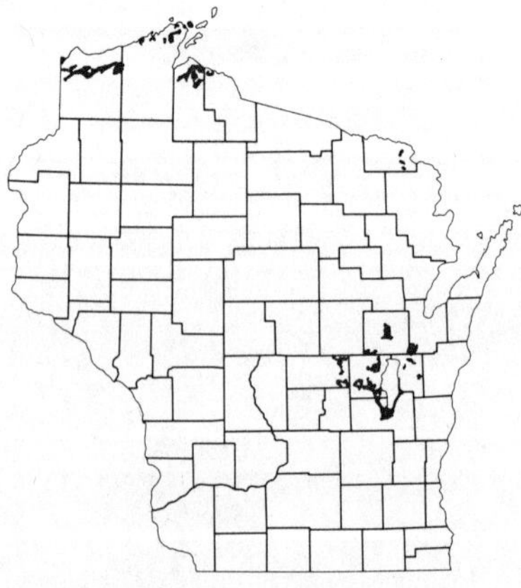

Figure 2-27. Surface horizons are commonly silty clay loam to clay loam in texture.

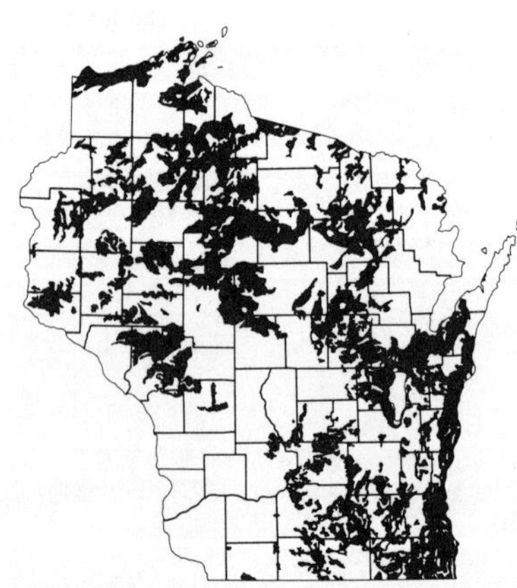

Figure 2-28. Silty soil is commonly 5 to 20 inches thick.

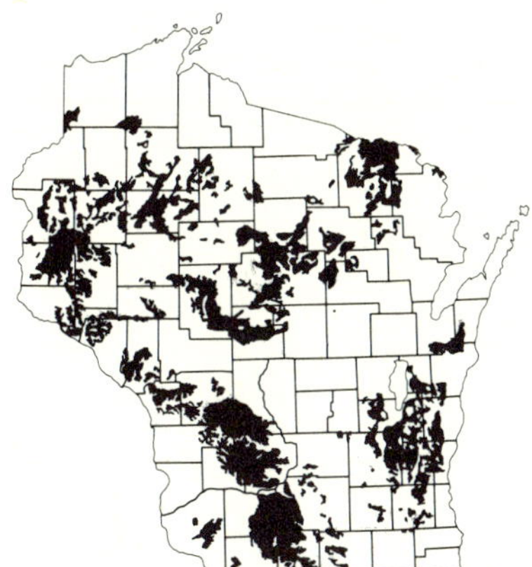

Figure 2-29. Silty soil is commonly 20 to 40 inches thick.

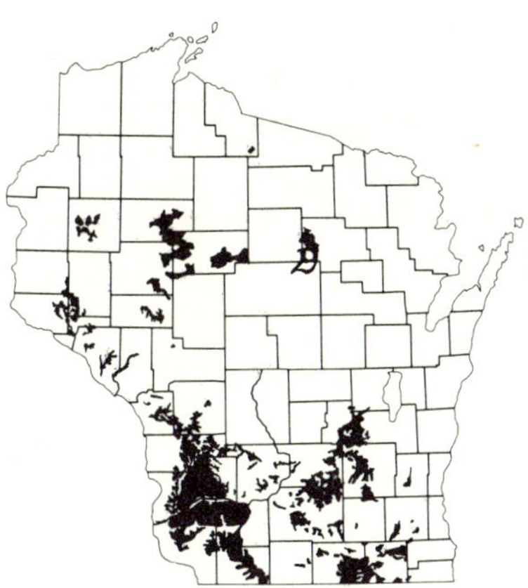

Figure 2-30. Silty soil is commonly 36 to 50 inches thick.

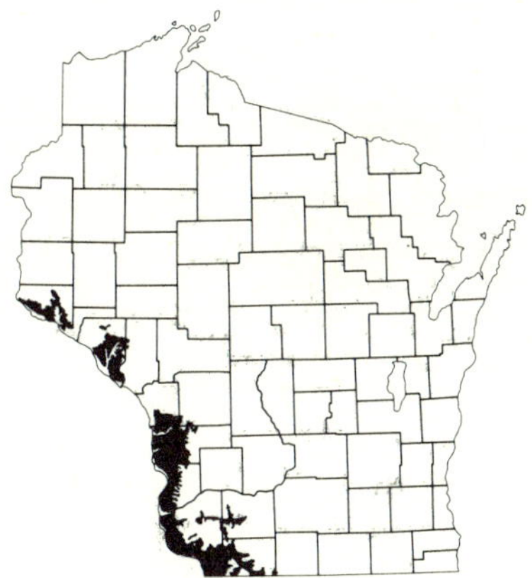

Figure 2-31. Silty surficial horizons commonly total more than 50 inches in thickness.

contact, and glacio-lacustrine sediments (Figs. 2-5, 2-6, 2-7, 2-13, 2-14, 2-15) and of unstratified drift, i.e., till (Figs. 2-8 through 2-12; 2-16 through 2-22). In about a third of the state the soils have formed from till, in another third from outwash, and in nearly 10% of the area, from glacio-lacustrine sediments and associated inwash deposits (Table 2-5). Textures of these Pleistocene substrata (proportions of sand, silt, and clay) range from sand and gravel to clay (Table 2-5), carbonate content from zero to more than 75%, and colors from reddish brown to gray and olive brown (Figs. 2-16, 2-18 through 2-22). Interlayering of different kinds of glacial drift within the depth of soil profiles accounts for some contrasts between topsoil and subsoil (Figs. 2-23 through 2-27). For example, the Borth silty clay loam is formed from lacustrine clay overlying outwash sand and gravel at a depth of 2 or 3 feet. However, the most common surficial layering of geologic origin is produced by another group of materials, the wind-laid or aeolian deposits, to be considered next.

Aeolian Deposits

There is no doubt that every square mile of Wisconsin holds some dust particles from other land areas of the world. Most important are coverings of less-traveled aeolian (windblown)

Table 2-5. Estimated proportionate extent of Wisconsin soils in which Pleistocene substrata are recognized[a]

Deposits subdivided by properties and with proportionate areas given in parentheses (%)				
Till (36)	Sand to sandy loam (3)	Acid (2)	without loess cover	(1.6)
			with loess cover	(0.4)
		Dolomitic (1)	without loess cover	(0.6)
			with loess cover	(0.4)
	Sandy loam to loam (25)	Acid (16)	without loess cover	(6)
			with loess cover	(10)
		Dolomitic (9)	without loess cover	(3)
			with loess cover	(6)
	Silty clay loam to silty clay (all dolomitic) (8)	Gray (2)	without loess cover	(1)
			with loess cover	(1)
		Red (6) (all without loess cover)	bordering Lake Superior	(2)
			bordering Lake Michigan	(4)
Outwash (30)		Acid (20)	without loess cover	(16)
			with loess cover	(4)
		Dolomitic (10)	without loess cover	(6)
			with loess cover	(4)
Lacustrine (9)				(9)
(75)				(75.0)

[a]Determined from sums of relevant categories of soils in the legend of the soil map (see Table 19-1). The remaining 25% of the area of the state is largely in the "Driftless Area" and in alluvium and colluvium. The table does not indicate that in the 8% of the area of the state covered by peat and muck the substratum is about 2% each of lacustrine deposits, colluvium, outwash, and glacial till.

Table 2-6. Estimated extent of aeolian deposits in Wisconsin (Plate 5)

Material (and depth)[a]	Area[b] mi²	%		Estimates from the soil map (%)
Silt[c] (8-16 ft.)	1,200	2	63	41
Silt[c] (4-8 ft.)	5,500	10		
Silt[c] (2-4 ft.)	8,900	16		
Silt[c] (0.5-2 ft.)	19,200	35		
Silt[c] (0-0.5 ft.)	12,600		24[d]	—
Silty clay loams	—	—		3 } 37[d]
Loams	—	—		34
Active dunes present	1,000	2	13	14
Sandy areas, subject to blowing	6,000	11		
Wetland soils	—	—		8
	54,400	100		100

[a]The map shows ridgetop conditions and this exaggerates the area covered by loess.
[b]Measurements are made by the cut-and-weigh method from Plate 5.
[c]Loess, leached and unleached.
[d]These two percentages probably refer to the same category of deposit.

silts and sands that are evident over nearly three quarters of the state. These silts and sands rest on a great variety of bedrock and glacial drift (Plate 5; Tables 2-5, 2-6, 2-7). Calcareous loess is present below a depth of 7 feet near the Mississippi River valley in western counties. This deposit is largely silt (0.002-0.05 mm dia. particles), containing quartz (about 60% by weight), feldspar (25%), pyroxene, mica, and numerous other minerals in small proportions, and carbonates (calcite and dolomite, 10%). Because of its variety of minerals and abundance of feldspar, which contains potassium, loess is a material from which fertile soils have developed during the approximately 12,000 years since the last major deposit. The thickness of the loess diminishes from somewhat more than 16 feet in the southwest to zero in eastern Wisconsin (Figs. 2-28 through 2-31). It is presumed that much of it blew onto the upland from glacial outwash in the Mississippi River valley and major tributaries to it during the outpouring of glacial meltwaters down those channels. Some additional loess must have been blown from states to the west. A small amount continues to fall even now. The work of Borchardt, Hole, and Jackson (1968) suggests that nearly a third of the state has soils that do not have a measurable silty surface layer and yet do contain admixed loess material. Some silty soils of northern Wisconsin have a relatively high content of coarse and medium silt that was probably derived from locally exposed glacial drift containing little fine silt. Northern aeolian silts were probably never calcareous, but were winnowed from acid materials.

Sandy and loamy coverings are also of local origin, sand dunes being the thickest. Stabilized sand dunes are extensive in parts of Soil Region C (Chapter 9). Dunes are active today in places near lake shores, river banks, and disturbed "barrens."

Table 2-7. Estimated proportionate extent of substrata of Wisconsin soils in which loess coverings are recognized[a]

Substratum		Approximate thickness of loess covering in feet 0.5-2	2-4	4-8	8-16	Totals	
Glacial outwash (and associated lacustrine deposits)		4.3	3.8	0	0	8.1	25.4
Glacial till		6.3	11.0	Tr	0	17.3	
Bedrock	Maquoketa Formation	Tr	0.3	0.2	0	0.5	15.4
	Sinnipee Group	2.0	1.0	2.0	1.0	6.0	
	Ancell Group	Tr	Tr	0.2	0.2	0.4	
	Prairie du Chien Group	2.0	1.0	2.0	1.0	6.0	
	Upper Cambrian formations	1.0	0.5	0.5	0.5	2.5	
	Precambrian rocks	Tr	0	0	0	0	
		15.6	17.6	4.9	2.7	40.8	40.8

[a]Extent determined from sums of soil association units (see Table 19-1).

Alluvium and Slope Wash

Each year runoff waters deposit alluvium (J1 and J2; Plate 1) and slope-wash sediment in numerous ravines and valleys; mass-wasting augments colluvial deposits on footslopes. These materials commonly consist of two principal layers: (1) a dark presettlement buried soil on sediments that represent centuries of relatively slow erosion and deposition, and (2) an overburden of lighter colored, less organically enriched sediments that eroded from fields during the past 130 years of agriculture in Wisconsin. These are initial materials for numerous bodies of fertile Entisols (Regosols) in the state.

Organic Deposits

Waterlogged remains of tamarack, black spruce, sedges, mosses, and other hydrophilic plants constitute initial materials for Histosols (Bog soils, peats, and mucks). Marl, diatomaceous earth, sand, silt, and clay are commonly interlayered with these. A unique characteristic of organic deposits is their combustibility when dry. As a consequence, numerous peat fires have smouldered for long periods in drained bogs of Wisconsin, leaving extensive pits and hollows. Even without actual burning, drained Histosols gradually oxidize and disappear as the organic matter is converted to water vapor, gases, and ash.

Table 2-8. Proportionate extent of major plant communities in the early nineteenth century in Wisconsin (Plate 6)[a]

Unit	Area mi²	%
Prairie (bluestem, composites)[b]	2,800	5
Oak savanna (bur oak, white oak, bluestem)[b]	11,300	21
Southern oak forest (white, black, and red oaks)	2,100	4
Southern mesic forest (sugar maple, basswood, elm)	5,600	10
Lowland hardwood (willows, soft maple, ash)	400	1
Sedge meadows (sedges, blue joint, cordgrass)	1,400	3
Pine barrens (jack pine, prairie grasses)[b]	3,500	6
Pine forest (white pine, red pine)	3,400	6
Northern mesic forest (maple, hemlock, yellow birch)	20,200	37
Conifer swamps (black spruce, tamarack, cedar)	2,800	5
Boreal forest (balsam fir, white spruce)	900	2
	54,400	100

[a]Determined by the cut-and weigh method on the basis of Plate 6.
[b]This is part of the 32% of the state that was especially affected by frequent burning before the advent of European settlers.

ORIGINAL COVER OF VEGETATION WITH ASSOCIATED ANIMALS

The major gross influences of the biota on the soil are (1) stabilization and protection against removal by water and wind erosion, and (2) development of characteristic layers (horizons) that are enriched in organic matter.

An S-shaped curve, delineated by the southern boundary of the northern mesic forest (Plate 6; Table 2-8; Figs. 1-9, 2-40), marks the transition between major areas of original northern mixed broadleaf-conifer forest, and southern broadleaf forest and prairie association (Curtis, 1959). Distribution of prairie on ridge tops (southwestern Wisconsin), on outwash plains (as in the vicinity of Janesville and Beloit in Rock County), and on irregular topography elsewhere has been explained as an effect of fires (Curtis, 1959) which burned over these areas often enough to suppress or eliminate forest growth. The presence of large areas of oak savanna in the south and the less extensive pine barrens and pine forest in the north is also attributed to the effects of major fires that either suppressed trees without eliminating them, as in oak savannas and pine barrens, or cleared the way for a succession of white and red pine on sandy loam soils. In northern counties fires created conditions locally favorable for bracken fern-sedge meadows. These are sometimes called stump prairies (Vogl, 1964), but are not true prairies. In places, the Rock and Crawfish rivers in Dodge and Jefferson counties in southeastern Wisconsin served as fire breaks that limited the spread of oak savanna into southern mesic forest (Zicker, 1955). This was important for soil formation because organic matter is abundantly incorporated into the soil under cover of prairie and savanna. Northern forests, especially hemlock stands, have favored the development of Spodosols (Podzols).

Animals associated with the plant communities have had their influence on soil directly or indirectly. Before European settlement, for example, beavers expanded the area of wet soils by constructing innumerable dams along drainageways throughout the state (Schorger, 1965; Hole et al., 1962). Recent build-up of the deer herd in Wisconsin, as a result of both the destruction of natural predators and particularly the increase in range following lumbering and burning, has damaged the forest reproduction in much of northern Wisconsin, with consequent interruption of the biocycling of nutrients by trees (Creed and Stearns, in Milfred, Olson, and Hole, 1967; Leopold, 1943). The vast numbers of smaller animals (rodents, worms, insects) have had a more direct and pervasive effect on soils of the state (Nielsen and Hole, 1964; Baxter and Hole, 1967; Salem and Hole, 1968). In general, fauna in Wisconsin have been more abundant and active in the soil of prairie and savanna lands than in forest lands, activity being proportionate to the intensity of solar illumination of the soil surface.

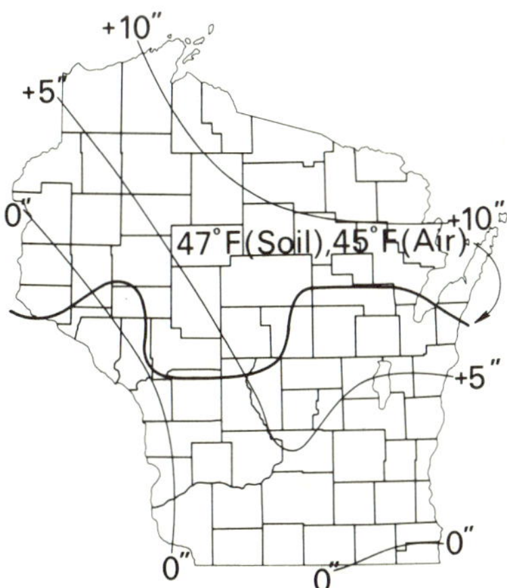

Figure 2-32. Transition boundary between mesic (to the south) and frigid (to the north) soils families; with isolines of precipitation minus potential evapotranspiration (after Bryson, 1957).

It is evident that the biotic overlay on soil and penetration into soil is dynamic. The soil responds to biologic events; sometimes rapidly, as in the case of biocycling of nutrients, and sometimes slowly. Long-term changes in prairie-border soil profiles, for example, have been very gradual, in response to the vegetative shift of the last few thousand years (Curtis, 1959) from grassland to encroaching forest. The fact that there must be a time lapse, called pedologic lag, between the conversion of vegetation and that of soil profile would lead us to expect the deep dark mineral soils to be more extensive in Wisconsin than the actual prairies that European settlers found here. It is possible that this is the real explanation for the fact that the soil map shows more of the state occupied by prairie soils (about 8% Hapludolls and Haplaquolls) than was occupied by prairie a century ago (5%; Table 2-8).

CLIMATE

Regional Climate

The air masses that pass over a region determine its climate. In Wisconsin the interaction is between continental arctic, continental polar, maritime tropical, and maritime pacific air masses. Wisconsin lies in a midlatitude (temperate) continental climatic zone (Dbf and Daf of Trewartha, in Finch et al., 1957) that also prevails over parts of eastern Europe and China. Borchert (1950) characterized the climate of northeastern Wisconsin as one that brings deep winter snows and reliable summer rains; that of southwestern Wisconsin as providing relatively dry winters and frequent severe summer droughts (Figs. 2-33 through 2-37). Fig. 2-32 shows the transition boundary between two soil-climatic regions of the state, one relatively cool (roughly corresponding to the Dbf zone) and the other warm (Daf). The new USDA soil classification designates these regions as frigid and mesic, respectively.[4] In terms of soils, these are the Spodosol (with widespread fragipans) and Alfisol-Mollisol (without fragipans) regions (Fig. 2-40). Precipitation is

4. These are defined as follows: Frigid soils are those with mean annual soil temperatures below 47° (8°C), at a depth of 20 inches (50 cm); mesic soils are those with mean annual temperatures between 47 and 59°F (8-15°C). Both groups of soils have a difference greater than 9°F (5°C) between mean summer and mean winter temperatures. The mean annual soil temperature at the depth indicated is about 2°F (1°C) warmer than the mean annual atmospheric temperature at the same site.

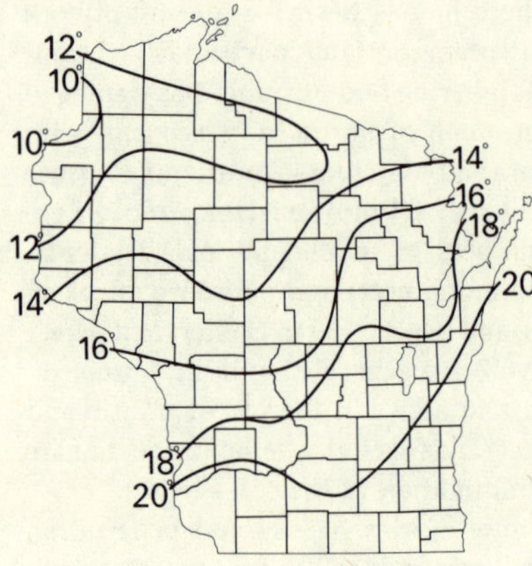

Figure 2-33. Mean air temperature (°F) for January (after Burley, 1964).

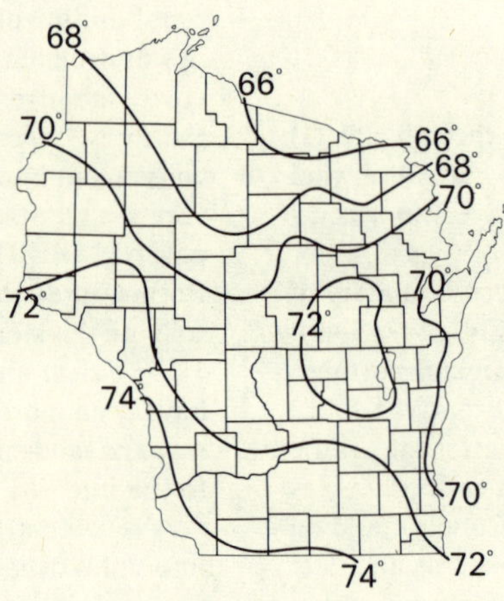

Figure 2-34. Mean air temperature (°F) for July (after Burley, 1964).

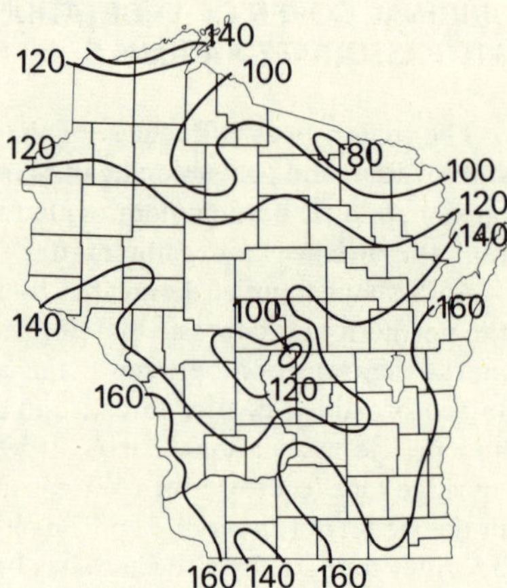

Figure 2-35. Average length (in days) of growing (frost-free) season (after Burley, 1964).

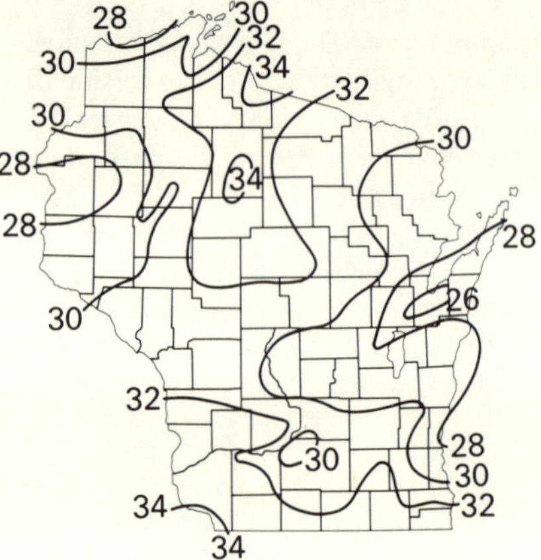

Figure 2-36. Mean annual precipitation (inches; after Holt, Young, and Cartwright, 1964).

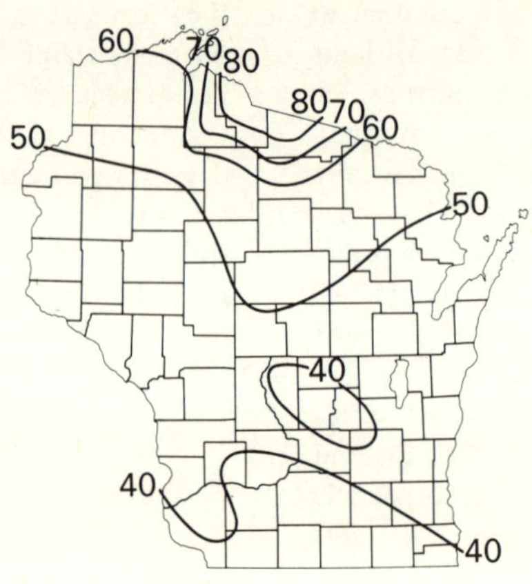

Figure 2-37. Average seasonal snowfall, 1930-1959 (inches) (after Burley, 1964).

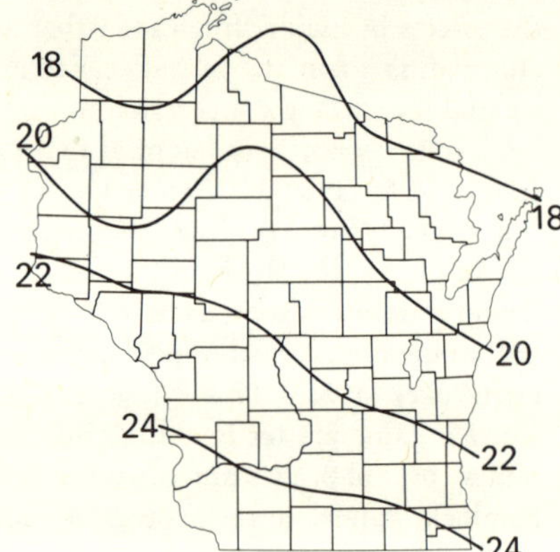

Figure 2-38. Estimated evapotranspiration in inches (annual precipitation, after Burley, 1964, minus annual total runoff, after Holt, Young, and Cartwright, 1964).

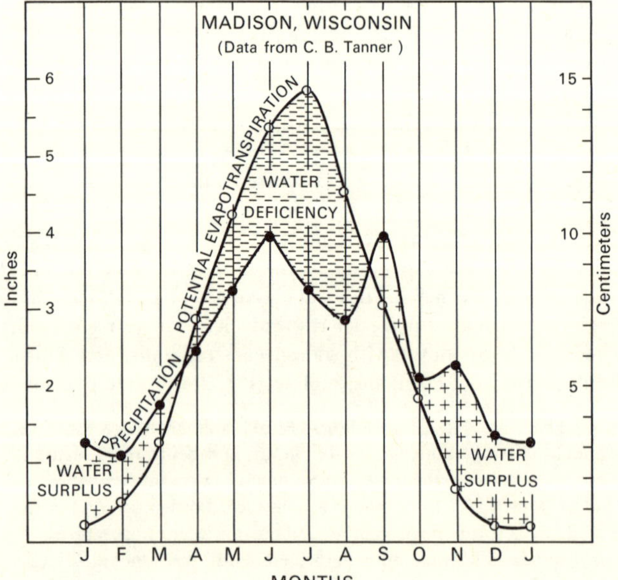

Figure 2-39. Water surplus and water deficiency periods at Madison, Wisconsin.

greatest in the growing season when it is needed for plant growth (Figs. 2-35 through 2-39); however, the moisture supply does not satisfy vegetation needs in most years. Hence, supplemental irrigation is beneficial on level areas (Figs. 2-53, 2-54), and is essential on sands in Adams and Juneau counties in the Central Plain. Three light lines in Fig. 2-32 indicate that Wisconsin might well be semiarid if evapotranspiration were to reach its theoretical potential.[5] Although the climate of the state is referred to as temperate compared with tropical and polar climates, it is actually rather severe (Figs. 2-33 through 2-37). During the year, changes in weather are rapid and marked. The northward march of the spring response of plant-

5. Out of the total solar radiation received in a year at Madison—that is, an amount sufficient to evaporate about 80 inches of water—about a 51-inch-equivalent is lost by reflection and long wave radiation. The remainder, the net radiation, could evaporate 29 inches, which is nearly equal to the 31-inch annual precipitation. But part of this net radiation is used in heating air and soil. Cattail plants in swamps transpire 40 inches a year. Upland conifers could transpire the same amount, if the moisture were available. Transpiration by corn is about 20 inches, and by alfalfa-brome is 23 inches (see Tanner, 1964).

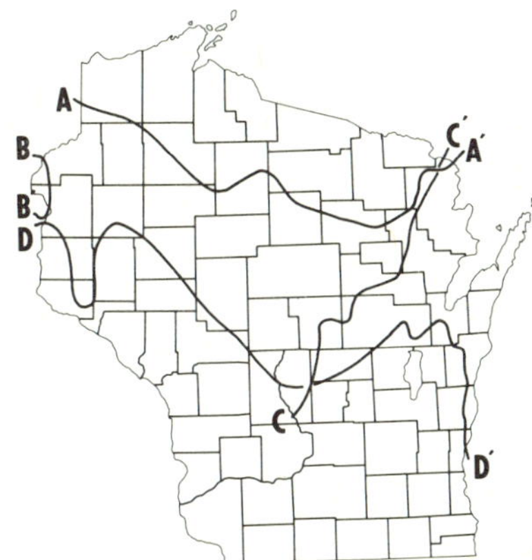

AA′ The southern limit of most strongly developed fragipans in Wisconsin

BB′ The eastern limit of carbonate-rich glacial drift of the Grantsburg glacial lobe (Wright and Ruhe, 1965)

CC′ The western limit of carbonate-rich glacial drift of the Green Bay glacial lobe (Thwaites, 1943)

DD′ The southern limit of the zone of bisequal soils (in part after Carroll, 1959, and Beaver, 1963)

Figure 2-40. Distribution of soil fragipans in Wisconsin in relation to calcareous and acid glacial drifts.

soil units proceeds at a rate of about 15 miles per day for a period of about three weeks, judging by phenological data for May blooming of lilacs (Dana, Zimmerman, and Lettau, 1963). Farmers frequently have difficulty in harvesting air-dried hay because of periodic thunderstorms in summer.

Local Climate

The climate is cooler and more moist on north-facing slopes and in low places than on southern exposures or moderate elevations. The growing season is not only notably short in wetlands of central and northern counties, but is likely to be abbreviated in small depressions (frost pockets), even in uplands. Soils adjacent to Lakes Michigan and Superior are subject to an ameliorated local climate, as in the case of shorelands of soil associations I18 and I19. Degree of exposure of the soil to sunlight and to wind affects the soil temperature and moisture content. Dark plowed soil and burned-over grassland heat up more quickly in the sun than soil protected by vegetation and mulch. Pavements and buildings bring about complicated local climatic regimes in adjacent and subjacent soil. These sometimes include accentuated freeze-thaw cycles. Local climates in nonurban settings must leave their impress on soils, and special studies of these are needed.

Paleoclimate

Students of climatic succession in Wisconsin and adjacent areas report evidence for alternation of relatively cool and warm, dry and moist climatic periods that began and ended abruptly (Bryson and Wendland, 1966, 1967). Fig. 7-6 presents a diagram of a postulated soil succession, corresponding to major climatic episodes. Freezing action on soils was so severe near the glacier during its long stay in the state that polygonal networks of cracks formed several feet deep in soft materials, including Cambrian sandstone. These cracks were filled and enlarged by growing wedges of ice. With later climatic warming the ice disappeared and sand blew into the empty cracks. The resulting ice-wedge casts are still present in the B and C horizons of a variety of soils in the state, from Pierce and Outagamie counties in the north to Rock and Green counties in the south (Black, 1965b).

The fragipan, so widespread in soils of northern Wisconsin (Fig. 2-40), may have been produced in part under permafrost conditions before and after late glacial advances, or during the thawing of frozen subsoil in post-Valderan (Holocene) times. Fitzpatrick (1956) noted soil structure and color in fragipans of northern Wisconsin similar to those in lower active layers over permafrost in Spitsbergen (see summary by Hole et al., 1962).

TOPOGRAPHY AND SOME SOIL-WATER RELATIONSHIPS

Topography

Wisconsin is in the Central Lowlands geomorphic province of the United States (Thornbury, 1965). Specifically it lies in (1) the Great Lakes Section characterized by lakes and glacial lake plains, conspicuous end moraines, cuestaform topography partially exposed, and poorly integrated drainage; and (2) the Wisconsin "Driftless Section" (approximately the Western Upland of Fig. 2-1), characterized by maturely dissected cuestas in Paleozoic rocks and by valley trains deposits of glacial outwash.

The Great Lakes Section is further subdivided into four subsections: (1) the Northern Highland, with a general elevation of 1,400 to 1,650 feet (Tim's Hill in Price County is at 1,952.9 feet, the highest point in the state); (2) the Lake Superior Lowland, an undulating to rolling plain, with elevations of 600 to 900 feet above sea level; (3) the Central Plain, with elevations commonly between 750 and 850 feet; and (4) the Eastern Ridges and Lowlands, with prominent end moraines and the Kettle Moraine. The Wisconsin "Driftless Section" is dissected

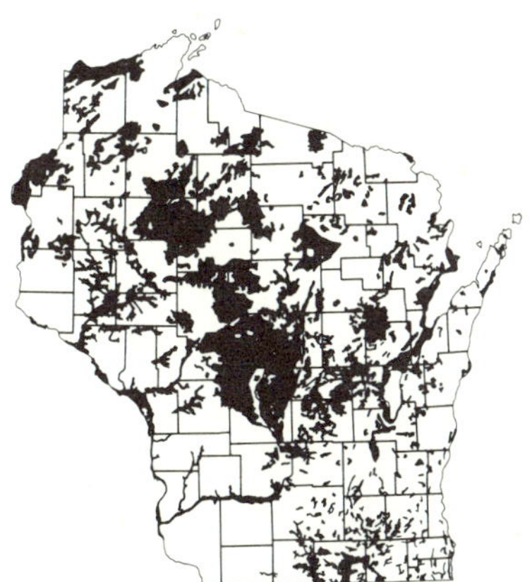

Figure 2-41. Soil association areas in which nearly level slopes (0 to 3% gradient) are common.

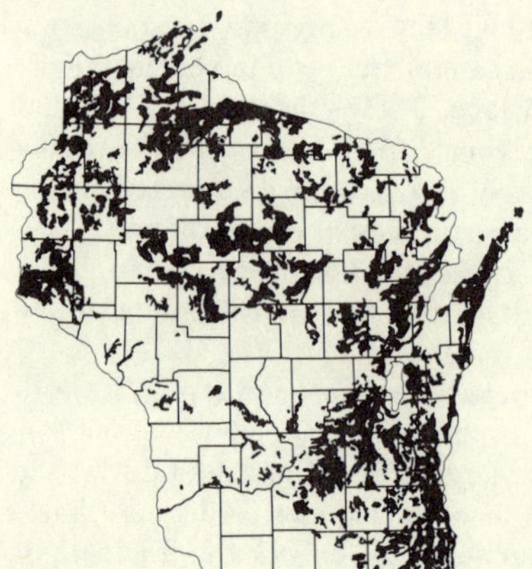

Figure 2-42. Soil association areas in which undulating to gently rolling topography (2 to 10% gradient) is extensive.

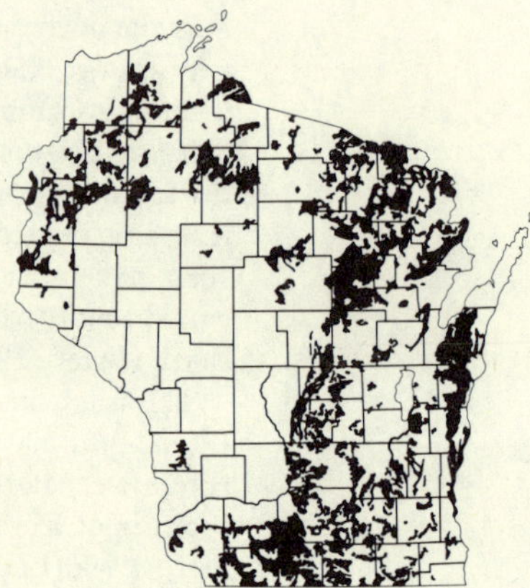

Figure 2-43. Soil association areas in which rolling topography (5 to 20% gradient) is extensive.

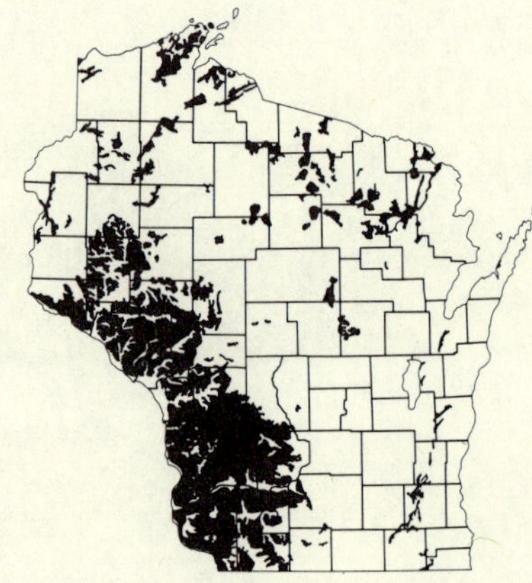

Figure 2-44. Soil association areas in which hilly and steep topography (slopes of 20% gradient or more) is common.

by valleys to a depth as great as 500 feet below the upland, whose surface stands at 1,000 to 1,450 feet above sea level.

The total range in elevation in the state is from 580 feet to 1,953 feet (Fig. 2-45). Local relief ranges from 600 feet at Wyalusing State Park at the confluence of the Wisconsin River with the Mississippi and about the same relief at Wausau where Rib Mountain stands boldly above the Wisconsin River, to less than a foot in a hundred feet of distance on old lake plains near Lake Winnebago. The relief in drumlin landscapes may be as much as 50 feet and in the Kettle Moraine reaches 300 feet in the vicinity of Lapham Peak in Waukesha County. The reader is referred to Martin (1932) for a full discussion of relationships between topography and geology.

The soil map shows that much of the land surface of Wisconsin is rolling, with slope gradients of 5 to 18%. In the eroded cuesta terrain (Thwaites, 1960) of southwestern counties and in scattered areas elsewhere many soils are hilly, with slope gradients of 18 to 30%. The remainder of the state is undulating to nearly level (0 to 5% gradient) (Figs. 2-41 through 2-44). The approximate distribution of slope (gradient) groups of soils of the state is as follows: 0 to 1% slopes, 6% of the area of the state; 1 to 3% slopes, 25%; 3 to 12% slopes, 48%; 12 to 30% slopes, 15%; and 30 to 45% slopes, 6%. Such generalizations do not do justice to the wide variety of soil slope patterns in Wisconsin landscapes.

Region A (see Plates 1 and 2) is characterized by mature topography. Valley fills, as much as 200 feet deep in the Mississippi and Wisconsin River valleys, have produced the broadest flats of the region. Here natural levees, oxbows, cutbanks, slip-off slopes, alluvial fans, and natural terraces are common features (Hole, Peterson, and Robinson, 1952; Hole, 1956a). Each has characteristic soils and soil patterns. Steep-sided tributary valleys, called coulees, are numerous, in accordance with the dendritic drainage pattern. Detached blocks of bedrock have moved down steep slopes by the process of creep. Soil patterns tend to follow the intricacies of knife-edge ridges, benches produced by resistant rock ledges, sinkholes, cliffs, natural bridges, rock "monuments" and crags, dissected cuestas, and escarpments (Martin, 1932). Even areas dotted by man-made pits and mounds in the lead-zinc district are given a soil name with a topographic term, "Fayette and Dubuque soils and Pits, eroded."

Region B can be subdivided first into three bands: (1) the Rock River-Lake Winnebago lowland characterized by many included bodies of wetlands (soils of Region J) and prairie soils; (2) the Silurian ("Niagara") upland to the east (see Plate 3), clearly delineated by soil association B1 just south of Lake Winnebago; and (3) the Prairie du Chien and Sinnipee cuestas (Fig. 1-2). These physiographic units are overlaid by glacial features, such as the Kettle Moraine (soil association B4) and end moraines diverging from it that trend northwest to the Baraboo Range and southeast to Lake Geneva (Plate 4). The Kettle Moraine is largely composed of gravel deposited in conical hills called kames, curvilinear ridges called eskers, straighter ridges called crevasse fillings, and deep pits called kettles. Glacial till is more abundant than outwash in the end moraines. Outwash plains (soil associations B32, B33, B34) are scattered throughout the region but are particularly notable at the borders of the major moraines. More than 1,400 oval, streamlined hills, called drumlins, are concentrated in a belt about 35 miles wide lying behind (up-ice from) and parallel to the prominent end moraines. The drumlins are largely composed of till (soil association B13) and they point in the direc-

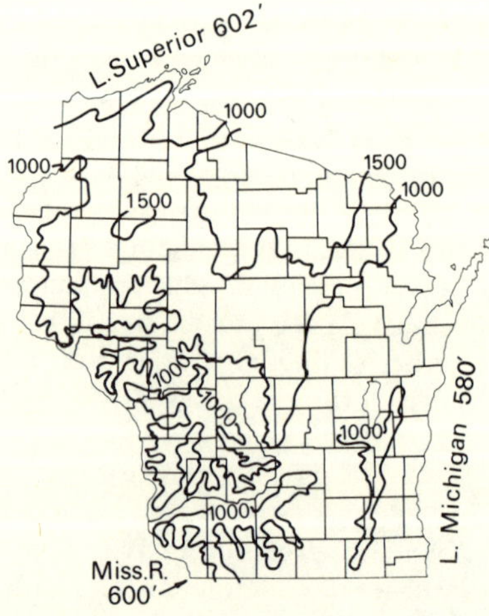

Figure 2-45. Generalized topographic map of Wisconsin (after Martin, 1932; elevations of contours are in feet above sea level).

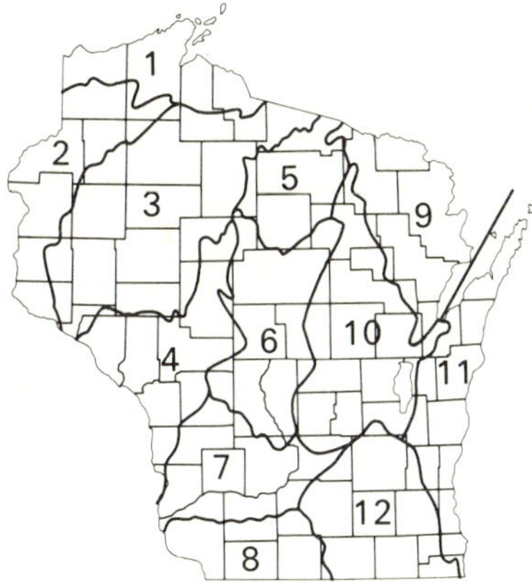

1. Lake Superior Streams
2. St. Croix River
3. Chippewa River
4. Trempealeau and Black Rivers
5. Upper Wisconsin River
6. Middle Wisconsin River
7. Lower Wisconsin River
8. Platte and Pecatonica Rivers
9. Menominee and Oconto Rivers
10. Wolf and Fox Rivers
11. Lake Michigan Streams
12. Rock and Fox Rivers

Figure 2-46. Principal river basins of Wisconsin (after Holmstrom, 1972).

tion of ice movement, which was in a spreading pattern in each of the two major ice masses, the Green Bay and the Lake Michigan lobes. Ground moraine occupies the remainder of the areas, interrupted by glacial lake plains (Plate 4).

In Region C, outwash and lake plains, with organic soils in depressions, are interrupted or bordered by glacial landforms, outwash terraces and escarpments, and cuestas. Areas of stabilized and active sand dunes are extensive. In the western part, sandstone pinnacles and hills are scattered here and there "like a child's blocks on the bedroom floor" (Black, 1964).

Region D is largely rolling, with characteristic conical hills of Cambrian sandstone and shale and buttes of sandstone. Between the hills are coulees and outwash and lake plains, with some associated wetlands.

Landforms of Region E include those already mentioned for Region B. Areas of shallow soils on limestone bedrock are extensive on the Door Peninsula, and the escarpment of Silurian dolomite is prominent on the west side of the peninsula.

Region F, in addition to aforementioned glacial landforms, has fluted landscapes that are composed of parallel ridges less than 20 feet high, separated by elongated bodies of poorly drained soils (Hole and Schmude, 1959). Undulating, relatively featureless till plains are extensive (soil association F21).

Region G includes silt-cover-free glacial terrain of northern Wisconsin. There are spectacular end moraine zones, festoons of eskers and kames with associated kettles, and vast areas of rolling ground moraine and pitted outwash (Plate 4). The region is one that contains many lakes, and small lake plains and outwash plains. "Reversal" of landscape occurs locally in Washburn County where surfaces of small lacustrine plains stand as depositional features a few tens of feet above surrounding ground moraine. Original retaining walls of stagnant glacial ice melted away. Bedrock hills and ridges protrude through the drift at the Penokee Ranges, Barron Hills, McCaslin Mountain (Fig. 1-2), and some other areas.

The landscapes of Region H are associations of eskers, kames, kettles, outwash plains, and stabilized and active dunes. Lakes and bogs occupy many of the countless kettles.

"Red clay" landscapes of Region I include lake plains, nearly level and rolling moraines, some drumlins, shallow kettles, and a few crevasse fillings. The region includes the Lake Superior Lowland and the northern part of the three topographic units mentioned for Region B, namely, two cuestas and the intervening Winnebago-Green Bay lowland (Figs. 1-2, 2-1). Limestone outcrops and bodies of shallow soil are numerous in places along the cuestas.

Region J has already been referred to in connection with mineral soil regions. Alluvial soils are prominent in the setting of Region A. Bogs and mineral soil wetlands are numerous in Regions B, C, E, F, G, H, and I. Local small-scale topographic features include burned-out or excavated pits in peat and masses of organic soils occurring naturally on slopes with gradients as steep as 15%.

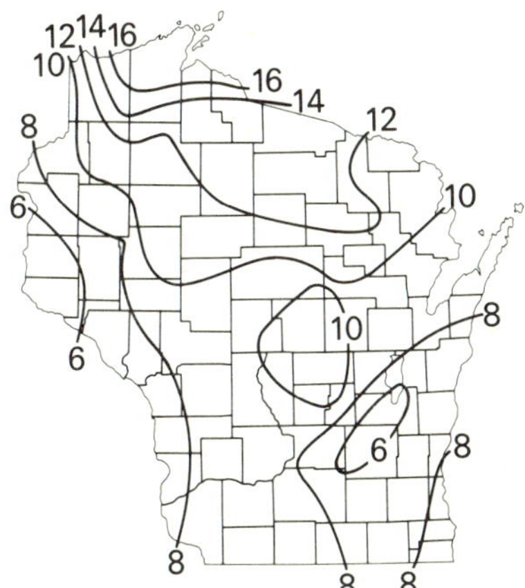

Figure 2-47. Average annual runoff (inches) (Holt, Young, and Cartwright, 1964).

Figure 2-48. Trout stream regions (after Threinen and Poff, 1963).

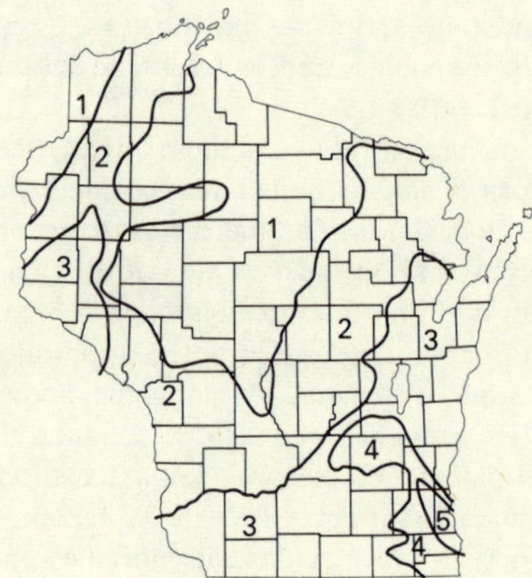

Figure 2-49. Quality of surface water; total dissolved solids (mg/liter) during periods of low flow (data from C. L. R. Holt, Jr., 1968).

The topography and the limited growing season of the state (Fig. 2-35) favor a predominant farming system that puts much of the cropland in legumes and mixed hay, large acreages in crop strips laid on the contour, and considerable areas in pasture. This system protects sloping soils from erosion by wind and water. Erosion-control practices have been widely applied in southwestern portions of the state, where many miles of terraces and grassed waterways are maintained.

Some Soil-Water Relationships

Soil-water relationships are closely tied to topography because of surface and subsurface movement of water from elevations to depressions and because of the down-slope decrease in depth to the water table.

The abundance of water in Wisconsin is indicated by the very name of the state, derived from an Indian word meaning "where the waters gather." In Wisconsin there are more than 9,000 lakes, 33,000 miles of rivers and streams, 800 miles of water frontage on lakes Superior and Michigan (Fig. 2-46), and two and a half million acres of wetland and alluvial soils (Fig. 2-50). Much of the water that falls on Wisconsin enters the soil (Tanner and Lemon, 1962). It may be returned to the atmosphere by evapotranspiration or may move below the root zone of plants to recharge the ground water. The average amount returned by evapotranspiration varies from more than 24 inches in southeastern Wisconsin to less than 18 inches in extreme northern counties (Fig. 2-38). Some water does not enter the soil but moves over the land as surface runoff (Fig. 2-47). Most of the runoff comes from snow melt and sustained heavy rains (Holt, Young, and Cartwright, 1964). The runoff has in the past produced serious erosion in many parts of the state. For best results in soil erosion control, combinations of practices are custom-made for each kind of soil association (Beatty and Peterson, 1963). The experience of Agricultural Experiment Station and Soil Conservation Service personnel has been invaluable in determining how strip-cropping, setting up grassed waterways and diversion terraces, and many other practices may

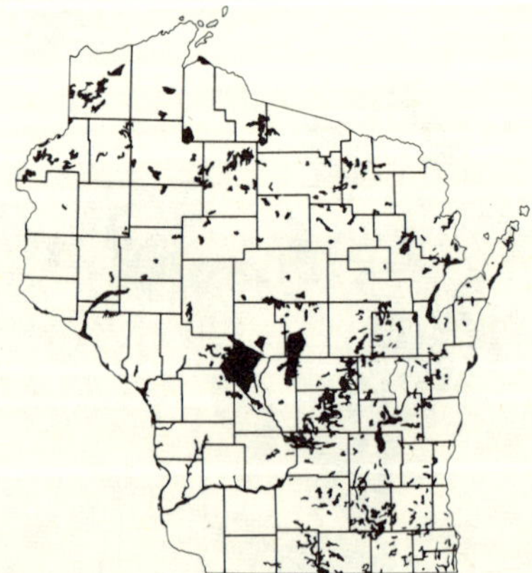

Figure 2-50. Distribution of major wetlands and alluvial soils.

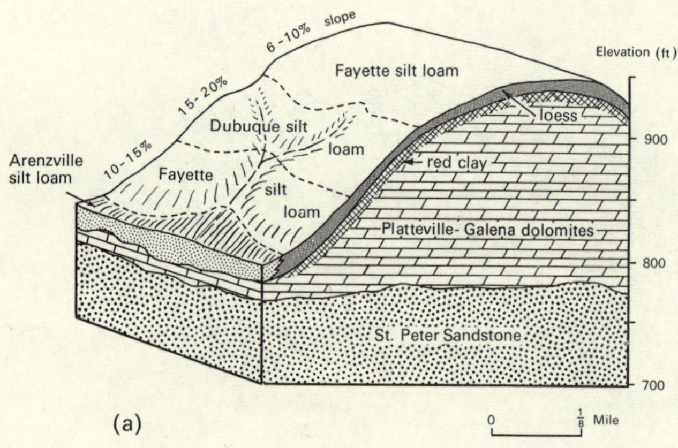

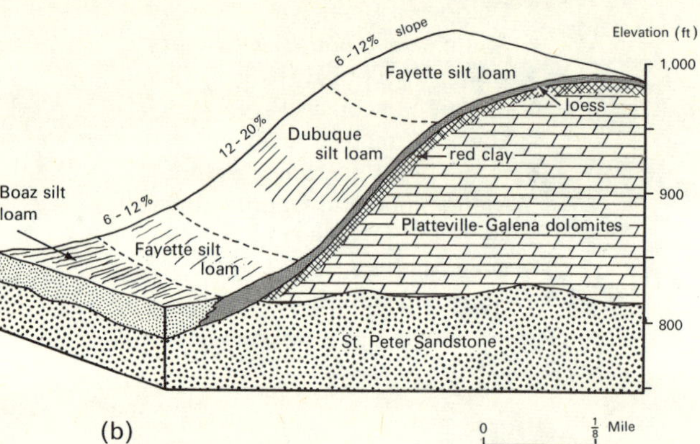

Figure 2-51. Block diagrams of a landscape representative of soil associations A7 and A9.

a. Fayette-Dubuque soil sequence in the SW ¼, Sec. 9, T32.N., R.3W., Grant County.

b. Fayette-Dubuque soil sequence in Sections 5 and 4, T.3N., R.5E., Lafayette County.

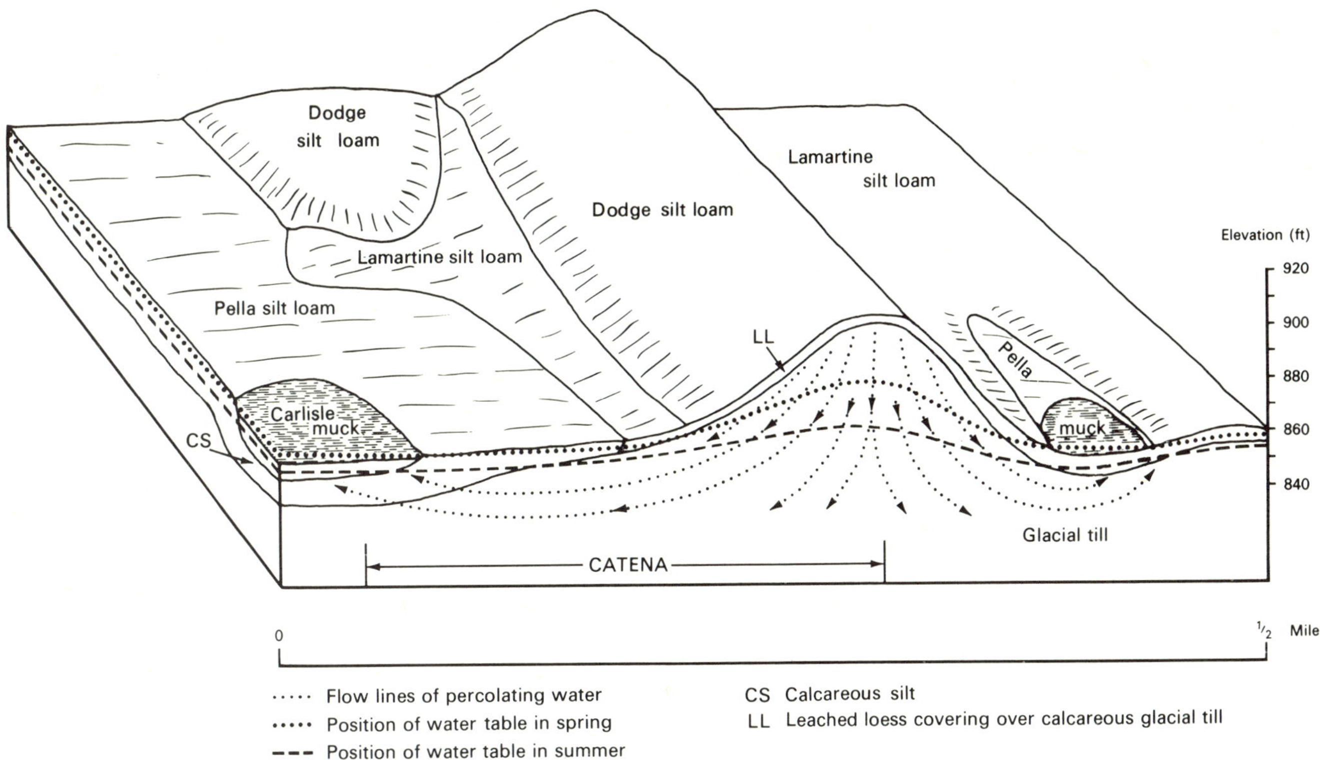

Figure 2-52. Block diagram showing landscape positions of soils on a drumlin and adjacent lowlands in soil association B13 in Dodge County. The glacial till is calcareous.

best be applied to specific areas to promote infiltration of water into the soil and removal of excess water from the fields, without notable incidence of erosion and sedimentation. Increased soil productivity has been the result (Figs. 2-53, 2-54). The dramatic increase in area of water-shedding surfaces—roofs and pavements—in urbanizing areas creates special problems for lowlands, lakes, and streams. Surface runoff and groundwater discharge combine to give the total discharge of streams. This total annual runoff varies from a low near 6 or 7 inches in southeastern and extreme western Wisconsin to more than 15 inches in the north. Threinen and Poff (1963) noted that trout streams are abundant in Soil Regions A and eastern C, D, and G, in which water-bearing sands and gravels, sandstones, and limestone promote high sustained low flow (Fig. 2-48). At low flow, quality of runoff water varies from soft water in north-central counties to very hard water in the southeast (Fig. 2-49).

Water which enters the soil and which is not returned to the atmosphere by evapotranspiration moves into the groundwater system. It moves to a point of lower groundwater elevation and may be discharged wherever the groundwater system intersects the land surface. This may be anywhere from a few feet to many miles from the point where it entered the soil. Except where artificially drained, many of the lowlands of the state have a high water table and wet soils (Fig. 2-50) that include peat, muck, and dark mineral soils. In most of these wet soil bodies, groundwater discharge (by evapotranspiration and surface runoff) exceeds recharge. Pockets of wetlands in elevated kettles and extensions of wetlands to upland drainage systems may be recharge areas and may contain bodies of soil that have the capacity to conduct percolating water faster, per unit volume, than do associated drier soils. Van Rooyen (1972, 1973) found this to be the case at the Marshfield Agricultural Experiment Station and in the University of Wisconsin Arboretum at Madison.

Soils are classified as to relative degree of dryness or wetness into six natural drainage groups: excessively drained, well drained, moderately well drained, somewhat poorly drained, poorly drained, and very poorly drained (Soil Survey Staff, 1951; Bouma, 1973). A sequence of soils from the top of a hill to the lowland at the footslope commonly includes three or more of these kinds of soils (Figs. 2-51, 2-52). Members of such a topohydrologic sequence, sometimes called a soil catena, are grouped together in soil keys. Each kind of soil has a characteristic landscape position (Hole, 1953) and soil catenas have specific geographic ranges.

The idealized pattern of flow of ground water in a common catena is shown in Fig. 2-52. Water enters and moves below the root zone of the well-drained Dodge silt loam when the frost melts in the spring, and periodically thereafter in rainy periods. Some of this water moves laterally down-slope near the surface and increases the amount of subsoil water in Lamartine silt loam (somewhat poorly drained) and Pella silt loam (poorly drained). Some water may move deeper and for longer distances laterally before being discharged. In the spring, the water table (surface of the ground water) may be at or slightly above the surface of Pella silt loam, may be only 15 to 30 inches below the surface in Lamartine silt loam, and, at the same time, several feet below the surface in the Dodge silt loam.

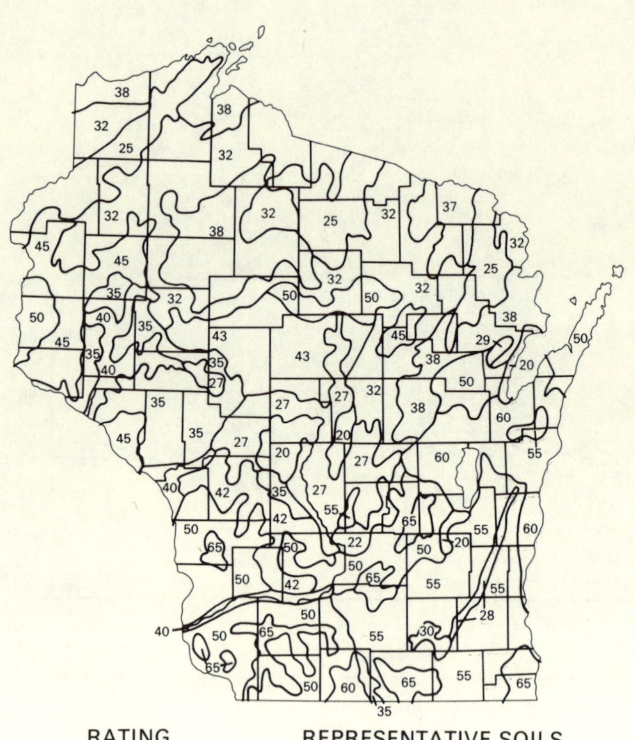

RATING	REPRESENTATIVE SOILS
Very Low - 20 to 30	(Plainfield, Vilas, Omega, Rodman, peat and muck soils)
Low - 31 to 43	(Hixton, Gale, Withee, Kennan, Gogebic, Santiago, Dakota, Onamia, Ontonagon, Pence, Iron River, Hibbing soils)
Medium - 43 to 50	(Fayette, Dubuque, Onaway, Antigo soils)
High - 51 to 64	(Kewaunee, Dodge, St. Charles, Fox soils)
Very High-65	(Tama, Dodgeville, Plano, Varna, Jewett soils)

Figure 2-53. Generalized soil productivity rating map for cultivated row crops, without irrigation or drainage. Ratings do not relate to yields of any particular crop, but are based on several soil properties and on length of growing season (generalized from Hole, 1955).

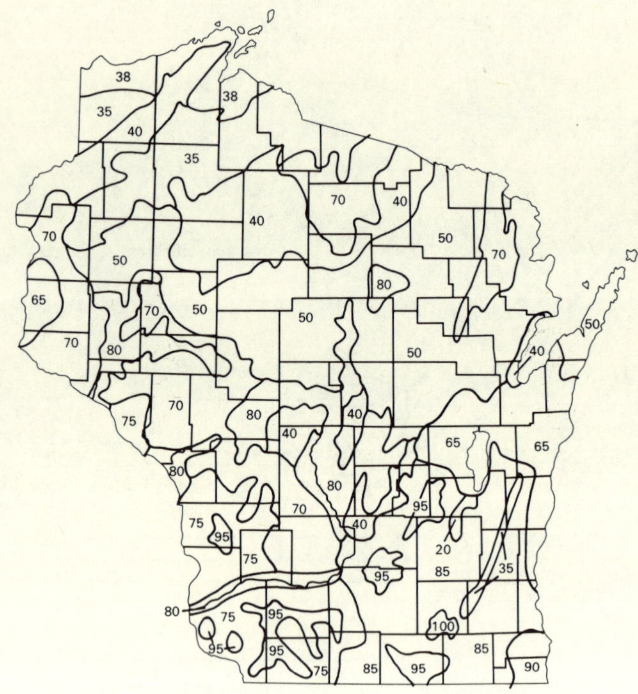

RATING	REPRESENTATIVE SOILS
Very Low - 20 to 39	(Ontonagon, Gogebic, Rodman, permanent wetlands)
Low - 40 to 49	(Pence, Iron River soils)
Medium - 50 to 79	(Kewaunee, Vilas, Omega, Withee, Stambaugh, Norrie soils)
High - 80 to 94	(Plainfield, Sparta, Antigo, Onamia, Dakota, Dodge, St. Charles, Fox soils)
Very High - 95 to 100	(Peat, muck, Tama, Plano, Dodgeville soils)

Figure 2-54. Generalized soil productivity rating map for cultivated row crops, with irrigation, drainage, and fertilization (as needed) on selected areas. The elevation of these ratings above those of Fig. 2-53 may be ascribed to intensive soil management. (Estimates are by R. D. Powell and F. D. Hole, with reference to Hole, 1955.)

The Dodge-Lamartine-Pella catenal sequence in soil association B13 shows the following features:

1. Increases down-slope of
   a) Organic matter content and thickness of the dark topsoil
   b) Mottling and gleying (bleaching) of the subsoil
   c) Content of the montmorillonitic clay
2. Decreases down-slope of
   a) Soil acidity
   b) Oxidation-reduction potential (Becking, Kaplan, and Moore, 1960)
   c) Distinctness of the clay-accumulation in the subsoil (Bt horizon) (Milfred and Hole, 1970; Weaver, 1970; Al-Rawi et al., 1969).

On sandy soils (Plainfield-Nekoosa) in central Wisconsin, growth of red pine improves down-slope as depth to water table decreases from 10 feet to 4 feet (Dosen, Peterson, and Pronin, 1950).

Each soil catena has a characteristic regime of movement of water over and through it, and therefore has a characteristic runoff potential (Chiang and Petersen, 1970). The movement of water into and through soils is related to such factors as texture, crusting, structure, biopores (Bouma and Hole, 1965), and permeabilities of soil horizons. During much of the growing season, water movement is slow and is limited to unsaturated flow in the soil mass lying between cracks and channels (Bouma et al., 1972), which conduct water only under conditions of saturation, such as are present during wet spring months.

Table 2-9. Approximate age of soil initial materials in Wisconsin

Period or System	Material	Age[a] (years before present)
Quaternary	Alluvium (surficial layers)	1 to 1,000
	Peat and muck	1 to 13,000
	Loess (surficial layers)	5,000 to 30,000
	(Peorian)[b]	(5,000 to 22,000)
	(Farmdale)[b]	(28,000 to 70,000)
	Glacial outwash and ice-contact deposits	5,000 to 30,000
	Glacial till (Wisconsinan)	11,000 to 70,000
	(Valderan)	(5,000 to 11,000)
	(Woodfordian: Mankato, Cary, Tazewell, Iowan)	(12,500 to 22,000)
	(Altonian: Farmdale)	(28,000 to 70,000)
	Residuum and duricrusts on bedrock	30,000 to 30 million
Devonian	Dolomites and shales	400 to 413 million
Silurian	Dolomites	413 to 425 million
Ordovician	Maquoketa (shales with dolomite)	425 to 475 million
	Sinnipee (dolomites with limestones and shales)	
	Ancell (sandstone with shale and conglomerate)	
	Prairie du Chien Group	
Upper Cambrian	Sandstones with dolomites and shales	475 to 500 million
Precambrian Groups	Igneous and metamorphic rocks	600 million to 3.5 billion

[a]Note that the age of the initial (parent) material is almost always much greater than the age of the soil (see text). These estimates are based on Black and Rubin (1968), Hogan and Beatty (1963), Dury and Knox (1971), Wascher et al., (1971), and Dott and Batten (1971).

[b]The boundary between these two loesses has not been distinguished in the field in Wisconsin.

AGE OF THE SOILS

The age of Wisconsin soils ranges from less than a year to approximately 24,000 years (Fig. 2-55). The very new soils are those forming in deposits made by the latest floods on river bottoms or on fill materials placed around new suburban homes by landscape contractors. The oldest soils are on upland crests in the "Driftless Area," soils of medium age in the Woodfordian Drift area, still younger soils on the Valderan Drift, and the youngest in accumulating deposits such as peat bogs and the alluvium just mentioned (see Table 2-9).

Some soils, including rather ancient ones, have been found lying under loess, glacial drift, and Paleozoic strata. Probably many more await discovery. Dury and Knox (1971) report horizon sequences more than 25 feet thick in the Ordovician Ancell (St. Peter Sandstone). A typical sequence consists of ferruginous and siliceous duricrusts, a mottled zone, and a pallid zone. Whether at the modern soil surface or lying under Sinnipee (Platteville-Galena) dolomites, the profiles are similar to ancient soils (paleosols) of Australia. Those of Wisconsin are estimated to be mid-Miocene in age, or about 20 million years old. Residual cherty reddish clay resting on limestone in southwestern Wisconsin may represent as much as 500,000 years of leaching of dolomite bedrock that contained 4% of insoluble residues (Black, 1970a). Black and coworkers (Black, 1970a; Black et al., 1970) report remnants of old soils, probably Sangamonian in age (about 70,000 years old; see Table 2-10) in several southern counties. A poorly drained Entisol, buried under 20 feet of glacial drift at the Marshfield Agricultural Experiment Station (Hole, 1943), has been dated at more than 45,000 years B.P. (Black and Rubin, 1968). Hogan and Beatty (1963) and Milfred (1966), who studied soil material that was buried under deep loess in southwestern Wisconsin, estimated the paleosol to be 29,400 years old. Paleosols on Altonian tills in southern Wisconsin were studied by Bleuer (1970, 1971) but cannot be dated because of the lack of old charcoal and wood material isolated from modern roots. The paleosol of the Two Creeks Forest Bed of Manitowoc County is only 11,840 years old (Black, 1970b). It has characteristics today of a poorly drained Entisol (Aquent), but may have originally been a Spodosol (Lee and Horn, 1972). A buried soil at the Aztalan village site is about 800 years old (Jaehnig, 1968).

Ages of bedrock (Table 2-9), determined by radioactive decay of various elements, are usually not related to ages of soils on them because these rocks, long buried, have been exposed only in relatively recent times. It is actually the age of the land surface that is most important in studies of soils. For example, the surface of a small, rapidly weathering sandstone butte near Camp Douglas in central Wisconsin is probably less than a century old, although the butte may be 12,000 years old and the rock itself is 500 million years old. Erosion is rapid on the steep sides of a butte, and hence the soil is thin or absent, except in cracks in which some fine material has collected and down which roots of pine trees have grown. In southwestern Wisconsin, shallow soils and stones on grazed grassy slopes (30 to 50% gradient) move at a rate of about 17 feet in 1,000 years (170 feet in 10,000 years), with a vertical lowering of the sloping surface of the land by 6 feet per thousand years. This process, by which cliffs have been exposed at the tops of valley slopes in southwestern Wisconsin, is slower in ungrazed, wooded areas (Black, 1969a, 1970a; Black and Hamilton, 1973). On less steep hillslopes of southeastern Wisconsin, much deeper and better developed soils (Fig. 8-8) have formed in an estimated

Table 2-10. Time-stratigraphic classification of the Pleistocene deposits of Wisconsin

C14 dates (years before present)	Time stratigraphy					Rock stratigraphy				Soil stratigraphy
	QUATERNARY SYSTEM	PLEISTOCENE SERIES	Holocene stage			Alluvium and lacustrine deposits				Modern soil
11,000			Wisconsinan stage	Valderan substage				Valders formation	Outwash & ice contact deposits[a]	
12,500				Twocreekan substage		Alluvium and lacustrine deposits				Two Creek Forest Bed soil
				Woodfordian substage	Cary	Peoria loess	Richland loess	Wedford formation	Outwash & ice-contact deposits[a]	
					Tazewell substages		Morton loess			
22,000					Iowa					
28,000				Farmdalian substage		Alluvium and lacustrine deposits				Farmdale soil
				(Rockian)[b] Altonian substage		Roxana silt?		Winnebago formation	Outwash & ice-contact deposits[a]	
			Sangamonian stage			Alluvium and lacustrine deposits				Sangamon soil?
			Illinoian stage			Loveland silt		Glasford formation	Outwash & ice-contact deposits[a]	
			Yarmouthian stage			Alluvium and lacustrine deposits				Yarmouth soil?
			Kansan stage					Banner formation	Outwash & ice-contact deposits[a]	

Source: Adapted from work in Illinois and Wisconsin as reported by Frye, Willman, and Black (1965), Willman, Glass, and Frye (1963), and Willman and Frye (1970). See Emiliani (1972) for evidence that fluctuations of climates during the Quaternary, from high to low temperatures, were abrupt and far more numerous than indicated by the classical climo-sequential model.

[a]Outwash and ice-contact stratified deposits are time-transgressive and are of much less use than till and loess in correlation of Pleistocene deposits.

[b]An old substage term (Flint, 1945). Note that D. M. Mickelson (personal communication, 1972) and R. F. Flint (1971) use the terms "Wisconsin," "Sangamon," and "Illinois" without the adjectival endings "an" and "ian."

period of 13,000 years (Milfred and Hole, 1970), with less mass movement.

Age of soil has two principal meanings: (1) actual age of soil in years, usually in terms of thousands of years, as has been noted; and (2) relative degree of maturity of the soil profile, as expressed in terms of minimal, medial, and maximal development (see Hole et al., 1962), which may not be strictly related to age in years.

Thickness of the soil solum (A plus B horizons) is commonly used as a measure of soil age. For example, in the absence of erosion, depth of an upland soil increases with age by downward advance of the lower boundary of the B horizon into the geologic material. Pedersen (1954) observed this in a comparison he made of somewhat shallow Elliott soils of southern Wisconsin with the deeper ones in Illinois. Accelerated erosion of uplands, since the coming of farmers from Europe to Wis-

Table 2-11. Estimates of age of soils developed from unconsolidated geological deposits on uplands in and near Wisconsin

Soil horizons	Estimated ages: Pedologically initial mineral deposits (years)	Oldest plants in vegetative cover, today (years)	Soil on completion of its present horizon(s) (years)	Texture of initial mineral material	Depth of soil (cm)	Reference adapted
A1 (mull, ochric epipedon) horizon in a Hapludalf (Gray-Brown Podozolic) soil formed from weathered loess in Wisconsin	13,000	200	265 ± 50	Silt loam	7	Van Rooyen (1973)
A1 (mollic epipedon) horizon of a Hapludoll (Brunizem) soil formed from loess in Iowa and Wisconsin	19,000	50	275 ± 140 to 400	Silt loam	33	Simonson (1959); Arnold and Riecken (1964); Van Rooyen (1973)
A and B horizons (solum) of a Spodosol (Podzol) soil formed from acid outwash in Wisconsin	12,000	200	1,000	Loamy sand	50	Tamm and Östlund (1960); Milfred et al. (1967)
A and Bt horizons (solum) of a Hapludalf (Gray-Brown Podzolic) formed from weathered loess in Iowa	19,000	200	4,000	Silt loam	100	Arnold and Riecken (1964)
A2 and Bhir/A'2 and B't sequence (Alfic Haplorthod) formed from acid outwash in Wisconsin	12,000	200	2,500	Sandy loam	32	Milfred et al. (1967); Arnold and Riecken (1964)
Two meter-thick Histosol (peat) in a bog Wisconsin	—	100	5,000	Peat	600	—

consin, has resulted in a rapid thickening of the young alluvial soils on bottoms (Riecken and Poetsch, 1967; Knox, 1972). Knox has pointed out that the man-induced acceleration of erosion in the landscape of southwestern Wisconsin is analogous to that associated with channel morphology adjustments produced naturally by Holocene climatic fluctuations, as recorded by a buried coarse deposit made 6,000 years B.P. near the end of a major drought. On the upland, the black topsoil of prairie soils has thickened slowly with time (see Fig. 7-6) (Hole and Nielsen, 1970). Van Rooyen (1973) estimated that about 275 years are required to form a mollic epipedon in the vicinity of Madison, Wisconsin. An ochric epipedon in a nearby forest requires about 265 years to form. The B horizons of forest soils have developed increasing clay accumulations with age (Muckenhirn et al., 1955). "Clay skins" (argillans) on the walls of cracks and channels are prominent in well-drained soils developed in loess on 14,000-year-old glacial drift in southeastern Wisconsin. By multiplying the per cent by volume of argillans in soil horizons times the thickness of horizons in centimeters, *profile clay illuviation indices* can be obtained that may be interpreted as indices of age—the larger the index, the older the soil. Miedema and Slager (1972) found that deciduous forest soils of north temperate regions (Typic Hapludalfs) have profile clay illuviation indices ranging from 450 to 850 (see Chapter 8, soilscape B25). Old soils are likely to be less fertile than young ones. The meaning of age of soil depends, therefore, on kinds of initial material and kinds of processes of soil formation that have operated, as well as the length of time involved.

Original lime (calcite, dolomite) has been leached out of some Wisconsin soils to a depth as great as 6 feet (Robinson, 1950). Clay minerals have been somewhat altered in that zone, but the bulk of the soil material is still little weathered (Hendrix in Hole and Schmude, 1959; Al-Rawi et al., 1969). As compared with soils of southeastern United States, which Thorp (1965) estimated to be hundreds of thousands of years old, most soils of this state are young indeed. Their mineralogic heterogeneity and geologic youthfulness account for their large reserves of the mineral nutrients on which plants depend. Some estimates of age of individual horizons and sequences of horizons, presented in Table 2-11, indicate that the surface horizons form more quickly than the subsoil horizons, and that clayey subsoils (Bt horizons) have required possibly four times as much time to form as have spodic horizons (Bhir of Podzols). There are many unknown variables involved in the processes (see Chapter 3) by which a soil horizon reaches equilibrium. Once this state of equilibrium has been attained, further changes in the soil may be so much slower than before that a vastly different time scale must be resorted to in evaluating them.

Several methods of estimating age have been used. Hypothetical correlation of duricrust soils with subtropical and tropical forest cover suggests the mid-Miocene age of the profiles studied by Dury. The extent of decay of the radioactive isotope of carbon, C^{14}, is useful to a maximum of about 60,000 years. This measurement was applied to buried spruce wood to obtain the dates mentioned for the Two Creeks Forest Bed. Robinson (1950) estimated ages of Wisconsin soils developed from calcareous materials. His calculations were based on (1) estimated amounts of carbonates that have been leached from the now acid upper horizons and (2) the amount of agricultural lime known to be leached annually from farmers' fields. Despite the uncertainties of his method, the results approximate those indicated by C^{14} dating (see Hole, Peterson, and Robinson, 1952).

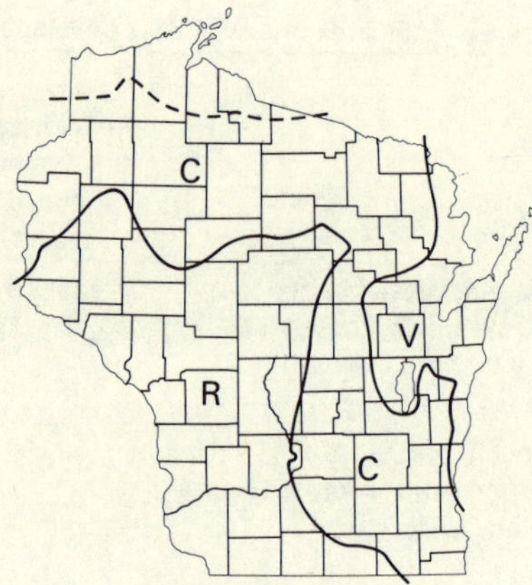

R 16,000 to 24,000 years
C 10,000 to 13,000 years
V 5,000 to 10,000 years

The dashed line represents a glacial drift boundary drawn by Leverett; it is possible that the territory north of it in Wisconsin is part of Region V.

Figure 2-55. Estimated ages of surfaces of Wisconsin landscapes (duricrust not considered).

MAN'S INFLUENCE ON THE SOIL

Before the coming in 1634 of Jean Nicolet, the first European explorer to visit the Great Lakes region, the area that we call Wisconsin was inhabited by about 12,000 American Indians of three tribes (Ritzenthaler, 1953) out of the approximately one to twelve million Indians who occupied the lands within the present limits of the forty-eight coterminous states of the United States (Farb, 1968). Small as their number in Wisconsin may have been, the Indians did over a period of many thousands of years make observable changes in soils. These are especially evident near lakes and streams. Archeological and paleoecological investigations have revealed Indian disturbance of soil at such sites throughout the state by digging, by ridging (Fox, 1959), and by accumulation of phosphates and other materials at camp and village locations and at effigy and other ceremonial mounds. It is quite possible that some prairie and forest fires were set intentionally by these native people in order to increase food supply for wildfowl and game animals (Curtis, 1959; Milfred and Hole, 1970). Dark soils formed under grasslands in burned areas.

In comparison with settlers from abroad and their descendants, the American Indians lived in immeasurably greater harmony with the natural environment. In fact, the Menominee Indians are still trying to continue this relationship on their tribal lands (Milfred, Olson, and Hole, 1967). In terms of degradation of soil from its original wild state (Hole, 1974a), postsettlement culture has taken considerable toll on the environment as the American Indians knew it. This is the price paid thus far for attainment of the present high living standards for over four million people in Wisconsin, most of whom reside in cities and villages. However, by systems of soil and water conservation, rural zoning, and regional planning, abuse of the soils can be kept to a minimum (Rolands, 1963; Bauer, 1964; Yanggen, Beatty, and Brovold, 1966).

Human beings have had a definite influence on the five factors of soil formation previously listed (Bidwell and Hole, 1965). In Wisconsin, this influence has been intensified since settlement by Europeans. In this section frequent reference will be made to some of the processes going on in soils, but the comprehensive treatment of these is reserved for the next chapter.

Mineral materials added to the soil by people may be thought of as new soil initial materials: pulverized limestone, mineral fertilizers, and wood ashes from stoves. In a sense these additives have rejuvenated[6] the soils by making them more like the original unleached calcareous loess and glacial drift. Additions of livestock manure and "green manure" have caused some forest soils to take on the appearance of prairie soils to a depth of 9 to 10 inches. Where soils have not been fertilized, although extensively cropped, removal of plant nutrients through harvesting has left the soils less fertile than they were in their natural state.

Human activity in Wisconsin has resulted in changes in topography of the soil surface, at least on a small scale. Some stream bottoms have been raised as much as 10 feet by deposition of soil material washed from exposed hillsides.[7] Uplands have been shaped by machinery to create drainageways and terraces in fields, and concrete structures have been built to check gullying (Zeasman and Hembre, 1963). Except where it has caused gullying, cultivation has tended to make fields smoother than they were originally. For example, tree-tip mounds have been erased in cropped areas of northern counties (Gaikawad and Hole, 1961). Wetlands have subsided as a result of drainage. Oxidation of drained peats and mucks over years of cultivation has gradually reduced the volume and lowered the surface of these soils.

The climate of the soil itself, that is, its temperature and moisture regimes, has been drastically changed in this state by removal of the natural forest and prairie cover. The cutting from 1860 to 1920 of an average of a billion board feet per year of the native forest, which originally totaled 200 billion board feet, exposed the soil to sudden temperature

6. Determinations of soil acidity (pH) in paired wooded and cultivated soils in southeastern Wisconsin in 1958 showed an increase of about one unit (as from 5.5 to 6.5) in the first 2 feet of soil in the fields (personal communication, John Cain).

7. Annual rainfall in the state delivers about 3,300 tons of water to each acre. In severe rainstorms raindrops fall as fast as 20 miles per hour and strike exposed soil with a force of about 25 pounds per square inch (personal communication, A. J. Wojta, 1955) (see also Ekern, 1950).

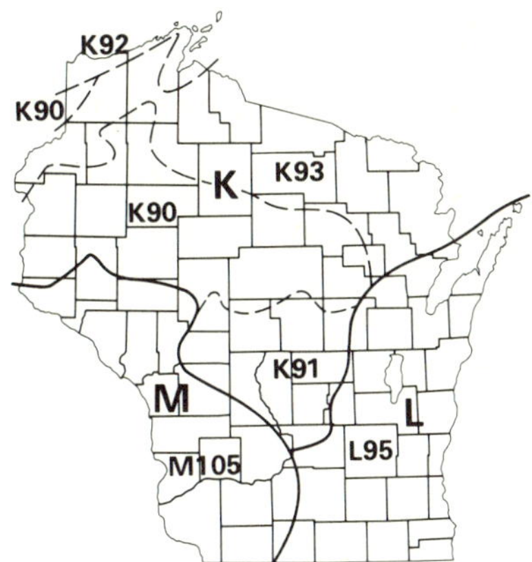

K Northern lake states forest and forage region
K90 Central Wisconsin thin loess and till area
K91 Wisconsin sand outwash area
K92 Superior lake plain area
K93 Northern Wisconsin stony, sandy, and rocky plains and hills area
L Lake states fruit, truck, and dairy region
L95 Southeastern Wisconsin drift plain area
M Central feed grains and livestock region
M105 Northern Mississippi Valley loess hills area

Figure 2-56. Land resource regions and major land resource areas (after Hobbs, 1963).

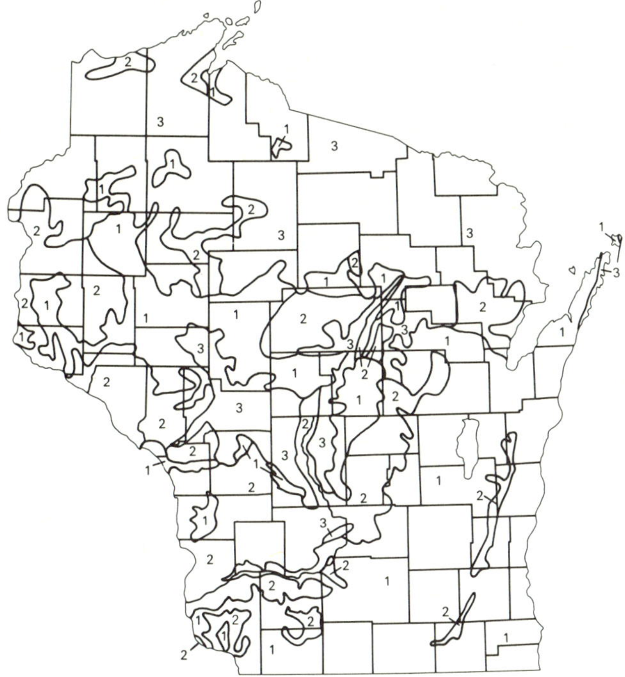

1. Largely cropland
2. Pasture, woodlots and cropland
3. Largely forest

Note: Cities and wetlands not shown

Figure 2-57. Generalized land use, 1956 (after Johnson and Carley, 1963).

changes. Early observers report that the forests dewatered the soils more than later crops and pasture grasslands did. Much of the charcoal that we find in the soil today is inherited from the lumbering days when great fires swept the cutover land. Natural deepening of the dark topsoil (A1 horizon), already noticeable in southern forests before Europeans came (Wilde, 1940), was extended to northern areas by fires and by the succeeding vegetation. This deepening was accelerated in many places by the importation of the large earthworm *Lumbricus terrestris* (Milfred, Olson, and Hole, 1967) and the spread of Kentucky bluegrass and quackgrass. Soil climate has been affected in lowlands by man-made changes in the position of the water table. Artificial drainage of wetlands has introduced a more variable soil climatic regime to formerly poorly drained soils. Impounding of water by dams along rivers and streams has caused a local rise in water table, converting some well-drained soils into poorly drained ones, with consequent changes in vegetation and land use. Planting of shelter belts on sandy soils has fostered accumulation of snow in winter and has ameliorated somewhat the dessicating effects of exposure of the soil to direct sunlight and winds in summer (Albert, 1951). Irrigation has made profound changes locally in soil climate.

The content and the distribution of organic matter in the soil have been changed by cultivators. In general, organic matter content of the surface soil horizons has been reduced by cultivation, but heavy manuring of sandy soils has had the opposite effect in places.[8] Plowing has tended to distribute organic materials evenly through the upper 7 inches of mineral soils in the approximately 11 million acres of cropland. In cultivated fields, compaction of the soil by machinery is counteracted somewhat in the surface layer by annual plowing, and by earthworms and other burrowing organisms. These are especially abundant wherever row crops are produced without plowing but rather with "trashy" farming practices (Peterson and Engelbert, 1959; Dixon and Peterson, 1971). As the area of cultivated fields has expanded (Anderson, 1967), there has been a reduction in the variety of plants and animals affecting the soil, as compared with the original array in natural ecosystems (Leopold, 1949). Where overgrazing has exposed soil of uplands and of stream and lake banks to erosion, contamination of waters with soil has developed into a serious man-made nuisance.

It is important, in connection with a discussion of man's influence on the soil, to differentiate between two uses of the word *development.* In this book the word is nearly always used to denote degree of soil profile formation. For example, Fayette silt loam has as much as 31% clay in the

8. The approximately 30 million tons (dry weight) of farm animal manure produced in the state each year and spread on the land has been estimated, in terms of plant nutrient value, to be worth $100 million.

subsoil (Bt) horizon, whereas the associated Seaton silt loam has only 24% clay in that horizon. The Fayette is therefore considered to have greater development than the Seaton (see Chapter 7, soil associations A5, A7, A8). This kind of comparison may be made between the thick topsoil (A1 horizons) of a prairie soil and the thin topsoil of a forest soil, with respect to pounds of carbon per acre (Russell, 1961). On the other hand, *development* is commonly used in an economic sense, as when commerical or residential use is made of agricultural or wilderness areas. We are used to saying that housing or industrial development is taking place on the fringes of cities.

Wisconsin's nearly 200 cities and 400 villages occupy about 6% of the land area of the state (Wisconsin Conservation Needs Committee, 1970), and most of them (along with cross-country highways) are expanding at the expense of agricultural lands (Durand, 1962). The disturbance, enrichment, and pollution of soils and the sealing over of the land surface with buildings and pavements is intense in urban areas and will have a lasting effect.

Three major land-resource regions and several land-resource areas have been delineated in Wisconsin on the basis of geologic features and agricultural enterprises (Fig. 2-56). The present land-use pattern may be seen in terms of encroachment of pasture and cultivated fields on forests, and of urban areas on rural lands (Fig. 2-57).

CHAPTER 3

Processes of Soil Formation

INTRODUCTION

The environmental factors of soil formation were more clearly defined early in the history of soil science by Dokuchaev (1879) and Hilgard (1892) than were the corresponding soil-forming processes. Factors and processes are inseparable, but the second category is the more difficult to analyze. It is still easier, even today, to describe the environment in which a soil is found, and to describe the soil itself, than to determine the soil-environment interconnections in terms of interplay of processes that formed and continue to form the soil. In Chapter 2 the customary approach of stressing factors first has been followed, though not without reference to processes, particularly in the section on man as a factor of soil formation. In the present chapter emphasis will be placed on processes of soil formation.

Processes have been at work in most of our soils for thousands of years to transform and combine geologic and organic materials (Fig. 4-1), to arrange them in units called peds and horizons (Fig. 4-2), and to evolve numerous chemical, physical, and biological properties. Chamberlin (1883) spoke of the glacial drift as being in "a crude uncongenial state" before agencies of soil formation had acted on it to produce "deep, rich and enduring soils."

Soil is formed wherever plants and associated animals and microorganisms congregate on and penetrate into rock and around mineral grains. Through the soil flows what Aldo Leopold (1949) called a "fountain of energy." Lichens and moss growing on bare rock cliffs of the Baraboo Range and Rib Mountain or on steep rock banks of the Wolf River have helped to weather the rock ever so slightly. However, most of the soils of Wisconsin have formed in unconsolidated (nonbedrock) deposits, including glacial till, outwash and lacustrine deposits, loess, and the original vegetable and animal ingredients of humus and peat.

Examples of subsoil features that may have formed millennia ago are fillings of animal burrows (krotovinas) made by rodent activity within a single season, and sesquioxide (certain oxides of aluminum, iron, and manganese) concretions precipitated very slowly in the course of centuries. We need not assume, however, that no living person has seen a pedogenic process in operation. Some surficial phenomena are readily observable, such as deposition of casts by earthworms, building of small mounds by ants, the opening and closing of cracks in soils with the succession of droughts and wet periods, and autumnal leaf fall that replenishes the O horizons of soils.

The total number of possible pedogenic events and combinations and interactions that occur within a depth of several feet in soil is beyond comprehension. In order to simplify the whole complex of soil-forming processes so that we can think about them, several general ecological terms were devised by the early Russian and American soil scientists. For example, *podzolization* is that combination of processes that produces Spodosols (Podzols). This is preceded by *leaching* of any carbonates that may be present and thereafter includes (1) the accumulation of an acid mat of forest litter and humus resting on the mineral soil (with little mixture into it), and (2) the downward movement from the A to the B horizon of organic matter and sesquioxides to produce the characteristic dark brown "spodic" horizon in the subsoil. Three other classic processes of soil formation have been at work in Wisconsin. One is melanization, the all-important incorporation of organic matter into the mineral surface soil, which is most evident in Mollisols formed under prairie vegetation. Another is *gleization,* the production of a bluish-gray color in the mineral subsoil at wet sites. *Lessivage,* the washing of particles of clay and fine silt from the topsoil (A horizon) into the subsoil (B horizon) and possibly farther, is the third process. Ranney and Beatty (1969) saw evidence of this even in soils of northern Wisconsin, where podzolization would be expected to decompose clays (leaving siliceous residues in the topsoil or A2 horizon) rather than permit them to be transported unchanged. Lessivage accounts in considerable part for the concentration of clay in the Bt horizons of Alfisols.

Leaching, podzolization, and lessivage are accomplished by percolating waters. In Wisconsin 70% of the annual rainfall (21 out of 31 inches) is evapotranspired, and of the remaining 10 inches that run off, about 6 inches of water pass through the soil and become part of subsurface runoff. This latter may be double the amount of water that percolated all the way through the soil prior to the development of agriculture, because the crops that were substituted for the native forests and prairies have a lower annual water consumption. Most of the water that evapotranspires passes through some part of the soil solum first. Percolating water has brought about many changes in Wisconsin upland soils during the 10,000 to 20,000 years of their formation.

Two general kinds of processes and conditions are observed in our soils (Hole, 1961). There are those that tend to produce profiles with many horizons and those that tend to produce profiles with few horizons. A Spodosol illustrates the first case. This soil may have seven distinct horizons above the C horizon (see Fig. 1-7). The Boone loamy sand (Fig. 1-3) is an example of the second case. In this soil, the thin organic horizons and dark topsoil (A1 horizon) are the only ones above the initial material (C horizon). The relative complexity of layering of a soil may not tell us much about its age. A Spodosol with five major horizons (O, A1, A2, Bhir, C) may be no older than 2,250 years (Franzmeier and Whiteside, 1963). A Mollisol with four major horizons (O, A1, B2t, C), with thickness of the A1 as great as 60 cm, may be 6,000 years old (Fig. 7-5) (Baxter and Hole, 1967).

Opposing processes at work in soils include those that tend to loosen it and increase its porosity as contrasted with those that cement it and form various kinds of "pans." Processes that leach nutrients out of the soil contrast with those that concentrate nutrients in it, as by biocycling in a deciduous forest (Milfred, Olson, and Hole, 1967). The balance between opposing processes determines the nature of a given soil horizon. According to the concept of entropy, soils tend to "run down," that is, to become infertile and uniform. But this theoretical "doomsday" condition is perpetually postponed by antientropic or rejuvenating processes of melanization, biocycling, aeolian enrichment, soil mixing (pedoturbation), and gradual migration of the lower boundary of the soil solum into fresh, fertile materials below.

In most soil profiles, both complex and simple, we can identify four principal kinds of processes: (1) alteration of materials in place, (2) removal of materials from one or more horizons, (3) additions of materials to one or more horizons, and (4) mixing or homogenizing of one or more soil horizons (Simonson, 1959; Hole, 1961). These will now be considered, along with processes of the plant-soil system that preserve soil.

ALTERATION OF MATERIALS IN PLACE

In road-bank exposures of till-derived soils we find occasional stones of crystalline rocks that have been softened by weathering and entry of plant roots. Such stones may have already been somewhat altered when deposited in a frozen state by glacial ice, but alteration has proceeded further in the course of soil development. Clay and silt-size particles of mica are gradually altering to more expansible kinds of clay (montmorillonite and vermiculite) in the soils of the state. Pictures have been taken through electron microscopes showing mica particles partially weathered to expansible clays (Borchardt, Jackson, and Hole, 1966).

In general, however, we can say that sands and coarser materials in Wisconsin soils have weathered very little during soil formation. Microscopic examination usually shows these coarse fragments to be virtually fresh and unweathered, despite a surface coating of iron oxide stains. This suggests that the vast amount of weathering of solid rock that produced these clastics was primarily physical and that few products of chemical weathering have survived in the coarse fractions. The clay and fine silt particles have undergone more change than the sands. Profound alteration of vast quantities of organic materials has been rapid at well-drained sites, particularly in the O horizons in which leaf litter is transformed into humus.

REMOVAL OF MATERIALS FROM ONE OR MORE HORIZONS

Weathering releases soluble and colloidal materials from minerals and organic matter. Leaching (eluviation) removes these materials from their place of origin in the soil.

The pale A2 (albic) subsurface horizon of Spodosols (Podzols; Fig. 1-7) is an example of a depleted soil horizon. Calcium, magnesium, and other cations have been partially leached out of this layer, along with significant amounts of iron, aluminum, organic matter and clay. Such depleted horizons are known as eluvial horizons.

Probably the most abundant mobile mineral constituents in Wisconsin soil profiles today are calcite ($CaCO_3$), commonly referred to as lime, and its associate, dolomite ($CaMg(CO_3)_2$). These carbonates of calcium and magnesium have been removed from 6 or 7 feet of the loess in which the Fayette soil has formed (Fig. 7-9). This entire depth of material is eluvial with respect to carbonates. The agricultural practice of liming fields is to renew the supply of carbonates in the surface soil. Growth of most agricultural crops (that is, excepting potatoes, watermelons, cranberries, and others that do best under acid conditions) is promoted by liming, because of resulting improvement in availability of plant nutrients, in favorable soil structure, and in beneficial activity of microorganisms, and also because of reduction in toxicity and in incidence of plant diseases.

ADDITIONS OF MATERIALS TO ONE OR MORE HORIZONS

Addition of organic matter to soil is a major process of soil formation. One to 5 tons of new organic material (dry weight) are added per acre to the soil each year, mostly on the surface under mesic forest, and mostly underground under prairie in Wisconsin (Hole and Nielsen, 1970). In southern Wisconsin enough of the forest litter decomposes each year to reduce the litter and humus layer from about 3 tons (one third of which is wood) in February, to about 2 tons (one half of which is wood) per acre, dry weight, in August. The prairie litter layer may be reduced from 5 tons (dry weight) in April to 2 tons in October. In cropped fields, aboveground production is routinely removed in harvesting operations. Therefore, annual additions of organic matter are chiefly by root growth.

The new production of roots by corn, alfalfa, and timothy grass is estimated to be between 1 and 2 tons, annually, about the same as under native forest and prairie. Because corn (maize) is an annual crop, the equilibrium organic matter content of the soil to a depth of 4 feet of a former forest soil is probably only about 30 to 40 tons, and of a former prairie soil, about 50 to 60 tons (interpreted from Soil Conservation Service, 1967a). Continuous alfalfa and timothy hay probably maintains somewhat more soil organic matter than this (possibly 90 tons, to a depth of 4 feet).

Under natural vegetation, the organic matter content of the soil is maintained at about 70 tons (in forest soils) to 120 tons (in prairie soils) to a depth of 4 feet. The darkened B (spodic, argillic)[1] subsoil horizons are zones of accumulation of materials moved down from the surface A horizon. These are largely colloidal mineral and organic substances that include layer silicate clays, hydroxides of iron, and aluminum, with some oxides of phosphorus and manganese along with humus. Dark clay films ("argillans" of Brewer, 1964) coat blocky units of soil in the subsoil (B horizon) of many Wisconsin soils. These argil-

1. Dark brown (7.5YR 4/4) soil is observed in B horizons of soils developed under prairie, hemlock, and hardwood forests (Hapludolls, Haplorthods, and Hapludalfs). This suggests that the same processes are at work in each case to produce the same color.

lans contain about 85% clay, much of it fine clay, 4% free iron oxide, and 3% organic matter, as compared with the host soil matrix, which has these corresponding analyses: 24%, 2%, 0.7% (see Buol and Hole, 1959; Ranney, 1966; Milfred, 1966).[2] Some vertical streaks or "tongues" of bleached silt and very fine sand extend from the A horizon down cracks in the B horizon in forest soils (Ranney and Beatty, 1969). Acid, grainy gray coatings occur on ped surfaces in B horizons of some Argiudolls (Arnold and Riecken, 1964). Massive deposits of calcite directly below the B horizon are rarely present in Wisconsin soils because rainfall is adequate to leach carbonates. However, the Rodman gravelly sandy loam does exhibit lime crusts and miniature stalactites on the undersides of stones in the C1 horizon (Gaikawad and Hole, 1965). In the thick loess-derived soils of southwestern Wisconsin, secondary calcite deposits begin as films on ped surfaces at about 6 feet. In soils over glacial till in southeastern Wisconsin, white seams of calcite appear in the glacial till at a depth of 3 or 4 feet in places. Black films of manganese oxides line the cracks to greater depths in southern Wisconsin. Except in bleached soil layers, iron oxide stains seem to be very common in soils, giving them the typical yellowish- and reddish-brown colors. Both iron and manganese are involved in the formation of concretions and mottles in soils (Van Rooyen, 1973).

Addition of material to surface horizons in lowlands is evident in alluvial and colluvial soils. F. F. Riecken (personal communication, 1960) has called these soils "the step-children of the landscape," whose parents were the upland soils. Peat and muck soils, which also receive annual surficial increments, may be termed "the major natural organic waste dumps" of the landscape.

MIXING OR HOMOGENIZING OF ONE OR MORE SOIL HORIZONS

The absence of a dark topsoil (A1 horizon) in many Spodosols (Podzols) (Figs. 1-7, 12-9) is ascribed to absence of organisms such as earthworms, ants, and rodents that can mix forest debris and humus into the mineral soil. A contrasting condition prevails in a forest dominated by maple trees, and even more so in a prairie. In these ecosystems the animals mentioned are active in churning organic matter into the soil profile. Bacterial gums produced from the organic materials stabilize soil aggregates (Harris, Chesters, and Allen, 1966). In the Udalfs (Gray-Brown Podzolic soils) the A1 is thus formed at the expense of the pale A2 horizon. In the Udolls (Brunizem soils) of the prairie, no A2 is present because of the deepening of the dark topsoil (A1), which may be as much as 24 or 30 inches thick in places. Some mixing of soil horizon materials is accomplished by freezing and thawing (Hole, 1961). In forest lands of northern Wisconsin, tree-tipping by strong winds has churned sand and stones up into silty surface soil over as much as 10% of the area (Jacobson, 1969). Chamberlin (1883) described "the great windfall of September, 1872" as being 40 miles in length in northwestern Wisconsin. Local churning of soils by tree-fall has been intense in this area.

2. As percolating water carrying fine sediments moves down cracks in soil, it may soak into the soil peds, leaving the clay behind as a film on the ped surface. Clay coatings are also common on carbonate stones in soils (see Milfred's description of the Underhill soil, in Milfred, Olson, and Hole, 1967), which indicates that calcium ions favor deposition of suspended clay particles.

PRESERVATION OF WISCONSIN SOILS

Human survival depends, among other things, on the restriction and control of antipedogenic (soil-destroying) factors and processes (Hole, 1971). Natural processes by which soils have been stabilized and preserved for thousands of years include the following:

1. Protection of the vulnerable soil material from the erosive action of raindrop impact and runoff water by a cover of
   a) Vegetation,
   b) Leaf litter and other organic debris,
   c) Mulch of gravel and other coarse mineral fragments,
   d) Fragile soil crusts (common in cultivated fields),
   e) Exhumed soil pans.
2. Safe (nonerosive) disposal of excess water through maintenance of a high capacity of the soil-plant system to absorb and transmit water by the development of
   a) Cracks that extend downward in a three-dimensional dendritic system,
   b) Vertical tubular channels and burrows that lead from the soil surface to considerable depths, possibly to sandy substrata and jointed bedrock,
   c) Spongy surface soil,
   d) Evapotranspiration process that moves large quantities of water from soil to air.

Wisconsin soils are tenacious. They have survived innumerable storms and droughts. They are locally protected by temporary surface crusts or more resistant lag gravel, boulder pavement, and duricrust. But the most common mode of preservation is by the process of vegetative growth that forms a live canopy over the soil and a mat of roots in it. Whenever the vegetation-soil continuum is breached by running water, waves, slump, or by human activity in ditching and earthmoving, vegetation soon spreads over the exposure again and over related terrestrial deposits of sediment, protecting them from further transport by wind and water.

Practices of soil and water conservation imitate natural processes of landscape stabilization. For example, mulches are used to protect soil, and contouring, terracing, and establishment of grassed waterways and wooded shelterbelts utilize vegetation to slow the flow of water and control soil erosion.

CHAPTER **4**

Properties of Wisconsin Soils

Marvin T. Beatty, coauthor

The properties of soils reflect the impact of the active factors of soil formation, climate, and organisms, and also exert an influence on the life of plants, animals, and human beings.

SOIL COMPONENTS

General Nature and Origin of Soil Particles

Soils are made of particles of crystalline and amorphous solid mineral matter, water (or ice), air, and organic matter (Fig. 4-1). The absolute and the relative amounts of each component as well as their arrangements (soil fabric; Fig. 4-2) strongly influence soil behavior (Figs. 4-3, 4-4).

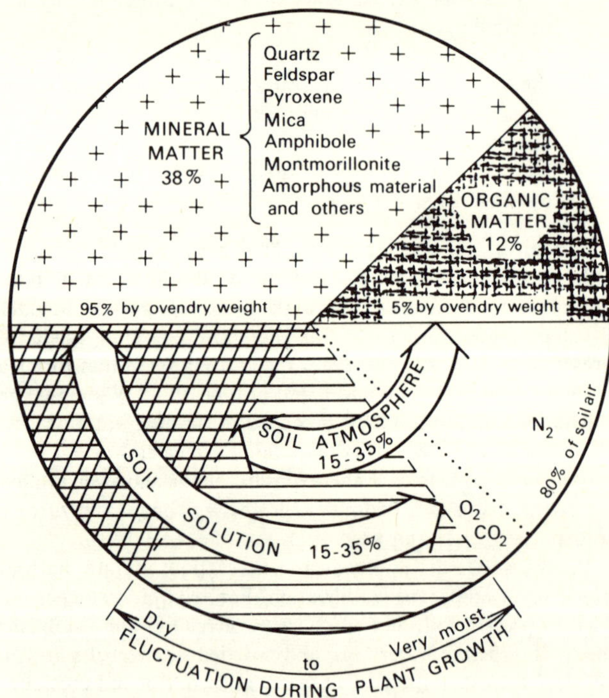

Figure 4-1. Percentage composition (by volume) of a representative stonefree A1 soil horizon. Fluctuation of water content during plant growth is shown as covering a range of 20% by volume (15-35%); contents of air and water, singly, do not fall below 15%.

Wisconsin mineral soils are for the most part rather quartzose and vary in per cent of total mineral matter, from approximately 4 to 45% on a volume basis or 70 to 99% on a dry weight basis.[1] The size distribution of soil particles strongly influences soil properties. So does the mineral composition. Quartz, for example, is chemically inert, whereas mortmorillonite has enormous capacity to interact with inorganic ions. Hence, a small amount of montmorillonite has much greater influence on active soil properties than a large amount of quartz. The mineralogical properties of Wisconsin soils are most closely related to the mineralogy of the initial deposits from which the soils formed (see Ciolkosz, 1964).

Figs. 4-5 and 4-6 show the striking contrast in clay mineral composition between clays of the Varna soil formed from clayey glacial drift and those of the Tama soil formed from thick loess in southwestern Wisconsin (Fanning and Jackson, 1966a,b). They also show how little the clay mineralogy of the soils has been modified by soil-forming processes. The suites of minerals each remain essentially as they were upon deposition.

The quartzose nature of many Wisconsin soils is illustrated in Fig. 4-7 for the Antigo silt loam. The mineralogical composition varies little within this soil even though the upper horizons (A2, B11, B12) are silt loams and the two lower horizons (IIB21, IIIC) are sandy loam and loamy sand, respectively. The Lapeer loam, Fig. 4-8, is similar in composition. The more clayey Saylesville silt loam has more layer silicate clays (mica, vermiculite, montmorillonite, chlorite, kaolinite) and less quartz because of its finer texture (Fig. 4-9). These layer silicates are most abundant in the subsoil (B horizon) where clay has accumulated.

A cubic foot of Wisconsin Montello granite, weighing about 175 pounds, has a surface area of 6 square feet or 1/7260th of an acre. The same block of granite, broken up and ground to a fine loam and mixed with some humus, provides about a thousand billion particles, the surfaces of which total about 640 acres or a square mile in area. The original granite block, with scarcely any capacity to hold water and plant nutrients, may thus be transformed into an artificial soil that can hold about 10 pounds of available water and 4 ounces of available plant nutrients, mostly calcium.[2] The 7-inch plow layer of 3 acres of dark loam soil has a total particle surface area of about 35 million acres, which is the land area of the state. The rock materials of Wisconsin have undergone a tremendous amount of physical weathering, and most of the resulting products have been transported by wind, water, and ice (Plates 4 and 5). Through leaching of these materials and admixture of organic matter under vegetative cover for thousands of years (Plate 6 and Fig. 7-6), Wisconsin soils (Plate 7) have formed from the residues of rocks, plants, and associated animals.

A cubic foot of peat is much lighter in weight than an equal volume of granite. Such a block of peat, when dried, weighs

1. This applies to surface soils. Subsurface horizons include some cemented "pans" with a higher content of mineral matter and correspondingly reduced porosity. Presence of significant amounts of gravel, stones, and other coarse fragments (> 2 mm in dia.) is associated with reduced porosity and increased bulk density (Gaikawad and Hole, 1965).

2. The artificial soil would occupy a volume of 2 cubic feet by reason of increased porosity.

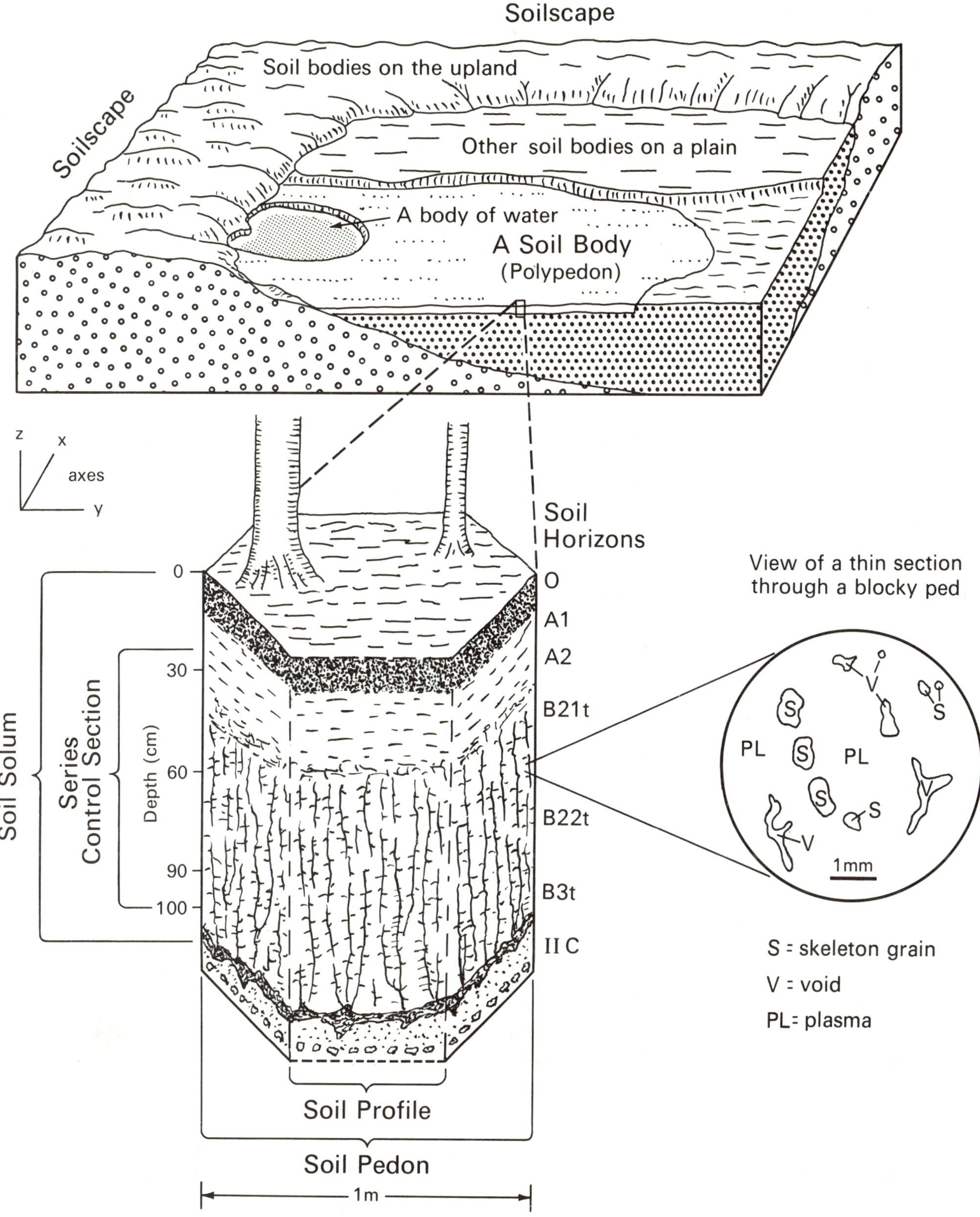

Figure 4-2. Diagram illustrating concepts of the soil pedon, polypedon, and related features. The two-dimensional front panel is the soil profile. The tree trunks symbolize a forest.

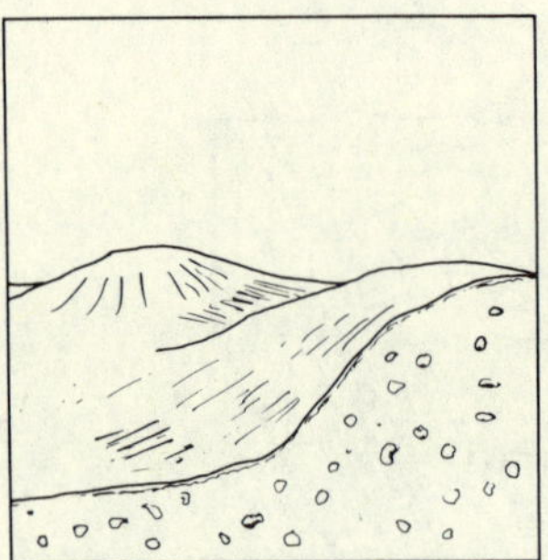
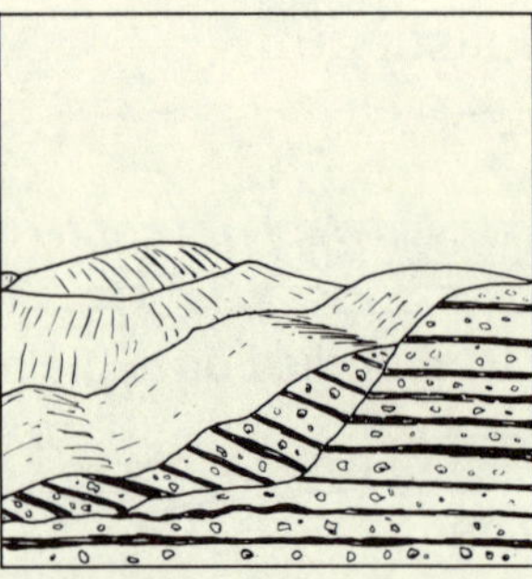

Figure 4-3. Cross section of two soil bodies with the same clay and sand contents but one (left) with homogeneous fabric and the other (right) with banded fabric. The latter is less pervious to water and more subject to landslip and piping. Illustration is conceptual.

about 8 pounds but when saturated with water weighs 70 pounds.[3] This organic soil is obviously low in bulk density (0.13) as compared with the dominantly mineral soil described in Fig. 4-1, a cubic foot of which weighs about 114 pounds saturated, 83 pounds dry, and has a bulk density of 1.32 (compared to water, which has a value of 1.0). Removal of water from saturated peat and mineral soils by artificial drainage allows the soil to change by processes that Dutch soil scientists call *ripening*. This accompanies the gradual collapse of the land surface as the soil becomes more dense, as raw fibers of sedge, moss, or wood decompose to make a very fine humic material called muck, as shrinkage cracks form, defining platy, blocky, and prismatic soil units (peds), and as worms work over the soil, producing granular peds.

There are many kinds of soils intergrading from sands through loams to peats, with various amounts of organic matter in them.

General Nature of Soil Horizons and Soil Bodies

The soil particles are organized into peds (Fig. 4-2; see Glossary) and the peds are organized into horizons. Principal horizons have developed in place since the time of deposition of the geologic materials. The A1 horizon of the soil sketched in Fig. 4-2 is a dark surface soil horizon formed by additions of humus and other organic material to the upper few inches of mineral soil. The A2 subsurface horizon is somewhat bleached by percolating water. The B2 subsoil horizon is a zone of accumulation of clay washed out of the upper horizons. The C horizon is the initial geologic material that at many sites is like that from which the A and B horizons (together called the solum) formed. However, this soil (shown in some detail in Fig. 4-2) is a deep Antigo silt loam that has formed from several feet of silt overlying sand and gravel. The lower stratum is designated IIC, meaning second (II) initial material, and the letter "C" is repeated by convention, even though the sand and gravel are very different from the initial material of the solum.

As noted earlier, farmers have plowed up the surface and subsurface horizons and converted the A1 and upper A2 into a plow layer, designated the Ap horizon (Table 1-1). People have also made additions to soil, such as organic and inorganic fertilizers, which can increase soil productivity. Some soils are actually mined, as on sod farms where the A1 horizon is periodically removed to be laid as new (instant) lawns in the cities. Unfortunately, mismanagement of soils has resulted in unintended removal of soils in many places. That is to say, serious erosion has taken one or more horizons from the upland and deposited them on low-lying soils, and in adjacent water bodies.

A landscape may be likened to a jigsaw puzzle consisting of soil bodies and included bodies of water, bedrock outcrops,

3. The dry peat, if wettable, would expand several-fold upon addition of water.

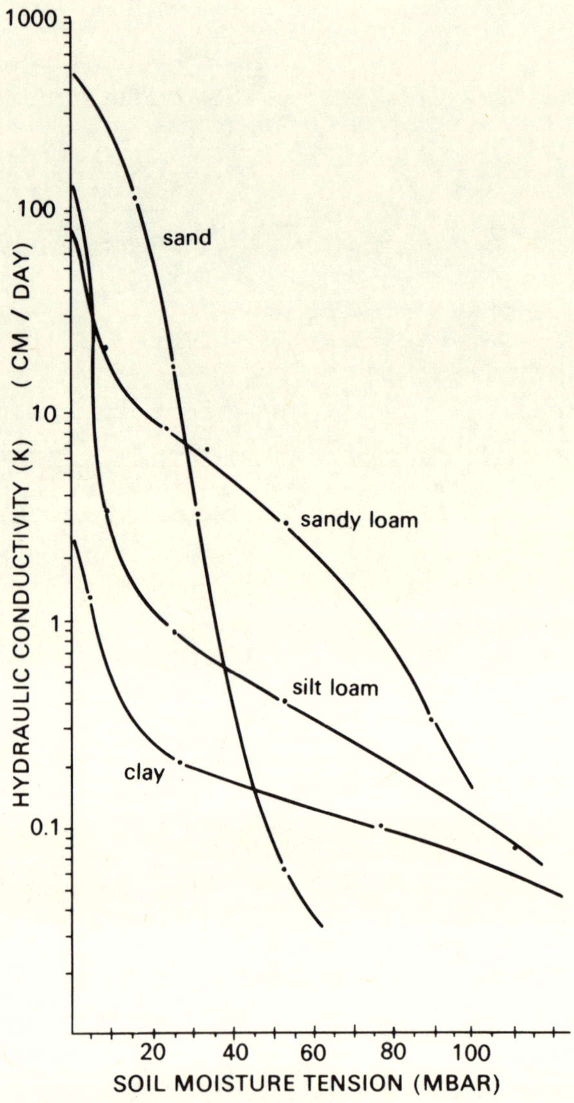

Figure 4-4. Hydraulic conductivity (K) versus soil moisture tension relationships for several soil horizons (Bouma et al., 1972). The left end of each curve represents the saturated condition of the soil, in which the water, being without tension, flows most freely, as indicated by the high K values. As the soil dries, tensions increase and flow of water (K) slows. At saturation the sand conducts water at a rate of about 16 feet (480 cm) per day, as compared to about 1 inch (2.5 cm) through clay. At 20 millibars tension, unsaturated flow of water is at a rate of about 12 inches (30 cm) per day through sand and 0.1 inch (0.25 cm) through clay.

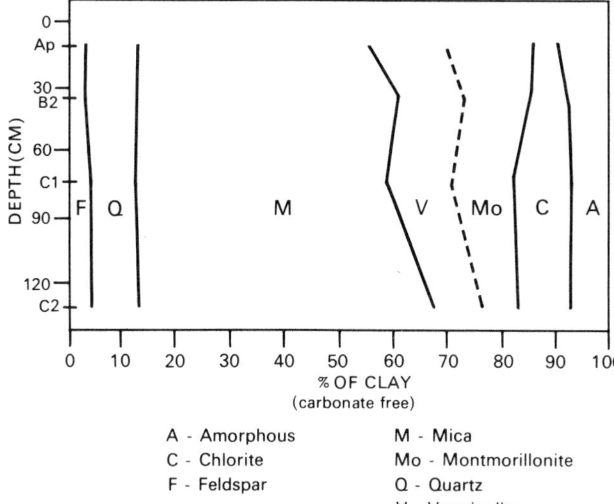

Figure 4-5. Mineralogy of the clay fraction of the Varna soil formed from clayey glacial till (after Fanning and Jackson, 1966b).

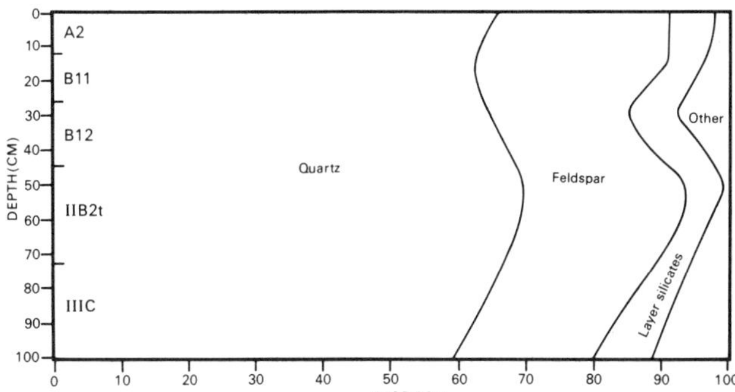

Figure 4-7. Mineralogy of the Antigo silt loam, whole soil (< 2mm) (after Al-Rawi et al., 1969). Layer silicates include chlorite, kaolinite, mica, montmorillonite, and vermiculite. "Other" refers to other minerals also present.

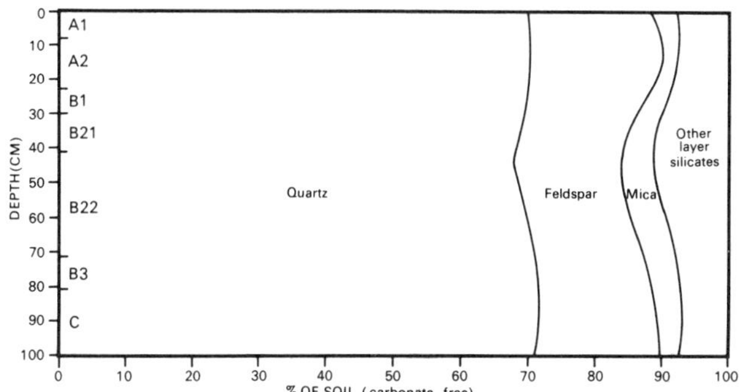

Figure 4-8. Mineralogy of the Lapeer loam, whole soil (< 2mm) (after Borchardt, 1966). Other layer silicates include chlorite, kaolinite, montmorillonite, and vermiculite.

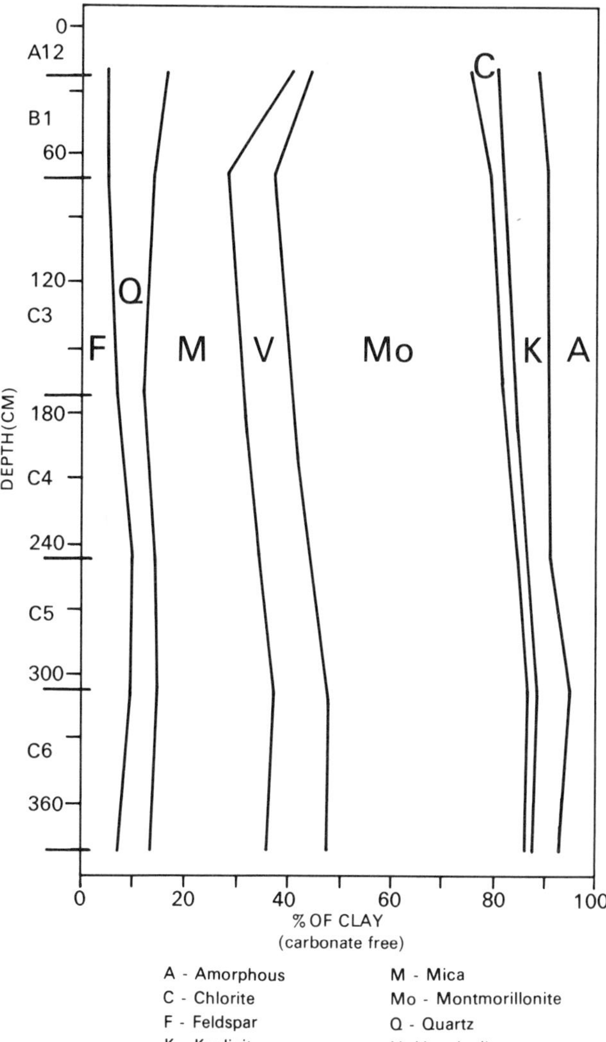

Figure 4-6. Mineralogy of the clay fraction of the Tama soil formed from thick loess (after Glenn, Jackson, and Hole, 1960).

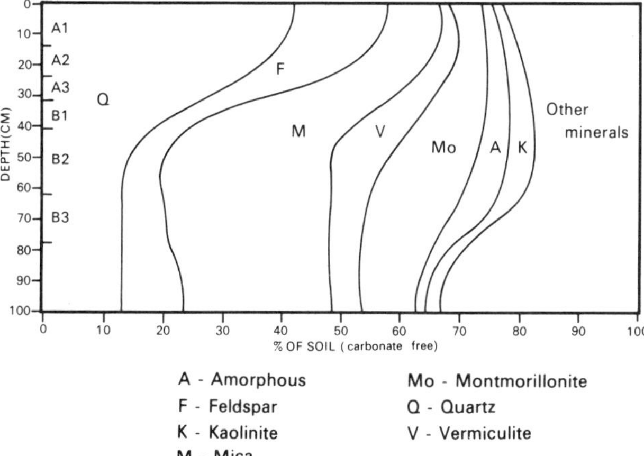

Figure 4-9. Mineralogy of the Saylesville silt loam, whole soil (< 2mm) (after Borchardt, 1966).

Table 4-1. Capacities of several Wisconsin soils to exchange, fix, or complex various chemical ions

Soils	Exchange capacity for cations Ca^{+2}; Mg^{+2}, K^{+}, Na^{+}, NH_4^{+} (me/100g)
Plainfield sand (Typic Quartzipsamment)	
A1 horizon	5
Miami silt loam (Typic Hapludalf)	
A1 horizon	23
Plano silt loam (Typic Argiudoll)	
Ap horizon	29
Omega loamy sand (Entic Haplorthod)	
Bir horizon	3

Source: Data from Soil Survey Investigations Report No. 17, Soil Conservation Service, 1967a.

Table 4-2. Summarized status of soil pH, soil organic matter, and plant-available phosphorus and potassium in the plow layer (0-7 inches) by major soil regions (in fields currently under cultivation)

Major soil region	pH	Available P (lb/A)	Available K (lb/A)	Organic matter (%)
A	6.7	30	180	3.0
B	6.8	40	166	3.3
C	6.4	76	133	1.6
D	6.5	71	122	2.2
E	6.7	38	115	2.8
F	6.1	40	120	3.2
G	6.0	53	127	3.2
H	5.3	200	225	2.8
I	7.2	26	143	3.5
J	6.8	30	150	10.0+

Source: Data from E. E. Schulte, State Soil Testing Laboratory, University of Wisconsin, Madison.

and many works of man. Some boundaries, such as those between dark wet soils and light-colored drier soils, are easily seen in freshly plowed fields, as previously noted. Rather than study in detail an entire soil body that may be more than a quarter of a mile (0.4 km) across, we customarily examine a soil profile that is a narrow vertical slice through it. This cross section is thought of as one face of a soil column (pedon) (Fig. 4-2). Like a plug taken from a watermelon or from a wheel of cheese, the pedon shows the layering and fabric characteristics of the soil body. A vertical face of a pedon is called a soil profile, in which we can see the soil horizons exposed. Soil sola (A plus B horizons) of Wisconsin soils range in depth from a few inches (Sogn silt loam, shallow to bedrock; Fig. 7-5) to 5 feet (Tama silt loam, in deep loess; Fig. 7-5). The number of major soil horizons in sola varies from three, the O1-O2-A1 horizon in Chelsea sand, to six, the O1-O2-A2-Bhir-A'2-B'2 sequence in the Goodman silt loam (Fig. 12-10).

SOIL CAPACITY TO HOLD AND EXCHANGE NUTRIENTS

Many Wisconsin soils have enormous capacities to retain, transform, and exchange a variety of nutrients (chemical ions). This property of soils is one which is of great importance in the total earth ecosystem: It allows most plant nutrients to be retained for long periods of time in the root zone for future plant use; it immobilizes many harmful and/or toxic substances; it allows for the orderly transformation of organic material through the use of energy originally received from the sun.

Table 4-1 gives the capacities of some representative Wisconsin soils to retain and supply, via exchange, several kinds of chemical ions. The reactive properties of Wisconsin soils come principally from the organic matter, which can retain up to 300 milliequivalents of cations per 100 grams of dry soil; from montmorillonite, which can retain up to 80 milliequivalents of cations per 100 grams of dry soil; and from hydrated oxides of iron and aluminum, which coat most soil particles and which can react with and immobilize ions such as phosphate and arsenate.

The major widely abundant ions that do not react strongly with soil are nitrate (NO_3^-), chloride (Cl^-), sulfate (SO_4^{2-}), carbonate (CO_3^{2-}), and bicarbonate (HCO_3^-). These may move through the soil, although all of them except chloride participate in many of the reactions in the water, plants, and soil as they move.

Since Wisconsin soils are young and relatively little leached, immense reserves of plant nutrients are present. This has permitted profitable cultivation of all but the most sandy or acid soils for long periods without use of fertilizers other than barnyard manure (Hole, 1974b). In 1967 about 812,000 tons of commercial fertilizer were added to surface soils of the state on farm and forest land, lawns, and gardens. Fertilizer consumption may double or triple in the next few decades, but the amounts involved will still be slight compared with the immense reserves of fertility in the surface soil and subsoil. Nearly half of the phosphorus needs of a corn crop are thought to be obtained by the plants from native material in the subsoil (Murdock and Engelbert, 1958).

Plant-available phosphorus and potassium and soil pH have been widely studied in Wisconsin. Estimates of available levels of phosphorus and potassium for cultivated fields in the major soil regions, based on summaries of recent soil tests, are given in Table 4-2.

The geography of trace element deficiencies is being investigated. For example, Beaton (1971) summarized reports of Walsh and Hoeft (1971) that Soil Regions C, D, H, and northeastern E and I contain areas that are deficient in sulfur. Formulae of commercial fertilizers may include additional trace elements, such as boron, copper, zinc, molybdenum, and manganese, to assure optimal growth of plants on mucks and other nutrient-deficient soils of Wisconsin. Animal nutritionists, in their studies of sheep and cattle, have noted deficiencies of cobalt in soils of northern Wisconsin, such as soil association F6 in northern Shawano County (Glover, 1952).

The capacity of soils to receive and degrade organic wastes from sewage sludge, sewage effluent, and solid organic wastes from cities, as well as animal wastes, appears likely to become increasingly important as society urbanizes and as demands for high-quality surface waters increase. Related to these special uses of soils are various natural processes by which heavy metals, such as zinc and mercury as well as phosphates and organic materials, are rendered alternately mobile and immobile in the ground.

SOIL CAPACITY TO STORE AND TRANSMIT WATER

The sizes of soil pores, and their continuity, determine the ability of soils to retain and transmit water. Retention is highest, i.e., transmission is slowest, in the absence of large continuous pores, such as earthworm channels, and in soils with an abundance of very fine pores but with few large pores. Pore size distribution varies widely among the soil series of Wisconsin and also substantially within soil series. Fig. 4-4 gives some conductivity curves of several soils which have been studied in detail (Bouma et al., 1972).

Soil water available to plants has for many years been considered to be the volume of water held between a state of low suction or tension called *field capacity* (⅓rd bar) and a very high water suction of 15 bars. Field capacity is now known to be nonexistent, and recent work shows that plants normally suffer from shortages of water at tensions far less than 15 bars (Gardner, 1971). The amount of water available to plants, then, depends upon (1) the rates at which soil pores drain (a continuous process), (2) the demand which the plants make for water, and (3) the volume and type of root system of the plants. Thus no one figure (percentage) can be given for available water for all soils, vegetation, and weather conditions. Soil water potentially available is related to that held at various suctions by the soil and to the rate of drainage of water at a given suction (Bouma et al., 1972).

Movement of water in soils is complicated by their fabric and horizonation (Fig. 4-2). At the time of the spring thaw, medium-textured surface soils of Wisconsin are commonly saturated, or nearly so. At sloping sites water usually flows laterally in the A horizon over a compacted plowsole or even over a natural Bt horizon, or in the solum over a contrasting substratum, even though underlying horizons are not as wet as those above. Under saturated conditions, channels made by worms and by roots and major intersecting cracks conduct water rapidly. Under unsaturated conditions, water does not flow through these large voids, but rather moves slowly through fine pores. These are chiefly inside peds in structured soils.

SOIL CAPACITY TO SUPPORT LOADS

The same block of granite considered at the beginning of this chapter can support a great weight, such as part of a highway or the foundation of a large building, but it cannot absorb sewage effluent or other liquid waste. A mass of soil of the same volume is usually a fairly good absorber of liquid, but less stable under foundations of structures. Well-drained sands commonly provide much more stable support for man-made structures than silt or clay. Seasonal freeze-thaw cycles affect sands the least. Because soils have many engineering uses besides agricultural, horticultural, and silvicultural ones, an increasing number of land users and community planners are taking advantage of soil survey information in their work. Engineering test data for representative soils of Wisconsin are available from the Wisconsin Department of Transportation and the U.S. Soil Conservation Service.

CHAPTER 5

Classification of Soils

Gerhard B. Lee

The purpose of soil classification is to organize our knowledge of soils by grouping similar soils into manageable classes. If intellectually possible, it would be better to deal with each soil individually, i.e., with each pedon or polypedon; however, individual soils are so numerous that in most cases this cannot be done. Because it is useful to deal with soils at different levels of generalization, classification systems have been devised that group soils at several categorical levels. These levels range from very general classes to very specific ones. In summary, soil classification is simply a device by which we order, or organize, taxonomic units of soil into usable classes.

Soils have been classified in several ways (Bidwell and Hole, 1963). A simple classification relates to use, as illustrated by the terms *corn soils, alfalfa soils, hardwood soils,* and *pine soils.* Such systems, while useful for some purposes, may change rather rapidly with changes in technology. More permanent, scientific[1] systems of classification are based on soil characteristics and their relationship to soil genesis. Such systems express universal relationships that exist in nature and enable one to understand, remember, and predict from the information obtained. Scientific systems of soil classification have the following advantages:

1. They enable us to identify pedons or soil individuals and place them in their proper class.
2. They make it possible for us to organize our information about soils.
3. They show natural relationships among soils.
4. They allow us to make predictions, i.e., extend to other soils information gained by study of and experience with one soil.

At the present time several systems of soil classification are in use in the United States. The New System of Soil Taxonomy (final version now in press) has been in official use by the USDA since 1965. However, much published material exists in which older systems or terms from older systems are used. The county maps and soil survey reports published prior to the early 1930s, for example, classify soils as *series* and *types.* The soil series unit dates back to the early 1900s and was first conceived of as a textural sequence of soils from a common source. Thus we might have Miami clay, Miami clay loam, loam, sandy loam, etc., all derived from glacial drift. With time the series concept narrowed. In the soil survey report for Green Lake County (Whitson et al., 1929), soils were divided into series on the basis of the amount of organic matter in them, topography, drainage conditions, and other factors. "Each series represents a definite combination of these factors" and each series is given a proper name. Soil types were reorganized on the basis of differences in textural class of otherwise uniform material.

In the early 1930s, United States systems of soil classification changed drastically under the influence of C. F. Marbut (see his *Soils of the United States,* 1935). In Marbut's concept, adapted from the Russian approach, the soil profile was envisioned as the unit of study. Soils were classified according to similarities in morphological features, such as kind and arrangement of soil horizons, their color, texture, structure, consistence, and depth, as well as composition and origin of the initial geologic deposits. Marbut's system was revised in 1938 (Baldwin, Kellogg, and Thorp) and again in 1949 (Thorp and Smith). The 1949 system was used until the "new" system was officially adopted in 1965 and appears in many soil survey reports and other soils publications of that period. For this reason a brief outline of the higher categories of this system is included (Table 5-1). As can be seen from this outline, soils in the order and suborder categories were classified essentially on the basis of assumed genetic factors of soil formation. At the Great Soil Group level, however, groupings were made on the basis of kind and arrangement of soil horizons.

In addition to the higher categories, the 1949 system had three lower categories: family, series, and type. An example of how a soil such as Miami silt loam would be classified in that system is shown in Table 5-2.

The "new" United States soil classification system (Soil Conservation Service, 1960) was developed over a period of many years, most intensively since the early 1950s. At that time it was realized that the older systems, patterned after Marbut, had limitations that did not allow the proper classification of all soils. As a result a new system was developed, with greater reliance on properties of soils that could be observed and measured. The new system placed less emphasis on soil genesis, although a genetic thread runs throughout it. Since its recent adoption, the new system has been in official use by the USDA Soil Conservation Service.

The new system has several unique features. One is the nomenclature used in the higher categories. To avoid confusion with old terms, new names, formed from elements derived mainly from the classical languages, have been coined. In the past, for example, dark-colored granular soils in southern Wisconsin were called Prairie soils, because they presumably formed under prairie vegetation. In the new system they are called *Mollisols* from *mollis,* Latin for "soft," referring to their good tilth, and *solum,* Latin for "soil," with *i* as connecting vowel. Lists of the formative elements used in the higher categories of Wisconsin soils in the new system are shown in Table 5-3.

The basis for classification of soils into classes in the new system includes morphological features that can be observed

1. The Land-Use Capability System of the U.S. Soil Conservation Service is a useful classification of soil landscape units on the basis of degrees of hazard to the soil and response of the soil under various managements. This classification is, therefore, not based solely on soil properties and is considered a practical system rather than a scientific one.

Table 5-1. Soil classification in higher categories of the 1949 system

Order	Suborder	Great soil groups
Zonal soils	1. Soils of the cold zone	Tundra soils
	2. Light-colored soils of arid regions	Desert soils Red Desert soils Sierozem Brown soils Reddish-Brown soils
	3. Dark-colored soils of semi-arid, subhumid, and humid grasslands	Chestnut soils Reddish-Chestnut soils Chernozem soils Prairie soils Reddish Prairie soils
	4. Soils of the forest-grass-land transition	Degraded Chernozem Noncalcic Brown or Shantung Brown soils
	5. Light-colored podzolized soils of the timbered regions	Podzol soils Gray Wooded, or Gray Podzolic, soils Brown Podzolic soils Gray-Brown Podzolic soils
	6. Lateritic soils of forested warm-temperate and tropical regions	Red-Yellow Podzolic soils Reddish-Brown Lateritic soils Yellowish-Brown Lateritic soils Laterite soils
Intrazonal soils	1. Halomorphic (saline and alkali) soils of imperfectly drained arid regions and littoral deposits	Solonchak, or Saline, soils Solonetz soils Soloth soils
	2. Hydromorphic soils of marshes, swamps, seep areas, and flats	Humic-Glei soils (includes Wiesenboden) Alpine Meadow soils Bog soils Half-Bog soils Low-Humic Glei soils Planosols Ground-Water Podzol soils Ground-Water Laterite soils
	3. Calcimorphic soils	Brown Forest soils (Braunerde) Rendzina soils
Azonal soils	....	Lithosols Regosols (includes Dry Sands) Alluvial soils

Source: 7th Approximation, Soil Conservation Service, 1960.

and measured in the field as well as some characteristics that must be determined by laboratory study. Also included are environmental factors such as soil temperature and moisture regimes. *Diagnostic horizons* are defined and used to order soils in the higher categories. Six diagnostic surface horizons, called *epipedons,* are recognized. Five of these occur in Wisconsin:

1. *Mollic*—Latin *mollis,* "soft." A thick, dark-colored surface horizon, saturated with bases, high in organic matter content, and having good tilth (structure). *Mollisols* are characterized by a mollic epipedon. Soils with a mollic-like horizon are recognized at the subgroup level, for example, *Mollic Hapludalf.*
2. *Umbric*—Latin *umbra,* "shade." A thick, dark-colored surface horizon having low base saturation (very acid). A few soils having an umbric epipedon have been noted in central and northern Wisconsin.
3. *Ochric*—Greek *ochros,* "pale." A light-colored surface horizon, low in organic matter, or a thin, dark-colored surface horizon. *Alfisols, Entisols,* most *Inceptisols,* and *Spodosols* are characterized by an ochric epipedon.
4. *Histic*—Greek *histos,* "tissue." Surface horizons formed from plant tissue. Histic epipedons are known to occur on some undisturbed northern mineral soils. This horizon, as presently defined, is difficult to identify over mappable areas.

Table 5-2. Categories of the 1949 scheme of soil classification

Name of category	Degree of inclusiveness	Example
Order	Most general or inclusive. All soils in 3 genetic groups.	Zonal soils
Suborder	Slightly less inclusive. Nine genetic groups are shown in the 1949 scheme. Azonal soils are not subdivided into suborders.	Light-colored soils of timbered regions
Great Soil Group	Thirty-seven groups are shown in the 1949 scheme: 21 zonal, 13 intrazonal, and 3 azonal.	Gray-Brown Podzolic soils: A1, A2, Bt, C sequence of horizons
Family	Less inclusive but not carefully defined or commonly used. A family would include several series of closely related soils.	Miami Family
Series	Much less inclusive. A soil series includes those soils which are alike in every respect, except for texture of the surface horizon.	Miami Series
Type	Very specific. Some series include only 1 type (monotype series). Others may include 3 or 4 types. A soil type includes soils alike in every respect including texture of the surface horizon.	Miami silt loam Miami loam Miami sandy loam

Source: Thorp and Smith, 1949.

5. *Anthropic*—Greek *anthropos,* "man." Refers to a thick, usually dark-colored horizon modified by man, and high in phosphorus content. Occurs locally in kitchen middens or other areas used by American Indians.

A number of subsurface horizons are also recognized. Some of these, important in the classification of Wisconsin soils, are as follows:

1. *Albic*—Latin *albus,* "white." This is a light-colored subsurface horizon (sometimes just below the litter layer) from which clays and oxide coatings on soil grains have been removed (in podzolic soils, for example, *Spodosols* and *Alfisols*). In some soils, e.g., *Glossoboralfs* or *Glossic Eutroboralfs,* tongues of albic materials extend into the B horizon.
2. *Argillic*—Latin *argilla,* "clay." This is a subsoil horizon enriched in clay by soil-forming processes (an important diagnostic characteristic in Wisconsin soils). Mollisols in Wisconsin, mainly *Udolls* because of the humid climate, can be separated into *Argiudolls,* having a clay-enriched subsoil, and *Hapludolls,* lacking clay enrichment in subsoil horizons. Not always used consistently, however; for example, *Aqualfs, Boralfs,* and *Udalfs* usually have an argillic horizon without its being noted in group names such as Hapludalf.
3. *Cambic*—Latin *cambire,* "to change." This is a subsoil horizon in which soil formation has occurred, resulting in leaching, change of color due to chemical reactions, and structure formation, but without intense weathering or enrichment by clay or iron. *Hapludolls* have cambic horizons. A common example in southern Wisconsin is the *Haplaquoll,* a wet Mollisol found in depressions, which has a very gray (gleyed) cambic subsoil horizon.
4. *Fragipan*—Latin *fragilis,* "brittle," and English *pan.* This is a dense subsoil horizon that is extremely hard when dry. When moist it breaks suddenly under stress. Fragipans are usually impermeable and do not allow root penetration. In Wisconsin, several soil series are classified as *Fragiorthods,* i.e., typical Spodosols except for the presence of a fragipan.
5. *Spodic*—Greek *spodus,* "wood ashes." A spodic horizon is a subsoil layer enriched in aluminum, iron, and/or organic matter, usually brown or reddish brown in color. *Spodosols,* formerly called *Podzols,* have a spodic horizon.

As in most older systems of classification, the new system orders soils at several categorical levels. There are several thousand very specific classes (soil series) in the lowest category, and ten very broad classes (orders) in the highest category. Between *order* and *series* are *suborder, great group, subgroup,* and *family* as follows:

Order—All soils are grouped into ten broad classes called *orders.* Within each order all soils have some important common property, e.g., all Mollisols have a mollic epipedon from which the order derives its name. In Wisconsin six orders are recognized, as shown in Tables 5-4 and 5-5.

Suborder—Orders are divided into several classes called *suborders* on the basis of common properties; for example, Aquolls (wet Mollisols) and Udolls (Mollisols of temperate climates). All suborders have two syllables consisting of a prefix indicating the character of the class plus an identifying part of the order name. Representative suborders of orders recognized in Wisconsin are also shown in Tables 5-4 and 5-5.

Great group—Suborders include varying numbers of *great groups* as indicated by the prefix(s) to the suborder name. An Argiudoll is a Mollisol of the temperate climates with a clay-enriched subsoil horizon.

Subgroup—Each group may consist of one or several *subgroups,* each of which has an additional word to describe its particular characteristics. For example, a *Typic Argiudoll* is typical of the group; a *Lithic Argiudoll* is shallow over bedrock.

Family—A soil *family* includes soils within a subgroup that are essentially similar in texture, mineralogy, and temperature regime, and in some cases, in other soil features (Table 5-3). The family classification is a pragmatic grouping and some very practical interpretations can be made at this categorical level. Family names consist of descriptive terms; for example, a soil such as Dodge silt loam is classified as a *fine-silty* (silty, with more than 18% clay), *mixed* (no single mineral dominates soil behavior), *mesic* (mean annual soil temperature above 47°F), *Typic Hapludalf.*

Series—The concept of the soil *series* has not changed appreciably from the 1949 system. A series consists of soil individuals (pedons or polypedons) that are similar in morphology, and in their chemical, physical, and mineralogical characteristics. Soil series have always been given geographical names, usually after the location where the series was first described. For example, Dodge silt loam was named after Dodge County, Wisconsin.

Several other terms are used in classification. One of these is *soil type,* a category in the 1949 and earlier systems, used to designate differences in surface texture. Soil type is not a taxonomic category of the new system but it may be used for its original purpose when convenient. The term *phase* is also used, although not in a taxonomic sense. It refers to a characteristic of soil that is primarily of importance in man's use and management. For example, we may speak of *eroded, sloping,* or *stony phases* of Dodge silt loam.

The term *variant* has a taxonomic connotation. It is used in conjunction with a series name to describe a soil of very limited geographical extent that is somewhat similar to the soil named, but different enough to be classified in a separate series if it were more extensive.

In the accompanying key to soil orders and suborders (Fig. 5-1) the wet soils are placed together in the central part of the key and the better drained soils at the periphery. By this device the topographic sequence of soils, so important in soil geography and cartography, is represented simultaneously with the somewhat arbitrary taxonomic partitioning of the soil categories in the classification system.

For more detailed information about the classification of Wisconsin soils the reader should consult Special Bulletin 12, *Classification of Wisconsin Soils* (Lee et al., 1968).

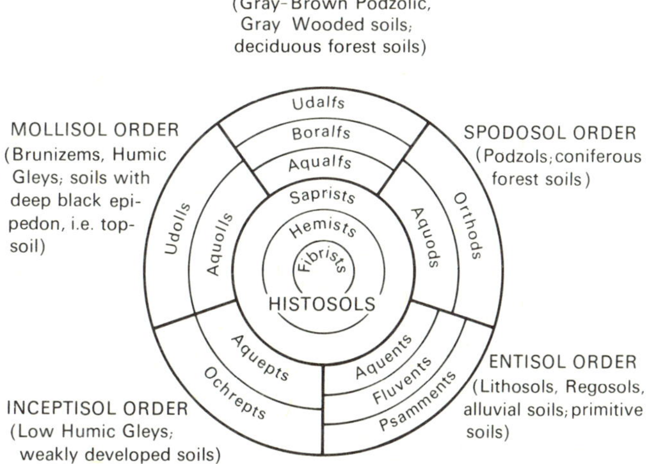

Figure 5-1. Circular key to the new USDA soil classification for Wisconsin. (See also color version in section of plates.)

Table 5-3. Formative elements[a] in names of soils of Wisconsin, USDA classification

Order	Suborder[b]	Great group	Subgroup	Family
Alf (Pedalfer)	Aqu (wet)	alb (white)	aeric (aerated somewhat)	Texture: fragmental, sandy-skeletal, loamy-skeletal, clayey-skeletal, sandy, coarse-loamy, fine-loamy, coarse-silty, fine-silty, fine-clayey, very fine-clayey
Ent (recent)	Bor (cool)	arg (clayey subsoil)	alfic (Pedalfer-like)	Mineraology: carbonatic, micaceous, mixed, illitic, calcareous
Hist (tissue)	Fibr (fibrous)	boro (cool)	aquollic (aquic and mollic)	Reaction: dysic or acid (pH<5.5); euic or nonacid (pH>5.5)
Incept *or* ept (incipient)	Fluv (fluvial)	calci (calcareous)	arenic (sandy)	Mean annual soil temperatures—at 20 in. depth: frigid (<47°F); mesic (>47°F)
Moll *or* oll (soft, dark)	Hem (half decomposed)	camb (changed slightly)	boric (cool)	Depth: shallow (<20 in. to bedrock)
Spod (ashes) *or* od (Podzol)	Ochr (pale)	dystr (infertile)	cumulic (accumulated)	
	Orth (true)	eutr (fertile)	dystric (infertile)	
	Psamm (sandy)	frag (fragipan)	entic (primitive)	
	Sapr (rotten)	gloss (tonguing)	eutric (fertile)	
	Ud (humid climate)	hapl (simple)	glossic (tongued)	
		medi (moderate temperature)	hemic (half decomposed)	
		moll (soft, dark)	histic (with histic epipedon)	
		ochr (pale)	humic (humus prominent)	
		quartz (quartz)	hydric (water layer present)	
		sphagno (sphagnum moss)	limnic (lacustrine)	
		ud (humid climate)	leptic (thin)	
			mollic (soft, dark)	
			pachic (thick)	
			psammentic (sandy, primitive)	
			terric (mineral layer present)	
			thaptic (buried soil at <20 in.)	
			typic (typical)	
			udic (humid climate)	
			udollic (udic and mollic)	

Example of a soil name

Typic Argiudoll, fine-silty, mixed, mesic; Tama silt loam

[A typical clayey-subsoil, humid-climate, dark soil that is silty (with more than 18% clay), of mixed mineralogy, under moderate climatic temperatures; a silty soil first studied near Tama (Iowa).]

Order	Suborder	Great group	Subgroup	Family
Mollisol	Udoll	Argiudoll	Typic Argiudoll	fine-silty, mixed, mesic

[a]See the text for an explanation of each element.

[b]Note that there is one member of the Ustoll suborder ("ust" means subhumid) in Wisconsin, namely the Sogn silt loam, a Lithic Haplustoll, an excessively drained soil shallow to limestone bedrock.

Table 5-4. Brief definitions of major Wisconsin soil categories

	Order[a]	Suborder[a]	Definitions
MINERAL SOILS	ENTISOLS[b]	All	Soils with little or no development of pedogenic horizons. Thin A1, A2, or peaty horizons may be present. Alluvial soils, steep soils, eroding dune sands, other sandy soils with a weak A and "color B," and simple wetland soils are included. (These are mainly Azonal soils in the old system.)
		Aquents	Wet Entisols. Gray and mottled Entisols with very thin A1 horizon. Wet sands may be brown.
		Fluvents	Entisols that are not usually wet and are not sands, and are developing in stratified stream deposits (alluvium). Content of organic matter decreases irregularly with depth. These soils may be flooded periodically by streams.
		Psamments	Entisols formed from well-sorted sands and not usually wet.
	INCEPTISOLS	All	Soils with weakly developed horizons, such as A1, A2, and B horizons with structure and/or gleying, but little or no clay accumulation. Many wetland soils and some upland soils with rather uniform texture in the solum are included. (Low Humic Gley and some Brown Podzolics of the old system.)
		Aquepts	Wet Inceptisols. Surface soil usually dark over a mottled gray subsoil that may have pedal structure (such as blocky structure).
		Ochrepts	Inceptisols that are not usually wet. A thin A1 horizon overlies an A2 and a pedal (usually blocky) B horizon without appreciable clay accumulation.
	MOLLISOLS	All	Soils with deep dark A horizons (mollic epipedons) that are relatively fertile. The subsoil (B horizon) may or may not have notable clay accumulation. (Prairie soils, Brunizems, Humic Gleys, Brown Forest soil.)
		Aquolls	Wet Mollisols. Deep, dark-colored surface soil over an olive-gray, mottled subsoil.
		Udolls	Mollisols that are not usually wet. Deep, dark-colored surface horizon over a yellowish-brown or brown subsoil with little or no mottling.
MINERAL SOILS	SPODOSOLS	All	Soils with a dark brown or reddish-brown subsoil (Bhir horizon) in which organic matter and iron (and aluminum) have accumulated, but usually not clay. Commonly a lighter colored horizon (A2) overlies this Bhir horizon, but a plow layer (Ap) or thin A1 horizon may replace it. (Podzols.)
		Aquods	Wet Spodosols. (Groundwater Podzols.)
		Orthods	Spodosols that are not usually wet. A fragipan may underlie the spodic horizon (Bhir).
	ALFISOLS	All	Soils with thin A1 horizons and definite A2 horizons (or plow layer), and with subsoils (Bt horizons) in which clay accumulation is distinct. (Gray-Brown Podzolics, Gray Wooded soils.)
		Aqualfs	Wet Alfisols. Under the thin A (A1 and/or A2) horizon is a drab and mottled subsoil (Bt horizon).
		Boralfs	Alfisols of the cooler northern portion of the state. These soils are not usually wet. (Gray Wooded soils.)
		Udalfs	Alfisols of the warmer, southern portions of the state. These soils are not usually wet. (Gray-Brown Podzolics.)
ORGANIC SOILS	HISTOSOLS[c]	All	Peaty and mucky soils more than a foot thick. These soils are wet unless artificially drained. These soils contain no less than 20 to 30% (by weight) of organic matter.
		Fibrists	Histosols that are largely fibrous. Original plant fragments (mainly moss and/or sedge) are clearly visible.
		Hemists	Histosols that are only partially fibrous. Original plant fragments are present, but the soils are largely developed in and break down easily to a black or brown paste.
		Saprists	Histosols that are almost without fibers. Most of the soil is a fine black or brown amorphous material. Many of these soils have good structure.

[a] Categories in the new system of classification. Adapted from the Soil Conservation Service (1960, 1967c, 1970).

[b] The suborder of Orthents is represented in Wisconsin by the Emmert soil series.

[c] When the organic material in these soils is examined without rubbing, fiber content is observed to occupy the following volume of the material: more than two thirds in Fibrists, one third to two thirds in Hemists, and less than one third in Saprists.

Table 5-5. Names of orders, suborders, great groups, subgroups, and families of representative soil series of Wisconsin

Order	Suborder	Great group	Subgroup	Family	Representative series
Entisols	Aquents	Psammaquents	Mollic Psammaquent	sandy, mixed, frigid	Roscommon
	Fluvents	Udifluvents	Typic Udifluvent	coarse-silty, mixed, nonacid, mesic	Chaseburg
	Psamments	Quartzipsam-ments	Typic Quartzipsamment	sandy, uncoated, siliceous, mesic	Boone
Inceptisols	Aquepts	Haplaquepts	Mollic Haplaquept	coarse-loamy, mixed, nonacid, mesic	Keowns
	Ochrepts	Eutrochrepts	Aquollic Eutrochrept	coarse-loamy, mixed, mesic	Shiocton
			Typic Eutrochrept	coarse-loamy, mixed, mesic	Salter
Mollisols	Aquolls	Argiaquolls	Typic Argiaquoll	fine-loamy, mixed, mesic	Brookston
		Haplaquolls	Typic Haplaquoll	fine-silty, mixed, mesic	Pella
	Udolls	Argiudolls	Aquic Argiudoll	fine-silty, mixed, mesic	Lisbon
			Typic Argiudoll	fine-silty, mixed, mesic	Tama
Spodosols	Aquods	Haplaquods	Aeric Haplaquod	sandy, mixed, frigid, ortstein	Saugatuck
	Orthods	Haplorthods	Typic Haplorthod	sandy, mixed, frigid, ortstein	Wallace
Alfisols	Aqualfs	Ochraqualfs	Udollic Ochraqualf	fine-loamy over sandy or sandy-skeletal, mixed, mesic	Matherton
			Aeric Ochraqualf	fine, illitic, mesic	Del Rey
	Boralfs	Eutroboralfs	Typic Eutoboralf	very fine, mixed, frigid	Ontonagon
	Udalfs	Hapludalfs	Aquollic Hapludalf	fine-loamy, mixed, mesic	Mosel
			Arenic Hapludalf	fine-loamy, mixed, mesic	Metea
			Psammentic Hapludalf	sandy, mixed, mesic	Spinks
			Mollic Hapludalf	fine-silty, mixed, mesic	Batavia
			Typic Hapludalf	fine-loamy over sandy or sandy-skeletal, mixed, mesic	Fox
Histosols	Fibrists	Sphagnofibrists	Hemic Sphagnofibrist	dysic, frigid	Lobo
	Hemists	Borohemists	Typic Borohemist	dysic, frigid	Greenwood
	Saprists	Medisaprists	Typic Medisaprist	euic, mesic	Houghton

Part II

Characteristics of Wisconsin Soil Associations

CHAPTER **6**

Introduction to the Soil Associations of Wisconsin

A soil association is a geographic assemblage of soils. Any account of the soil associations (i.e., soilscapes, page 4; see Buol, Hole, and McCracken, 1973) of a region must take the soil series of that region into consideration. The list of kinds of soils present in Wisconsin is still incomplete, for several reasons: First, detailed observations have not been made in many areas, particularly in northern counties; second, the process of classifying the soils and correlating them with those of adjacent states is still in progress; finally, changes in soils brought about by land use have not been adequately observed, recorded, and evaluated. Even so, the current list of soil series is impressive, as represented in Part III. A soil association name, such as Tama, Ashdale, Downs, and Muscatine silt loam (A1), tells what kinds of soils are grouped in a repeating pattern in a particular landscape.

In this discussion several soil terms that are somewhat analogous to plant ecological terms are used.

Some Analogous Terms

Pertaining to plant ecology	*Pertaining to soil ecology*
botany (plant science)	pedology (soil science)
phytosphere	pedosphere
vegetation of a region	soil continuum of a region
flora of a region	total list of soil series, types, and phases ("pedota") of a region
sequence of plant communities down-slope { xeric, mesic, hydric }	soil association (soilscape) { excessively drained, well drained, poorly or very poorly drained }

The pedosphere is the continuum of soils on the land portions of the lithosphere. Within a particular region the soil continuum is a mosaic of soilscapes in which the soil species, called soil series, types, and phases, are present in certain proportions and arrangements. We might coin a new collective term, *pedota,* for the soil species of a region. Soil continuum and pedota differ from each other in a quantitative way. The pedota is the total list of soil species present, regardless of the numerical abundance of each.[1] The soil continuum, on the other hand, has to do with combinations of species present in a given region and with the relative abundance of each species (Hole, 1953). For this purpose, the common species are far more important than the rare ones. The presence or absence of the latter may be of little significance in the functioning of the landscape, though it may be highly significant as a record of past environments. The basic task of identifying and describing the soil series is arduous, and is still in progress.

1. One body of a rare species can contribute as much to the pedotal list as thousands of bodies of a common species. In actual practice, soil species are not listed in county soil survey reports unless they occupy more than 500 to 1,000 acres. Therefore, considerable information about unusual soils is not generally available. For example, in soil association G25, in Sec. 16, T.37N., R.10E. in Oneida County, a small body of a red (2.5YR 4/4, moist) silt loam 50 cm thick over a sandy loam fragipan resting on acid outwash sand was observed along Gudegast Creek. The soil map (Hole and Schmude, 1959) includes the area in Peat-Au Gres soils, nearly level. The red soil remains unnamed and is not described in the soil survey report.

Most soilscapes in Wisconsin include soils of both good and poor natural drainage status, and were originally covered by a vegetative sequence from a mesic or even xeric plant community to a hydric one, i.e., a fen, wet meadow, or bog. Hence, soils that are very different occur side by side on the land. Fig. 5-1 was designed to indicate this. Crop production is known to vary from one soil member of a soil association to another. Natural plant communities are so rare, particularly in well-drained uplands, that correlation of properties of soil profiles and native vegetation has been difficult. A serious attempt to do this was made in the forests of the Menominee Indian lands (Milfred, Olson, and Hole, 1967).

The soil map (Plate 1) is cartographically generalized, but with a legend that is detailed to the soil series level. This map shows the major broad pedological features of the state and serves as an introduction to detailed soil maps (see list, Appendix 5) and to the actual soil bodies themselves.

The 190 associations (A1 through J15) listed in the legend of the soil map are distinctive, though in varying degrees. Several examples will illustrate this. The level landscape of the Antigo Flats in Langlade County is occupied by an association of soils, the principal three of which are named Antigo silt loam, Brill silt loam, and Onamia loam. This grouping is labeled F25 on the soil map, not only in Langlade County, but wherever it occurs in northern Wisconsin (Figs. 12-2, 13-3, 13-5). The Baraboo silt loam and Skillet silt loam soil association (A10) is extensive in portions of the Baraboo Hills of Sauk County. The Horicon Marsh in Dodge County is a large body of peat rimmed with wet mineral soils. This soil association (slightly acid to alkaline sedge and woody peat and muck soils; Pella, Poygan, and Brookston silt loam and silty clay loam) is labeled J15 on the soil map. In the drumlin field of Dodge County, well-drained soils of drumlins are associated in the landscape with linear bodies of wet soils between the hills. This is soil association B13, Miami, Dodge, and Pella silt loam association. The exact proportions of soils present in each soil association are, with few exceptions, unknown at present and therefore cannot be tabulated in this report.

Each soilscape has a characteristic fabric. Soilscape fabric analysis is under study by the author and coworkers, and results will be reported elsewhere. In brief, the study concerns soil body patterns as they occur in loops, whirls, and stripes in the landscape.

Many of the soilscapes, like the four examples just given, are recognizable from the window of a vehicle during a trip across the state. Others, such as the associations dominated by Kennan (G5, G6, G15, G24) and Cloquet (G12) soils, are not so easily differentiated. In any case the usefulness of the soil association names lies in the considerable body of relevant information that is available about topography and soil profiles to a depth of several feet. The approximately 200 soil series mentioned in the legend to the soil map (Plate 1) and in the text are defined in Chapter 17, in terms of representative field and laboratory characterizations.

Even though a soil association map like Plate 1 is useful in showing major landscape patterns and serving as a basis for regional planning for land use, it has limitations. The chief of these is that the map cannot, because of scale, show exact locations of individual soil bodies, any more than a state forest cover map can show particular stands of hemlock, pine, or aspen. As a result, small bodies of Antigo soils, for example, were omitted from the map, although these productive soils are of great importance on individual farms that are dominated by soils less suited to agriculture. Very different excessively drained, well-drained, and very poorly drained soils, such as Vilas, Kennan, and peat soils, may be associated together (see G5, Chapter 13), and the general soil map enclosed them all within a soil boundary. The exact location of each component soil condition cannot be shown.

In the following ten chapters, the soil associations are grouped more on the basis of soil series than on the topographic units of the map legend. This is because our knowledge of soil profile characteristics is much more advanced than our knowledge of geographic patterns of soils on the land.

The reader is directed to detailed soil maps (see Appendix 5) of individual counties, towns, and farms for the exact location of specific soil bodies.

CHAPTER 7

Soil Region A: Soils of the Southwestern Ridges and Valleys

This region presents one of the most beautiful arrays of landscapes in the state. One does not need to climb the observation towers on Blue Mounds or stand on the Mississippi River bluffs at Wyalusing State Park to find impressive views of pleasing patterns of valley bottoms, steep forested bluffs, and cropped fields. The picturesque is to be seen almost everywhere. The intimacy of coves and amphitheatre-like heads of small valleys is striking. The soil surveyors who crisscrossed this area on foot got their feet wet in the marshes of the river bottoms and perspired profusely while climbing the bluffs. To speed their work, these men commonly mapped soils by physiographic units, namely, the ridge tops of a section first, then the valleys below. Modern analysis of landscape resources shows a high concentration in this region of scenic patterns, called *environmental corridors* (Lewis, 1964). The scarcity of lakes in the natural landscape is being remedied by creation of artificial bodies of water.

The southwestern region of ridges and valleys overlaps the "Driftless Area" (Fig. 7-1) and includes those portions of the state in which landscape of considerable relief has developed on limestone bedrock overlying sandstone (Plates 2 and 3).

About 15% of the area of this scenic region is occupied by steep rockland (Fig. 2-2) that separates the gently rolling to moderately steep upland ridges from the valley floors and footslopes (Figs. 7-2, 7-3, 7-4). The nearly level valley bottoms account for another 15% of the area. Gently sloping and sloping soils (2 to 12% gradient) cover a third of the landscape on ridge tops and footslopes. Strongly sloping and moderately steep soils occupy the remainder of the landscape, immediately above and below the steep, rocky valley walls. Within the major landforms are smaller ones, including coulees, spurs, and coves (see Hole, 1956a, Fig. 7).

Relief amounts to as much as 600 feet; bluffs along the Wisconsin River valley are commonly 200 to 400 feet high. The region is traversed by the Wisconsin, Black, and Chippewa rivers and is rather completely dissected by a dendritic system of smaller streams (west portion of Fig. 7-3; also see Hole, 1956a, Fig. 9).

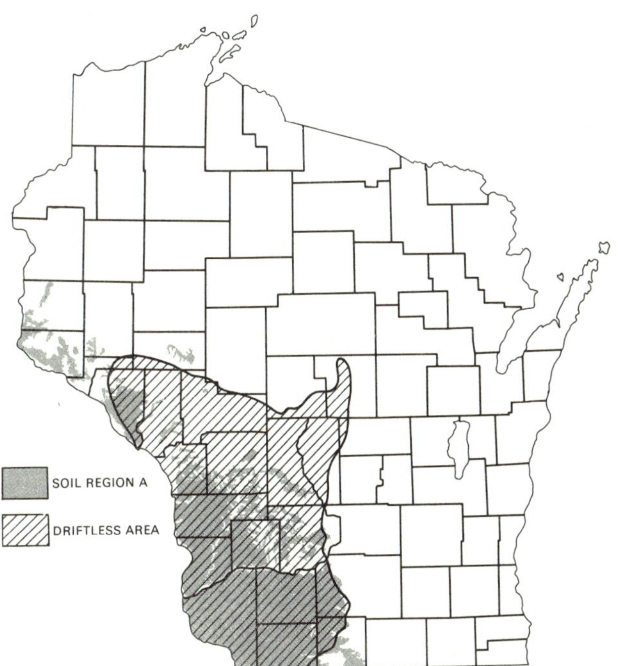

Figure 7-1. Index map showing the geographic relationship between Soil Region A and the "Driftless Area."

Most of the region was originally in oak savanna (Plate 6). Dry prairies, so-called goat prairies, occupied southwest-facing slopes, now dotted with bushy juniper trees. Prairie vegetation dominated on broad ridge crests. A body of sugar maple trees and a basswood forest were centered in Richland County. Woodland is the predominant cover today on valley slopes, with oak, birch, aspen, and black cherry in evidence.

Soils of the region are largely well drained, with only about 5% of the area occupied by wet soils that are concentrated in marshes, poorly drained river bottoms, and at occasional seepage spots and springs on footslopes. Droughty conditions prevail where bedrock comes close to the surface at the borders of the ridge tops (Sogn stony soils; Fig. 7-5) and on the deep sands (Plainfield, Sparta, Chelsea) that were distributed by water and wind over valley floors and against footslopes many centuries ago.

About 30% of the area is in woodland, pastured and nonpastured. Cropland occupies 40% of the land and cleared pasture about 25%. Ahlgren et al. (1946a,b) established that pastured woodland produces only one fifth as much forage per acre as open pastures, and one tenth as much as renovated pastures in Richland County.

The landscape surface is estimated to be about 250,000 years old (Palmquist, 1965). The shape of the upland surface and the pattern of streams show influence of rock structure (joints and folds). Geologic erosion of the entire landscape in the ensuing 250,000 years may have been at the average annual rate of about 2 tons per acre in order to carve out the deep valleys that we see today. This rate is half of the "maximum allowable soil loss for good farming" as defined by the U.S. Soil Conservation Service (Beatty, 1960).

There is some evidence that part of the region was glaciated more than 30,000 years ago. During this early period, considerable reddish clay residuum and illuvium may have been removed from the dolomite surfaces of the uplands, boulder trains were formed, and some gravel deposits were made (Palmquist, 1965; Black, 1964). Dury's discovery of duricrust in this region (Dury and Knox, 1971) suggests a much greater age for some of the land surfaces than hitherto recognized.

Figure 7-2. Sequence of maps showing distribution of soil associations in Soil Region A.

Soils of this region have formed primarily from windblown silt overlying discontinuous, cherty, reddish-brown clay, dolomite (Figs. 2-28 through 2-31; 7-5, 7-7, 7-9) and other bedrocks on upland ridges and in silt and loam material over sandstone on the valley slopes (Fig. 10-6). Fayette silt loam alone occupies almost a million acres, 85% of which is on upland ridges and the remainder on valley footslopes (Klingelhoets, 1959). Shale influences the soils in several places (Fig. 7-8). Less extensive are silty deposits on valley benches and natural terraces (Richwood silt loam; Fig. 7-10) that stand 20 feet or so above the floodplains. In a few places springs and seepage create poor drainage conditions even on the benches.

Prairie soil bodies occur prominently on ridges south of the Wisconsin River and less extensively in Vernon, Pierce, and St. Croix counties to the north; and on terraces in river valleys in Eau Claire, La Crosse, Juneau, and Richland counties. The development of nearly black topsoils as much as 2 feet thick and of somewhat darkened subsoils is in large part a result of the penetration of roots of big bluestem and little bluestem to depths of 4 to 7 feet (Weaver, Hougen, and Weldon, 1935). Intergrades[1] between prairie and forest soils developed over much of the region under oak savanna (Riecken, 1965). Some dark soils in Richland and Grant counties formed under maple-basswood vegetation. Alluvial soils (J1, J2) occupy about 6% of the area of the region.

Neary four fifths of the land has been plowed at least once. The average slope of cultivated land is about 12% gradient. It is little wonder that erosion of cultivated fields has been serious in many places. As a result, alluvial soils such as Orion silt loam (Figs. 7-9, 7-10) have about 3 feet of light-colored eroded soil

1. Colors of the three kinds of soil profiles are useful in identification: Tama silt loam, prairie soil: A1, 0 to 21 inches, black (10YR 2/1, moist); A3-B1, 21 to 33 inches, dark grayish brown (10YR 4/2); Downs silt loam, prairie-forest intergrade: A1, 0 to 8 inches, very dark grayish brown (10YR 3/2); A2, 8 to 12 inches, brown (10YR 5/3); B1, 12 to 16 inches, dark brown (10YR 4/3); Fayette silt loam, forest soil: A1, 0 to 2.5 inches, very dark grayish brown (10YR 3/2); A2, 2.5 to 14 inches, pale brown (10YR 6/3); B1, 14 to 17 inches, yellowish brown (10YR 5/4).

material overlying the nearly black original surface soil that was formed before European settlers came to the area. Modern soil surveys show that soils of about half of the area of the region are eroded, most of them (about 45% by area) moderately so, but some (about 5%) severely.

In 1870, August Kramer, a farmer in Mormon Coulee in La Crosse County, laid out narrow field strips on the contour to control erosion of soil and reduce loss of water. Many years later, Professor O. R. Zeasman promoted this practice in the region, and seeing the need for additional safeguards against erosion, advocated the use of drop-inlet dams to stop catastrophic gullying in Bertrand silt loam flats (A12) on the Buffalo River and at the McDonald farm near North Bend in Jackson County (Zeasman and Hembre, 1963). For thirty years (1933-1963) the U.S. Department of Agriculture and the Wisconsin Agricultural Experiment Station of the University operated the Upper Mississippi Valley Conservation Experiment Station (Hayes, McCall, and Bell, 1948; Anderson and Hembre, 1955) to measure the actual rate of soil loss from fields with and without erosion-control practices. It was found that most losses occur during four rains each year, sometime between April and October, and usually in June, July, and August (Beatty, 1960). A six-year study showed that in grain fields annual soil losses in tons per acre were 5, 11, 23, and 29 on slopes of 3, 8, 13, and 18% gradient, respectively. Terracing, grassed waterways, contour strip-cropping, diversions, and emphasis on productive legume-rich hay in oats-corn-hay rotation were found to be effective erosion-control practices, reducing the annual soil loss to less than 4 tons per acre per year.

Soils of the region are grouped on the map in fourteen associations (Fig. 7-2) on the basis of topography, soil initial material, and dominant soil types.

In keeping with the two-story topography of this part of the state, the soil map legend starts with the most level parts of the upland ridges, proceeds through the steeper terrain, and concludes with silty benches (natural terraces) of valley floors. The soil associations are numbered in that order from A1 to A14. In the following discussion they have been grouped on the basis of soil series rather than topography.

A1, A2. *Tama and associated deep silt loams of former prairies on Military Ridge and similar uplands.*

A1. The undulating and rolling Tama, Ashdale, Downs, and Muscatine silt loam association.

A2. The undulating and rolling Dodgeville, Ashdale, and Sogn silt loam association.

Undulating and rolling land occupies most of the ridge crests of southwestern Wisconsin (Fig. 7-5). Over the last several thousand years (Fig. 7-6) these loess-derived soils have formed, for the most part, under prairie vegetation. The faintly light gray A2 horizon in the Downs soil indicates oak forest influence. Some areas of Tama and Ashdale soils in southeastern Grant County developed under maple-basswood forest (Plate 6). The Muscatine soil occurs where Maquoketa Shale underlies the silty covering (Fig. 7-8).

The loess deposit is not of uniform thickness in Region A. It usually thins from ridge crests toward valley slopes, as in the

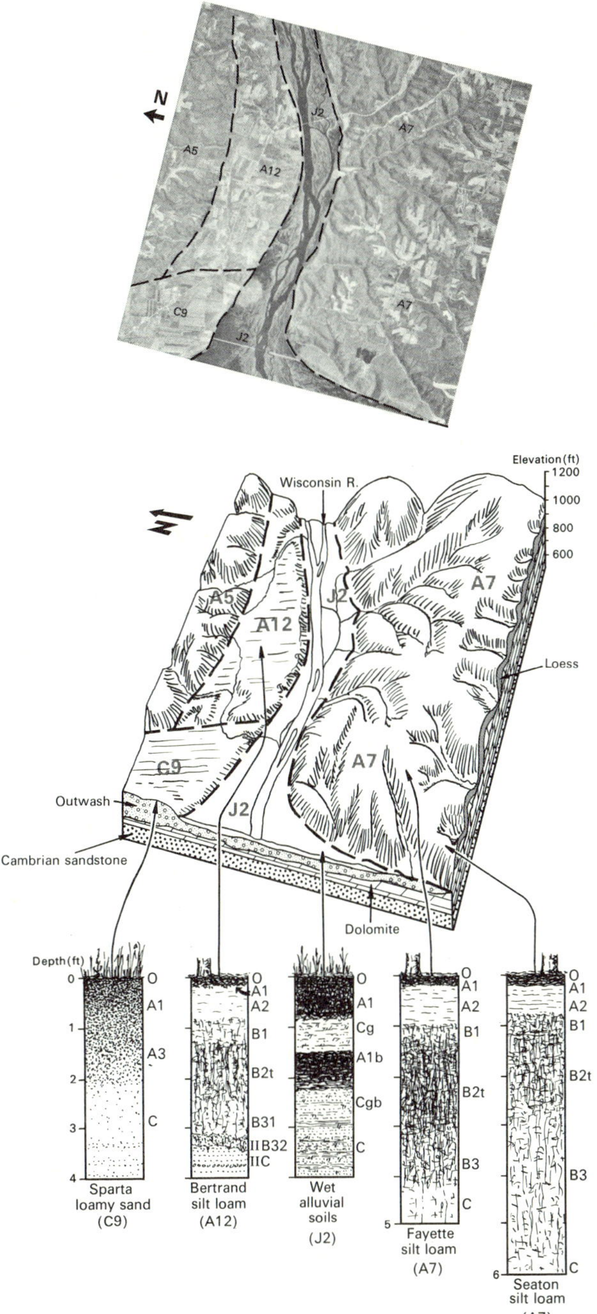

Figure 7-3. Aerial photo map of T.6N., R.6W. in Grant and Crawford counties. The area shown is 6 miles on a side.

Figure 7-4. Block diagram showing landscape positions of major soils of T.6N., R.6W. in Grant and Crawford counties.

Tama-Ashdale-Dodgeville-Sogn sequence of Fig. 7-5, in which depths of original loess were, respectively, more than 50 inches, 30 to 50 inches, 20 to 30 inches, and less than 12 inches. However, unusually thick deposits lie on some slopes, as in the case of the body of Tama silt loam shown in Fig. 7-8. Thickness of loess

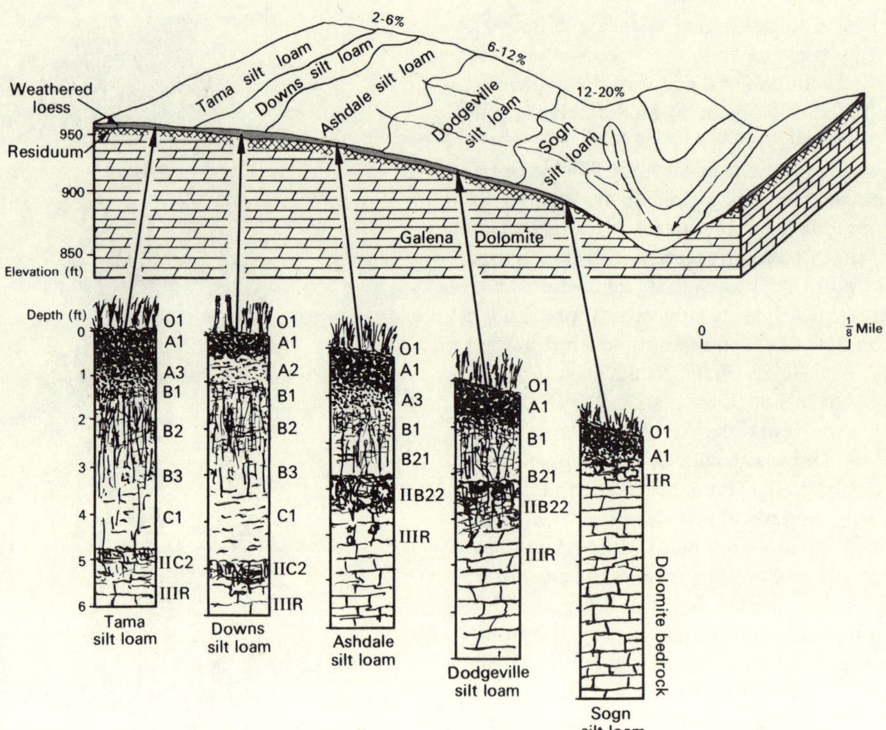

Figure 7-5. Block diagram showing landscape positions of representative soils of soil association A1 in Section 35, T.1N., R.5E., Lafayette County.

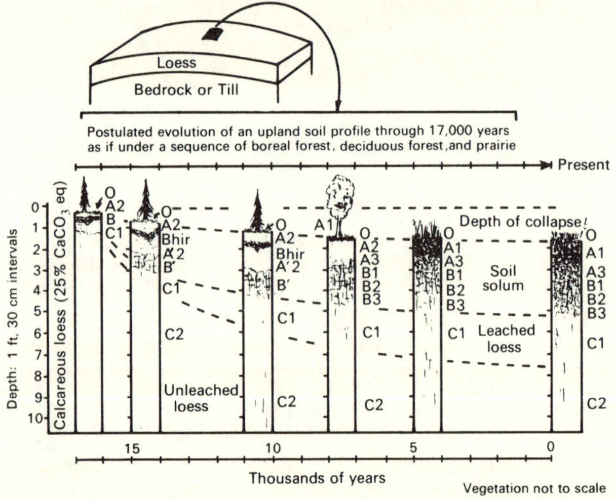

Figure 7-6. Postulated evolution of an upland soil profile through 17,000 years under a vegetative sequence of boreal forest, deciduous forest, and prairie (Hole and Nielsen, 1970). The soil succession is Entisol (not shown), Spodosol, bisequal Spodosol-Alfisol, Alfisol, Mollisol.

also decreases eastward from the Mississippi River (Plate 5), which accounts for the distribution of the shallower silty soils of soil association A2 to the east of soils of soil association A1 (Fig. 7-2).

Excavation of a Tama silt loam and underlying loess at an elevation of 1,032 feet near Glenhaven (Grant County) to a depth of 18 feet revealed a Udoll (Brunizem) soil profile developed in 17.5 feet of loess overlying about a foot of colluvium resting directly on Sinnipee (Platteville-Galena) dolomites (Milfred, 1966). The loess was calcareous below 116 inches, gleyed below 125 inches, laminated (stratified?) between 101 and 210 inches, and had a gleyed and concretionary zone at 155 to 210 inches. This zone may be interpreted (Ruhe and Scholtes, 1956) as a relic of a former saturated zone in which oxidation of iron has taken place around root channels subsequent to lowering of the water table in the course of landscape evolution. About three quarters of the total phosphorus of the A1 horizon of this soil is in organic form and is unavailable to plants (Glenn, 1959) until released by the slow decomposition of the organic materials. Even the subsoil (6 to 24 inches) contains only about 35 pounds per acre of available phosphorus (Soil Science Department, 1958).

The fairly uniform clay content of the combined A and B horizons in Tama silt loam in southwestern Wisconsin (Glenn, 1959; Fanning and Jackson, 1966b; Milfred, 1966) has been difficult to understand. One possibility is that clay formed faster in topsoil under prairie than under forest cover, and was not washed down rapidly into the B horizon. The discovery that the formerly common prairie mound-building ant (Fig. 7-7) constructs its mound largely out of subsoil (Baxter and Hole, 1967) offers an alternative explanation. Microscopic study of ant-mound material (Milfred, 1966) has shown that clay skins from the subsoil have been carried to the surface by the ants. The mound-building ants cannot survive in cultivated land and are therefore now confined to idle lands, such as edges of cemeteries, railroad rights-of-way, and wetlands (for example, on the Calamine silty clay loam on Blue Mounds, soil association A4; see Denning, Hole, and Bouma, 1973).

Figure 7-7. Ant mounds at an undisturbed Tama silt loam site, built by the western prairie ant, *Formica cinera Montana emery* (Baxter and Hole, 1967). The horizonal bar is 1 meter in length.

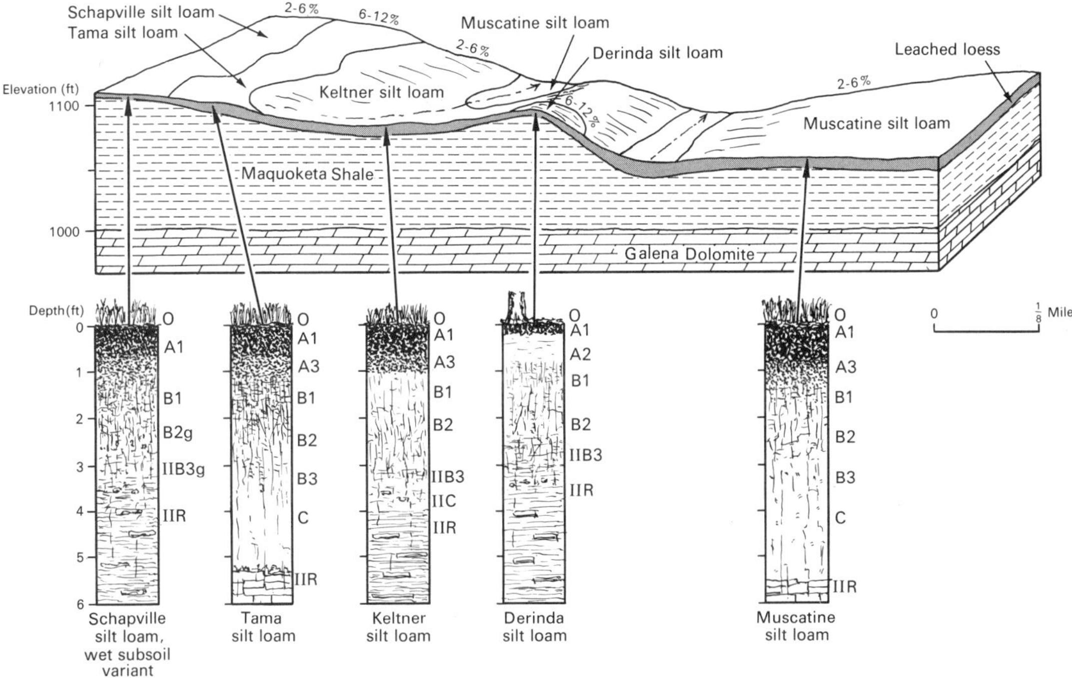

Figure 7-8. Block diagram showing landscape positions of representative soils of soil association A4 in Section 30, T.1N., R.3E., Lafayette County.

The somewhat poorly drained Muscatine (Aquic Argiudoll) and poorly drained Garwin (Typic Haplaquoll) soils occupy a small proportion of the A1 landscape. These soils lie in slight depressions in the broad uplands of southeastern Grant County, and more often, on seepage slopes just off the ridge crests. In the latter position the leached loess C horizon is commonly so nearly saturated that a hand-operated soil auger can be pushed almost effortlessly through it, once an access hole has been drilled to it through the overlying solum. In places, presence of clayey residuum at a depth of about 10 feet may account for the impeded subsoil drainage.

The Argiudoll (Brunizem) Tama soil of southwestern Wisconsin is considered to be about 5,000 to 6,000 years old (Fig. 7-6), and to have formed in a preceding Hapludalf (Gray-Brown Podzolic) soil. The loess deposit collapsed about 15%, or 8 inches, as leaching removed carbonates from the soil profile.

The depressing effect on soil productivity of limited depth of soil to bedrock is shown by the following shift in annual alfalfa-brome hay yields in tons per acre, dry weight, under good management: Tama (48 inches or more), 4.75; Ashdale (36 to 45 inches), 4.0; Dodgeville (15 to 30 inches), 3.5; and Sogn (4 to 20 inches) 2.0 (Beatty et al., 1966).

Although shown on the map as distinct and separate, the two soil associations (A1 and A2) may be found in the same landscapes, as shown in Fig. 7-5. In bodies of soil association A2 the approximately 3-foot Dodgeville soil is predominant and occupies many ridge tops in the absence of the deeper Ashdale silt loam. Fencing between woodland and plowland-pasture usually coincides with or closely parallels the Dodgeville-Sogn soil boundary. The IIB and IIC horizons (Fig. 7-5) are formed in cherty red clay over dolomite bedrock. This clay seems to be highest in content of chert gravel near cherty zones of the Galena and Oneota formations.

A3, A6, A9. *Dubuque and associated silt loams, moderately shallow to limestone and formerly forested.*

- A3. The undulating, rolling, and hilly Dubuque, Palsgrove, Sogn, and Dodgeville silt loam association.
- A6. The gently rolling to very steep Palsgrove, Dubuque, and Fayette silt loam association, with steep rocky land.
- A9. The gently rolling to very steep Dubuque and Palsgrove silt loam association, with steep rocky land.

Down-slope as well as eastward from the prairie soil ridge tops of soil association A2 are these A3, A6, and A9 soilscapes dominated by forest soils (Typic Hapludalfs, Gray-Brown Podzolics) (Figs. 2-51, 7-9). Moderate depths of loess predominate in the A6 soilscape, which forms a north-trending belt as much as 40 miles wide and half that distance from the Mississippi River valley which it parallels. Common to all three soil associations is the Dubuque series formed in shallow loess over cherty reddish-brown clay and limestone bedrock and most extensive in the southeastern part of Region A. In places, the clayey subsoil is greenish yellow and free of chert where the shaly members of the Sinnipee group are immediately under the silty topsoil.

Original depths of loess for these soils are as follows: Fayette, more than 48 inches; Palsgrove, 36 to 45 inches, Dodgeville and Dubuque, 18 to 36 inches; Sogn, 4 to 20 inches.

The dendritic pattern of dissection of the landscape is evident in Figs. 7-3 and 7.4. In the vicinity of Blanchardville, density of the natural drainageways is greatest (Palmquist, 1965) on soils of the lowest soil permeability (Klingelhoets, 1962). Gravelly and stony spoil material from old lead and zinc mine diggings, as described under A5, occurs on these soils in patches 2 to 20 acres in size in Lafayette County. In northwestern Richland County the cherty reddish-brown clay substratum (IIB2 horizon) of the Palsgrove and Dubuque soils is sandy in places. This probably represents residuum from St. Peter Sandstone, some striking outcrops of which occur over Prairie du Chien dolomites on ridge crests of the area. In parts of Crawford County, where the Platteville Dolomite has wasted away, the cherty reddish clay rests on mildly alkaline Glenwood Shale. Of minor extent is the dark soil formed under prairie sod, the Dodgeville silt loam.

As in other silty forest soils of the region, phosphorus in forms available to plants occurs in the subsoil of Palsgrove and Dubuque silt loam in amounts about three times those in analogous soils developed under prairie (soil associations A1, A2),

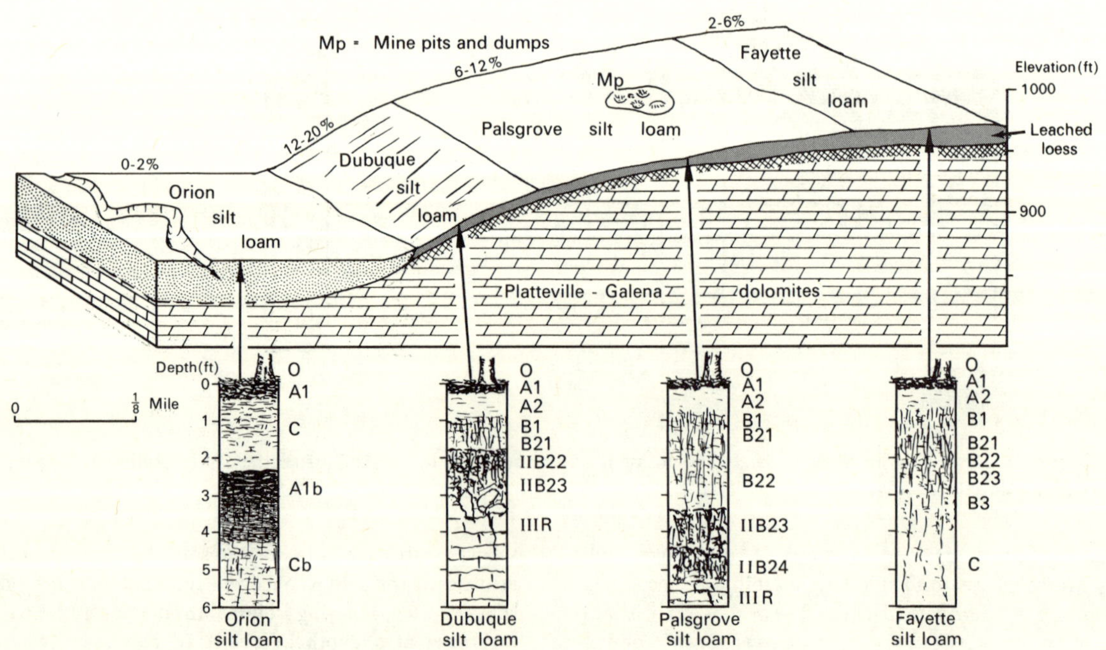

Figure 7-9. Block diagram showing landscape positions of representative soils of soil association A5 in Section 16, T.2N., R.1E., Lafayette County.

because of the lower content of organic matter in the forest soils. Evidences of mass movement of materials that have been observed on slopes of 3 to 10% gradient in soilscapes between Stitzer and Beetown and near Kieler in Grant County (Hogan, 1961) include (1) interlayering and intertonguing of brown silty B2t soil with cherty red clay IIB2t soil, (2) convolutions of chert bands in the cherty red clay residuum, and (3) abrupt vertical deflections of chert bands in red clay as if by miniature faulting (by a few inches) of material on slopes less than 10% in gradient. It is likely that these features developed under periglacial conditions (Smith, 1949; Hamilton, 1963). The content of chert by volume in the residual clay ranges from 1% between bands to 80% in bands.

Nearly all of soilscape A3 is cleared and in cropland and pasture. Upland ridge soils, mostly cultivated and pastured, occupy nearly a third of soilscape A6; wooded steep stony land nearly a third; farmed footslope soils nearly a third, and bottomland soils about 10% by area. Over half of soilscape A9 is occupied by cleared uplands of Dubuque silt loam with some Palsgrove silt loam; over a fourth by wooded steep stony land with cherty Sogn, sandy Hixton, Norden, and Hesch, and silty Fayette; and another 10% consists of bottom soils (Arenzville, Chaseburg, Ettrick; see Chapter 16).

In a study by Kaddou (1960), clay impurities in Prairie du Chien dolomites immediately below a Dubuque solum were found to be composed largely of mica with some interstratified layer silicates. Clay species in the overlying IIB2 soil horizon ("residuum") were montmorillonite (about 35%), vermiculite (27%), amorphous material (14%), with small amounts of mica and chlorite. The less clayey A horizons contained a somewhat larger proportion of montmorillonite, vermiculite, and chlorite. It is not known to what extent the reddish-brown subsoil is residual from dolomite bedrock and to what extent it is aeolian in origin. It has been presumed for many years that the reddish-brown IIB material is entirely residual, but a significant proportion of it may be windblown, as concluded by Ballagh and Runge (1970) after a study of an Ashdale soil in Stevenson County, Illinois, just south of the Wisconsin state line. These authors described the cherty reddish-brown clay IIBt horizon as a paleo- and modern illuvial B horizon ("beta" horizon of Bartelli and Odell, 1960a,b).

In soilscape A6 near Bosstown, Richland County, Akers (1964) observed (1) a 400-foot-long boulder train of St. Peter Sandstone extending south from a 20-foot-high sandstone pinnacle, and (2) sand and gravel of the controversial Windrow formation on ridges. These may be products of early glaciation.

Close relationships exist between slope and soil properties. Small (1973) studied this phenomenon in soil association A6 near the town of Mineral Point. He noted that slopes from ridge crest to valley bottom consist of a chain of convex, plane or straight, and concave units over which soil properties vary. Soil associations have a characteristic range of properties, based on slope length, slope steepness, sharpness of slope curvature, and stream order (Strahler, 1957). In soilscape A6, solum and subsoil thickness decrease over the steepest part of the slope, but thicken in either the up-slope or down-slope direction. Texture of the A horizon is related more to the geologic origin of the materials across which the slope is developed than to slope gradient. The amount of organic carbon in the A horizon increases with increasing distance down-slope, probably because of the moderate to severe erosion which has occurred on ridge tops in this soilscape.

A4. *The undulating and rolling Schapville, Derinda, Vlasaty, and Calamine silt loam association.*

This soil association has developed where loess is underlain within a few feet by Maquoketa Shale or clayey glacial till (Fig. 7-8). This condition exists at Blue Mounds in Dane and Iowa counties, in the southern part of the Wisconsin lead and zinc mining district (Lafayette County) (Fig. 7-2), and in Pierce and St. Croix counties in the northwest. Limited areas of somewhat similar soils occur in soil association B1. The shale of the IIC horizon is calcareous to slightly acid in many places. In the Plum Creek valley of eastern Pierce County an acid variant of the Derinda soil is recognized where the shale contains some sulfuric acid as a result of oxidation of the mineral pyrite (Wurman, 1961). The stratification of the shale persists into the solum in this valley area. A thin solum variant of Gale silt loam (bedrock at about 16 inches) is associated with Derinda soils in Pierce County. The somewhat poorly drained Stronghurst silt loam is mapped where silty material is deep (50 inches or more) on somewhat poorly drained benches on nearly level land.

Depths of original loess are as follows: Derinda and Schapville, 15 to 30 inches; Vlasaty, more than 24 inches; Calamine, 30 to 50 inches. Clay loam till of Rockian Wisconsin age (Foss and Rust, 1968) forms the C1 horizon of the moderately well drained Vlasaty and poorly drained Calamine soils in Pierce County. The upper solum of all four soils is developed from a blanket of leached loess. The Schapville soil developed under prairie cover, the Derinda and Vlasaty under forest, and the Calamine under sedge meadow. Somewhat poorly drained areas occur on nearly level benches and at seepage spots on hillsides. Schapville silt loam, wet subsoil variant, is representative of such sites.

The shale substratum of the Derinda soil is fertile with respect to plant growth, but has physical characteristics that impede rooting (Jones et al., 1967). Creep of the shale and of soils and chert blocks on shale is evident on many slopes, particularly at Blue Mounds.

A5, A7, A8. *Fayette, Seaton, and associated silt loams developed on uplands under forest near the Mississippi River valley.*

- A5. The rolling Fayette, Palsgrove, and Dubuque silt loam association, with steep rocky land.
- A7. The rolling Fayette and Seaton silt loam association, with steep rocky land.
- A8. The rolling Seaton, Palsgrove, and Dubuque silt loam association, with steep rocky land.

These rolling to hilly soilscapes stretch for more than 200 miles along or near to the Mississippi River valley (Fig. 7-2). The soils have developed under forest cover in loess of Woodfordian (Cary) age (Foss and Rust, 1968) (Table 2-10), ranging in thickness over cherty reddish-brown clay from about 18 inches or even less to as much as 30 feet (Figs. 7-9, 7-10): Seaton and Fayette, more than 50 inches; Palsgrove, 30 to 50 inches; Dubuque, 20 to 30 inches. The effect of variation in thickness of silty soil is apparent in annual alfalfa-brome hay yields (tons, dry weight, per acre) under good management: Fayette, 4.5; Palsgrove, 3.8;

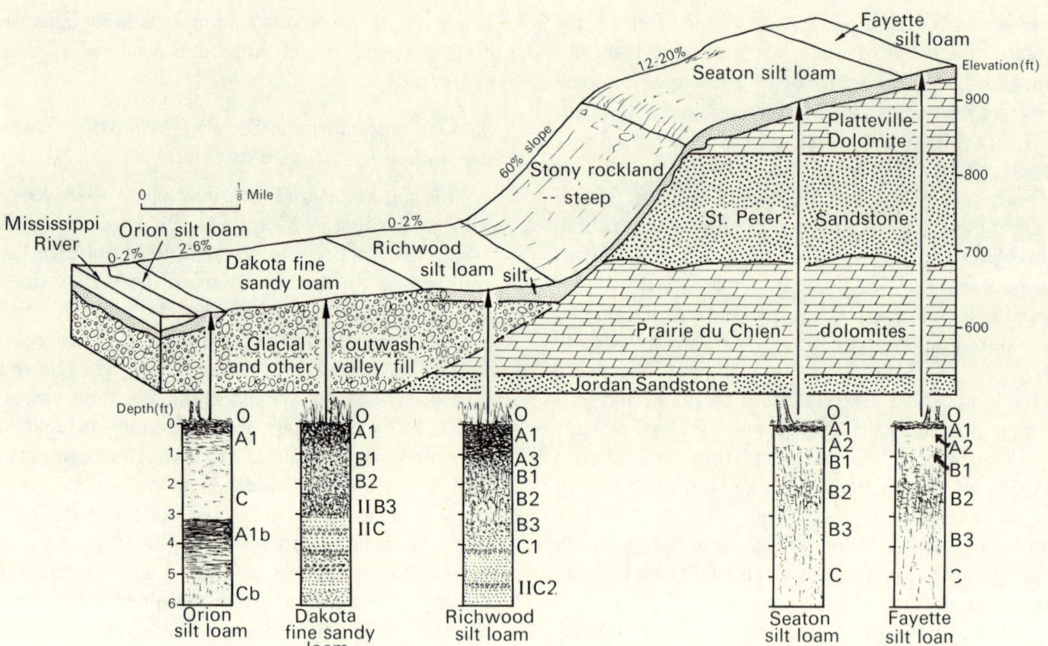

Figure 7-10. Block diagram showing landscape positions of representative soils of soil association A7 in Sections 10 and 11, T.2N., R.3W., Grant County.

Dubuque, 3.5 (Beatty et al., 1966). The hay yield for Seaton is 4.0 tons. This is less than for Fayette because of the lower water-holding capacity of the slightly coarser textured Seaton. Steep rocky land includes slopes of 30 to 60% gradient on which a wide range of stony, clayey, silty, and sandy soils are found, along with outcrops of dolomite, sandstone, and siltstone.

Fayette silt loam occurs in two principal positions in the two-story landscape of southwestern Wisconsin: above and below the steep stony land zone. The lower-lying Fayette is referred to as Fayette, valley phase (Fig. 2-51). It contains stonelines and seams of sandy material. In a few places, where boulders have crept down from overlying cliffs in such volume as to make the soil uncultivable, it is classified as boulder, valley-phase Fayette silt loam. The distribution of phosphorus and other plant nutrients is heterogeneous in the valley phase as compared with the normal phase.

The loess is about 20 feet thick on ridge crests near the Mississippi River valley and thins logarithmically eastward. It is leached of carbonates to a depth of 6 to 7 feet. Laminae about 0.5 mm thick have been observed in the calcareous loess (about 15% $CaCO_3$ equivalent) in the Cg horizon at a depth of about 16 feet (Milfred, 1966). Concentric bands of iron and manganese oxides in carrot-shaped volumes, as much as 3 feet long, with root channel axes, are present below a depth of 10 feet in a gleyed paleophreatic zone in calcareous loess.

Two distinct kinds of loess are recognized in soil classification in these soilscapes: (1) the coarse silt deposits in which Seaton soils (soilscapes A7, A8; Fig. 7-4) are developed, and (2) the medium and fine silt deposits on which Fayette, Palsgrove, and Dubuque soils are mapped (A5 and parts of A7). Adjacent to the glacial outwash terraces in the Mississippi River valley, wind-blown deposits of fine sand and silt (loess) are thickest and coarsest. Chelsea fine sand and Lamont fine sandy loam have formed from aeolian sand deposits located respectively on foot-slopes of the Mississippi River valley eastern bluffs and on the immediate crests of the bluffs (Robinson and Klingelhoets, 1961). Eastward from there is the narrow zone (Fig. 7-10) of Seaton silt loam and beyond that a wide zone of Fayette silt loam. The Seaton soil has in the B2t horizon a maximum of about 24% clay and about twice as much coarse silt as fine silt. The Fayette soil has in the same horizon a maximum of about 31% clay, only slightly more (1.2 times) coarse silt than fine silt, and in comparison to the Seaton, slightly lower base saturation and more exchangeable magnesium relative to calcium.[2]

Substrata observed over dolomite bedrock include reddish-brown clay and paleocolluvium. It is likely that deposition of the loess from which the Seaton and Fayette soils developed was principally between 20,000 and 29,000 years ago (Hogan and Beatty, 1963).

About half of the A5 soilscape consists of cultivated and pastured upland ridges, a third is wooded steep rocky land, and the remaining fifth is nearly equally divided between farmed foot-slopes and valley bottoms.

2. The Clinton soil of west-central Illinois, also loess-derived, has a maximum clay content of 38% in the B2t horizon (Muckenhirn et al., 1955, 1960). Presumably it is an older soil than Fayette and completes the chronosequence of the three soils. The coatings (argillans) on the surfaces of peds in the Bt horizons of these soils commonly have the following moist colors: 7.5YR 5/7 in Seaton, 10YR 5/6 in Fayette, and 10YR 4/4 in Clinton.

A10. *Baraboo and associated shallow silt loams over quartzite.*

The rolling Baraboo and Skillet silt loams and associated steep rocky land are largely under forest cover (Figs. 7-11, 7-12) and used for recreation and wildlife. This soilscape lies on the south limb of the Baraboo Range (Dalziel and Dott, 1970), west of the Woodfordian (Cary) terminal moraine. In places, silty soil overlies blocks of quartzite mixed with finer materials in what appear to be deposits of glacial drift and products of periglacial mass wasting (Black, 1965a, 1968). The silt loam soils have formed from 20 to 40 inches of leached loess that rests on quartzite bedrock (Fig. 7-12) instead of on dolomite, as in the case of Dubuque and associated soils farther west. Where not interrupted by rock outcrops, Baraboo silt loam may be cultivated profitably on gentle slopes, as for example on one of the most ancient land surfaces in the state, at Happy Hill, in Section 35, T.11N., R.5E. in Sauk County (Thwaites, 1958; Geib et al., 1925). The somewhat poorly drained Skillet soils occupy seepage spots.

A11, A12, A13, A14. *Richwood, Bertrand, Tell, Dakota, and associated soils on terrace benches of large valleys.*

A11. The level Richwood, Toddville, and Bertrand silt loam association.

A12. The level Bertrand, Curran, and Arenzville silt loam, and Dakota and Meridian loams, association.

A13. The level Tell and Curran silt loam, and Ettrick silty clay loam, association.

A14. The level Dakota and Onamia loams, and Waukegan and Antigo silt loam, association.

Dark soils (Argiudolls) that formed under prairie occupy much of the soilscapes of associations A11 and A14. Deep silty soils on benches of outwash and on alluvial fans stand 10 to 30 feet above adjacent streams in soilscape A11 and are rarely flooded (Fig. 7-10). One narrow strip of these soils is found in the appropriately named Black Earth Creek valley in western Dane County. Some bodies of these soils are locally referred to as "prairies," as is the case with the West Salem Prairie of La Crosse County. Included in a body of A11 in southern Richland County are areas of Downs silt loam (Mollic Hapludalf), which show incipient A2 horizon development and "silica flour" coatings on surfaces of peds in the Bt horizon. In the northwest corner of Region A (Fig. 7-2) there is a body of dark silty soils (soilscape A14) on a bench of Prairie du Chien dolomites capped with sand and gravel outwash that surrounds some picturesque drift-capped hills of St. Peter Sandstone and Platteville Dolomite. These Dakota and associated soils lie nearly 200 feet above the Kinnickinnic River and Lake St. Croix (Fig. 12-4). Predominantly light-colored soils, formed under forest cover, occupy the other two soilscapes. Bertrand (Typic Hapludalf) and associated soils (soilscape A12; Fig. 7-4) are on benches that are separated in most places from the nearby floodplain by a distinct escarpment 10 to 25 feet high. Bertrand soils appear to have somewhat higher pH and a stronger argillic horizon than Fayette (personal communication, G. H. Robinson, 1952). Silty soils are present on all six bench (terrace) levels in the Wisconsin River valley (Hole, Peterson, and Robinson, 1952), but occur chiefly above the lowest sandy one. In places the terraces are crossed by shallow abandoned stream channels that

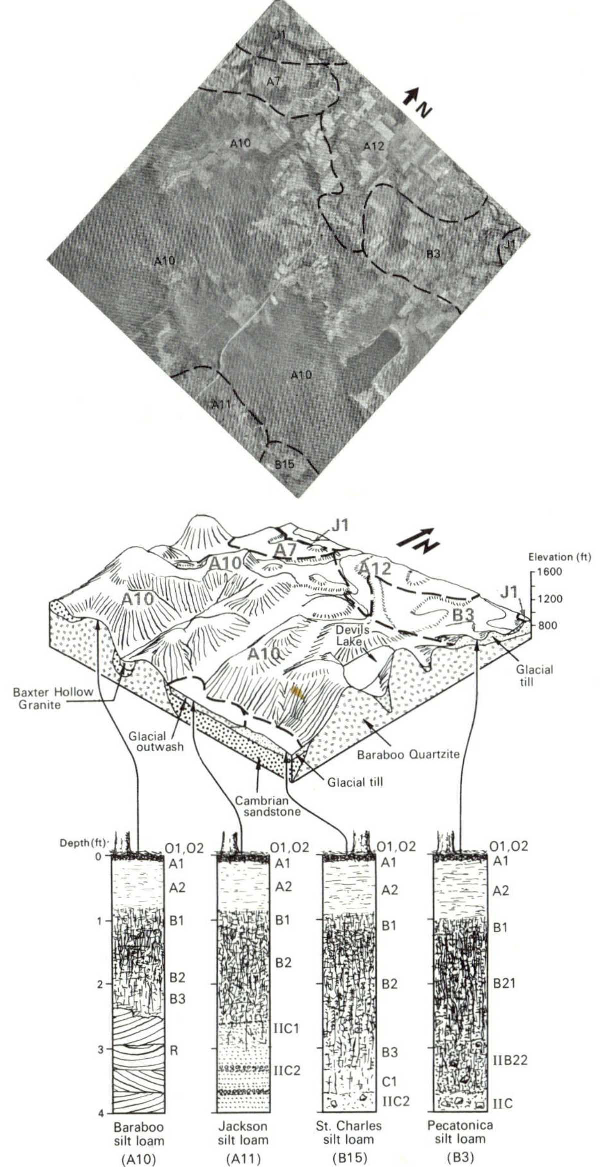

Figure 7-11. Aerial photo map of T.11N., R.6E., Sauk County. Part of the city of Baraboo shows in the northeast corner. The area shown is 6 miles on a side.

Figure 7-12. Block diagram showing landscape positions of major soils of T.11N., R.6E., Sauk County. The Jackson silt loam is a member of the Bertrand catena.

provide sites where, in dry seasons, corn grows unusually tall and stays green longer than on adjacent flats. Between the Lemonweir River and the base of the escarpment of the Cambrian sandstone-Prairie du Chien dolomites is a strip of nearly level land (Fig. 7-2) occupied by the association of well to poorly drained silty soils listed for soilscape A13.

Depths of original medium-textured deposits are 36 to 50 inches of silt for Arenzville, Bertrand, Jackson, Curran, Richwood, and Toddville soils; 20 to 40 inches of silt for Antigo, Ettrick, Tell, and Waukegan soils; 20 to 40 inches of loam in

Dakota, Meridian, and Onamia soils. The substratum at a depth of 6 to 10 feet is commonly outwash sand, acid or calcareous (Figs. 2-5, 2-6). But some benches are rock-controlled, as in the A11 soilscape of southern Richland County.

Associated with these soils (particularly in Richland, La Crosse, and Grant counties) is the Medary-Zwingle-Perrot catena (Typic Hapludalf-Aquollic Hapludalf-Mollic Haplaquept) that has developed on reddish-brown calcareous clays. These materials appear to be glacio-lacustrine in origin, probably derived both from Soil Region I and from residual cherty clay of Soil Region A.

Included in soil association A11 in La Crosse County is the Port Byron silt loam (Typic Hapludoll), a prairie equivalent of the Seaton silt loam, on gently rolling land.

CHAPTER 8

Soil Region B: Soils of the Southeastern Upland

Prominent topographic features of this region include the spectacular Kettle Moraine, the less bold End Moraine, the glacially smoothed east end of the Baraboo Range, the Silurian ("Niagara") Escarpment, and several lower cuestas. Even the ground moraine presents a rolling landscape that is in many places corrugated by "swarms" of drumlins and intervening depressions. In this setting lie hundreds of lakes, marshes, bogs, streams, ancient lake beds, and level outwash flats. The soil surveyors who walked these hills and lowlands were impressed with the extent of the wetlands, the variability of soil depth and character, and the high level of natural soil productivity.

This region includes the southern portions of the Eastern Ridges and Lowlands (Fig. 2-1), in which the glacial drift is yellowish brown to grayish brown in color (the Munsell color notations are 10YR 5/4-5/2 for moist material) (Fig. 2-18). Two patches of soil are separated from the main body of Soil Region B—a large one in Brown, Manitowoc, and Kewaunee counties (mostly B17), and a small one in southern Adams County (mostly B15) (Figs. 8-1, 8-2). The pattern of ridges and lowlands is irregular, because of glaciation (Figs. 8-3, 8-4), in contrast to the well-developed dendritic pattern of Soil Region A (Plate 2). Strip-cropping, so characteristic of Region A, is much less common in Region B (compare Figs. 7-3 and 8-3), except in drumlin country (Fig. 8-12). In most of Region B the bedrock is buried under tens of feet of glacial drift (Fig. 8-7) (Alden, 1904, 1918b) of Woodfordian and Altonian (Cary and pre-Cary) age (Fig. 8-1; Table 2-10). Poorly drained swales are far too numerous to show on the general soil map of the state. Nearly two thirds of the soils of the region are formed on glacial till (Fig. 8-6), about a fourth on glacial outwash, and the remainder on glacio-lacustrine deposits. Nearly all of these materials are calcareous at depth

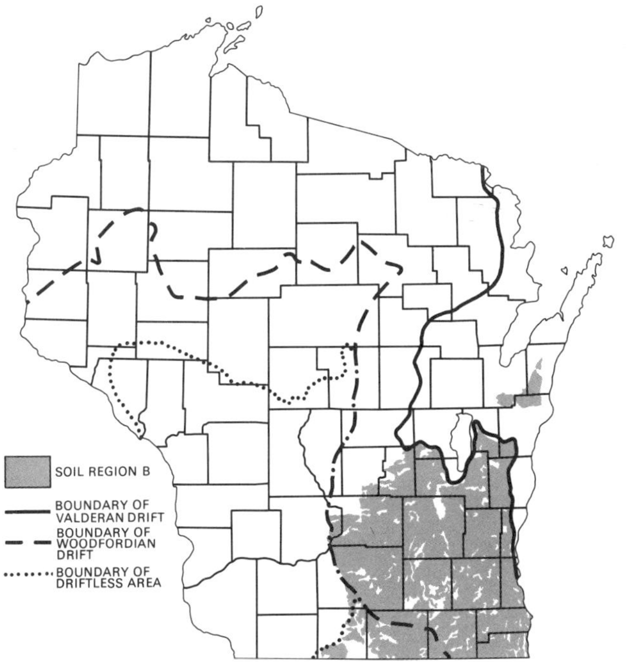

Figure 8-1. Index map showing the geographic relationship of Soil Region B to major glacial boundaries.

(Figs. 2-6, 2-8, 2-10, 2-15, 8-5). The local relief of the land surface is usually less than 100 feet, and in many areas less than 50 feet.

About three fourths of the landscape is occupied by soils without important impediments to natural downward drainage. More specifically, about a third of the area is in well-drained soils, another third is in excessively drained (droughty) soils, and about 12% of the landscape is in moderately well drained soils. Nearly 10% of the region is in peat and muck soils. The remaining 15% is wet mineral soil of the numerous lowlands.

The usual rectilinear pattern of fields and woods is modified or interrupted in places by several features already mentioned: (1) nearly parallel glacial flow traces where there is an abundance of drumlins (Figs. 8-12, 8-13) in central Dodge and Jefferson counties; (2) sinuous or curved bold ridges of the Kettle Moraine (Plate 4; Figs. 8-3, 8-4; see Black, 1969b) and other prominent moraines; (3) natural drainageways such as those of the Rock River system; and (4) bedrock escarpments, particularly where the drift covering is slight. Included in the fourth category are the Silurian ("Niagara") Escarpment of eastern Fond du Lac County and lesser escarpments to the west.

Because of urban expansion in the region, notably in southeastern counties, scarcely two thirds of the land is in farms and this proportion is steadily decreasing. The farmland consists mainly (62%) of cropland harvested, with about 17% in pasture and 10% in woodland. Some of the woodland is pastured.

The landscape surface is estimated (Bryson and Wendland, 1966) to be about 13,000 years old, except for a section in the southwest corner of the region where the age may be 20,000 years (Black et al., 1970). Large blocks of stagnant ice probably stood in depressions, some of which are now occupied by lakes and peat bogs. Soils of the uplands can be expected to be the oldest, wherever they have not been truncated by erosion or rejuvenated by surficial deposits of aeolian material. Soils of depressions have been cumulative, chiefly of organic materials where peat and muck have formed, or of mixed mineral and organic matter in the extensive loamy and clayey wetlands. Acceleration of erosion by land clearing and farming since about 1850 has resulted in siltation of lowlands and bog fringes. This process continues in the vicinity of paved surfaces along highways and in urbanizing areas. The area of "made land" delineations on soil maps is increasing and there is need for detailed mapping of these young soils.

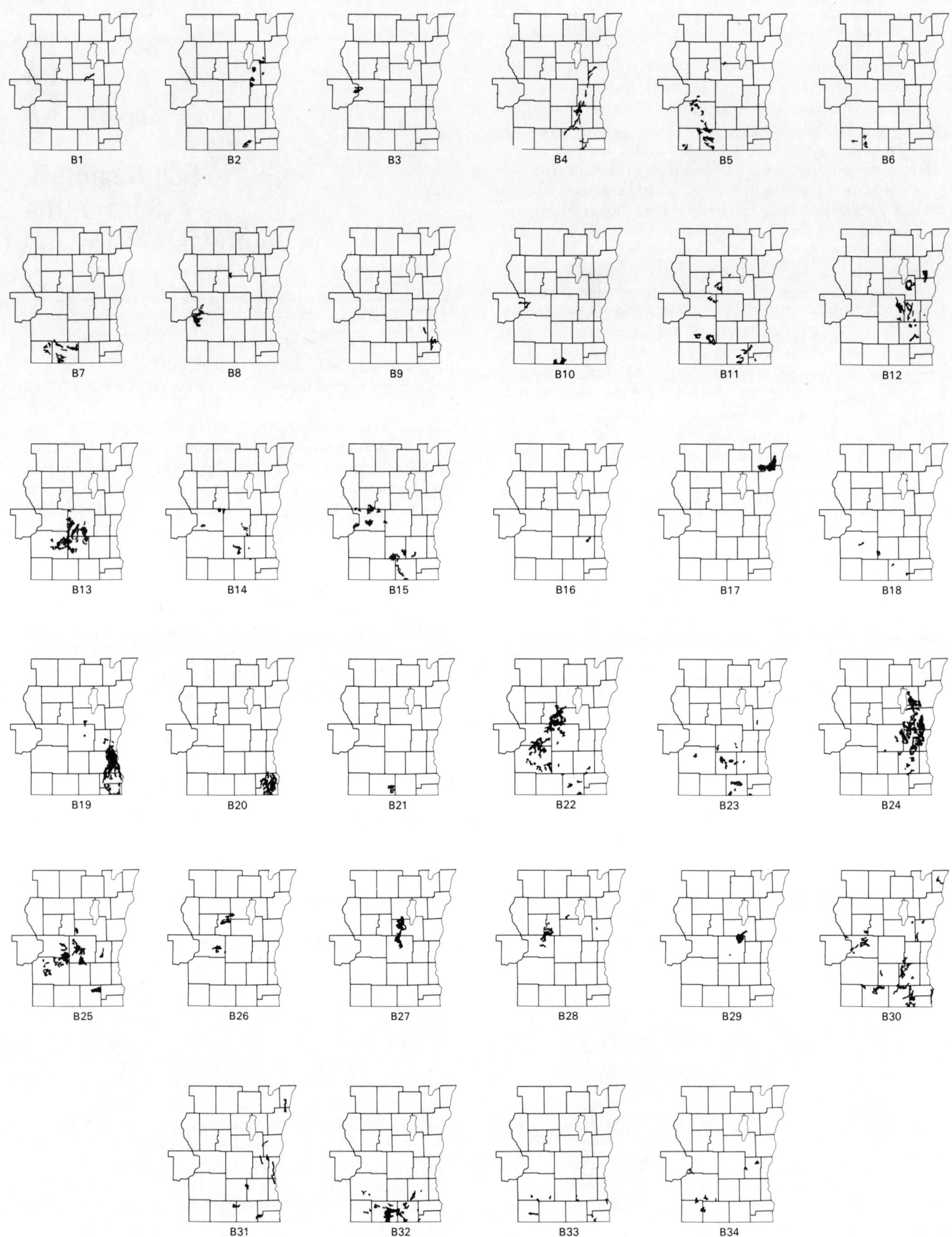

Figure 8-2. Sequence of maps showing distribution of soil associations in Soil Region B.

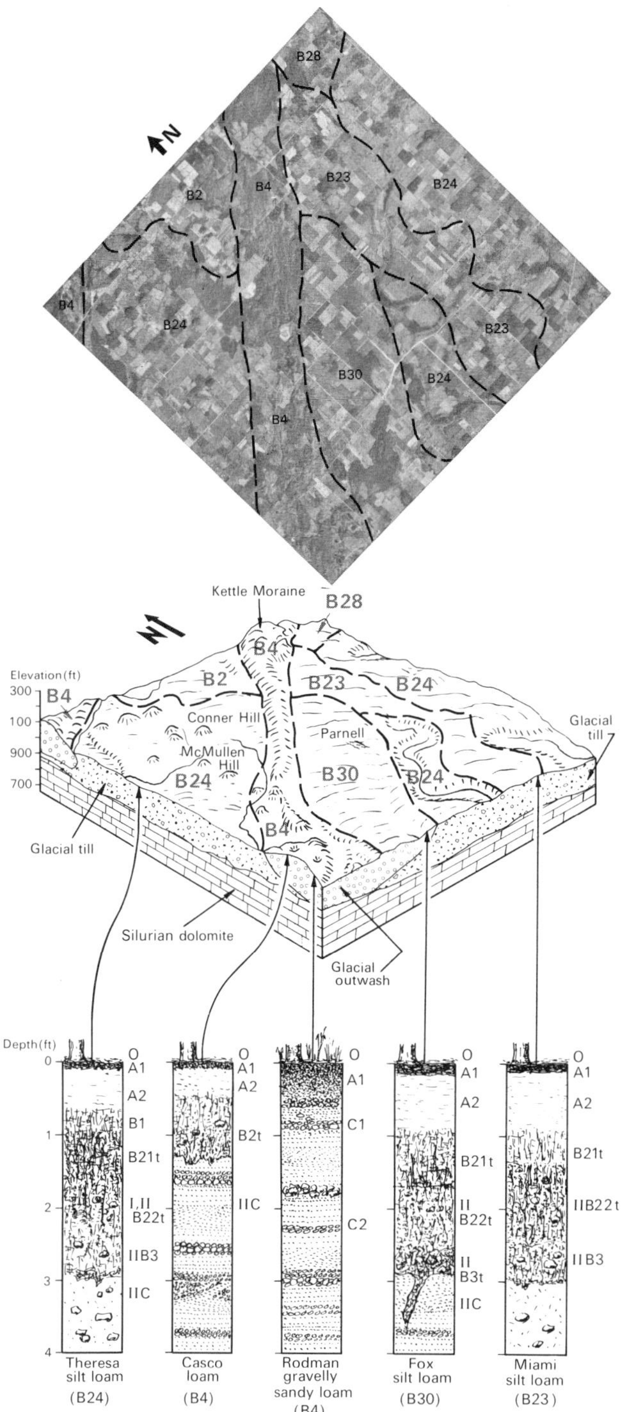

Figure 8-3. Aerial photo map of T.14N., R.20E., Sheboygan County. The area shown is 6 miles on a side.

Figure 8-4. Block diagram showing landscape positions of major soils of T.14N., R.20E., Sheboygan County.

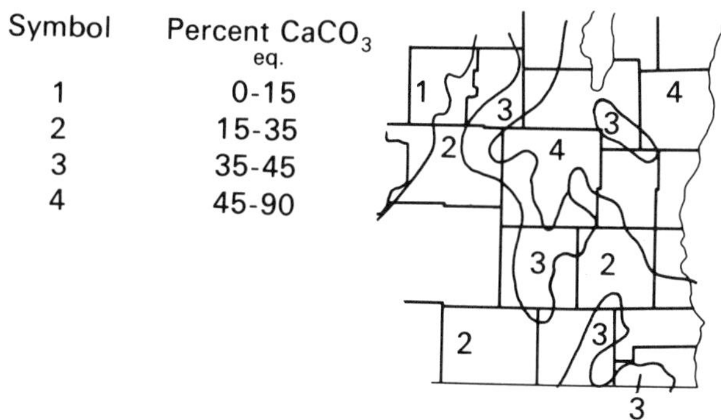

Figure 8-5. Map showing distribution of glacial tills of different carbonate contents in southeastern counties (after Owens, 1968).

The proportion of soils formed, at least in part, from aeolian silty coverings (leached loess) decreases northeastward in the region (Figs. 2-28, 2-29, 2-30; Plate 5). The detailed pattern of loess deposits is not fully understood. For example, the preponderance of silty upland soils in northern Jefferson County and the scarcity of them in southern sections of the same county have not been satisfactorily explained. Some observations indicate that loess was deposited most thickly on southwest-facing slopes in Columbia County (Rieger, 1947). Windblown sandy coverings are also notable in places, as in the vicinity of Columbus and Rio, northeast of Madison.

Available moisture-holding capacity of the soils varies with texture, particularly with respect to depth of fine-textured material over coarse deposits. In the nongrowing seasons when vegetation uses almost no soil moisture, soils become relatively moist throughout the landscape. Perched water tables appear over the Bt horizon in many medium-textured upland soils during winter and early spring.

The carbonate content of the glacial till decreases westward from well over 50% north of Milwaukee to less than 15% in Marquette County (Fig. 8-5). Clay content in the till also shows a regional trend of decrease from more than 50% on the east to less than 5% on the west (Fig. 8-6). Watson (1963) prepared a map showing the distribution of four major Woodfordian tills in Columbia County, which are, from east to west: (1) heavy sandy loam and loam with 60% or more $CaCO_3$ equivalent; (2) similar till but with 28 to 44% $CaCO_3$ equivalent; (3) sandy loam with 18 to 31% $CaCO_3$ equivalent; and (4) loamy sand

Symbol	Percent Clay
1	0-5
2	5-10
3	10-15
4	15-25
5	25-40
6	40-80

Figure 8-6. Map showing distribution of glacial tills of different clay contents in southeastern counties (after Owens, 1968).

and light sandy loam, with 5 to 19% $CaCO_3$ equivalent. Depth of leaching of soils is greatest in till of lowest carbonate and clay contents. G. B. Lee (personal communication, 1959) reported a lithosequence of soil series from southeastern Wisconsin in which thickness of solum increases with decrease in carbonate content of the underlying glacial till: LeRoy, 20 inches; Lomira, 32 inches; Theresa, 34 inches; Dodge, 38 inches; McHenry, 40 inches; and Pecatonica, 60 inches. The corresponding differences in percent $CaCO_3$ equivalent in the till range between 75 (LeRoy) and 20 (Pecatonica). Loess coverings of various thicknesses are all leached of carbonates in Region B.

Most of the well-drained soils are Hapludalfs (Gray-Brown Podzolics), but at sites of "ancient maple forests" (Curtis, 1959; Harper, 1963; Lee, 1949) are Argiudolls (Brown Forest soils; Pierce, 1951), with nearly black A1 horizons as thick as 9 inches.

It may be noted that in southeastern Wisconsin the "beta B" (Bartelli and Odell, 1960a,b), which is the argillic horizon at or near the upper boundary of calcareous glacial drift, occurs in various positions in the pedon: (1) as the lowest subhorizon of the solum (see dark horizon at a depth of 3 feet in Fig. 1-6), (2) as a subhorizon between the main B2 and the B3 horizons, and (3) as a subhorizon between the C1 and IIC2 horizons. The beta B subhorizon marks the deep subsoil zone to which percolating waters flush solutes and colloids seasonally, probably in the spring of the year.

The thirty-four soil associations (Fig. 8-2) in this region are grouped in the soil map legend (Plate 1) in order, from those with steepest slopes to those on nearly level terrain. In the following discussion these associations are regrouped with more emphasis on soil series than on topography.

B1, B16. *Knowles, Ripon, Morley, Casco, and Sisson silt loams and loams, with rocky land.*

B1. Knowles and Morley silt loam association, with rocky land.

B16. Knowles, Ripon, Casco, and Sisson loams association.

The Knowles silt loam, a Typic Hapludalf that formed from 20 to 36 inches of loess overlying limestone bedrock (Fig. 2-3), is prominent in two soilscapes: (1) a narrow strip of land (B1) along the Silurian ("Niagara") Escarpment in Fond du Lac and Dodge counties, overlooking the Horicon Marsh and other lowlands (Fig. 8-7); and (2) a rolling to undulating area in northeastern Waukesha County famous for its quarries from which Lannon building stone comes. These two soilscapes occupy only about 10,000 acres each, but they are distinguished from adjacent associations on the basis of presence of soils shallow to bedrock.

Associated with the Knowles on the escarpment (B1) are two soils formed from less than 20 inches of silt cover over calcareous shaley (Maquoketa Shale-derived) glacial till, namely, the Neda (Mollic Hapludalf), formerly mapped as Morley, and the Ashippun (Auqollic Hapludalf). The Sogn series (Lithic Hapludalf) is also present, where bedrock is as close as 15 inches to the surface.

In the Waukesha area a wider variety of soils is associated with the Knowles. The Ripon silt loam (Typic Argiudoll) is the dark equivalent of Knowles. Casco (Typic Hapludalf) has loamy sola 12 to 20 inches thick over calcareous outwash sand and gravel. The Sisson (Typic Hapludalf) is formed in calcareous lake-laid fine sand and silt.

These soilscapes are largely used for pasture and field crops.

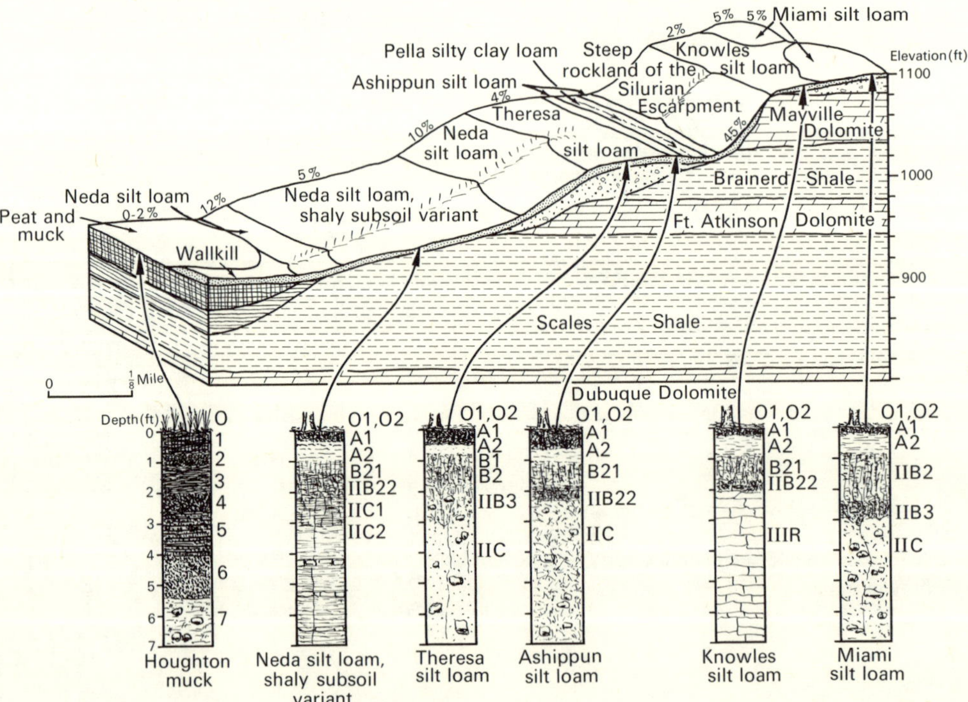

Figure 8-7. Block diagram showing landscape positions of representative soils of soil associations B1 and B8 in Sections 4 and 5, T.13N., R.16E., Dodge County.

B2, B12, B17, B24. *Theresa, Hochheim, and associated soils, shallow over stony, dolomitic glacial till.*

- B2. The hilly to rolling Hochheim, Theresa, and Brookston silt loam association.
- B12. The rolling to undulating Theresa, Hochheim, and Nenno silt loam association.
- B17. The rolling to undulating Theresa, Onaway, Fox, and Salter silt loam and loams association.
- B24. The undulating to rolling Theresa, Hochheim, and Nenno silt loam association.

Theresa (Typic Hapludalf; Figs. 8-7, 8-8) and Hochheim (Typic Argiudoll) soils are about 36 and 20 inches deep, respectively, over highly calcareous glacial till (see Fig. 8-4). They occur on convex slopes of drumlins and moraines, including some ridges in the Kettle Moraine complex. The Theresa is on slopes of 2 to 12% gradient; the Hochheim on slopes of 2 to 30% gradient. The Theresa soil has formed in 20 to 30 inches of loess covering. The Hochheim may be silty to a depth of 20 inches or may be somewhat sandy throughout. The subsoil in both cases is a dark clay loam. The sola are alkaline to neutral for the most part. The till adjacent to the Kettle Moraine becomes coarser in texture, differing from outwash in fabric more than in particle size distribution. The presence of lenses of stratified sand and gravel makes disposal of liquid waste hazardous in the C horizon. The native vegetation was dominantly maple-basswood forest, some of which still survives on steep lands, as in the Kettle Moraine area near Dundee in Fond du Lac County (Scholz and Trenk, 1959). Most of the area is now used for production of corn, oats, and hay. Alfalfa does particularly well on these soils.

The four soilscapes, totaling 880,000 acres or 2.4% of the land area of the state, are listed in order of decreasing relief.

The B2 association includes parts of the Kettle Moraine and other hilly land in which the two soils just discussed occur, along with depressional bodies of poorly drained black Brookston soils (Typic Argiaquolls). Around Lake Geneva in Walworth County these soils lie on long uniform slopes of 4 to 15% gradient. In Fond du Lac County, they are on shorter, steeper slopes.

The more extensive B12 soil association is characterized by shallower depressions, and these are commonly occupied by Nenno loam and silt loam (Aquic Argiudoll), a somewhat poorly drained associate of the Hochheim.

Onaway (Alfic Haplorthod) and some Ozaukee (Typic Hapludalf) soils are associated with Theresa soils in the northern outlier of this region, surrounded by reddish-brown clay soils in Brown, Manitowoc, and Kewaunee counties. The glacial till is pink to brown (the Munsell color notations are 7.5YR—10YR 5/4 for moist material). In lake basins the moderately well drained Salter loam (Typic Eutrochrept; Fig. 1-4) has formed on calcareous silts and fine sands. On small outwash plains is found the Fox silt loam (Typic Hapludalf; see Fig. 1-6). An interesting pattern of contrasting depths of leaching of carbonates has been found in this area (personal communication, Professor David M. Mickelson, Department of Geology and Geophysics, University of Wisconsin-Madison, 1975). The average depth in well-drained soils over till is about 44 inches in the B17 soil association and in associations of red clays (Region I) lying to the south thereof as far as the B region in southern Calumet and Manitowoc counties. In contrast, the depth of leaching averages about 23 inches in the red clay region (I) bordering the B17 soilscape on the Two Rivers and Denmark moraines and immediately beyond to the east, north, and west.

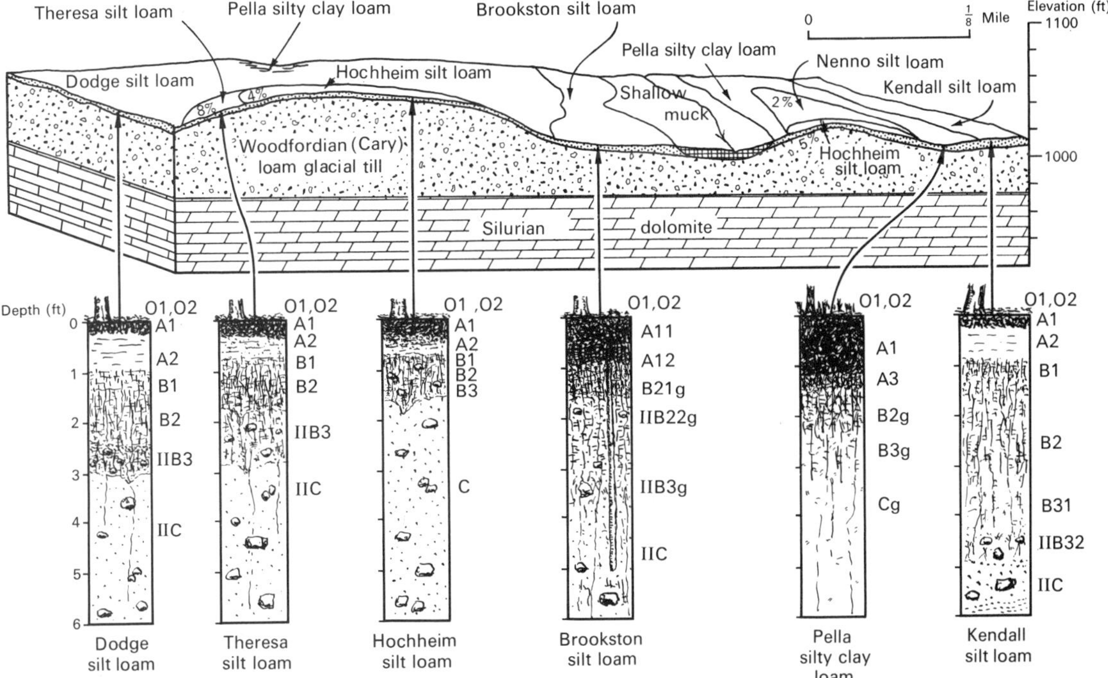

Figure 8-8. Block diagram showing landscape positions of representative soils of soil associations B24 and B25, Dodge County.

The most extensive soilscape in this group is the B24 association, composed chiefly of the same three soils as association B12.

B3, B6, B7, B10. *Pecatonica and associated silt loams, with clayey subsoils and some rocky land.*

- B3. The hilly to rolling Pecatonica and Flagg silt loam, and Baraboo stony silt loam, association.
- B6. The rolling to hilly Dubuque, Pecatonica, McHenry, and Whalan silt loam association.
- B7. The rolling to hilly Pecatonica, Dodge, McHenry, and Whalan silt loam association.
- B10. The rolling to hilly Flagg, Pecatonica, and Mingo silt loam association.

These soilscapes are interesting because they include truncated remnants of some of the oldest soils in the state (Fig. 2-17). Soils with and without glacial drift components are associated, which indicates that processes of mass wasting have stripped uplands of glacial deposits locally. Bleuer (1970) classified the glacial drifts under the Flagg and associated soils as Altonian and Illinoian in age (see Table 2-10). Young soils (on Woodfordian drift) are associated with the old soils (paleosols) in these soilscapes. The paleosols themselves are blanketed by aeolian silt in which the upper, younger horizons have formed. The Flagg and Pecatonica silt loams (Fig. 7-12) developed in 36 to 50 inches and 20 to 30 inches, respectively, of loess overlying paleosolic Bt horizons. These soils are from 4 to 8 feet deep over calcareous sandy loam to loam glacial till. The Mingo soils differ from the Flagg in being somewhat poorly drained.

Most of the soils listed in the soil map legend for these associations are well-drained Typic Hapludalfs. They occur in front of the end moraine of the Woodfordian ice advance (south part of Fig. 8-1) and in the vicinity of the east end of the Baraboo Range in eastern Sauk County. The Baraboo silt loam has formed in 20 to 40 inches of loess over quartzite bedrock. The Dubuque (and associated New Glarus, which has more than 6 inches of residuum on the limestone) and Whalan silt loams are commonly about 2 to 3.5 feet deep to limestone bedrock. The last-named soil has weathered glacial till in the Bt horizon (Figs. 2-2, 2-3). On the younger glacial drift are the Dodge (over loam till) and McHenry (over sandy loam till) silt loams.

Most of these areas are used for production of corn, soybeans, oats, and hay, but steep slopes, especially where rocky, are forested, as in the scenic Baraboo range. Soil association B3 is dominated in Columbia County by St. Charles (20% by area), McHenry (15%), Baraboo (15%), and Pardeeville (15%) soils (McColley, 1971).

B4, B18, B30, B31, B33, B34. *Fox and associated gravelly and silty soils over sand and gravel.*

- B4. The hilly to rolling Casco, Rodman, Fox, and Lapeer loams association.
- B18. The rolling to undulating Fox, Casco, and St. Charles (stratified substratum) silt loam association.
- B30. The undulating to rolling Fox and Casco loams, and Boyer sandy loam, association.
- B31. The undulating to rolling Fox, Will, Casco, and Fabius silt loam association.
- B33. The nearly level and gently undulating Fox, Hebron, and Del Rey loams association.
- B34. The nearly level and gently undulating Fox silt loam, and St. Charles (stratified substratum) and McHenry loams, association.

The Fox silt loam, a Typic Hapludalf with shallow silty cov-

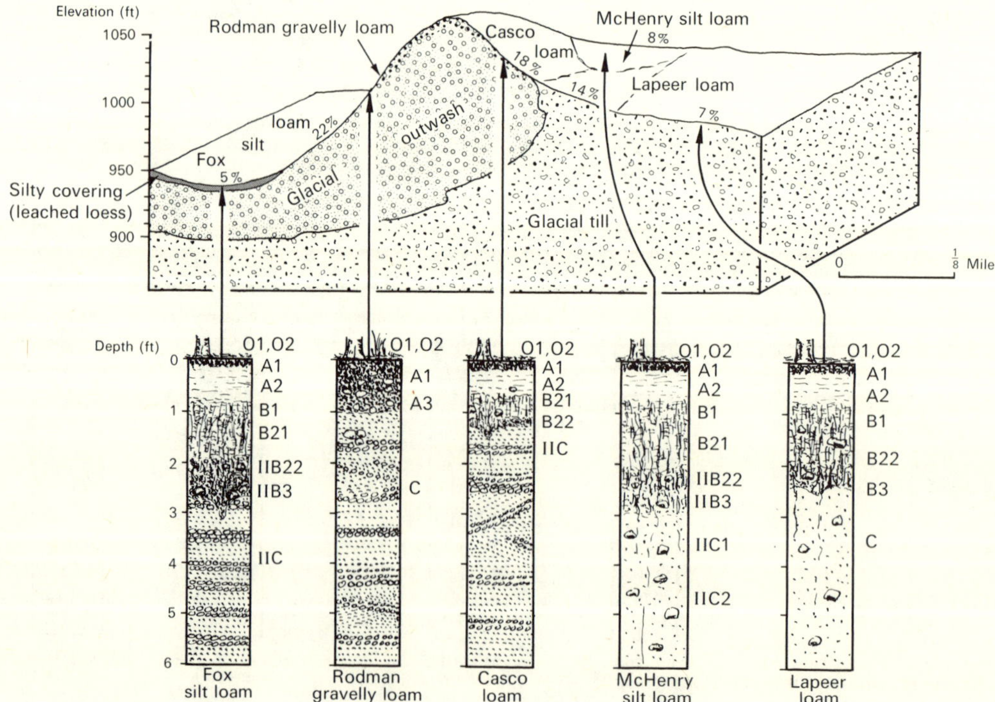

Figure 8-9. Block diagram showing landscape positions of representative soils of soil association B4, Dodge County.

ering (less than 20 inches thick, but the solum is 20 to 42 inches thick) over calcareous outwash sand and gravel (Figs. 1-6, 8-4, 8-9), is the most extensive soil in this group of soil associations that occupy about 2% of the land area of the state. In Rock County, a variant of Fox is underlain by calcareous sand in which is found a thick, weakly developed B3 horizon. The range of landscape conditions is considerable, from Kettle Moraine topography (B4) to nearly level outwash and lake plains (B33). A sequence of soils over calcareous outwash sand and gravel, in order of increasing thickness of silty covering, is Rodman (Typic Hapludoll; Figs. 8-4, 8-9), no silt; Casco (Typic Hapludalf; Figs. 8-4, 8-9) and Fabius (Aquic Argiudoll), 0 to 12 inches of silt; Fox (Typic Hapludalf) and Will (Typic Haplaquoll), 0 to 20 inches of silt; St. Charles, stratified substratum (Typic Hapludalf; Fig. 7-12), 36 to 50 inches of silt. The Boyer (Typic Hapludalf) soil has no silty covering, but has a sandy solum 20 to 40 inches deep over calcareous sand. Two soils in the association, both Typic Hapludalfs, formed over calcareous glacial till: Lapeer, with no silty covering, and McHenry with 20 to 36 inches of silty covering. Two other soils formed in calcareous lake-laid silts and clays with loamy coverings: Hebron (Typic Hapludalf) and Del Rey (Aeric Ochraqualf). Hebron and associated soils have free carbonates as shallow as 30 inches below the surface. Studies by Borchardt, Hole, and Jackson (1968) and Gaikawad and Hole (1965) indicate that the same suite of minerals is present in all of these soils, although the proportions of size-separates vary from one series to another. South of Milwaukee along the shore of Lake Michigan soil associations B30 and B31 are dominated by Boyer and Granby (Typic Haplaquoll) sandy loams.

Production of alfalfa-brome hay in tons (dry weight) per acre per year on these soils (under good management) is an index of available moisture-holding capacity of the soils: St. Charles (4 feet to coarse material), 4.75; Fox (30 inches), 3.0; Casco (15 inches), 2.75; Rodman (6 inches), none.

In the Kettle Moraine of Waukesha, Jefferson, and adjacent counties, Spinks sand with banded B horizon (mentioned by Robinson and Rich, 1960) occupies a higher position on flanks of ridges of Rodman soils than do the band-free Oakville and Tedrow sands. The Spinks soil may have formed from stabilized aeolian sands and the Oakville-Tedrow soils from aeolian and beach deposits that were periodically reworked by wind and water.

B5, B21, B22, B32. *Ringwood, Ogle, Plano, and associated dark, well-drained soils developed under former prairie.*

- B5. The rolling to undulating Ringwood, Durand, and Ripon silt loam association.
- B21. The gently undulating to rolling Ogle, Durand, and Pella silt loam association.
- B22. The gently undulating to rolling Plano, Saybrook, Ringwood, Elburn, and Pella silt loam association.
- B32. The nearly level and gently undulating Plano and St. Charles (stratified substratum), Warsaw, and Fox silt loam association.

The influence of the original prairie vegetative cover (Green, 1950) is shown by the dark color of the thick surface horizon of these soils. Most of them are Typic Argiudolls: Ogle, Plano; Ringwood, Saybrook, Ripon; Durand; Warsaw. They are listed here in order of decreasing thickness of silty covering, from as much as 50 inches (Ogle, Plano) to less than 20 inches (Warsaw), over calcareous glacial drift. Corresponding annual yields (tons, dry weight, per acre) of alfalfa-brome hay, under good management, range from 4.75 to 3.5. Depths of 2 to 4 feet of silty soil predominate in these landscapes. The other soils are the somewhat poorly drained Elburn (Aquic Argiudoll) and poorly drained Pella (Typic Haplaquoll), both associated with Plano and the deep silty St. Charles, stratified substratum phase, and with the shallower Fox silt loam. The last two soils have developed under forest cover (Typic Hapludalfs). The landscape position of the St. Charles, stratified substratum, is along the contact between till upland and outwash flat. The Ripon soil (Typic Argiudoll) is 20 to 42 inches deep over limestone bedrock. This entire group of productive soils occupies about 2% of the land area of Wisconsin, or 700,000 acres.

The deep (36 to 50 inches in the Ogle) and moderately deep (20 to 30 inches in the Durand) silty upper sola are the dark equivalents of the Flagg and Pecatonica, discussed in a preceding section. All four silt loams are underlain by old truncated subsoils formed in early glacial drifts (Altonian and Illinoian).

Named prairies under which some of these soils formed are Walworth and Rock prairies in the two counties of the same names, and the Arlington Prairie (also called the Empire or "High" Prairie; Engel and Hopkins, 1956) in Columbia and Dane counties.

In Columbia County, soil association B22 is dominated by Plano (50% by area) and Ringwood (15%) soils.

B8, B15, B26, B28. *Lapeer, Metea, and associated loams on sandy dolomitic glacial till, with some rocky land.*

- B8. The rolling to undulating Lapeer, McHenry, and Miami silt loam association, with rock outcrops.
- B15. The rolling to undulating Lapeer, Pardeeville, Boyer, and McHenry loams association.
- B26. The gently undulating to rolling Metea, Puchyan, Miami, and Lapeer loams association.
- B28. The gently undulating to rolling Lapeer, Pardeeville, and McHenry loams association.[1]

On glacial ground and end moraines of calcareous till on which patches of distinct loess covering are not extensive, the Lapeer (Typic Hapludalf) and Metea (Arenic Hapludalf) soils predominate. The Lapeer and its somewhat dark associate, Pardeeville (Mollic Hapludalf), overlie sandy loam till, whereas the Miami (Typic Hapludalf; Figs. 8-4, 8-7) overlies loam till. All three soils have less than 20 inches of silt cover. The McHenry (Typic Hapludalf; Fig. 8-9) has 20 to 36 inches of silty soil. This soil is underlain by pink till on the Marengo ridge, classified as a Tazewell moraine by Thwaites (Table 2-10) and so published by Flint (1945). Sandy coverings over these same tills characterize the Puchyan (15 to 36 inches of sandy deposit over silt over till) and the Metea (18 to 36 inches of sandy deposit over till) soils, both classified as Arenic Hapludalfs. It is apparent that considerable blowing of both silt and sand took place over these landscapes after deglaciation.

1. On the soil map, Plate 1, in T.16N., R.11E., B should read B28.

The Boyer soil (Typic Hapludalf) consists of 20 to 40 inches of sandy soil over calcareous outwash sand. These are all agriculturally productive soils, even though somewhat droughty as compared to deep silt loams of the region. Rock outcrops in soil association B8 indicate that excavation and even cultivation are not possible in places. Mixed into the sandy loam and loam Lapeer soil are some silt and clay particles of aeolian origin.

A detailed study in a gravel pit in Boyer and associated soils 2 miles southwest of Dekorra in Columbia County showed the presence of tongues of B2t horizon 5 to 6 feet deep and as much as 7 feet long and 2 feet wide at the tops, having the forms of cones and linear and branching wedges (as much as 10 feet long) (Yehle, 1954). Stratified gravel beds cross the tongues, with downward deflection and loss of carbonates in the tongues. These are not to be confused with ice-wedge casts (see page 19), but are the result of intense leaching localized by unknown factors that controlled patterns of percolation of water into the soil.

A Seaton-like soil is included in soil association B15 in Columbia County (McColley, 1971).

B9, B19, B20. *Morley, Varna, and associated silty soils over gray clayey glacial till.*

B9. The rolling to undulating Morley, Blount, and Varna silt loam, and Ashkum silty clay loam, association.[2]

B19. The gently undulating to rolling Morley, Blount, and Ozaukee silt loam, and Ashkum silty clay loam, association.

B20. The gently undulating to rolling Varna and Elliott silt loam, and Ashkum silty clay loam, association.

The dolomitic, clayey (about 30% clay) substratum of these soils is glacial till (Figs. 2-12, 2-22, 8-10) containing considerable shale, derived from the Devonian Milwaukee Shale[3] in Racine and Kenosha counties and adjacent basin of Lake Michigan, and from the underlying Silurian ("Niagara") dolomites. The glacial drift is commonly brown and gray (10YR and 2.5Y colors) but under the Ozaukee soils is slightly pinker (7.5YR-5YR colors, transitional toward the 5YR-2.5YR colors of till under Kewaunee soils) (Fig. 8-11). The Ozaukee C horizon contains 40% calcium carbonate equivalent and in many places is thin (6 to 12 feet) over outwash or loamy till. Where silty clay loam till rests on loam till a white horizon of calcium carbonate accumulation occurs at the contact (Watson, 1961). All of the soils in these associations are formed in less than 20 inches of loess covering over the clayey till. Both forest and prairie vegetative covers have left their impress. The dark Varna (Typic Argiudoll) and Elliott (Aquic Argiudoll) soils formed under native grasslands. Barnes Prairie was the name given to a large body of these soils in Racine County. The lighter colored Morley (Fig. 8-10), Ozaukee (both Typic Hapludalfs), and Blount (Aeric Ochraqualf) soils were originally forested. The Morley soil solum is about 7 inches thicker and is developed in slightly thicker loess covering than the Ozaukee soils. Associated with Ozaukee are the somewhat poorly drained Mequon loams (Udollic Ochraqualfs) and in Ozaukee County, Martinton (Aquic Argiudoll) and Saylesville (Typic Hapludalf) loams on lacustrine deposits. In the lowlands are bodies of Ashkum (Typic Haplaquoll), many of which have been drained for crop production. The low permeability of the glacial till makes these landscapes unique in that in places farmers have installed both tile drains and runoff diversion terraces on the same slope, where the gradient is about 9%. This unusual combination of drainage and erosion-control practices accelerates removal of subsoil water and at the same time slows removal of surface water.

Pedersen's investigation (1954) of the Varna soil catena extended into Illinois and covered the entire geographic province of the association. He found the sola of the Elliott-Varna members of the catena in Wisconsin to be slightly shallower, more silty in the A horizon, and more clayey in the B horizon than corresponding soil profiles in Illinois. The differences he attributed to the age difference (possibly as much as 10,000 years) between the early Woodfordian (Tazewell) till landscape in several Illinois counties and the late Woodfordian (Cary) till landscape in Wisconsin. Pedersen suggested that the increasing proportion of the landscape occupied by the poorly drained Ashkum soil (Watson, 1961) from north (about 26%) to south (about 34%) may have been caused in part by gradual collapse of borderlands of depressions, as a result of removal of colloidal material from sola by lateral seepage waters moving through the soil just above the dense calcareous silty clay loam glacial till (bulk density, 1.8; 35% $CaCO_3$ equivalent).

In Sheboygan County a Theresa-like soil is associated with the Ozaukee series on reddish-brown till.

Illite and chlorite are the predominant clay minerals in both the Paleozoic shales and silty clay loam tills of southeastern Wisconsin (Dixon and Jackson, 1960) (Fig. 4-5).

B11, B13, B14, B23, B25, B27, B29. *Miami, McHenry, Lomira, Dodge, St. Charles, and associated silty soils on sandy, dolomitic glacial till.*

B11. The rolling to undulating Miami, McHenry, Clyman (Crosby), and Brookston silt loam association.

B13. The rolling to undulating Miami, Dodge, and Pella silt loam association.

B14. The rolling to undulating McHenry, Lapeer, Miami, and Brookston silt loam association.

B23. The gently undulating to rolling Miami, McHenry, and Brookston silt loam association, with peat and muck,

B25. The gently undulating to rolling St. Charles, Dodge, Miami, and Pella silt loam association.

B27. The gently undulating to rolling Lomira, LeRoy, and Knowles silt loam association.

B29. The gently undulating to rolling Dodge, Lomira, and Knowles silt loam association.

These productive agricultural soilscapes occupy about a million acres, or nearly 3%, of the land area of the state. Soil mineralogy is mixed, providing an ample reserve of plant nutrients (Batson, 1940). Soils of the uplands are most extensive and are all underlain by calcareous glacial till. Bedrock comes close to the surface in a few places. Lowlands contain fertile black

2. A blank strip on the soil map, Plate 1, in T.6N., R.20E., interrupts a body of soil association B9, boundaries of which should be continued due south.

3. Watson (1961) reported finding in the glacial till some fossil spores of a Devonian plant, *Tasmites Chicagoensis.*

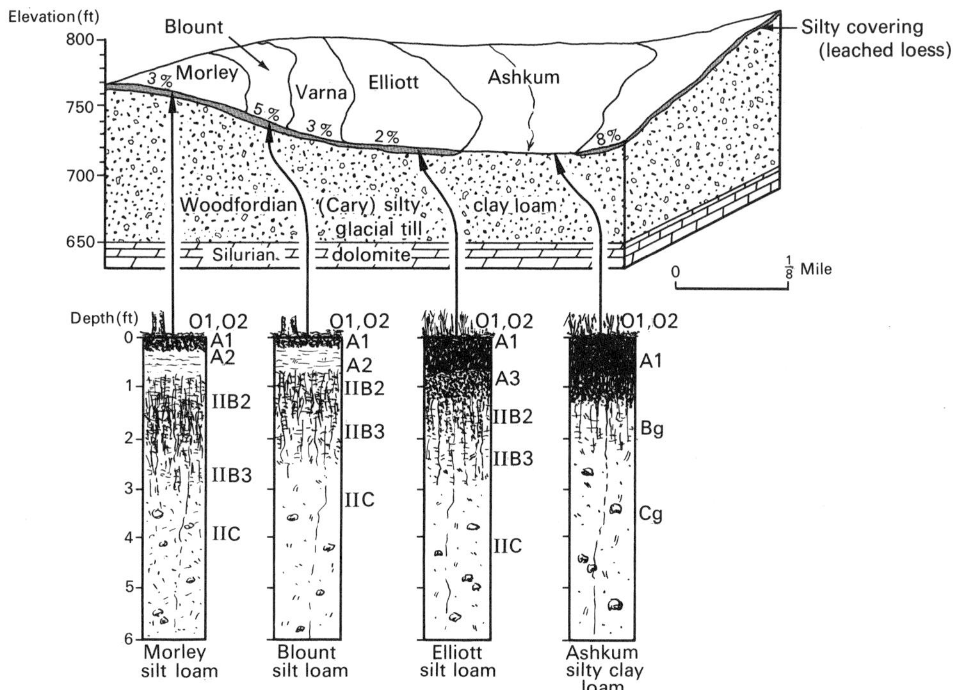

Figure 8-10. Block diagram showing landscape positions of representative soils of soil associations B19 and B20 in Section 29, T.3N., R.21E., Racine County.

soils, both mineral and organic. A dark, thick A1 horizon may extend from the lowlands to soils of footslopes. Thus, in some sections of Dodge County, Elburn soils occupy the place of Kendall soils in the St. Charles catena (see catenas numbers 47 and 49 in Table 18-1). Landforms are glacial moraines for the most part, but drumlins are prominent (Alden, 1905) (B13; Figs. 8-12, 8-13), particularly in Dodge County, where many field boundaries run parallel to them (Collins, 1971). This Woodfordian glacial drift is calcareous (dolomitic). Loess coverings are thought to have been deposited between 14,000 and 8,000 years ago, silts having been blown in part from the Mississippi and Wisconsin River valleys, in part from southwestern states, and in part from local till and outwash surfaces.

The naturally well-drained soil series (Typic Hapludalfs) are differentiated on the basis of (1) depth of loess covering and (2) texture of the underlying glacial till. The Miami (on loam till; Fig. 8-13), LeRoy (on highly calcareous channery till), and Lapeer (on sandy loam till; Fig. 8-9) soils have less than 20 inches of silty coverings. The Dodge (on loam till; see Ciolkosz, 1967) (Figs. 8-8, 8-12, 8-13), Lomira (on highly calcareous channery till), McHenry (on sandy loam till; Fig. 8-9), and Knowles (over limestone bedrock; Fig. 8-7) soils have loess blankets 20 to 36 inches thick. The St. Charles and Kendall soil series (Figs. 7-12, 8-8, 8-13) formed in 36 to 50 inches of loess over sandy loam or loam till. In southern Dodge County the B2t horizon of a pedon has clay films containing about four times as much clay and organic matter and twice as much free iron as the natural soil blocks that the films coat (Buol and Hole, 1959). Miedema and Slager (1972) reported a profile clay illuviation index (see page 27) of 550 for a silty soil in this area. Depressions in the landscapes are occupied by the very poorly drained Pella (Typic Haplaquoll; Figs. 2-52, 8-8), bordered by the poorly drained Brookston (Typic Argiaquoll; Fig. 8-8) and Lamartine (formerly called Clyman and Crosby; Aquic Hapludalf; Fig. 2-52).

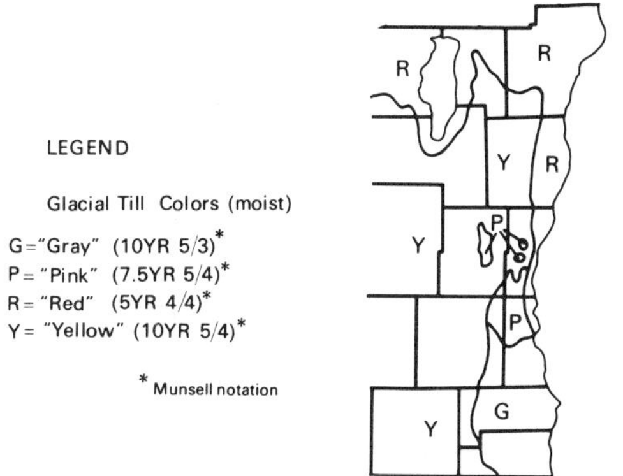

Figure 8-11. Map showing the distribution of glacial tills of four different colors in southeastern Wisconsin (after Watson, 1961).

Soil texture influences available moisture-holding capacity as indicated by these annual yields (tons, dry weight, per acre) of alfalfa-brome hay: Lapeer, 3.0; McHenry, 3.5; Miami, 4.5; and St. Charles, 4.75.

These soils have been leached of carbonates to depths of 20 to 50 inches. White seams of secondary calcite along joints in the till attest to the translocation of carbonates by water down cracks. Soil reaction in the upper sola of these soils ranges from very strongly acid (pH 4.5) in deep silty upland soils, to neutral in soils overlying channery till, to calcareous in alkaline phases

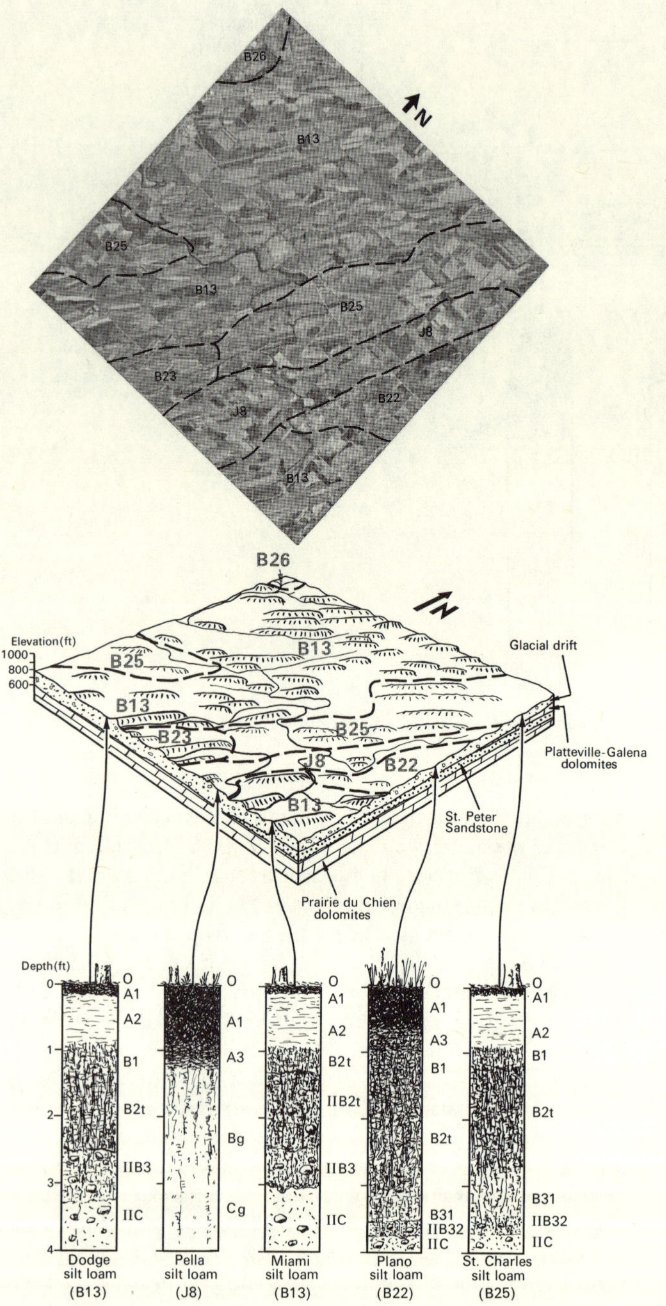

Figure 8-12. Aerial photo map of T.10N., R.13E., Dodge County. The area shown is 6 miles on a side.

Figure 8-13. Block diagram showing landscape positions of major soils of T.10N., R.13E., Dodge County.

of Pella of lowlands in which snail shells are abundant all the way up to the soil surface.

In Columbia County, soil association B25 includes depressional soils (Typic Haplaquolls) that are more silty than the Pella series, which they otherwise resemble.

In soilscape B13, erosion on sides of drumlins has stripped the silty solum from the glacial till locally (Hole, 1956b), removing the solum of the original Miami silt loam and leaving in its place a primitive soil on stony, dolomitic sandy loam or loam (a soil like Hennepin sandy loam; Typic Eutrochrept). Alfalfa can do well on such a soil, but oats and corn cannot.

In southern Walworth County some bodies of Miami soil are underlain by pink (7.5YR 5/4) and reddish-brown (5YR 4/4) glacial tills—ranging in texture from sandy loam to plastic light clay loam—over a far more extensive area than that occupied by the older Wisconsinan drift called Tazewell by Thwaites (Flint, 1945). Pink till underlies Hochheim and Theresa soils in western Washington County.

CHAPTER 9

Soil Region C: Soils of the Central Sandy Uplands and Plains

The central sandy area of Wisconsin has special charms and limitations. The charms have been well documented in *Sand County Almanac* by Leopold (1949), and are capitalized on by summer home developments around recently created artificial lakes. The limitations were experienced by early farm families who suffered severe crop losses in dry years.

This region responds very quickly to seasonal changes. The well to excessively drained soils warm up earlier in the spring than do the finer textured soils that have been considered in the previous chapters on Soil Regions A and B. When the soils are still moist from snow melt and spring rains, vegetative growth is good. The lupine, the wild rose, and many other flowers make the landscape attractive. Farmers can cultivate early. The sandy soils are never sticky. Yet drought can strike the area with a vengeance. Except where silt and clay beds underlie the sands, uplands tend to dry up early in the summer. The "scrub oak barrens," also called oak savannas (Curtis, 1959), are extensive, though far less open today than they were before forest fire protection became systematic.[1] The prickly pear cactus is a natural component of some plant communities. Wind erosion produced sand dunes in ages past and continues to do so today around some cultivated areas. The sand flats and the associated wetlands of the Central Plain constitute a "cold spot," with a shorter growing season than in the immediately surrounding terrain (Fig. 2-35). Yet with proper land management under irrigation, large acreages of these soils can support impressive crop yields. As a result, some agriculturalists apply the term *golden sands* to the soils of this region.

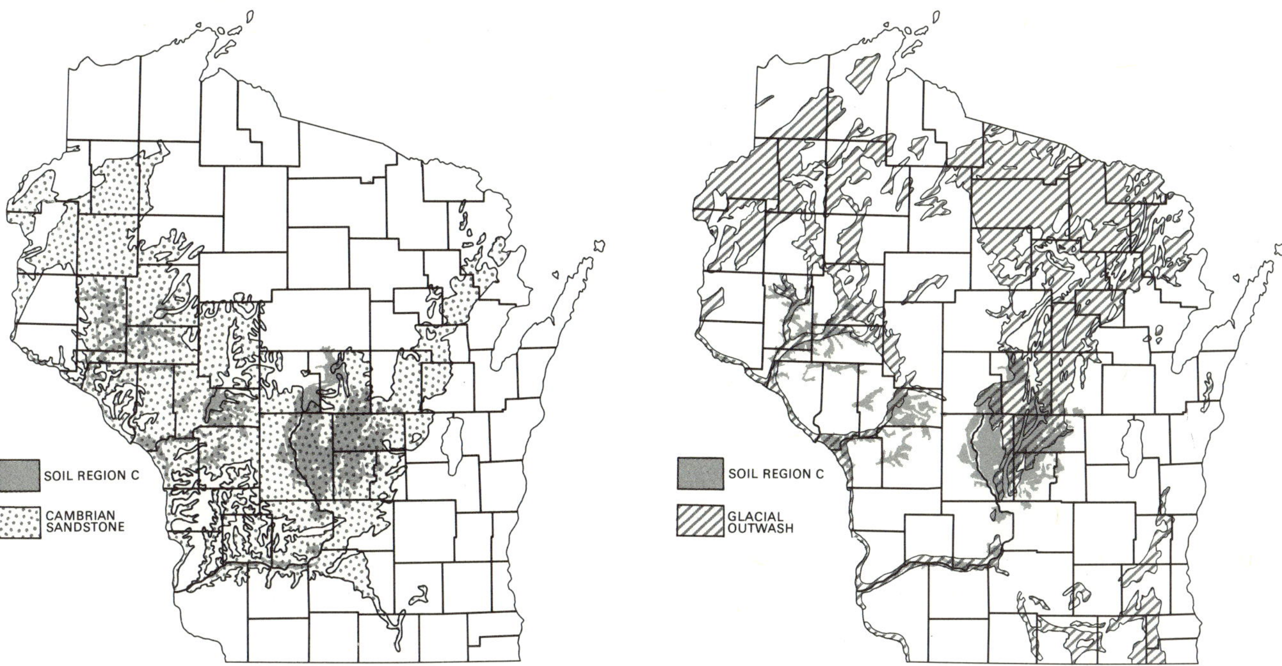

Figure 9-1. Index maps showing the geographic relationship of Soil Region C to bodies of Cambrian sandstone and glacial outwash.

1. Curtis (1959, p. 327) states that "an oak savanna with an intact groundlayer is the rarest plant community in Wisconsin today." The sharptail grouse and other fauna are threatened by the gradual disappearance of the savanna in this region.

The soils of the central sandy uplands and plains, occupying about two and a half million acres (7.1% of the area of the state), are nearly level to undulating for the most part (80% of the area). A number of butte-like sandstone mounds, such as Friendship Mound in Adams County, are prominent but isolated features of the landscape. On the east are uplands of sandy glacial end moraines, till-capped sandstone hills, and pitted outwash. These soil associations are distributed throughout more than a dozen counties and are largely on stratified Pleistocene deposits (Plate 4) in the Cambrian sandstone province of the state (Figs. 9-1, 9-2).

Glacial drift including extensive outwash sand (about two thirds of the area) and less extensive, weakly calcareous, sandy glacial till constitutes the principal substratum of the region. Old, stabilized sand dunes are abundant in some townships. Major bodies of lacustrine silts and clays, laid down in extinct glacial Lake Wisconsin, are inclusions of Regions E and I. Also included are some extensive wetlands of Region J. The sandy

Figure 9-2. Sequence of maps showing distribution of soil associations in Soil Region C.

soils on sandstone mounds are shallow to bedrock. Excessively drained (droughty) soils occupy 95% of the area.

The soils are weakly developed as a result of the instability of the region, both in terms of repeated burning of the vegetation and wind-blowing of sand before settlement. Fires are fewer and better controlled now than under natural conditions before the pioneers came. Agricultural activities have accelerated wind erosion in recent decades, even though considerable effort was at one time put into development of shelter belts of trees to reduce speed of winds across fields.

The sands are of mixed mineralogy except on extremely quartzose sandstone bedrock. Glacio-fluvial sands of central Wisconsin contain up to 15% feldspar grains and 1% heavy minerals by volume, whereas sands weathered from Cambrian and Ordovician sandstones have about a tenth of these amounts of the two mineral groups (Madison, 1963; Madison and Lee, 1965). Some soils with reddish-brown (5YR 4/4) sandy subsoils (Trempe, Trempealeau, Seckler, and reddish subsoil variants of Crown and Chetek) have been observed in Waushara, Juneau, Iowa, Dunn, and Trempealeau counties.

A periodically high water table does not produce the degree of mottling in these sandy soils that it does in the silty soils of Regions A and B. Therefore, soil colors that would indicate moderately good drainage in silty soils probably represent somewhat poor drainage in soils like the Nekoosa loamy sand (Typic Udipsamment). But since quantitative data are not available at present to substantiate this impression, the classification of sandy soils having impeded drainage still results in about the same interpretations of soil colors for all textures. Restrictions on installation of private septic systems should be especially severe on mottled sands that commonly have high water tables in the spring season (Bouma et al., 1972).

Major native plant communities on these soils were the pine barrens, oak savanna, prairie, and pine forest. Extensive forests of jack and other pine and scrub oak are held by paper manufacturing companies in the area.

Sparta loamy sand, an Entic Hapludoll that developed under prairie vegetation in La Crosse, Monroe, Trempealeau, and Dunn counties, has a very dark brown A1 horizon that may be as much as 2 feet deep. This particular sandy soil has a high soil development rating (800) on the Menominee scaling system (Milfred, Olson, and Hole, 1967). However, the amount of natural ecological achievement represented in terms of incorporation of organic matter in the sandy soil is possibly about one hundreth of that represented by a similar depth of A1 horizon in a Tama silt loam (Hole and Nielsen, 1970; Baxter and Hole, 1967; Bouma and Hole, 1971).

Soil management of unirrigated sandy soils includes emphasis on legumes and somewhat drought-resistant crops, such as rye, barley, peas, flax, cabbage, potatoes, and other root crops (Albert, 1951), accompanied by adequate fertilization according to soil test. The practice of irrigation farming on 160-acre blocks of land has spread in recent years, particularly in Adams County. This has caused some disturbance of the soil through leveling and brush removal, and notable changes in vegetative cover, microtopography, and soil climate. Cranberry beds in sand and peat near Wisconsin Rapids and Black River Falls are protected from frost by overhead irrigation or flooding.

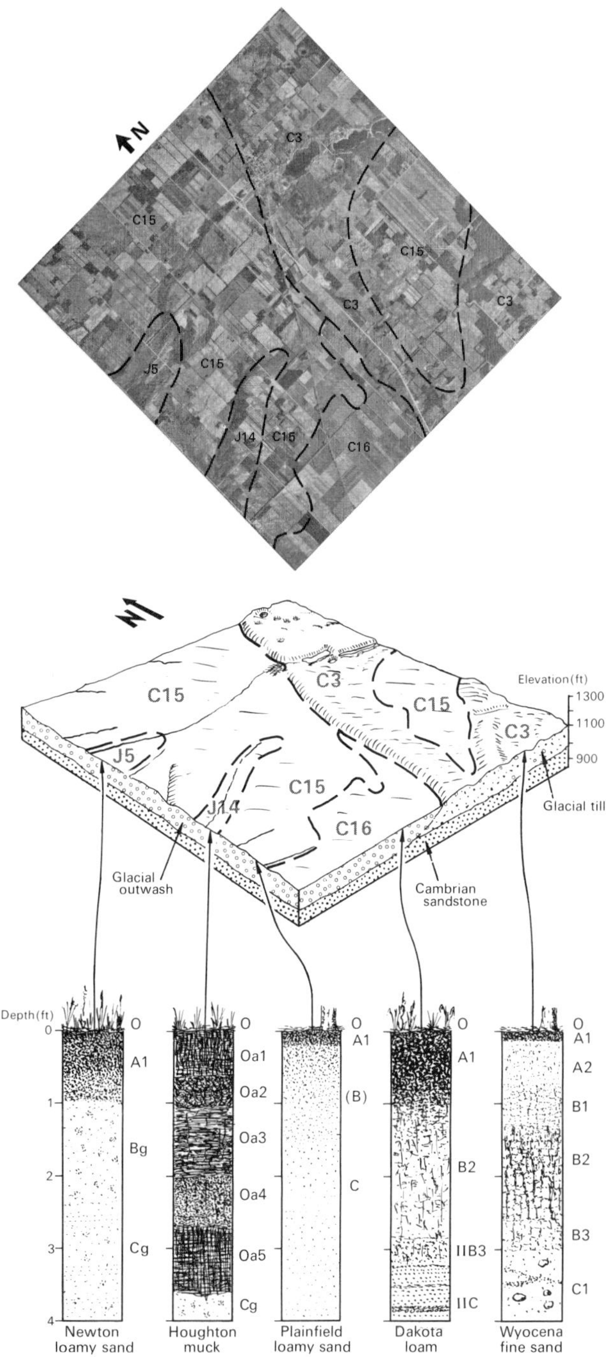

Figure 9-3. Aerial photo map of T.19N., R.8E., Waushara County. The area shown is 6 miles on a side.

Figure 9-4. Block diagram showing landscape positions of major soils of T.19N., R.8E., Waushara County.

The variation in texture of the soils of Region C, from loamy sand to loams (Figs. 9-3, 9-4), in combination with differences in depth to water table, is reflected in crop yields. For example, under high levels of management that include adequate fertilization and drainage, annual yield per acre of alfalfa-brome hay ranges, according to Beatty et al. (1966), from 1.5 to 3.0 tons (dry weight), as follows: 1.5, Boone loamy sand; 2.25, Plainfield, Sparta, and Morocco loamy sands; 2.50, Oshtemo, Burkhardt, and Wyocena sandy loams, Gotham, Newton, and Nekoosa loamy sands; 2.75, Meridian, Dakota, Onamia, and Mecan sandy loams; 3.0, Meridian, Dakota, and Onamia loams, Granby and Delton sandy loams. Irrigation on a soil as droughty as Plainfield loamy sand can raise yields to 4 to 6 tons per acre.

The eighteen soil associations of this region (Fig. 9-2) are listed in the legend of the soil map (Plate 1) under four major landscape headings, from the hilliest to the broad, nearly level flats. In the following paragraphs the soil associations are regrouped with more emphasis on soil series.

C1, C2. *Oshtemo, Mecan, and other loamy and sandy soils on sandy glacial drift.*

C1. The hilly to nearly level Oshtemo and Gotham loamy sand, and Plainfield loamy sand and sand, association.

C2. The rolling and hilly to nearly level Mecan and Wyocena loamy sand and sandy loam, and Plainfield and Gotham loamy sand and sand, association.

These hilly to nearly level sandy lands occupy about 270,000 acres, or nearly 1%, of the land area of the state. Developed over outwash sand (with some gravel) are the deep Oshtemo sandy loam (Typic Hapludalf) and the shallower Gotham loamy sand (Psammentic Hapludalf), and the Plainfield loamy sand (Typic Udipsamment; Figs. 9-4, 9-6). Slightly calcareous glacial tills underlie the moderately deep Mecan sandy loam and the shallow Wyocena loamy sand (both Typic Hapludalfs). The till has a distinct reddish cast under the Mecan soil, as may be seen on the Arnott Moraine in Portage County. Many of the till-mantled hills are sandstone-cored (Peck and Lee, 1961). Oak savanna was the original dominant plant community. The landscape is largely droughty and nearly half of it is left in woodland and pasture. Wind erosion is a problem on cultivated areas. In Marquette County some depressions associated with Gotham loamy sand may contain somewhat poorly drained Spodosols (Au Gres).

C3, C6, C10, C11, C14, C15. *Plainfield and associated sands on glacial outwash sand.*

C3. The rolling and hilly to nearly level Plainfield, Gotham, and Wyocena loamy sand association.

C6. The nearly level to gently undulating Sparta, Plainfield, and Gotham loamy sand and sand association.

C10. The nearly level and undulating Plainfield, Kellner, and Newton sand and loamy sand association, with peat and muck.

C11. The nearly level and undulating Plainfield loamy sand and sand, and Gotham loamy fine sand, association.

C14. The nearly level and undulating Plainfield, Nekoosa, and Newton loamy sand and sand association.

C15. The nearly level and undulating Plainfield, Richford, and Kellner loamy sand, and Dakota sandy loam, association.

Plainfield loamy sand (Typic Udipsamment) is the dominant soil in the 800,000 acres of these six soil associations that range from hilly in the Woodfordian glacial end moraine to nearly level on the adjacent irrigable outwash plain (Figs. 9-3, 9-4,

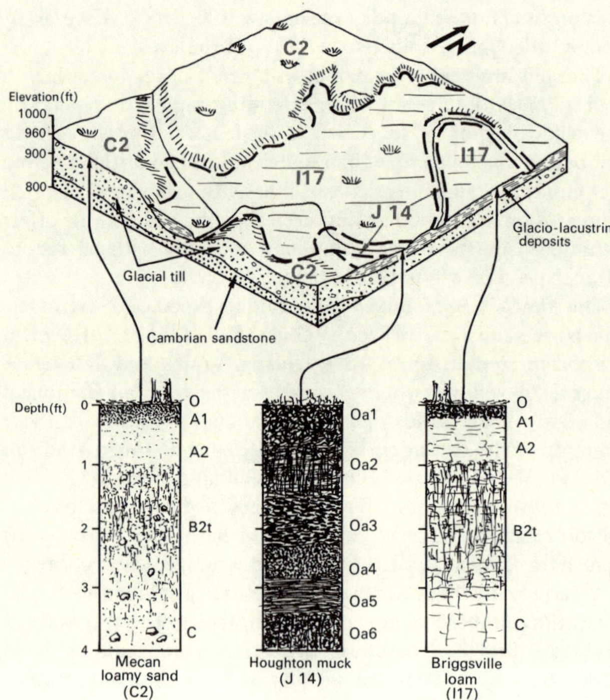

Figure 9-5. Block diagram showing landscape positions of major soils in parts of Sections 15 and 16, T.16N., R.8E., Marquette County.

9-6). Stabilized ancient sand dunes are present. The 10% of silt and clay in the surface soil, probably largely aeolian in origin, and the mixed mineralogy of the sands raise the natural productivity of these soils above the extremely low level of the Boone sands (Fig. 10-4) of scattered sandstone mounds.

In central Wisconsin very sandy soils occupy 30% of the five-county area (Adams, Juneau, Marquette, Portage, and Waushara counties), and moderately sandy, poorly drained sandy, and organic soil groups each occupy about 8% of the area (Beatty et al., 1964). The other soils associated with the Plainfield in these soilscapes contain more colloidal material, either in organic and inorganic forms or both. Three taxonomic soil sequences can be distinguished on the basis of (1) increasing influence of prairie, or of cultivation with good management, (2) increasing wetness, and (3) increasing development of a textural B (Bt) horizon. The first sequence is Plainfield (Typic Udipsamment)-Kellner[2] (Mollic Udipsamment)-Sparta (Entic Hapludoll). In southeastern Dunn County cultivation has actually converted Plainfield soil into Sparta, and the boundary between the two follows fence lines. The second sequence includes Plainfield-Nekoosa (Typic Udipsamment)-Newton (Typic Humaquept). The third sequence is Plainfield-Gotham (Psammentic Hapludalf)-Richford (Arenic Hapludalf)-Wyocena (Typic Hapludalf). Such arrays of soils rarely if ever occur on the landscape exactly in the order named because of the complexities of the factors of formation of soils and soilscapes.

Oak savanna, pine barrens, and prairie were the original

2. Greenhouse experiments with flooding of this soil and several others have indicated that manganese toxicity, which has a long-lasting, deleterious effect on alfalfa, results from brief periods of poor aeration, e.g., during rainy weather (Graven, Attoe, and Smith, 1965).

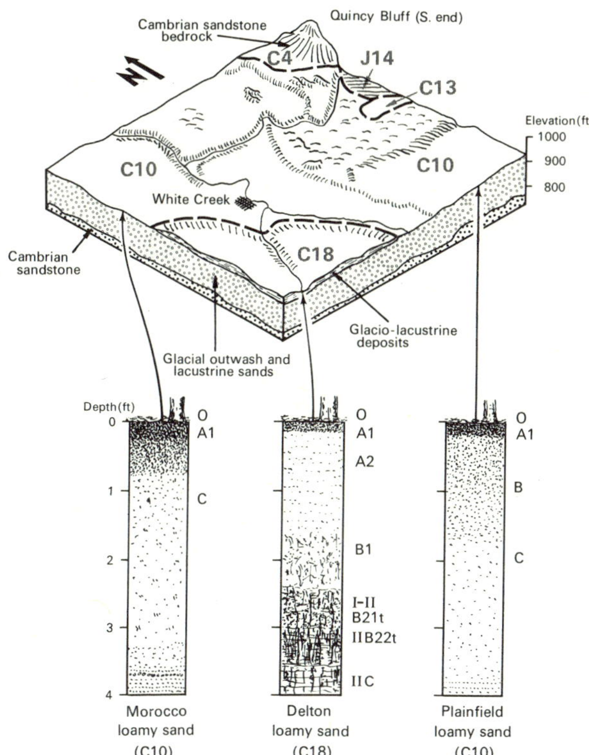

Figure 9-6. Block diagram showing landscape positions of major soils in Sections 25, 26, 35, 36, T.16N., R.5E., Adams County.

plant communities. Fire protection has allowed oak and pine forests to close somewhat, in uncultivated and unpastured areas. Irrigation by large self-propelling mobile systems has brought about the clearing of many acres of nearly level, formerly idle land.

Serious disturbance of vegetation on Sparta and Plainfield soils exposes them to wind and results in blowouts and sand drifts. Natural healing of these exposures may take place with invasion of spike moss and lichens, followed by sedges, prairie grasses, and forbs. Red cedar and jack pine appear on stabilized dunes.

Coloma loamy sand (Alfic Udipsamment) is differentiated from other soils of the C3 soilscape by the presence of 0.25- to 2-inch bands of finer material between the depths of 3 and 5 feet.

Sandy soils appear to be subject to wide variations in profile developments. For example, in northern Juneau County a Mollic Haplorthod (Brunizem-Podzol intergrade) and a Spodic Udipsamment (Podzol-Regosol intergrade) have been observed. Chetek (Eutric Glossoboralf) and its somewhat poorly drained associate, Crown sandy loam (Aquic Eutroboralf), are occasionally found in the same soilscape with Plainfield in Adams County. A Typic Haplorthod (Podzol), Wallace loamy sand (Fig. 1-7), level phase, was found (Lambert, 1970; Lambert and Hole, 1971) in this association near a domain of stabilized sand dunes in T.19N., R.6E. in Adams County. The ortstein pan, with a bulk density of nearly 2.0, did not impede water movement under saturated conditions but did so when the soil was unsaturated. In Portage County, soil association C15 includes the Anson sandy loam (Mollic Hapludalf), a prairie-forest intergrade (Brunizem, Gray-Brown Podzolic) between the Meridian and Dakota series. Also present is a Plainfield loamy sand, loamy substratum (probably in part residuum from igneous bedrock) phase, with angular rock fragments throughout the solum. At the boundary between soil associations C6 and D8 in eastern Eau Claire County is an unnamed Spodic Lithic Udipsamment soil which intergrades between Northfield and Humbird loamy sands. Gotham loamy sand (Psammentic Hapludalf) is prominent in soil association C6 in Dunn County, where it is associated with an Arenic Dystrochrept (lacking the usual Bt horizon) and a Typic Hapludalf formed from thin sand over silty material.

C4. *Boone and associated sands on rather infertile Cambrian sandstone, nearly level to undulating.*

Boone loamy sand (Typic Quartzipsamment), which developed largely from quartzose sandstone of Cambrian age (Figs. 1-3, 9-6, 10-4) is considered the least fertile soil in Wisconsin. Madison (1963) studied the mineralogy of a representative site in Jackson County and found the soil to contain more than 95% quartz by weight. Some sandstone fragments and an occasional chert stone are found in the soil, which is loose to a depth of 1 to 5 feet and is underlain by weakly cemented sandstone bedrock (Fig. 2-3). Rock outcrops are numerous, and some have brittle surfaces that are case-hardened with silica. The subsoil has a reddish-brown color in places. Stabilized and active sand dunes are present. Because of the droughtiness and low fertility of this soil, it has been little cultivated. Much of it is left in jack pine and scrub oak. Cranberry bogs have been developed in associated wetlands in Jackson County.

A shallow variant of Northfield sandy loam is recognized where the underlying sandstone is cemented close to the surface. Near the Wisconsin River are small bodies of Wautoma sandy loam (Mollic Haplaquept). Vilas (Entic Haplorthod), Hiawatha (Typic Haplorthod), and Kinross (Typic Psammaquent) loamy sands are reported from the soil association south and east of Black River Falls in Jackson County. In Eau Claire County some bodies of Boone loamy sand are actually Lithic Entic Haplorthods (Podzols).

C5, C8. *Sparta and associated dark, excessively drained sands of former prairies, on glacial outwash.*

C5. The nearly level to gently undulating Sparta, Plainfield, and Gotham loamy sand and sand association.

C8. The nearly level to gently undulating Sparta loamy sand, and Dakota and Meridian sandy loam, association.

The Sparta loamy sand (Entic Hapludoll; Figs. 7-4, 16-4) is the predominant soil in about twenty outwash flats scattered from Rock to Dunn County, and in the Mississippi and Wisconsin River valleys. This soil developed under prairie vegetation. The dark topsoil is 1.5 to 2 feet thick over brown sand, becoming yellower with depth, but remaining acid. Several associated soils constitute a taxonomic sequence toward finer textured forest soils: Sparta-Plainfield loamy sand (Typic Udipsamment)-Gotham loamy sand (Psammentic Hapludalf)-Meridian sandy loam (Typic Hapludalf). The Dakota sandy loam is a productive prairie soil about 3 feet deep to loose sand and gravel. Extensive old wooded and grassed dunes predate

European settlement, but wind is active locally today in making small blow-outs and dunes. Shelter belts of trees are widely used to reduce wind erosion on cultivated level fields.

In La Crosse County, Sparta loamy sand occurs in well-drained depressions between dunes that are occupied by Plainfield loamy sand. In some places the Sparta A1 horizon lies buried under the Plainfield, indicating that the latter soil formed on dune sand that had previously moved onto and buried the Sparta soil.

Some inclusions in bodies of Sparta sand in Pierce County have a somewhat thin A1 horizon and are actually Typic Udipsamments. Presence of bands of brownish (7.5YR 5/5) sandy loam at a depth of about 4 feet is reported for some Sparta loamy sand soils from Crawford County. In Trempealeau and Iowa counties two dark sandy soils, the Trempe (Entic Hapludoll) and Trempealeau (Typic Argiudoll), with distinguishing reddish-brown B horizons (5YR hue), are included in association C8. The origin of the reddish color is not known. Soil association C8 in southern Trempealeau County also includes some small bodies of clayey soils formed from old terrace remnants of slack-water silts and clays: Denrock (Aquic Argiudoll) and Perrot (Mollic Haplaquept) silty clay loams.

Centuries ago bodies of sand were blown against footslopes of the Mississippi and Wisconsin River valleys and Chelsea soils (Alfic Udipsamments) subsequently formed on them. These soils occupy rolling to steep sandy lands bordering C5 areas (Hole, 1956a). In places the sand is banded below a depth of 4 feet with reddish-brown sandy loam streaks 2 to 5 inches thick. Akers (1964) observed these sands near Soldiers Grove in Crawford County.

C7, C9, C16. *Dakota, Meridian, and associated loams developed on glacial outwash sand and gravel.*

C7. The nearly level to gently undulating Meridian loams, Plainfield and Sparta loamy sand and sand, and Shiffer loams, association.

C9. The nearly level to gently undulating Dakota, Onamia, Meridian, and Burkhardt loams association.

C16. The nearly level to undulating Dakota and Onamia sandy loam association.

These moderately deep (2 to 3 feet) loams and shallower associated sands and sandy loams occupy nearly a dozen widely scattered glacial terraces of acid outwash sand and gravel (Fig. 2-5). The dark Dakota soil (Typic Argiudoll; Fig. 9-4) formed under prairie. Associated with it are the forest soils, Meridian and Onamia loams (Mollic and Typic Hapludalfs, respectively). The Shiffer loam (Aquollic Hapludalf) is a somewhat poorly drained associate of the Meridian soil. The Burkhardt soil (Typic Hapludoll) is shallower (about 16 inches deep) than the Dakota. Plainfield (Typic Udipsamment) and Sparta (Entic Hapludoll) loamy sands have been discussed in previous sections. The Plainfield in Trempealeau County is reported to be slightly finer in texture than is usual. These three soil associations are more productive than those dominated by the more sandy, droughtier Sparta and Plainfield soils.

The Burkhardt is predominant in Pepin County, where it is associated with Dakota. Some bodies of Dakota loam in southwestern counties contain brownish bands of loam and sandy clay loam in the sand and gravel substratum, features that improve the water-holding capacity of the soil.

In soil association C9 in south-central Dunn County some double soil profiles were found consisting of Mollic Udipsamments in sandy wind-deposits overlying truncated sola of Tell and Meridian soils.

C12, C13. *Nekoosa and associated poorly drained sands, and peat.*

C12. The nearly level to undulating Nekoosa and Plainfield loamy sand and sand, and Newton loamy sand and sandy loam, and peat and muck, association.

C13. The nearly level to undulating Nekoosa and Morocco loamy sand, Granby sandy loam, and Plainfield loamy sand and sand, and peat, association.

The predominant soil is the moderately well drained Nekoosa loamy sand (Typic Udipsamment), which is in the upper part of the following toposequence of soils: Plainfield (Typic Udipsamment)-Nekoosa-Morocco (Aquic Udipsamment)-Newton (Typic Humaquept)-Granby (Typic Haplaquoll)-peat (Histosol). The sand and gravel outwash immediately under the Granby sandy loam is neutral to weakly calcareous. The soil pattern of these soilscapes is an irregular one. Poorly drained soils (Newton, Granby, peat) and somewhat poorly drained soil (Morocco) interrupt the areas of better drained soils.

Near the Castle Rock Flowage in Adams County, an Aquic Eutroboralf (somewhat poorly drained Gray-Brown Podzolic; Crown sandy loam), with a reddish-brown Bt horizon, was observed to be associated with the Plainfield and Nekoosa soils. Both the Castle Rock and the Petenwell Flowage impoundments have raised the water table locally so that soil properties are changing under the altered moisture regimes.

C17, C18. *Guenther, Delton, and associated loamy sands over fine sandy lacustrine deposits.*

C17. The nearly level to undulating Guenther, Dancy, Nekoosa, and Newton loamy sand and sandy loam association.

C18. The nearly level to undulating Delton loamy sand, Alban loams, and Wyeville loamy sand, association.

These few soil association bodies, along with the more clayey Delton-Wyeville-Poygan-peat-muck association (E10), include soils formed from lacustrine fine sands and also sandy coverings overlying clayey substrata. Among these are the well-drained Guenther and Delton loamy sands (Alfic Haplorthods and Arenic Hapludalfs), the somewhat poorly drained Wyeville (Aquic Arenic Hapludalf), and poorly drained Dancy (Aeric Glossaqualf). The clayey substratum is acid. The Alban fine sandy loam (Typic Glossoboralf) is a well-drained soil over lacustrine fine sands. Associated with these soils are two sands, the moderately well drained Nekoosa (Typic Udipsamment) and the poorly drained Newton (Typic Humaquept). The variability in drainage conditions complicates the landscape and adds to the difficulties of farming.

CHAPTER 10

Soil Region D: Soils of the Western Sandstone Uplands, Valley Slopes, and Plains

Soil Region D, like Soil Region C, is underlain by sandstone of Cambrian age (Fig. 10-1), but it is generally more hilly, is much less blanketed by glacial drift, and has less wetland. Region D is a scenic area with hundreds of coves, sharp ridges, pointed hills (called *tepee buttes* by Martin, 1932), angular mesa-like prominences, and valleys with low cliffs along streams. The Dells of the Wisconsin River are much publicized features of this part of the state. The relief diminishes from the hilly dissected borderlands on the southwest stretching from Sauk County to Pierce County, to level lands of southern Clark and Wood counties on the north. The region has received a moderate covering of loess and, in the northwestern reaches, some glacial till that is less sandy than that of eastern parts of Region C.

In this soil region of 3.3 million acres (9.4% of the land area of Wisconsin) the landscape is dominantly hilly (80% by area, hilly to rolling). Rock outcrops are numerous on steepest slopes (Fig. 2-2). The nearly level summits of the occasional buttes and the stairstep or bench features on hillsides reflect the presence of some relatively resistant strata in the bedrock sequence. Cementation along joints and stratification planes and on remnants of ancient land surfaces accounts for some of this resistance to geologic erosion. Soils that are well to excessively drained occupy 90% of the region. Gently rolling and nearly level soils are most extensive on the northeastern borders of the region. Only there do soils of impeded drainage occupy much more than 5% of the area of the landscape.

Cambrian siltstones, fine sandstones, and some shales influence the soil properties. A thin loess covering is discontinuous (Figs. 2-28, 2-29). Very thin glacial drift is present in places, as in southern Barron County. A certain amount of down-slope creep of silty and loamy soil has taken place in hilly terrain over a period of thousands of years. But this movement has usually not blurred the soil horizon sequence, A1-A2-Bt. Soft "greensand" (glauconite) is abundant both in extensive silty and loamy subsoils and in small bodies of soil over coarse, less friable glauconitic strata. Glauconitic sand is notably high in content of available potassium. Twenhofel (1936) published a general map of the distribution of these greensands in Wisconsin.

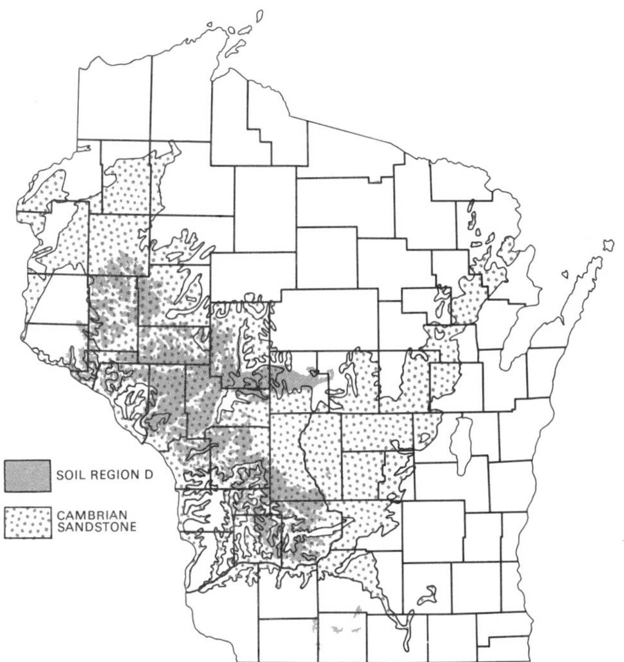

Figure 10-1. Index map showing the geographic relationship of Soil Region D to bodies of Cambrian sandstone.

The separation by field soil scientists of Hixton soils (on brown silts and sands of Cambrian age) from Norden soils (on greensands and silts) has been arbitrary, admittedly, and has been done on the assumption that the presence of glauconite gives soil added fertility. This may not always be the case. Future classification might better be based on soil properties as determined by time-consuming laboratory analyses and in field crop trials.

These soils have supported a wide range of native plant communities: prairie, oak savanna, southern oak forest, and southern and northern mesic forests. Dark, well-drained Mollisols are commonly found on footslopes where favorable moisture conditions have fostered growth of vegetation. A third of the region is in woodland, pastured and unpastured. Cropland occupies 40% of the land and cleared pasture about 25%.

The highly dissected topography so characteristic of Richland and La Crosse counties fostered the compartmentalization of rural life in the early days. Ridges were effective barriers between farm communities (Hole, 1968). Modern transportation and communication have overcome this isolation.

The thirteen soil associations (Fig. 10-2) in this region are grouped in the soil map legend in order from those with steepest slopes to those on nearly level terrain. In the following discussion these associations are regrouped with more emphasis on soil series than on topography.

D1, D2, D3, D4, D5, D7, D9, D10. *Norden, Hixton, Gale, and associated loams over moderately fertile Cambrian siltstones and sandstones.*

- D1. The steep to rolling rocky land, Gale silt loam, Norden and Hixton loams, and Fayette and Seaton silt loam, association.
- D2. The hilly to rolling Norden, Gale, and Fayette silt loam, and Hixton loams, association.
- D3. The hilly to rolling Gale, Norden, and Fayette silt loam association.
- D4. The hilly to rolling Norden, Hixton, and Northfield loams, and Boone sand, association.

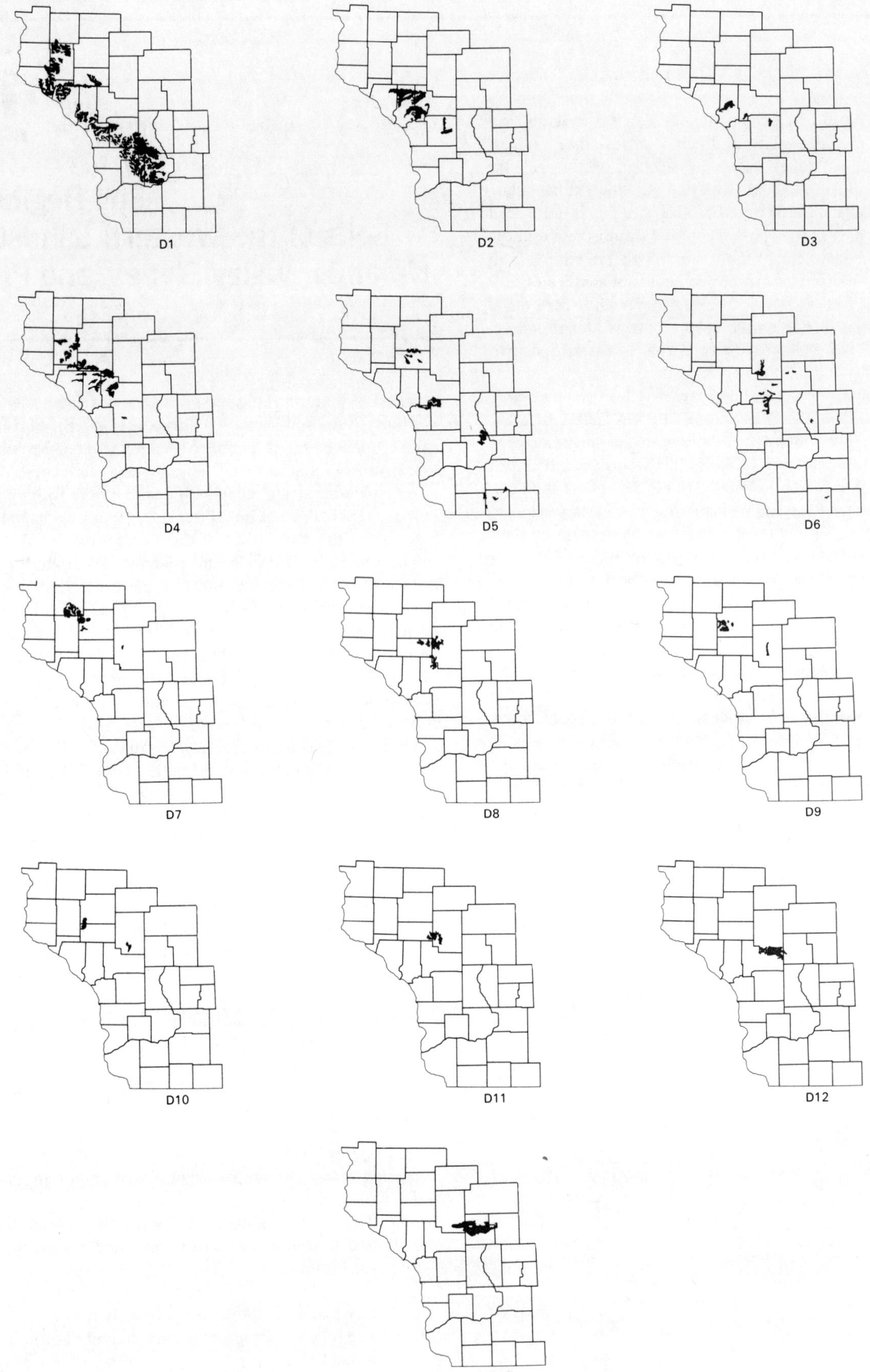

Figure 10-2. Sequence of maps showing distribution of soil associations in Soil Region D.

D5. The hilly, rolling, and steep Hixton and Northfield loams, Gale silt loam, and Boone sand, association.
D7. The hilly to rolling Hixton and Arland loam and silt loam, Gale silt loam, and Norden loams, association.
D9. The gently rolling and rolling Hixton loams, Gale silt loam, and Northfield, Arland, and Milaca loams, association.
D10. The gently rolling and rolling Hixton, Onamia, and Chetek loams association.

Norden and Hixton soils are representative of these eight soilscapes which together occupy more than 80% of the region. These two loamy soils are 2 to 3 feet deep over weakly cemented sandstones. Norden soils (Typic Hapludalfs; Figs. 10-3 through 10-6) are underlain by glauconitic green siltstone and sandstone, whereas Hixton soils occur over brown strata. At some locations highly weathered glauconitic greensand has weathered to a reddish brown in the Norden soils (Wurman, 1961). In places bedrock ledges outcrop. Where silty coverings are much deeper than 3 feet, other soils are recognized, such as the coarse-silty Seaton or the fine-Silty Fayette soils (both Typic Hapludalfs). The Seaton is more extensive than Fayette in Dunn County. The Gale silt loam soil (Fig. 10-6) has developed in 15 to 36 inches of loess overlying soft bedrock which is brown, the same as the substratum of the Hixton silt loam. The substratum of the Northfield sandy loam, loam, and silt loam (Lithic Hapludalf) is hard-cemented sandstone. Boone (Typic Quartzipsamment; Fig. 10-4) soils are deep, quartzose, infertile sands over soft sandstone. Where glacial drift is present, as in southern Barron County, several other soils are present: the Arland loam (in glacial till underlain at about 3 feet by sandstone) and the Milaca loam (with till substratum usually tens of feet thick); the Onamia and Chetek loams (Eutric Glossoboralfs) formed, respectively, in more than 20 inches and less than 20 inches of loam over acid glacial outwash.

Over many years, the need for more cropland has caused the boundaries of cultivated fields to be pushed gradually uphill, beyond the reach of the deepest and most productive valley soils, to slopes as steep as 28% gradient. Still steeper lands are left in forest or pasture.

Perrot State Park near Trempealeau includes many of these soils and associated sandstone cliffs. Vegetation ranges from patches of dry prairie to dense deciduous forests (with Kentucky coffee trees) in narrow valleys. A great variety of microclimates and corresponding pedons are present. Labrador tea occurs on acid sandstone cliffs in areas of D1, as well as in peat bogs.

In Pepin and Trempealeau counties sandstone debris is abundant in the lower silty soil profiles of Fayette and Seaton of soil association D1. In soilscape D4 at the foot of slopes occupied by Norden soils is an unnamed prairie analogue of Norden. Up-slope from Norden soils are very shallow sandy loams (classified in the Urne series; Dystric Eutrochrept; Lithosol) over shattered green sandstone. A red variant of Norden sandy loam was found to have 32% clay and 11% free iron oxide (Fe_2O_3) in the Bt horizon.

In Dunn County bodies as large as 3 acres of a Typic Argiaquoll (Humic Gley) are associated with the Norden series in seepage spots on concave slopes. In the same landscape, a windblown sandy deposit on Norden silt loam is classified as a loamy sand variant. Where the sandy covering is deeper than 3 feet, the Lamont loamy sand (Typic Hapludalf) is recognized.

Associated with the Hixton soil (Typic Hapludalf) are some small bodies of moderately well and somewhat poorly drained soils. The Gale silt loam occurs both in ridgetop and footslope positions. Boone sand is associated with Hixton in Pepin County even in cultivated fields.

The D1 soil association in Richland County includes some small bodies of Rockbridge silt loam (Typic Hapludalf) developed on benches from thin loess over colluvial debris of ancient cherty residuum (Hole, Peterson, and Robinson, 1952).

Some bodies of Dubuque silt loam are delineated where cherty red clay subsoil rests on sandstone. The absence of dolomite bedrock at such sites is explained as (1) a result of complete removal of the carbonate rock by solution in percolating water and (2) creep of the clayey residuum down-slope over the sandstone.

D6. *The hilly to rolling Boone and associated sands and Northfield loams over rather infertile Cambrian sandstone.*

Boone sand (Typic Quartzipsamment; Fig. 10-4) and Northfield sandy loam (Lithic Hapludalf) combine in this hilly terrain to provide an acid, excessively drained, relatively infertile landscape. The Boone sand contains 95% or more of quartz, with less than 5% of feldspar and other minerals containing plant nutrient reserves (Madison and Lee, 1965). This is a very strongly acid soil (Wurman, 1961). Only hardy pine, oak, and grassland species can survive on these hills. Erosion of unconsolidated sandstone is accelerated where poor land management has allowed removal of plant cover.

Associated with the Northfield series is a Lithic Dystochrept, the Elkmound, which lacks an argillic B horizon. Gale and Hixton loams are also present on some of the lower slopes. Akers (1964) observed a patch of reddish till on sandstone (Arland soil) 6 miles west of Tomah in Monroe County.

The Boone series is also mentioned under C4 in Chapter 9.

D8, D11, D12. *Merrillan and associated somewhat poorly drained loamy sands over shaly sandstone.*

D8. The gently rolling and rolling Merrillan loamy sand, Boone sand, Northfield loams, Elm Lake loamy sand, and Arland loams, association.
D11. The nearly level and undulating Elm Lake, Merrillan, and Humbird loamy sand and sandy loam, Boone sand, and Northfield loams, association.
D12. The nearly level and undulating Merrillan, Elm Lake, and Humbird loamy sand association.

This is a somewhat poorly drained soilscape (Figs. 10-1, 10-2), except for areas affected by artificial drainage systems. The Merrillan loamy sand (Aqualfic Haplorthod) and the Elm Lake loamy sand (Typic Haplaquent) are the most extensive soils, and like the better drained Humbird loamy sand (Alfic Haplorthod), are formed from sandy sediments overlying acid greenish- or reddish-brown sandstone that has some siltstone and shale layers (Figs. 2-3, 2-4). Some workers have distinguished shallow (18 to 24 inches) and deep (24 to 42 inches) phases of these soils over bedrock. The low natural level of fertility of the landscape is due in part to the presence on elevated areas of Boone and Northfield sandy soils that have been discussed previously. The Arland loam (Typic Hapludalf) is a moderately productive soil, even though shallow (30 inches) over sandstone. In much of the area the native vegetation of oak, elm, maple, and basswood on the well-drained sites and of white pine, spruce, hemlock, red maple, red oak, white birch, elm, and ash on the somewhat poorly drained sites has been replaced by silage corn, small grains, pasture, or second-growth oak and aspen. Poorly drained soils are still largely covered by lowland hardwoods and conifers.

Some small areas of Rudolph and Altdorf soils have been reported from the D12 association in northeastern Jackson County. Some bodies of Merrillan loamy sand in soil association D11 in Eau Claire County have the greenish- or reddish-brown clay (shale) in the Bt horizon rather than in the C horizon, below a depth of 40 inches, as is usually the case. Erratic cobbles have been seen on the Merrillan and associated soils near the village of Merrillan, indicating a glacial influence on the soil initial materials in that vicinity.

D13. *The nearly level and undulating Kert, Vesper, and Veedum silt loam association.*

A 2-foot silty covering over acid shaly sandstone (Fig. 2-4) constitutes the materials for these somewhat poorly and poorly drained soils (Figs. 10-7, 10-8). The Kert silt loam is an Aquic Glossoboralf and the Veedum and Vesper silt loams are a Typic Humaquept and a Humic Haplaquept, respectively. The soilscape is a wet one, but because of the silty covering, it is more fertile than the previously discussed Merrillan-Elm Lake association. The original vegetation was lowland hardwoods and conifers. Woodlots and wooded pasture are extensive now, and some land has been cleared for pasture and for production of forage, small grains, and silage corn.

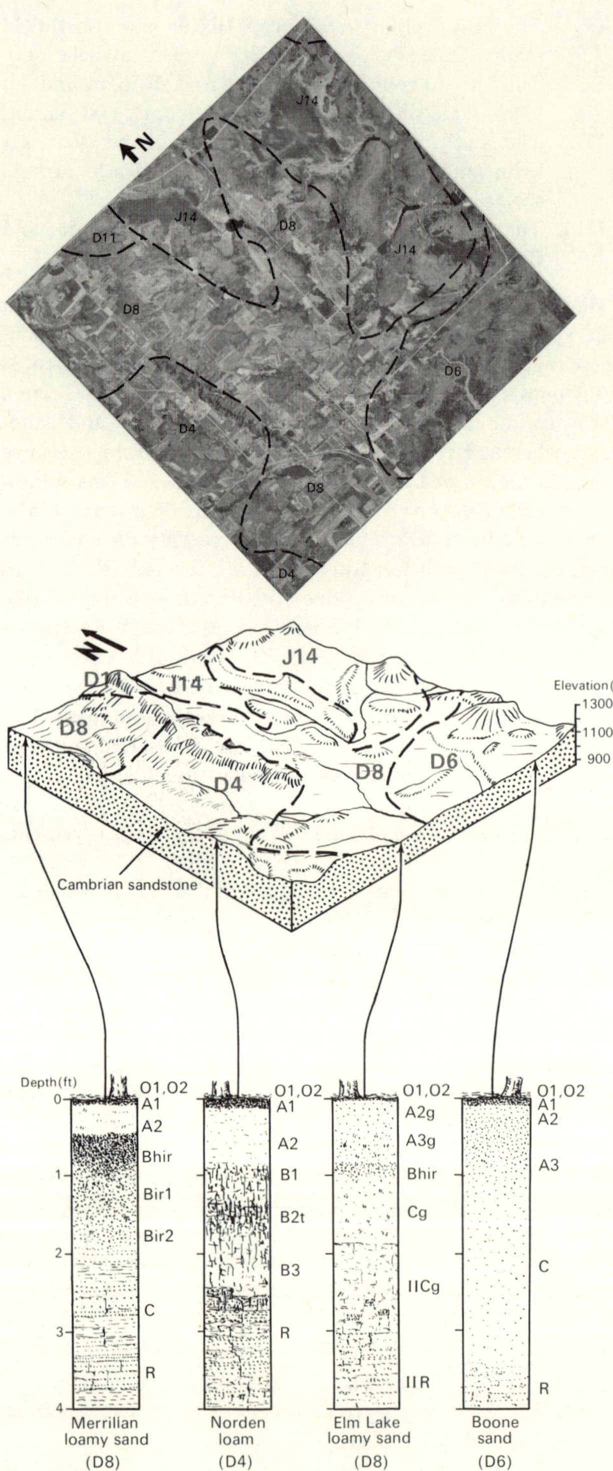

Figure 10-3. Aerial photo map of T.24N., R.4W., Clark County. The area shown is 6 miles on a side.

Figure 10-4. Block diagram showing landscape positions of major soils of T.24N., R.4W., Clark County.

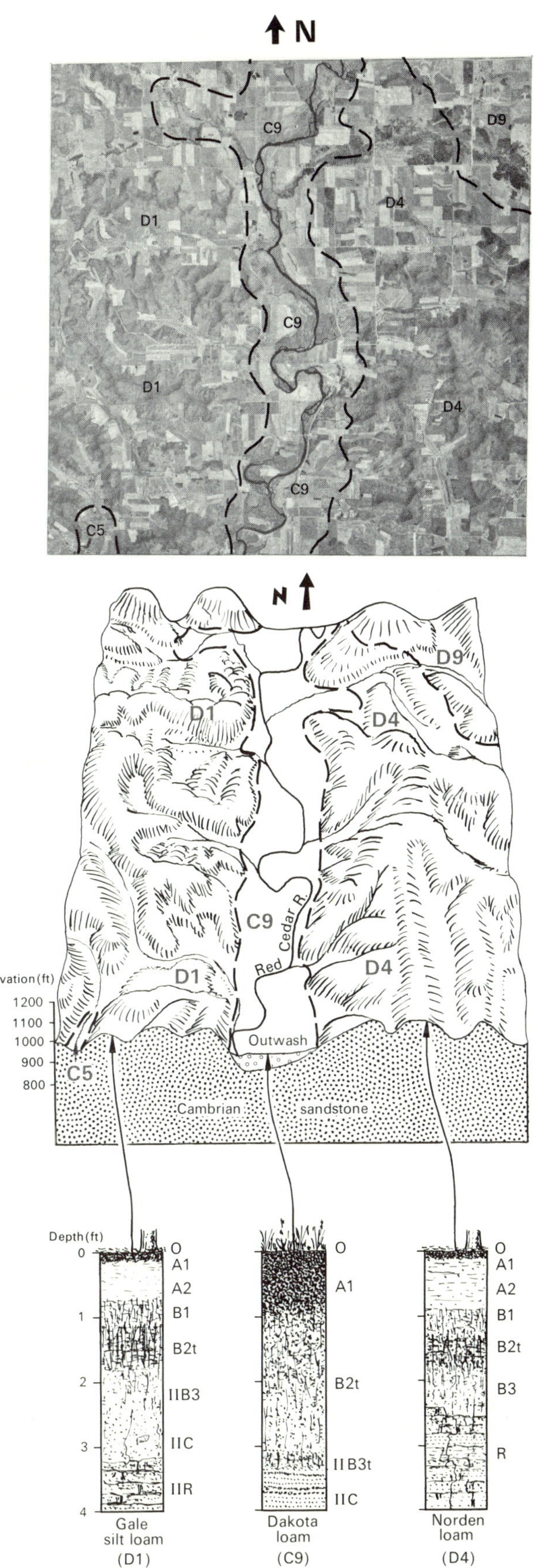

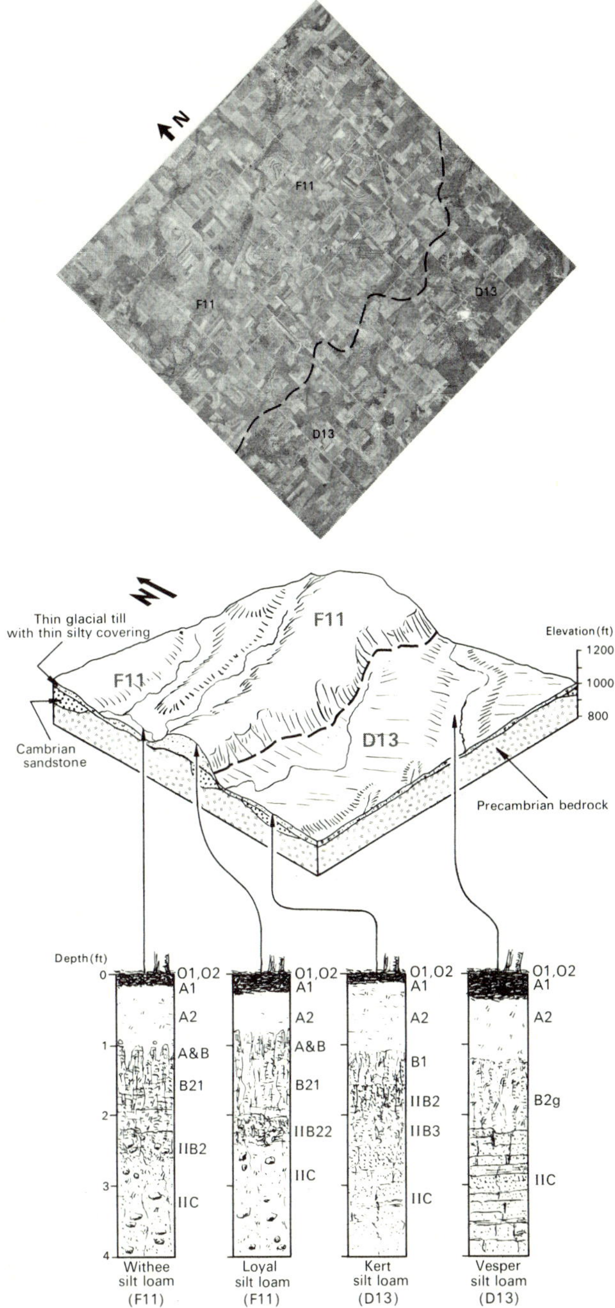

Figure 10-7. Aerial photo map of T.24N., R.1W., Clark County. The area shown is 6 miles on a side.

Figure 10-8. Block diagram showing landscape positions of major soils of T.24N., R.1W., Clark County.

Figure 10-5. Aerial photo map of T.30N., R.11W., Dunn County. The area shown is 6 miles on a side.

Figure 10-6. Block diagram showing landscape positions of major soils of T.30N., R.11W., Dunn County.

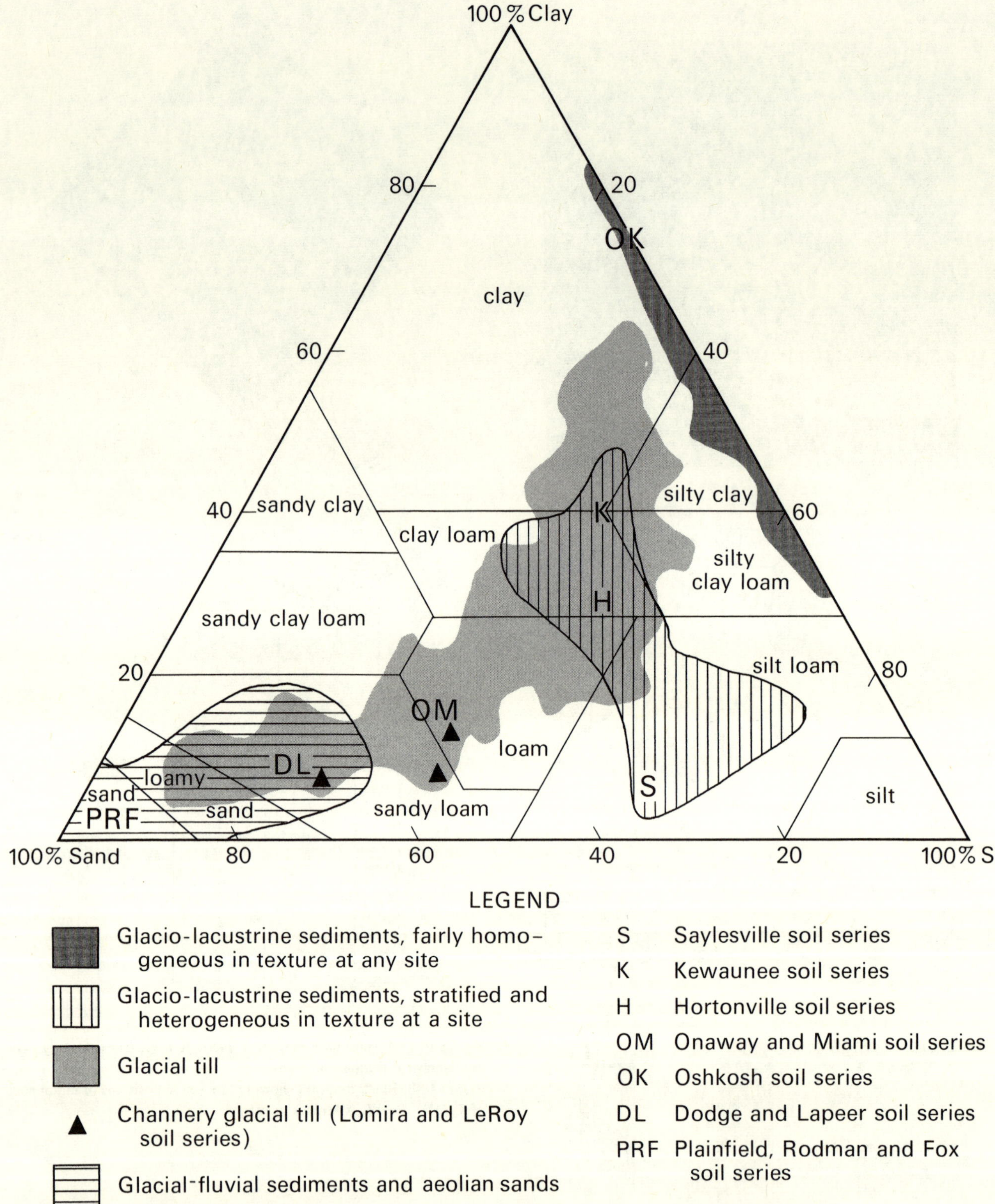

Figure 11-1. Particle-size distribution chart, characterizing common soil initial materials in eastern portions of Soil Regions E and I. (After Lee, Janke, and Beaver, 1962.)

CHAPTER **11**

Soil Region E: Soils of the Northern and Eastern Sandy and Loamy Reddish Drift Uplands and Plains

On both sides of Green Bay are Valderan Drift-covered cuestas rising from wetlands at the water's edge, to northwest-facing escarpments. The Silurian ("Niagara") dolomite escarpment on the east shore of Green Bay is bold, standing 100 feet high in many places. The cuestas formed by lower Paleozoic strata to the west of Green Bay are subdued and indeed blanketed by 100 to 200 feet of glacial deposits. The drift is calcareous, and decreasingly so toward the northwestern boundary of the region, where the land surface lies more than 300 feet above the level of Lake Michigan and Green Bay.

The region is distinguished from others by the pink (rather than red or brown) color of the glacial till and by its loamy texture, which is intermediate between that of the clayey drift of Soil Region I and that of the more sandy drift of Soil Regions G and H. The "pink" colors of the initial materials in Soil Region E (Fig. 2-21) are, according to the scientific soil color chart (Soil Survey Staff, 1951), actually light brown (7.5YR 6/4) and brown (7.5YR 5/4) and reddish yellow (7.5YR 6/6) in a moist condition. These brown to reddish colors are contributed by iron oxide in the component of glacial drift consisting largely of finely ground Precambrian iron formation (see Chapter 15). The textural range centers around the boundary between sandy loam and loam (OM in Fig. 11-1). The soils of the region exhibit a rather wide range of textures, carbonate content, and intensity of reddish color.

Spodosol (Podzol) soils are quite well developed in northeastern portions of the region, regardless of the presence of carbonates in the initial materials. The cool, moist climate, the original mixed forest cover, and the medium to coarse texture of materials have favored the development of these eye-catching soil profiles.

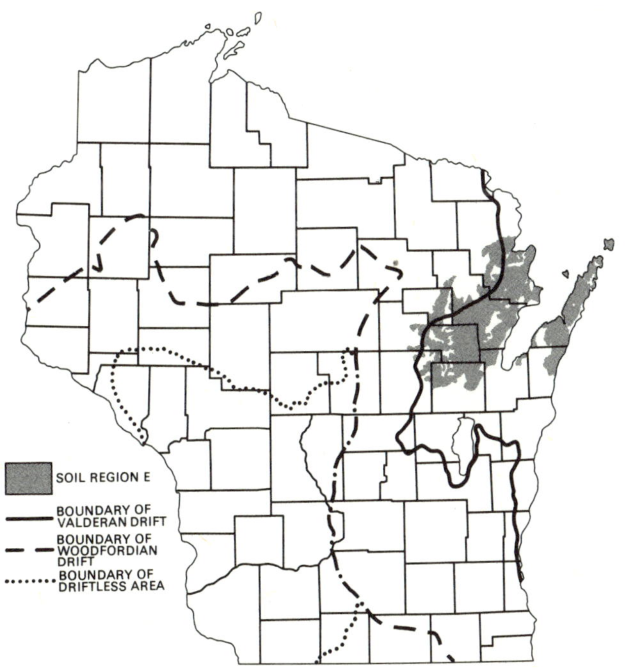

Figure 11-2. Index map showing the geographic relationship of Soil Region E to major glacial boundaries.

This region lies chiefly in the northern extension of two provinces, the Eastern Ridges and Lowlands Province and the Central Plain (Figs. 2-1, 11-2, 11-3), within the boundaries of the Valderan glacial ice advance and extinct glacial lakes as delineated by Thwaites (Plate 4; Flint, 1945). Small bodies of the pink sandy and loamy soils, including some acid variants, are also present in Dunn, Monroe, and Juneau counties.

All of this region of one and a half million acres as shown in Plate 1 is less sloping than 12% gradient; half of it has less than 6% gradient, and one fourth of the area, less than 3%. About three quarters of the region is underlain by glacial till and stratified inclusions and the rest by glacio-lacustrine deposits. In Door County the soil is shallow in many places over dolomite bedrock. The determination of the exact extent of such soils is hindered by the difficulty of distinguishing bedrock from channery glacial till by hand auger. Three quarters of the region is naturally well drained.

The region crosses several major plant communities (Plate 6): swamp conifers, sedge meadow, oak savanna, pine barrens, pine, boreal forest, and southern mesic forest. The corresponding soils range from the Entic Hapludolls to the Spodic Udipsamments, and from the dark Mollic Hapludalfs to Alfic Haplorthods. Included wetlands are occupied chiefly by Typic Haplaquolls and Histosols.

The naturally well-drained soils of the northeastern two thirds of the area (north of line D-D' in Fig. 2-40; show "double" (bisequal) profiles (Beaver, 1963, 1966) of varying degrees of distinctness. They are thus classified as Spodic and Glossic Hapludalfs or Alfic Haplorthods, transitional between Gray-Brown Podzolics, Gray Wooded soils, and Podzols (Table 11-1).

In the northeastern area about 55% of the landscape is cropland, 25% pastureland, and 20% woodland and wetland (Region J).

A number of soils that are common in southern Wisconsin are present in Door and Marinette counties as frigid taxadjuncts (see Fig. 2-32) of their series. These include the Kewaunee, Manawa, Casco, Salter, and Sisson soils.

For the sake of clarity, soil units are grouped in the following discussion by soil series affinities. This is a reflection of the fact that we know more about soil series than about soilscapes.

Table 11-1. Classification of major soils of soil associations E1 through E5, and E9

Texture of A horizon	Depth of solum (in.)	Character of substratum	Natural drainage condition of soil		
			Well to moderately well drained	Somewhat poorly drained	Poorly to very poorly drained
Loamy	24-36	Calcareous reddish-brown silty clay loam to clay loam till	Hortonville sl-sicl (Glossoboric Hapludalf)		
	>30	Calcareous reddish-brown clayey drift			Wauseon sandy loam, loam (Typic Haplaquoll)
	>30	Pink calcareous sandy loam to loam till	Underhill sandy loam silt loam (Typic Eutroboralf)		Angelica loam, silt loam (Aeric Haplaquept)
	15-30	Pink calcareous sandy loam to loam till	Onaway sandy loam, loam (Alfic Haplorthod)	Solona sandy loam, loam (Aquic Eutroboralf)	
	20-42	Calcareous silts and fine sands		Shiocton sandy loam, silt loam (Aquollic Eutrochrept)	
Sandy	20-40	Acid silts and fine sands	Alban fine sandy loam, silt loam (Typic Glossoboralf)		
	20-40	Pick calcareous sandy loam to loam till	Emmet sandy loam, loam (Alfic Haplorthod)		
	<15	Sand and gravel outwash	Emmert sandy loam, loam (Typic Udorthent)		
	>40	Acid pink sand outwash	Omega sand, loamy sand (Entic Haplorthod)		

E1, E2, E3, E4, E5, E9. *Emmet, Onaway, and associated loams on pink dolomitic glacial till.*

- E1. The rolling to undulating Emmet loamy sand, Onaway loams, and Omega loamy sand, association.
- E2. The rolling to undulating Onaway and Solona loams, Emmet and Underhill sandy loam, and Angelica loams, association.
- E3. The undulating Emmet and Onaway sandy loam, Solona and Angelica loams, and Omega loamy sand, association.
- E4. The undulating Onaway, Underhill, Emmet, Alban, and Solona loams association.
- E5. The undulating Solona, Onaway, Hortonville, Shiocton, and Angelica loams association.
- E9. The nearly level Underhill, Onaway, Angelica, and Wauseon loams association.

Prominent soils in this group of soil associations are the Emmet, Onaway, Underhill, and Solona series. The first two are Spodosols (Podzols; see Table 11-1). The other two are Alfisols: the Underhill, a Eutroboralf (Gray Wooded soil), and the Solona, an Aquic Eutroboralf (somewhat poorly drained Gray Wooded soil). The Onaway and Shiocton soils are illustrated in Fig. 11-5. Calcareous glacial drift (Fig. 2-8) is extensive under these soils, and much of it is pink (7.5YR 6/4). The Hortonville (Glossoboric Hapludalf) and Wauseon (Typic Argiaquoll) soils are developed over the finest textured and reddish materials. In these soilscapes highly fertile medium-textured soils are associated with bodies of acid sands and gravels (Emmet, Emmert soils) suited to coniferous forest.

Many other soils, well to poorly drained, are associated with the above-named series. The soil diversity is particularly marked along major soil boundaries, as between E4 and G14 in Menominee County (Milfred, Olson, and Hole, 1967).

On stabilized beach sands along the Lake Michigan shore is found a complex of sandy soils, identified in Michigan soil surveys as Wallace (Typic Haplorthod), Weare (Entic Haplorthod), and Oakville (Typic Udipsamment).

E6, E7. *Longrie and associated loams, shallow to limestone.*

- E6. The undulating Longrie, Summerville, Onaway, and Bonduel loams association, with rock outcrops.
- E7. The undulating Onaway, Longrie, and Detour loams and sandy loam association.

These soils of the Door Peninsula are neutral to alkaline, loamy, and shallow over dolomite (Summerville, 10 to 20 inches; Longrie and Bonduel, 20 to 40 inches; Fig. 2-3) or over calcareous, sandy loam till (Onaway, 15 to 30 inches deep to till; Figs. 11-4, 11-5). The Institute sandy loam is shallower than the Onaway soils and is fairly common in these landscapes. The abundant carbonates have favored the development of a deep, dark surface soil in the shallow, flaggy Summerville (Entic Lithic Haplorthod). In the other four soils, which are deeper, the cool, humid climate and the northern

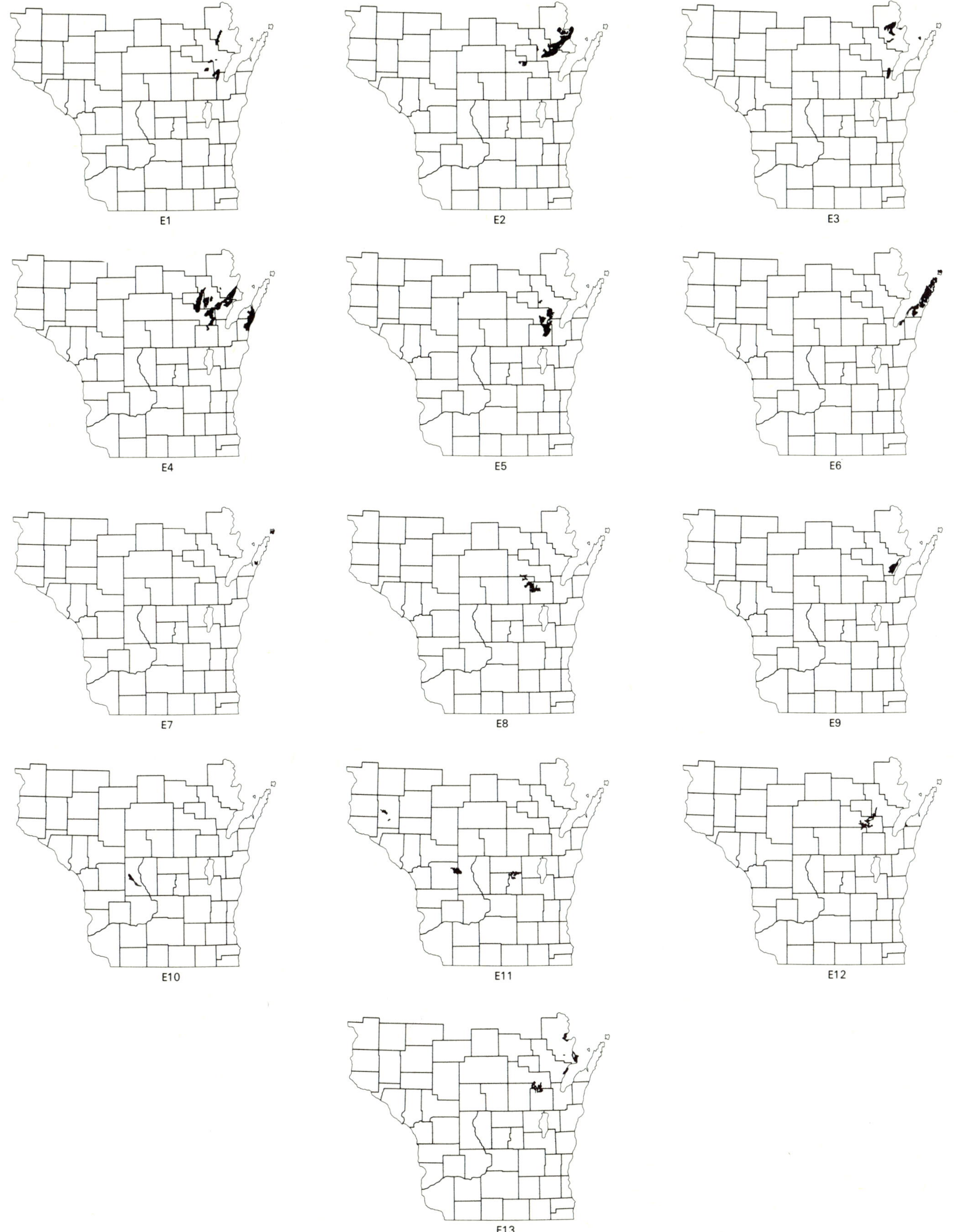

Figure 11-3. Sequence of maps showing distribution of soil associations in Soil Region E.

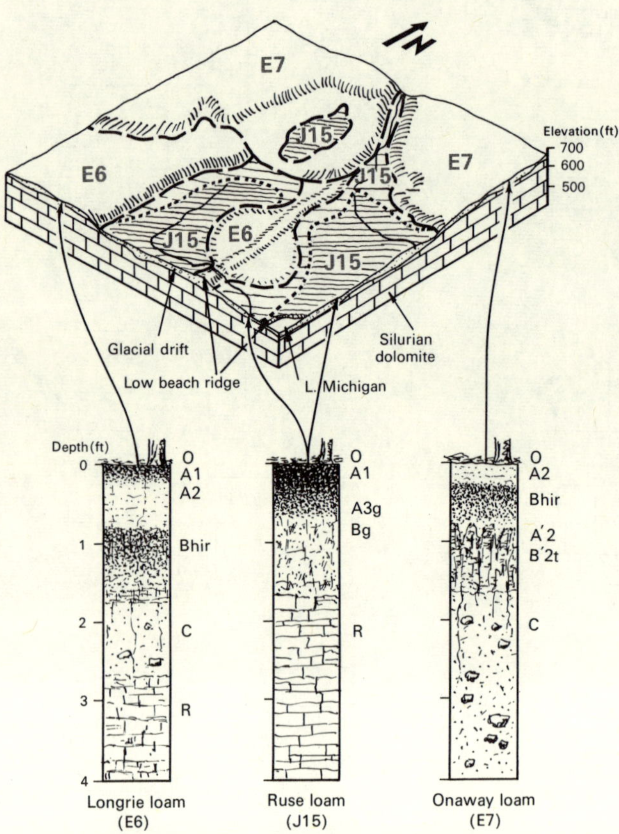

Figure 11-4. Block diagram showing landscape positions of major soils in 4 square miles of the Door County Peninsula (T.28N., R.26E., Sections 35 and 36; T.27N., R.26E., Sections 1 and 2).

Figure 11-5 (*below*). Block diagram showing landscape positions of representative soils of soil associations E8 and E11 in an east-west transect across Outagamie County through the Wolf River drainage basin. The reddish clay Ewen (moderately well drained DePere) and Stinson (somewhat poorly drained) Fluvents (Alluvial soils) occupy the floodplain. The soils labeled in the diagram are classified as follows: Shiocton, Aquollic Eutrochrept; Shawano, Typic Udipsamment; Tustin, Arenic Hapludalf; Oshkosh, Typic Hapludalf; Borth, Typic Hapludalf; Leeman (Oakville), Typic Udipsamment; Onaway, Alfic Haplorthod; Spinks, Psammentic Hapludalf. (After Beaver, 1966.)

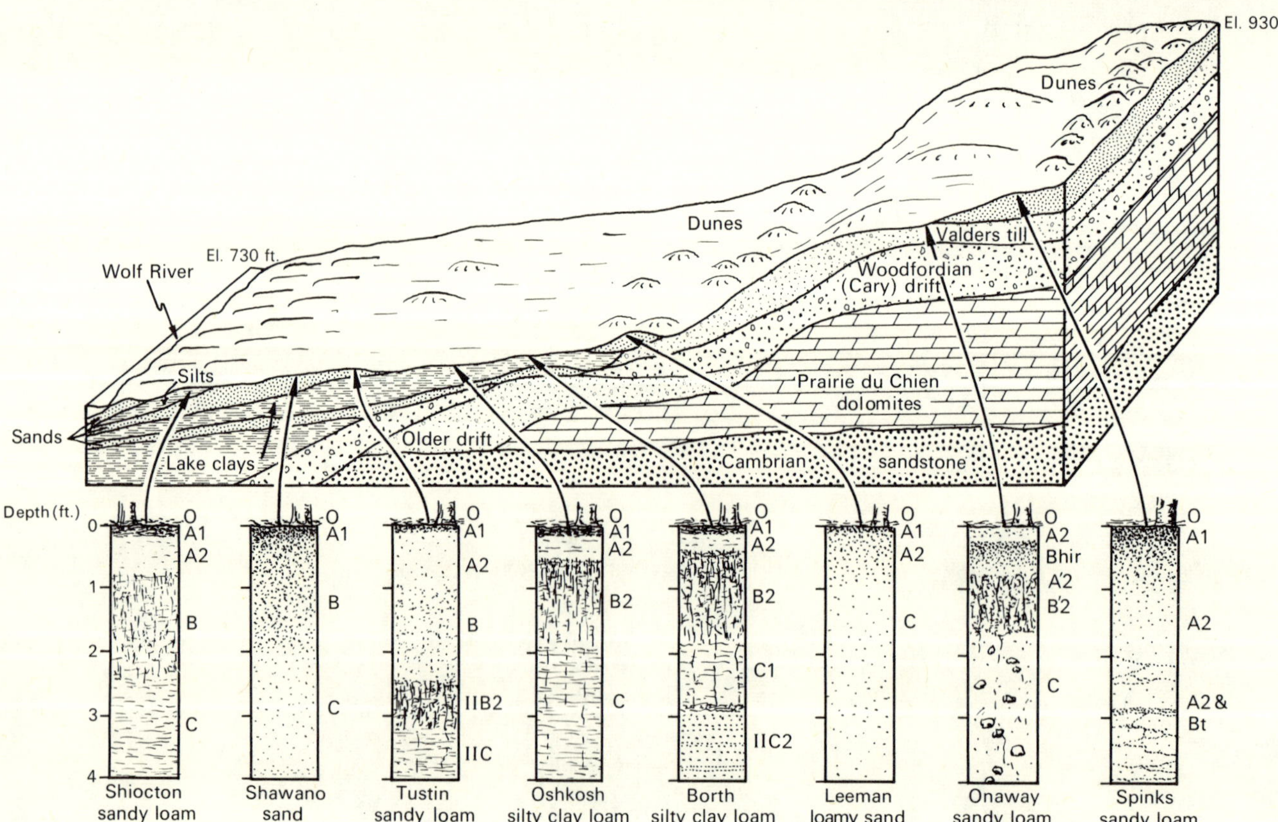

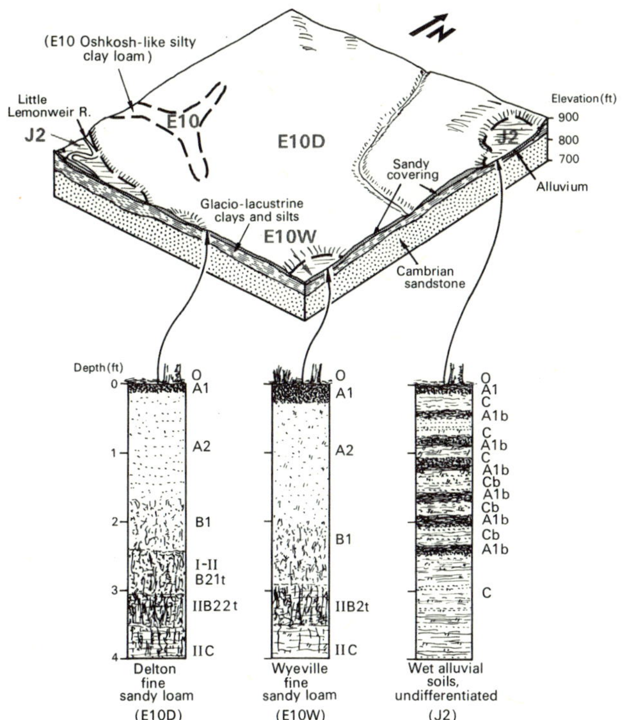

Figure 11-6. Block diagram showing landscape positions of representative soils of soil association E10 in Section 31, T.17N., R.3E., Juneau County.

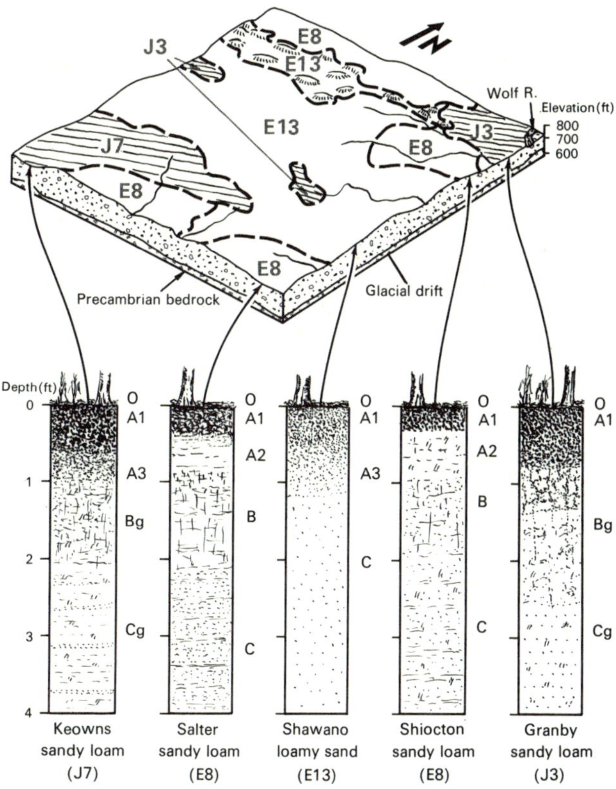

Figure 11-7. Block diagram showing landscape positions of representative soils of soil associations E11, E12, and E13 in Section 6 and parts of Sections 5 and 7, T.24N., R.16E., Outagamie County.

mixed forest vegetation have favored the development of a marked A2 horizon (albic) under the A1 horizon. The albic horizon is underlain by a spodic horizon (Podzol Bhir), namely in the Longrie (Entic Haplorthod), Onaway (Alfic Haplorthod), and Ruse (Entic Haplaquod) soils.

The shallowness of the soils over limestone limits agriculture on them. Cherry orchards have been successful, however, even where holes for tree planting had to be blasted into the underlying dolomite. Tonguing of B2t horizon material down cracks and cavities in the bedrock improves the capacity of these soils to support tree fruit production.

Safe disposal of liquid wastes in these soils is made impossible at many sites by the shallowness of the natural soil filter over channels leading down joints in the bedrock, and because of perched water tables in the depressional soils called Bonduel (Aquic Hapludalf) and Detour (Lithic Haplaquept). However, placement of private septic tank absorption fields in artificial sandy mounds under carefully controlled conditions makes subsurface disposal of effluent feasible (Bouma et al., 1972).

Rock outcrops are most numerous at and near the Silurian Escarpment that overlooks the east shore of Green Bay.

Some low drumlins are present in T.31N., R.28E. and T.32N., R.28E. (Thwaites and Bertrand, 1957; Kowalke, 1952). In T.27N., R.23E. in Door County a shale layer is associated with small patches of Shullsburg (Aquic Argiudoll) and Calamine (Typic Argiaquoll) soils of impeded drainage. Alpena (Typic Rendoll) and Kiva (Entic Haplorthod) sandy loams (northern counterparts of Rodman and Casco, respectively) are present in Door County. A Rousseau sand (Entic Haplorthod) is reported from stabilized sand dunes and has the unusual combination of properties of a spodic horizon and an alkaline reaction (pH 8.0) (Sec. 28, T.26N., R.20E.).

E8, E11. *Shiocton, Tustin, and associated soils on fine sandy and silty lacustrine deposits.*

E8. The nearly level Shiocton and Tustin sandy loam, Shawano loamy fine sand, Oshkosh and Poygan silty clay loam, and peat and muck, association.

E11. The nearly level Tustin, Shiocton, and Kibbie loams association.

Bodies of these soil associations, scattered throughout six counties (Fig. 11-2), are dominated by the well-drained Tustin (Arenic Hapludalf; Fig. 11-5) and the less clayey and somewhat poorly drained Shiocton (Aquollic Eutrochrept; Figs. 11-5, 11-7). These sandy loams have developed in calcareous glacial drift. Associated in the northeast are moderately well drained Oshkosh (Typic Hapludalf) and poorly drained Poygan (Typic Haplaquoll) soils of calcareous lacustrine clay deposits. The Kibbie sandy loam (Aquollic Hapludalf) resembles the Shiocton, but has a clay-enriched B horizon. Artificial drainage is a common practice in fields with soils whose natural drainage is impeded.

The E11 soil association in Dunn County includes Delton (Arenic Hapludalf) and Wautoma (Mollic Haplaquept) loamy sands and loams.

E10. *The nearly level Delton and associated Wyeville sandy loam and Poygan silty clay loam on lacustrine clays and silts.*

This soil association constitutes a body in Juneau and Monroe counties at the foot of the Franconia Escarpment that borders the Western Upland. Sand forms a surficial covering 20 to 40 inches deep over acid lacustrine clays in the well-drained Delton (Arenic Hapludalf) and the somewhat poorly drained Wyeville (Aquic Arenic Hapudalf) (Fig. 11-6). The wet Poygan soil has formed in calcareous reddish clay till with a covering of silt less than 20 inches thick. This soil is more acid in this landscape than it is in Region I. The Oshkosh-like soil of Fig. 11-6 is an acid variant that has the solum and upper C horizon leached to a depth of 5 or 6 feet.

E12, E13. *Shawano and associated sandy soils on lacustrine sands and silts.*

E12. The nearly level Shawano, Keowns, Granby, and Au Gres loamy sand and sandy loam association.

E13. The nearly level Shawano and Granby loamy sand and sandy loamy, and peat and muck, association.

These soils on glacial-lake plains are dominated by the well-drained Shawano loamy sand (Typic Udipsamment) formed in fine and medium sands that are neutral to slightly calcareous below a depth of 3 feet. Associated soils are mostly in depressions and in places where they are affected by a higher water table. The acid Au Gres (Entic Haplaquod) sand is somewhat poorly drained. The Granby (Typic Haplaquoll) loamy sand has formed in calcareous sand and is poorly drained, as is the finer textured Keowns (Mollic Haplaquept) sandy loam (Fig. 11-7). Some narrow bodies of Leeman fine sand represent old stabilized sand dunes.

Small amounts of charcoal in the surface soil are common throughout northern Wisconsin as a result of numerous forest fires, many of them connected with the major period of logging activity, particularly in the decades just before and after 1900. The Peshtigo fire of 1871 (Lapham, 1873) must have contributed to the charcoal in the soils of association E13 in southeastern Marinette County.

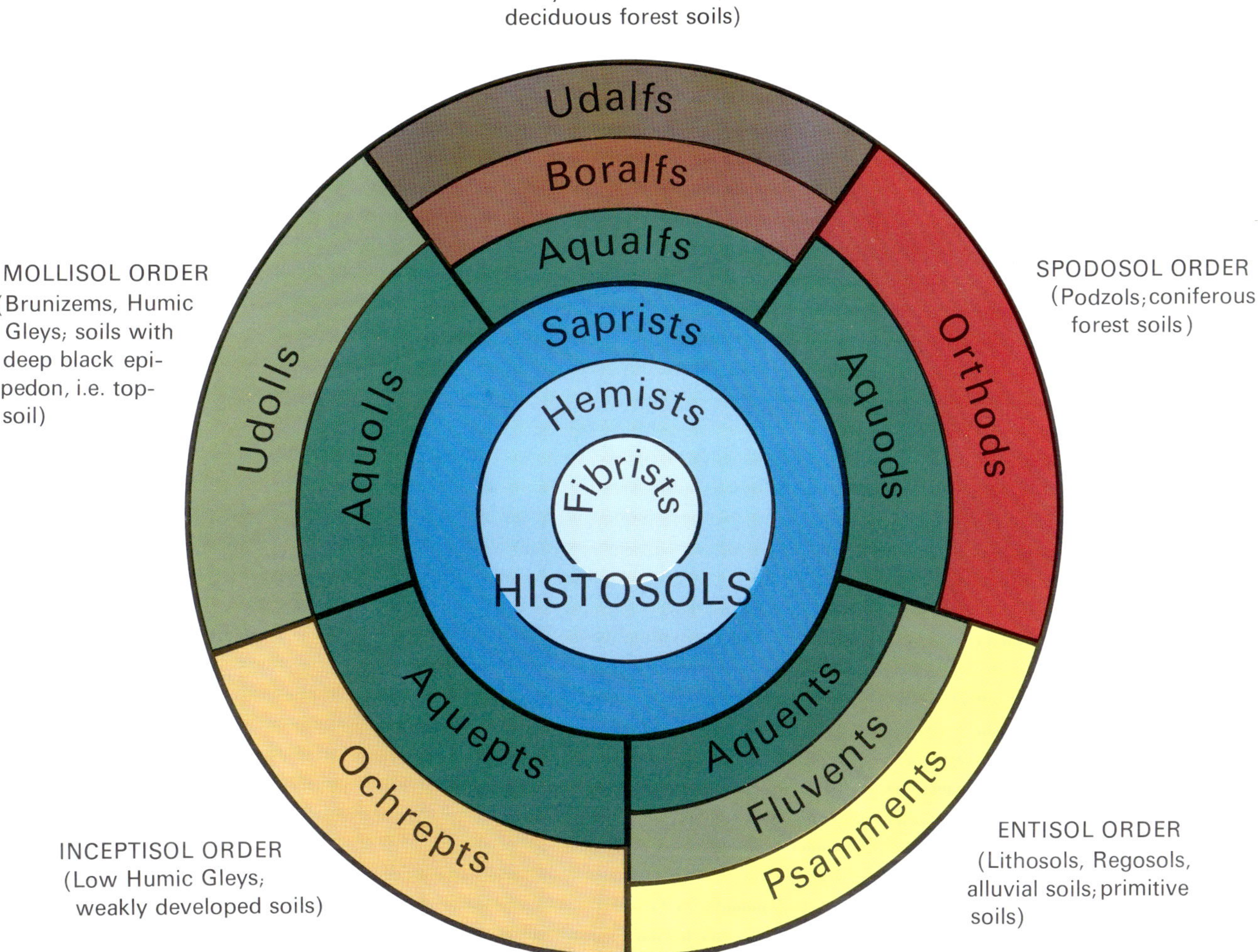

Figure 5-1, in color. Circular key to the new USDA soil classification for Wisconsin, coded in color to indicate, in each instance, a key feature of the soil or environment. The very poorly drained organic soils, represented here in blue, are normally found saturated with water in bogs and marshes; the Fibrists are assumed to be the wettest of the three suborders of Histosols. The poorly drained mineral soils (Aqualfs, Aquods, etc.) are colored green here to indicate the abundance of vegetation on them. Alluvial soils (Fluvents) and upland prairie soils (Udolls) are less wet, but even so support rather vigorous plant growth. The sands (Psamments) are droughty; Ochrepts may be thought of as a degree less droughty. Important upland forest soils are portrayed in the three remaining suborders: Podzols (Orthods) commonly have reddish-brown B horizons, Boralfs have less of a reddish tinge in clayey B horizons, and Udalfs have brown to yellowish-brown clayey B horizons. The chart may be visualized as a funnel with the wettest soils at the center and the driest ones at the periphery.

GENERALIZED SOIL REGIONS AND LANDFORMS OF WISCONSIN
UNIVERSITY OF WISCONSIN - EXTENSION
Geological and Natural History Survey
1971
0 30 60 Mi.
0 50 100 Km.
Legend
Soils of the . . .
A Southwest Ridges and Valleys
B Southeast Upland
C Central Sandy Uplands and Plains
D Western Sandstone Uplands, Valley Slopes and Plains
E Northern and Eastern Loamy Reddish Drift Uplands and Plains
F Northern Silty Uplands and Plains
G Northern Loamy Uplands
H Northern Sandy Uplands and Plains
I Northern and Eastern Clayey Reddish Drift Uplands and Plains
J Major Wetlands
Soil regions by Francis D. Hole Relief by David A. Woodward
MADISON

PLATE 2

BEDROCK GEOLOGY OF WISCONSIN
UNIVERSITY OF WISCONSIN - EXTENSION
Geological and Natural History Survey
1971
0 30 60 Mi.
0 50 100 Km.
LEGEND
SILURIAN DEVONIAN
ORDOVICIAN
CAMBRIAN
PRECAMBRIAN
Devonian Formations (dolomite and shale)
Silurian Formations (dolomite)
Maquoketa Formations (shale and dolomite)
Sinnipee Group (dolomite with some limestone and shale)
Ancell Group (mainly St. Peter Fm.) (sandstone, shale and conglomerate)
Prairie du Chien Group (dolomite)
Upper Cambrian Formations (sandstones with some dolomite and shale)
Upper Keweenawan Formations (sandstones with some shale and conglomerate)
Quartzite, Slate and Iron Formations
Gabbro and Basalt
Granite and Undifferentiated Igneous and Metamorphic Rocks
Granite and Undifferentiated Igneous and Metamorphic Rocks (sparse data)
Border of Wisconsin (Cary) Drift
Border of Older Drift
A
A′
MADISON
Elevation Above and Below Sea Level In Feet
1000
800
600
400
200
S.L.
200
455
600
800
1000
255
0 30 60 Mi.
0 50 100 Km.
P€ Basement
A
Cassville
A′
Racine

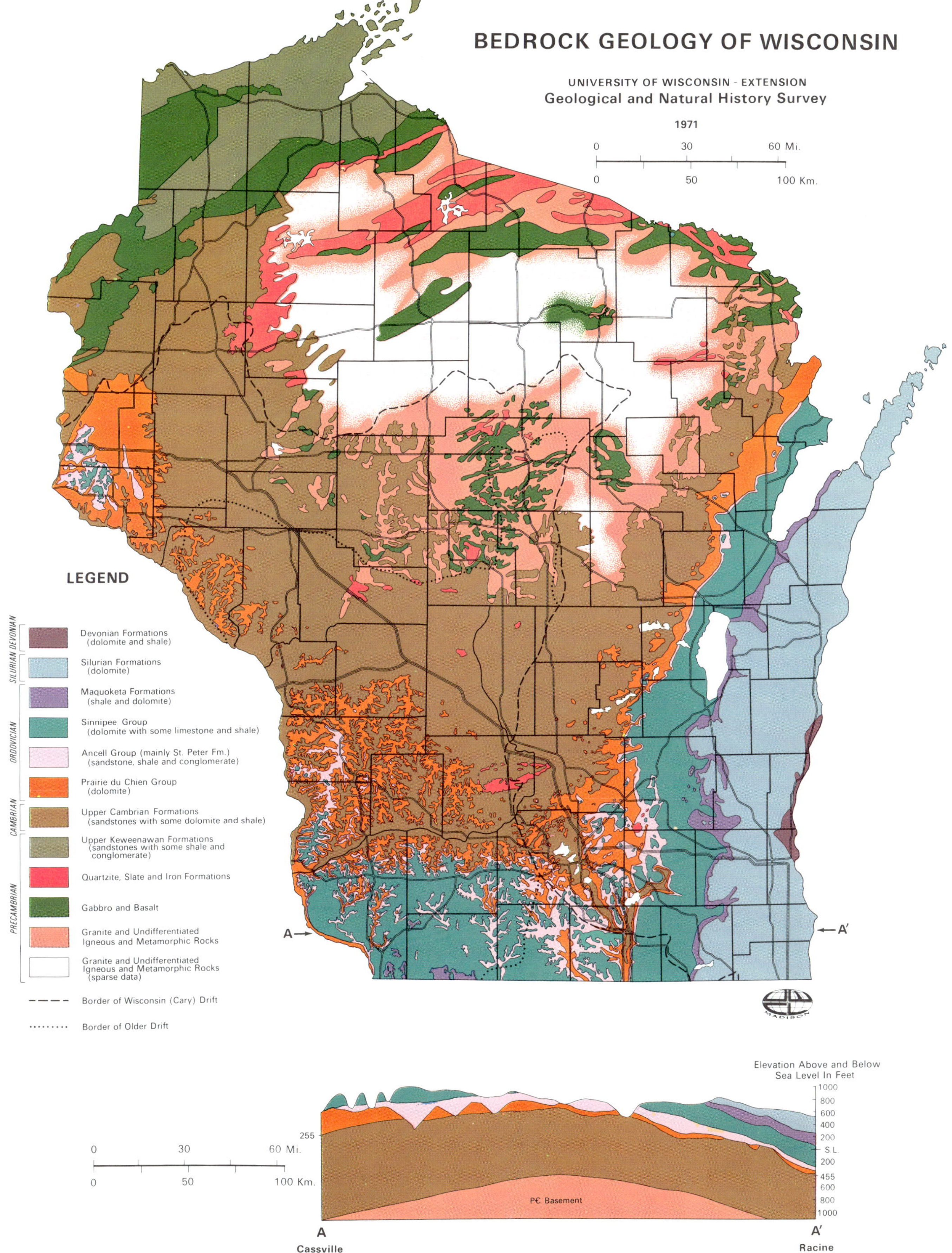

PLATE 3

GLACIAL DEPOSITS OF WISCONSIN
UNIVERSITY OF WISCONSIN - EXTENSION
Geological and Natural History Survey
1964
0 30 60 Mi.
0 50 100 Km.
"Driftless Area"
LEGEND
End Moraines
Ground Moraine
Outwash, unpitted
Outwash, pitted
Lake Basins
Drumlin Trends
After Thwaites, 1956
MADISON

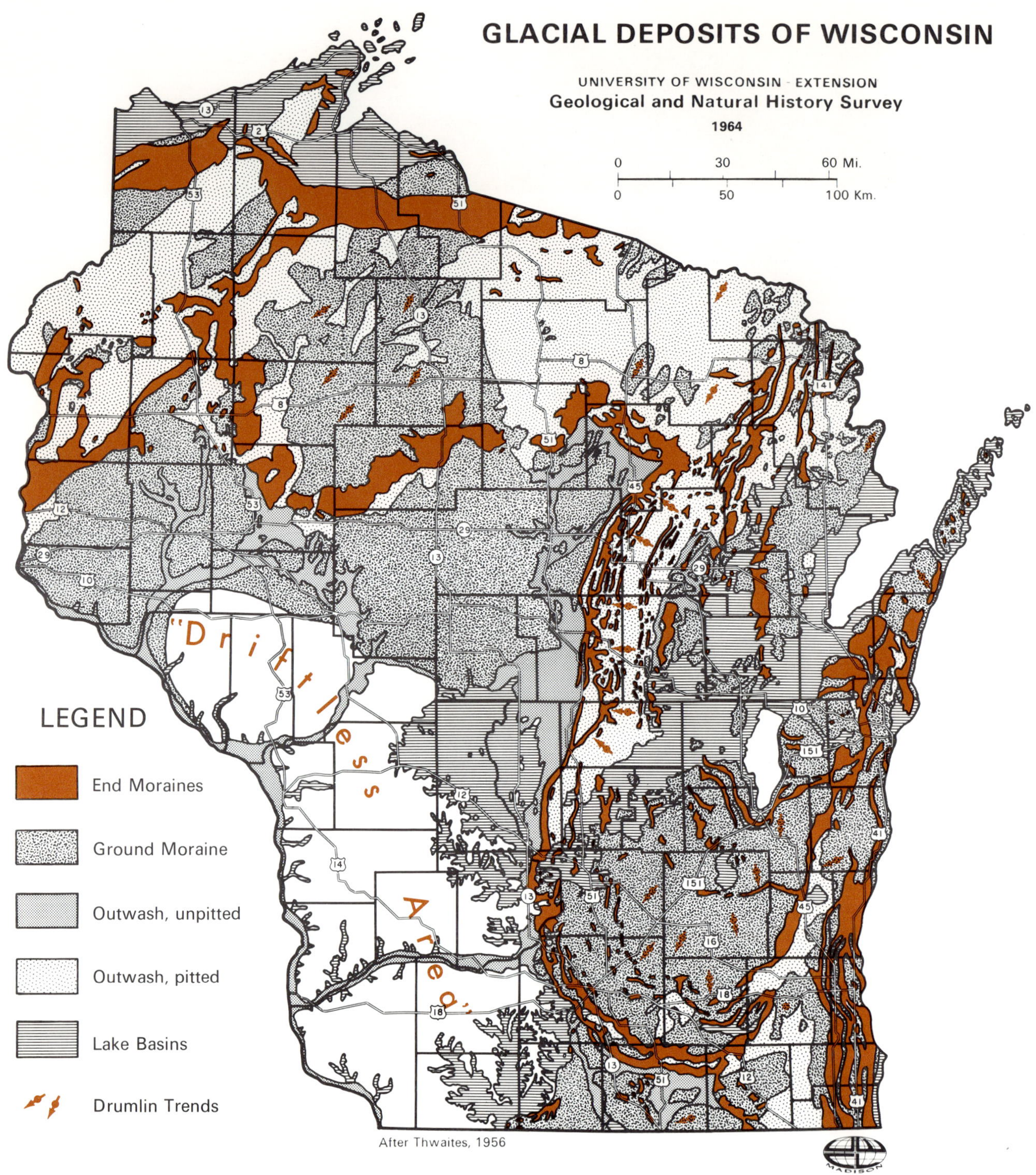

PLATE 4

AEOLIAN SILT AND SAND DEPOSITS OF WISCONSIN
UNIVERSITY OF WISCONSIN - EXTENSION
Geological and Natural History Survey
1968
0 30 60 Mi.
0 50 100 Km.
LEGEND
DEPTH OF AEOLIAN SILT
8′-16′, Silt
4′-8′, Silt
2′-4′, Silt
1/2′-2′, Silt
0′-1/2′, Silt
SAND AREAS
Active Dunes Present
Sandy Areas Subject to Blowing
Beach Sand
B
Compiled by Francis D. Hole
MADISON

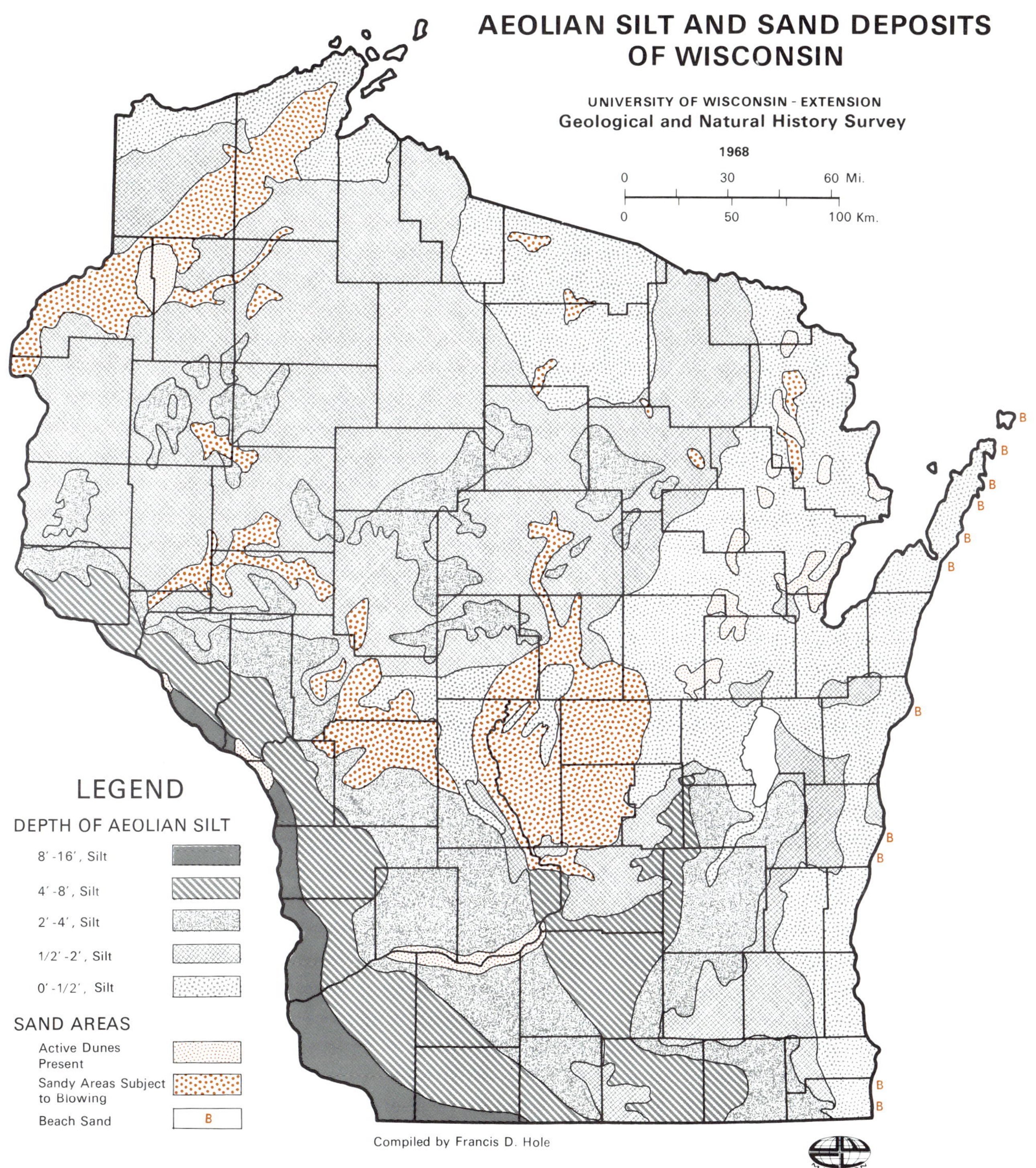

PLATE 5

EARLY VEGETATION OF WISCONSIN
UNIVERSITY OF WISCONSIN - EXTENSION
Geological and Natural History Survey
1965
0 30 60 Mi.
0 50 100 Km.
LEGEND
PRAIRIE
Bluestem, Composites
1
OAK SAVANNA
Bur Oak, White Oak, Bluestem
2
SOUTHERN OAK FOREST
White, Black and Red Oaks
3
SOUTHERN MESIC FOREST
Sugar Maple, Basswood, Elm
4
LOWLAND HARDWOOD
Willows, Soft Maple, Ash
5
SEDGE MEADOWS
Sedges, Blue Joint, Cordgrass
6
PINE BARRENS
Jack pine, Prairie Grasses
7
PINE FOREST
White Pine, Red Pine
8
NORTHERN MESIC FOREST
Maple, Hemlock, Yellow Birch
9
CONIFER SWAMPS
Black Spruce, Tamarack, Cedar
10
BOREAL FOREST
Balsam Fir, White Spruce
11
MADISON

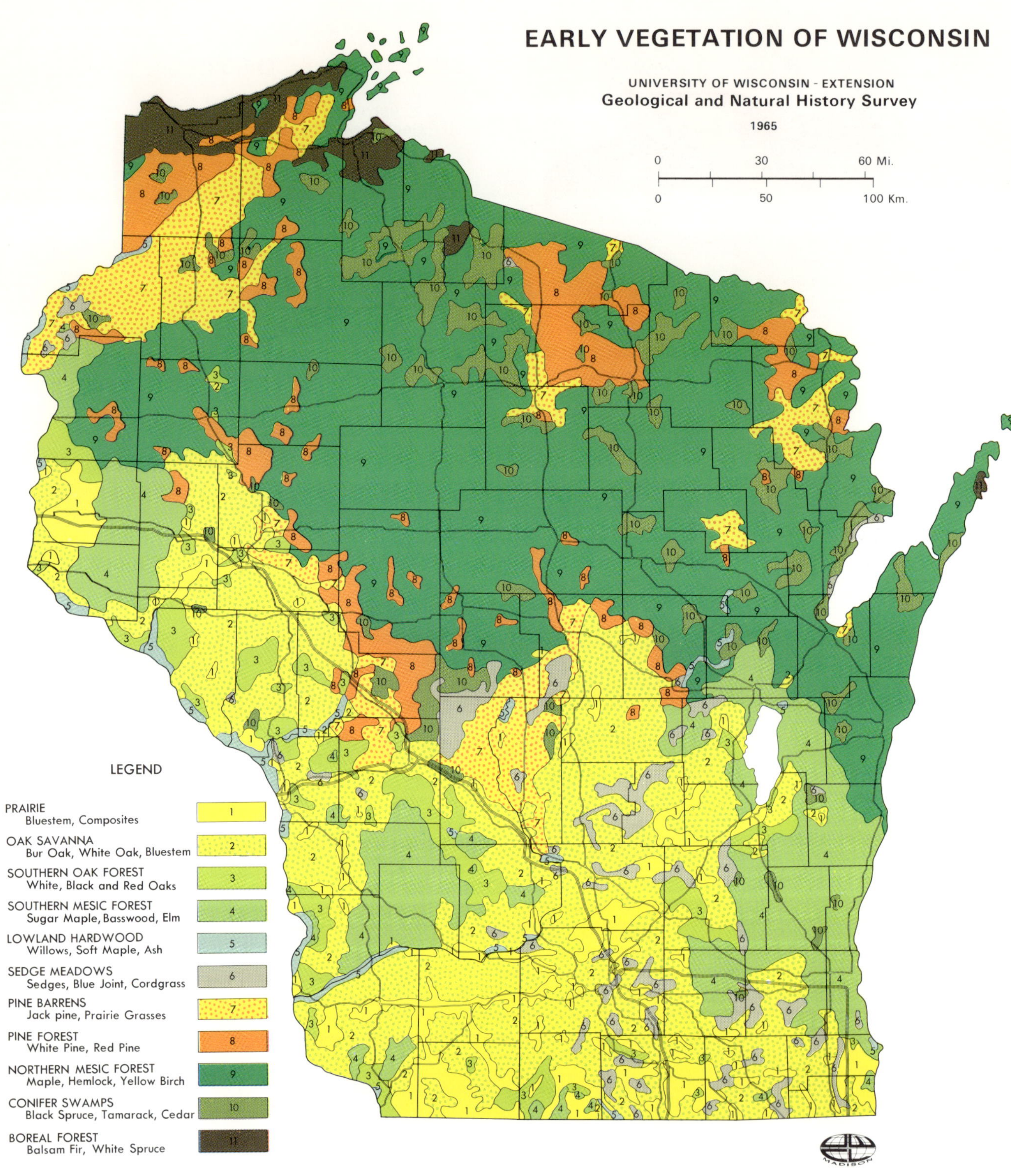

PLATE 6

SOIL REGIONS OF WISCONSIN
UNIVERSITY OF WISCONSIN - EXTENSION
Geological and Natural History Survey
1974
0 30 60 Mi.
0 50 100 Km.
MAJOR SOILS
A Fayette, Dubuque
Ap Tama, Dodgeville
B Dodge, Miami, Morley, Casco
Bp Plano, Warsaw, Varna
C Plainfield, Wyocena, Boyer
Cp Sparta, Gotham
D Hixton, Norden, Gale, Boone
E Onaway, Emmet, Shawano
F Withee, Santiago, Amery, Antigo
Fp Jewett, Pillot
G Iron River, Gogebic, Kennan
H Omega, Vilas, Hiawatha
I Kewaunee, Hibbing
J Pella, Poygan, Newton, Houghton, Arenzville
Subscript p (i.e.Ap,Bp) indicates prairie soil
Compiled by Francis D. Hole
MADISON
U.S.Department of Agriculture,
Soil Conservation Service and U.S.Forest Service
University of Wisconsin-Madison,
College of Agricultural and Life Sciences, Department of Soil Science

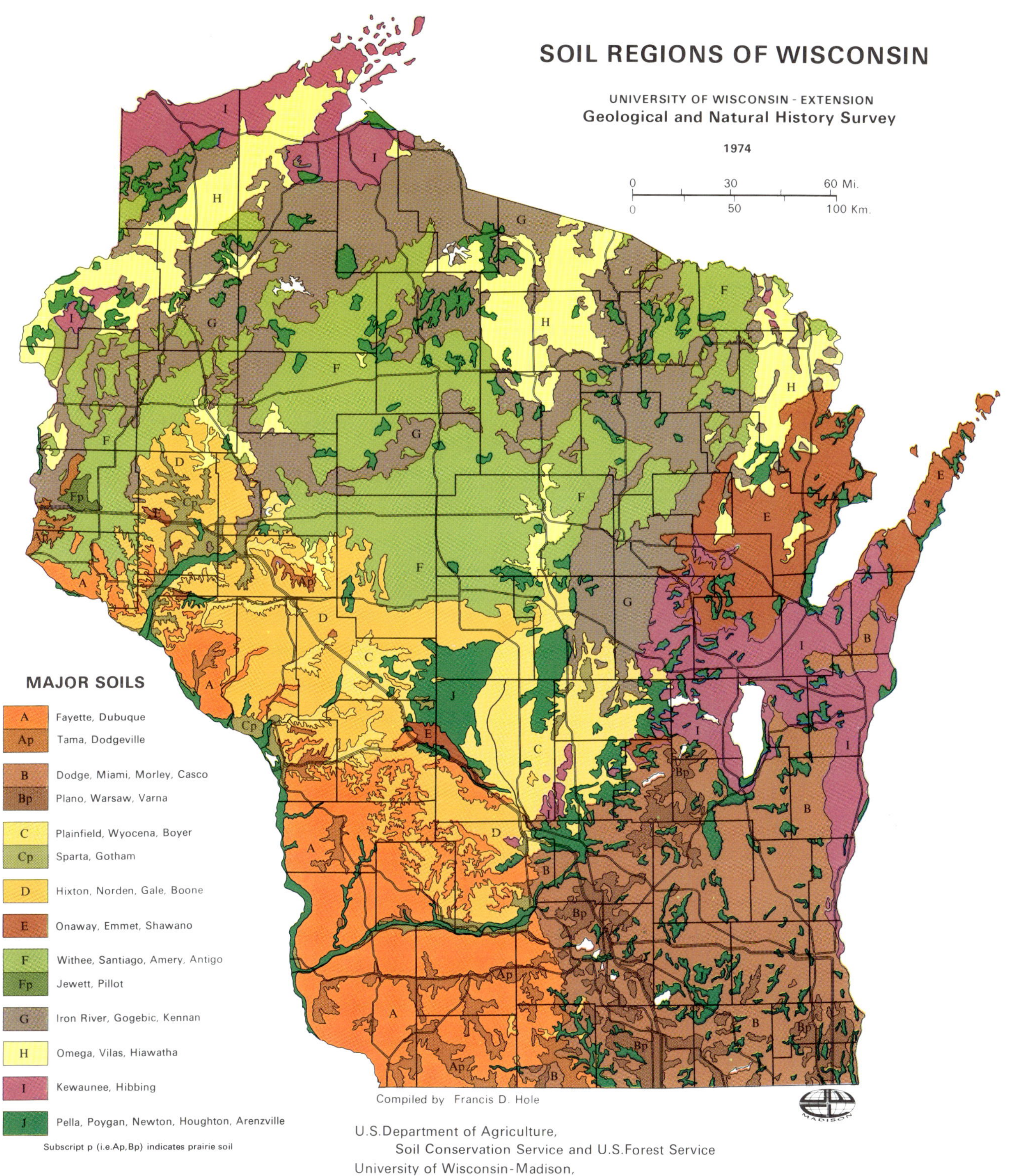

PLATE 7

Wisconsin soils: ecological motif. The stylized forms depict flowering plants rooted in mineral soil. From the top, down, the four horizontal rows of circles represent (1) flower heads; (2) cycling of nutrients between vegetation and soil, through the leaf litter, the dark A1 horizon, and the platy A2 horizon, all of which appear in the circles; (3) cavities in the subsoil occupied by hibernating mammals and insect nests; and (4) stratified gravel substratum, with illuvial clay fillings between the stones. The vertical lines aboveground represent plant stems bearing leaves. Below the surface soil are cracks of different widths that divide the soil into units ranging from fine blocky under the platy A2 horizon to prismatic just above the gravel layer. (Designed by F. D. Hole and executed by G. W. Stanley.)

PLATE 8

CHAPTER **12**

Soil Region F: Soils of the Northern Silty Uplands and Plains

A traveler going northward from Juneau County into Wood County is impressed with the change in landscape from wild sand barrens (Soil Region C) and wetlands (Soil Region J) to prosperous dairy farmland (F11, F21[1]). Only slightly less dramatic is the contrast in central Langlade County between the productive fields of the Antigo Flats (F25) and the pasturelands and forestlands of the rolling moraine to the north (G14). Without its silty soil covering, much of Soil Region F would have coarse-textured soils much like those of Soil Region G.

Tightness of the subsoil in 60% of this region severely limits two important land uses: production of field crops and disposal of liquid waste. Special practices are applied in order to cope with these limitations. For example, successful tillage and cropping are possible where excess water is removed quickly after the spring thaw and, indeed, throughout the growing season. Landforming practices that create surface (grassed) drainageways have largely taken care of this situation (Wojta et al., 1960), although droughty conditions in August still pose a problem. The second land use, soil absorption of liquid wastes, has encountered serious difficulties for many years. The subsoils are too impervious to permit absorption of wastes, as shown by the common phenomenon of surfacing of septic tank effluent and its seepage into roadside ditches, wetlands, and streams. Recent experiments with placement on these soils of soil absorption systems in artificial sandy mounds 4 to 5 feet high and 70 to 100 feet long have been successful where carefully controlled (Bouma et al., 1972).

About 40% of the area is well to moderately well drained. Irrigation is practiced on some level croplands, particularly on the Onamia and Antigo soils (F25).

With increasing use of lime and fertilizers since about 1950, good stands of alfalfa have been obtained on even the most acid silty soils, with the result that the dairy economy has greatly improved in many counties (Stauber, 1956).

The soils of this region (Figs. 12-1, 12-2) occupy nearly six million acres of northern Wisconsin that are blanketed with 2 or 3 feet of silt (Figs. 2-28, 2-29, 2-30) resting on a wide variety of substrata, mostly acid: stony glacial till (Figs. 2-9, 2-11, 2-20), glacial outwash (Fig. 2-5), glacio-lacustrine deposits (Fig. 2-14), and weathered Cambrian and Precambrian bedrock (Fig. 2-3) (Hole, 1943). Most of the silt may have blown and washed from local till and outwash materials. The thinness of the silty covering has made possible admixture of sand and stones from the underlying glacial drift through tree-fall, rodent activity, frost action, and mass movement (Al-Rawi et al., 1969).

The silty soils of this region show a range of intensity of development of the Spodosol profile (spodic and albic horizons) from moderate, in the Alfic Haplorthods of northeastern counties, to little or none in the more extensive Glossoboralfs and Hapludalfs farther south and west (Figs. 12-3, 12-4). Most of the soils have good to somewhat poor natural drainage, and commonly have clay-enriched subsoils (argillic horizons). Tonguing of A2 material down cracks in the Bt horizon is common. Where a shallow Spodosol solum is present in the upper horizons of these soils, two sequa (sola) are recognized (Figs. 12-9, 12-10).

The presence of a loess covering across much of Wisconsin (Plate 5), from Grant County to Florence County, provides as close an approximation to a uniform blanket of medium-textured initial material as exists in the state. It is interesting to note a trend of progressive weakening of the development of the argillic horizon notheastward across a distance of about 250 miles from the southwest. The ratio of percent clay in the argillic horizon to the percent clay in the A2 horizon diminishes from 3 in some Fayette pedons of southwestern counties (Muckenhirn et al., 1955), to 2 in Dodge and Withee pedons of Dodge and Wood counties, to between 1 and 2 in some LaFont pedons of northeastern counties (Soil Conservation Service,

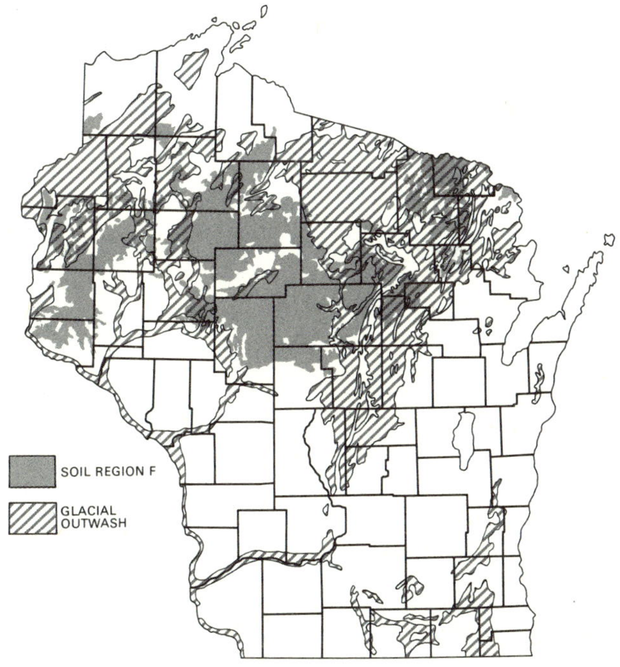

Figure 12-1. Index map showing the geographic relationship of Soil Region F to bodies of glacial outwash.

1. The Withee and the related Almena (F22) soils have been likened to Planosols (Albaqualfs) by pedologists visiting north-central Wisconsin from overseas or from other parts of the United States. The ratio of clay contents between the B and A horizons is not high enough, however, to support such a classification. Yet the field appearance and water relationships do suggest functional affinities of these soils to Planosols (see Van Rooyen, 1972).

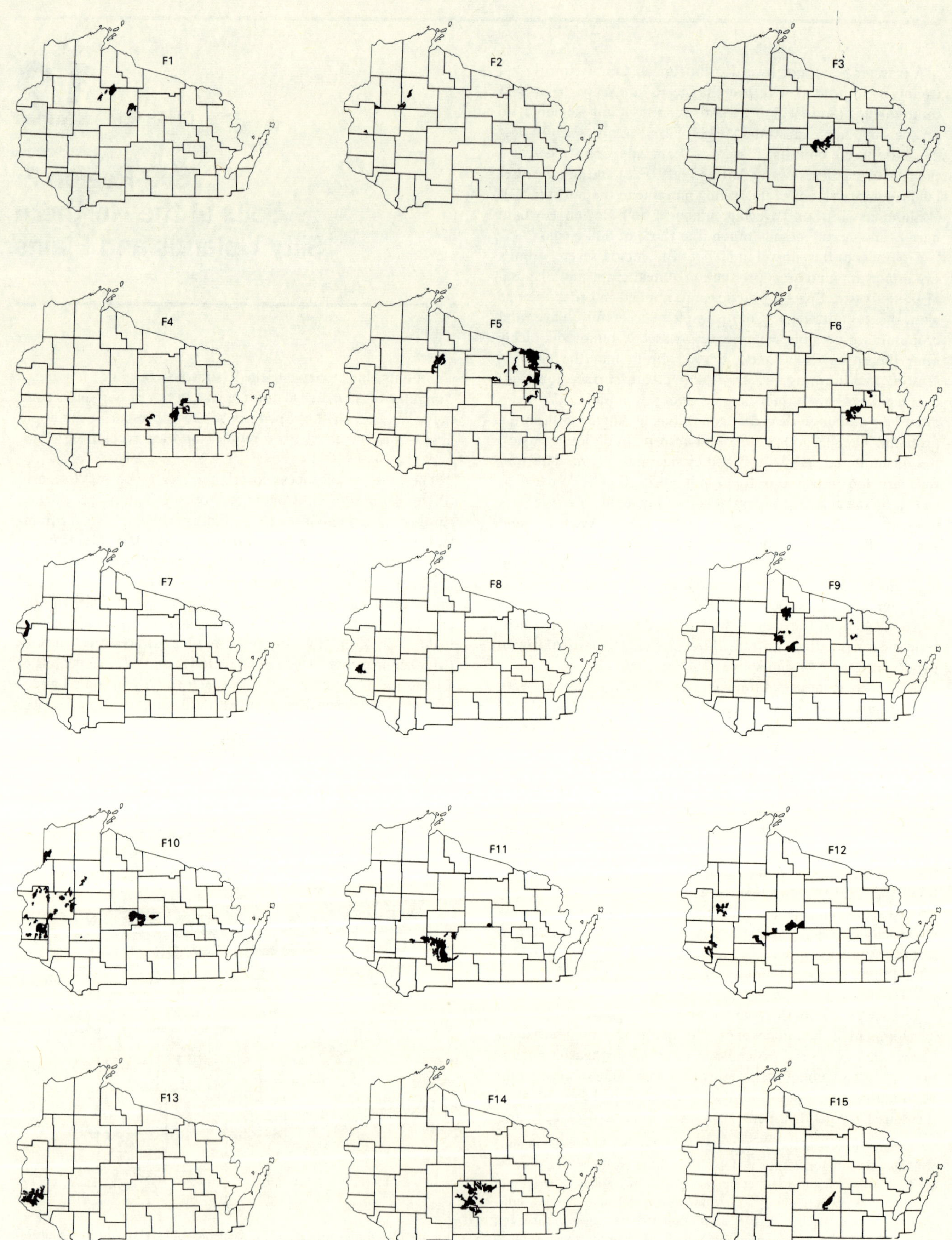

Figure 12-2. Sequence of maps showing distribution of soil associations in Soil Region F.

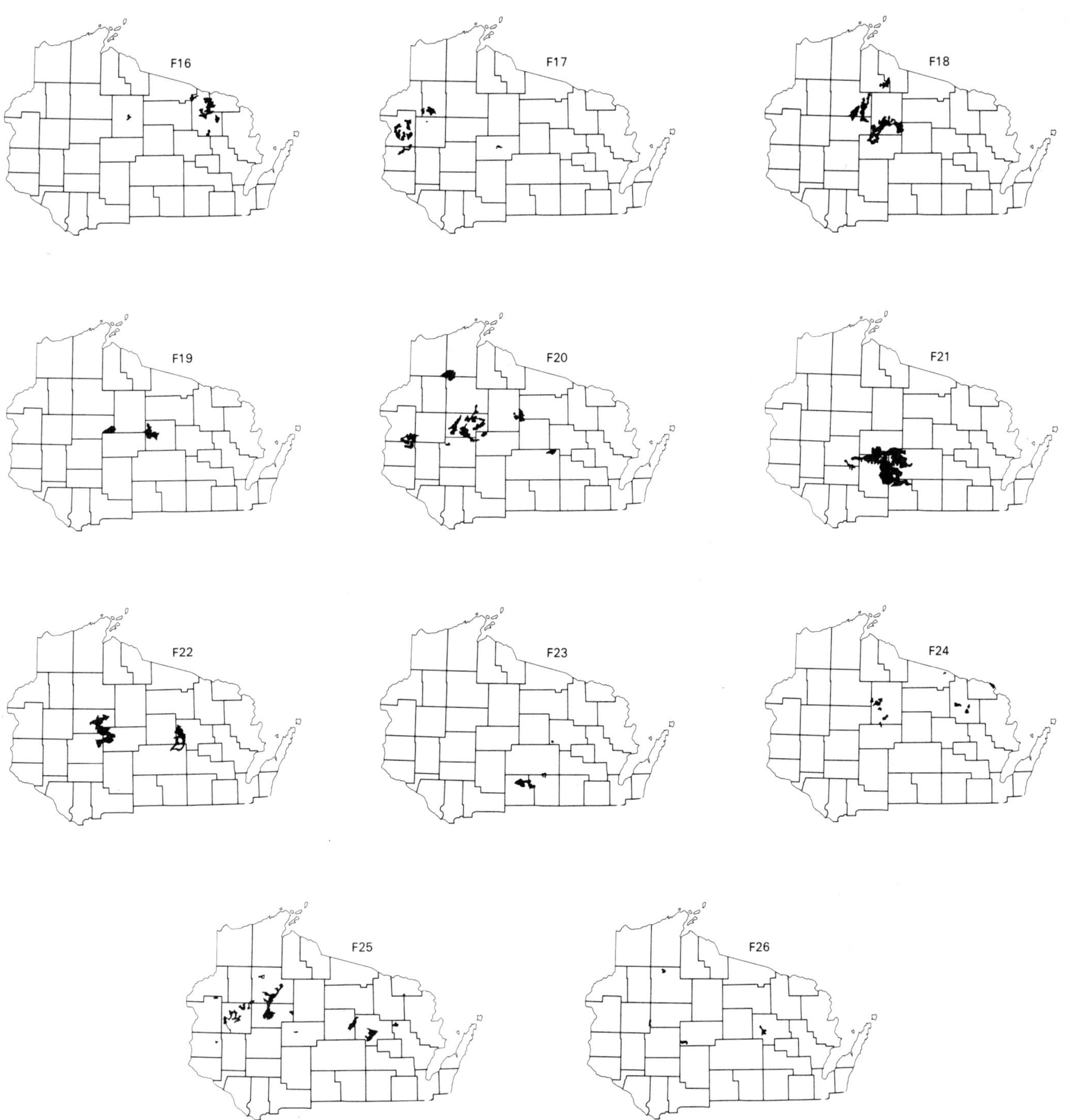

Figure 12-2, *continued.*

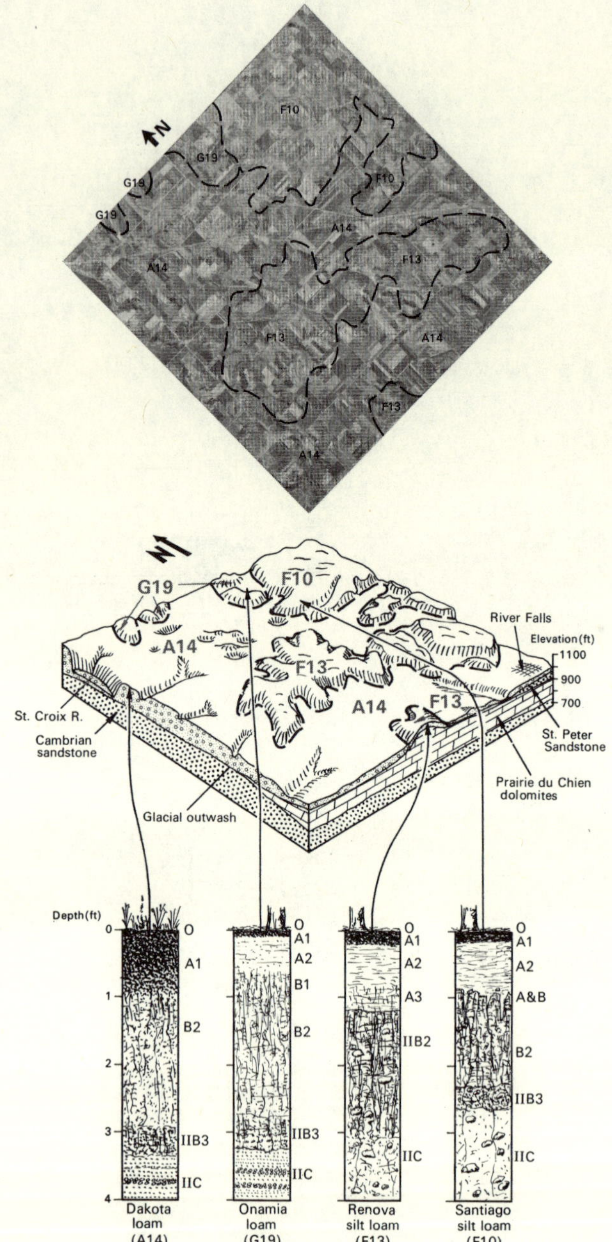

Figure 12-3. Aerial photo map of T.28N., R.19W., St. Croix County. The area shown is 6 miles on a side.

Figure 12-4. Block diagram showing landscape positions of major soils of T.28N., R.19W., St. Croix County.

1967a). Even where the ratio remains at 2 along the northern part of this transect, the actual amounts of clay involved diminish. For example, the B2t horizon of the Withee may contain 22% clay and that of the Lafont 12% clay. In each case, the amount in the Bt is double that present in the overlying A2 horizon. These ratios may be compared to the ratio of 4 which is typical for Miami silt loam in Indiana (personal communication, O. C. Rogers, 1952). The explanation for the diminishing degree of development of the argillic horizon in northern areas may lie in the shorter period of soil development (Figs. 2-35, 2-55), the coarser texture of the original loess and its dilution by coarser substrata through pedoturbation, and the lower frequency in the soils of wet-dry cycles that are thought to favor downward translocation of material within the solum.

The region is nearly level (41% of the region) or undulating (38%), but it has some rolling areas; even in these rolling areas, however, the slope gradients rarely exceed 20%.

Mottling of silty soils that have impeded drainage may be pronounced in the A2 and upper B2t horizons, but subdued in underlying horizons. This phenomenon seems to be peculiar to soilscapes of Region F, as contrasted with the condition in Region B, where moderately and somewhat poorly drained soils show continuous mottling in the lower horizons. The perched mottling in nothern silty soils is an expression of pseudogley conditions reported from European landscapes (Kubiena, 1953). These conditions result from seasonal perching above the main Bt horizon of a temporary water table, which is held up by the relatively impermeable subsoil and substratum. The persistence of frost in the subsoil into late spring, long after the surface soil layers have thawed and have received spring rains, may be an important factor in the genesis of this mottling pattern (personal communication, I. J. Nygard, after his study of similarly mottled soils in Alaska, 1946). Mottling in northern soils is not as reliable an indicator of poor conditions for crop growth as it is in soils of southern counties.

The region is nearly all in the northern mesic forest zone. Prairie and southern mesic forest formerly occupied only small areas of these soils in northwestern counties.

Present-day land use ranges from forest utilization (timber, maple syrup, recreation, and wildlife) in state and federal forest preserves and on much private land in northern counties, to utilization for pasture and fields cropped to hay, oats, corn (mostly for silage), potatoes, and other special crops.

The role of soil geography is illustrated by the advantage to a dairy farm of the presence of a body of 80 or more acres of arable silty soils in the midst of hilly loams (Region G). The sizes and slopes of soil bodies and their pattern of arrangement determine landscape fabric, which is as important to a region as is texture of soil horizons.

Crop yields reflect the combined influence of soil profile characteristics related to thickness of silty cover and to depth to water table (Beatty et al., 1966) (Fig. 12-5). For example, annual yields of alfalfa-brome hay under a high level of management that includes adequate fertilization and drainage range from 2.75 to 4 tons (dry weight) per acre: 2.75, Cloquet; 3.0, Dakota, Padus, Pence, Onamia, Stambaugh loams; 3.25, Clifford silt loam; 3.5, Altdorf, Antigo, Auburndale, Dolph, Fence, Freer, Goodman, Iron River, Kennan, Marshfield, Milaca, Poskin, Rudo, Rudolph, Waukegan silt loams and loams; 3.75, Almena, Comstock, Fenwood, Freeon, Lafont, Loyal, Withee; 4.0, Alstad, Bluffton, Brill, Campia, Crystal Lake, Cushing, Floyd, Jewett, Norrie, Ostrander, Rozell, Rozellville, Santiago, and Spencer silt loams.

F1, F9, F18. F19. *Clifford and associated somewhat poorly drained silt loams, shallow over compact loamy acid glacial till.*

F1. The rolling to undulating Lafont, Clifford, and Auburndale silt loam association.

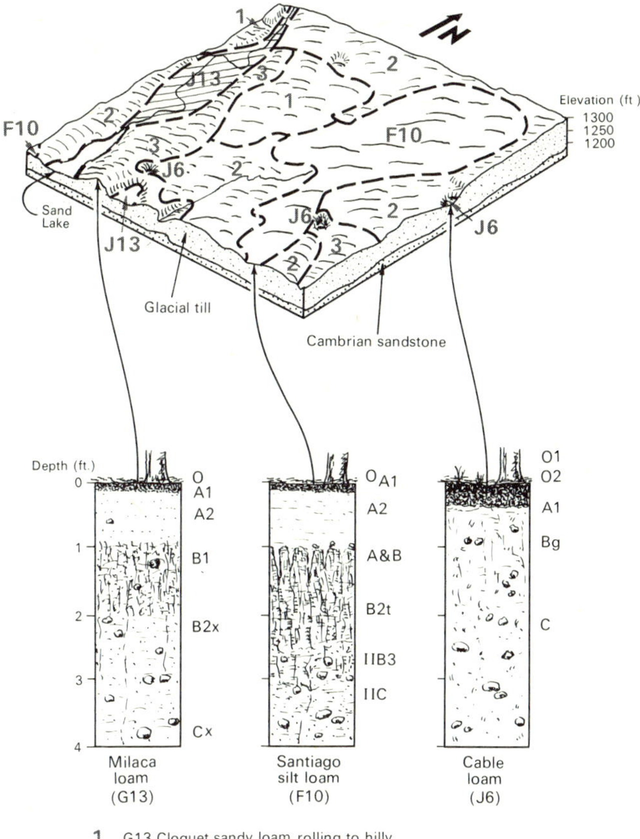

Figure 12-5. Block diagram showing landscape positions of representative soils in parts of Sections 8, 9, 16, and 17, T.36N., R.14W., Barron County.

F9. The undulating Clifford and Auburndale silt loam, and peat, association.

F18. The nearly level Clifford and Auburndale silt loam, and peat, association, with some stony areas.

F19. The nearly level Clifford, Lafont, and Auburndale silt loam association.

The somewhat poorly drained Clifford silt loam (Aqualfic Haplorthod) dominates these associations, which are extensive in Rusk, Price, and Sawyer counties (Figs. 12-6, 12-7). The mottling of the 1 or 2 feet of silt loam and underlying reddish-brown (5YR-2.5YR 4/4) sandy loam glacial till indicates the degree of wetness of this soil, which is intermediate between that of the better drained Lafont silt loam (Alfic Haplorthod) and the poorly drained Auburndale silt loam (Typic Glossaqualf). These soils are acid throughout. The Clifford and Lafont soils are bisequal with shallow Spodosol (Podzol) sola overlying weakly developed boralfic (Gray Wooded) sola (Carroll, 1959), the upper parts of which are fragic. Much of the landscape is cutover forestland. Where clearing and drainage have been effective, dairy farming is practiced.

In northwestern Taylor County the landscape is characterized by a catenal mixture of F19 and F20: Lafont-Spencer-Almena-Adolph. In places the reddish till (2.5YR 4/4) charac-

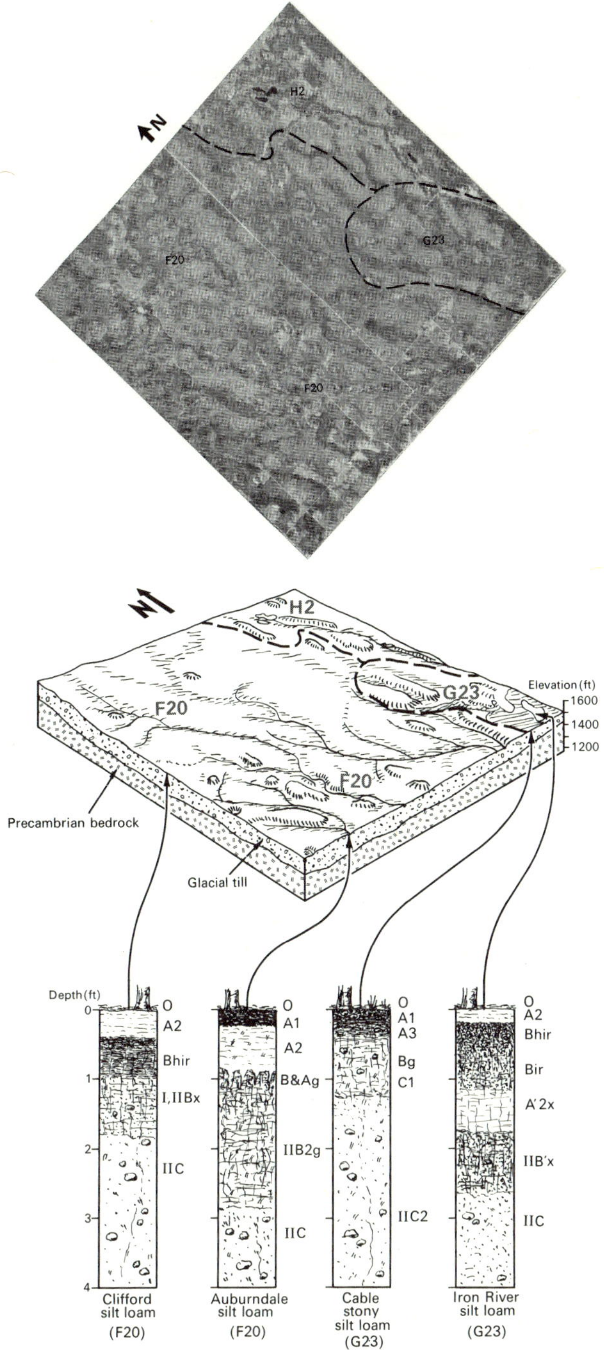

Figure 12-6. Aerial photo map of T.35N., R.4E., Oneida County. The area shown is 6 miles on a side.

Figure 12-7. Block diagram showing landscape positions of major soils of T.35N., R.4E., Oneida County.

teristic of the Lafont series is overlain by brown (10YR-7.5YR 4/4) till in which Iron River soils have formed.

In northern Price County some bodies of Goodman silt loam (Alfic Haplorthod) have a distinct pinkish cast in the silty material. This may indicate importation of red clay by wind and water, and possibly some staining of the soil with forest-derived substances.

F2, F10. *Santiago, Milaca, and associated somewhat poorly drained silt loams over sandy loam acid glacial till.*

F2. The rolling to undulating Santiago, Freer, Milaca, and Cable silt loam association.

F10. The undulating Santiago, Freeon, Freer, Milaca, and Cable silt loam, and peat, association.

The Santiago silt loam (Typic Glossoboralf; Figs. 12-3, 12-4, 12-5, 13-5) is well drained and acid throughout, and dominates these two associations, which are mapped in northwestern counties (Robinson et al., 1958). The rolling lands are much less extensive than the undulating areas. The associated Milaca (Typic Hapludalf) loam is also well drained but lacks the 15- to 36-inch silt covering over the glacial till which characterizes the Santiago toposequence. This sequence includes the moderately well drained Freeon silt loam (Typic Glossoboralf), the somewhat poorly drained Freer silt loam (Aeric Ochraqualf), the poorly drained Cable silt loam and loam (Typic Haplaquept), and very poorly drained peat (Fibrists and Hemists).

F3, F11, F20, F21. *Freer, Withee, and associated poorly drained silt loams, shallow over compact acid loamy glacial till.*

F3. The rolling to undulating Freeon, Freer, Almena, and Adolph silt loam association.

F11. The undulating Loyal, Withee, Arland, and Marshfield silt loam association.

F20. The nearly level Freer, Freeon, Almena, and Auburndale silt loam association.

F21. The nearly level Withee, Marshfield, and Adolph silt loam, and peat, association.

The silty soils of these soilscapes of north-central and northwestern counties have an impeded natural drainage to the extent that the somewhat poorly drained soils are dominant. These soils exhibit the pseudogley mottling in the A2g and A&Bg horizons (Kubiena, 1953; Van Rooyen, 1972). The substratum is acid, reddish-brown (5YR 4/4) sandy loam and loam glacial tills. This is represented by the three acid Aeric Glossaqualfs: Freer silt loam (in 15 to 30 inches of silt over sandy loam till), Withee silt loam (in 15 to 30 inches of silt over heavy loam till; Fig. 12-8), and Almena (in 36 to 50 inches of silt over sandy loam till). The slightly better drained Typic Glossoboralf, Freeon silt loam, is an associate of the Freer. There are three prominent poorly to very poorly drained soils in the soilscapes: the Marshfield, Auburndale, and Adolph. The Marshfield is a Typic Ochraqualf and the Auburndale is a Typic Glossaqualf, associated, respectively, with the Withee and the Almena soils. The Adolph and Mann series are Typic Haplaquolls (with a dark A horizon 10 to 18 inches thick) (Van Rooyen, 1972). In west-central Marathon County, over gray compact till, are found the Cassel (Aeric Glossaqualf) and Wien (Mollic Ochraqualf) silt loams. In northeastern Eau Claire County where the silt covering is discontinuous, some soils that have developed entirely in loam till are included with the Withee.

The topography of these soilscapes is largely inherited from the Precambrian bedrock surface that is only thinly mantled with glacial drift. Some stony drift was mixed by natural processes into the silty soil in presettlement times, possibly by tree-tipping and frost action. Occasional stone piles indicate that farmers have found an abundance of stones in the surface soil in places. Soil association F11 includes the "Marshfield Moraine" that extends eastward from Neillsville and consists of a drift-covered sandstone ridge (Figs. 10-7, 10-8).

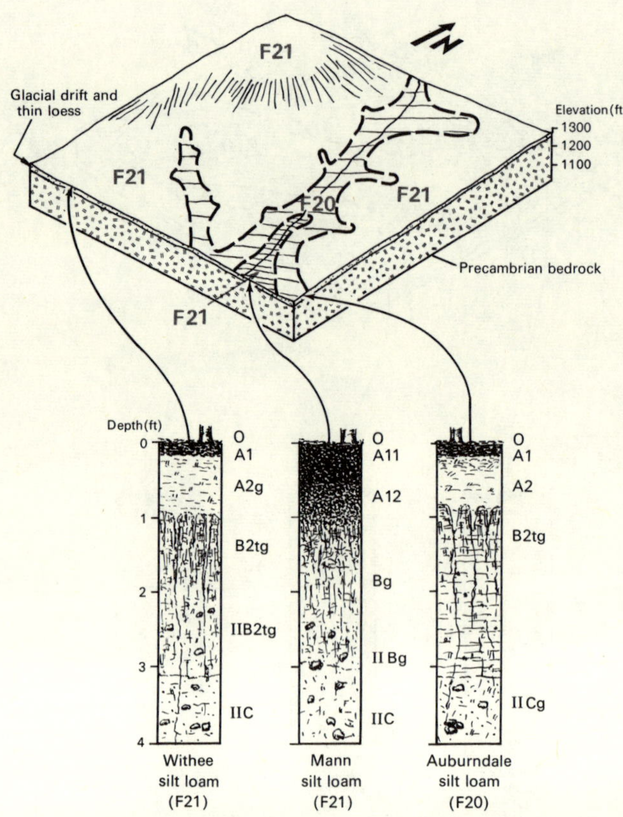

Figure 12-8. Block diagram showing landscape positions of representative soils of soil association F21 in 1 square mile of Wood County at the University of Wisconsin Agricultural Experiment Station (portions of Sections 15, 22, T.25N., R.3E.). The maximum slope gradient is 4%.

A low stream-divide that parallels Mill Creek Valley for 20 miles between Marshfield and Milladore is underlain by silty till containing carbonates (20% $CaCO_3$ equivalent) from an unknown source (Hole, 1943). Alfalfa production is unusually high in some fields in this area.

Early farmers harvested good crops from soils of Region F because they were well supplied with nutrients from the original forest humus. With the exhaustion of this reserve, soil productivity underwent a serious decline. This was finally remedied, beginning in the 1950s, by adequate applications of lime and fertilizers, which on the Withee silt loam have raised annual yields of alfalfa-brome hay from 1.6 tons (dry weight) per acre to as much as 3.75 tons (Love, Peterson, and Engelbert, 1960; Beatty et al., 1966).

F4. *The rolling to undulating Norrie silt loam, and Kennan and Onamia loams, and peat.*

This soil association is distributed through Langlade, Menominee, Oconto, and Marathon counties. The soils are well drained for the most part, and are underlain by acid sandy loam glacial till in the Norrie (Typic Glossoboralf) and Kennan

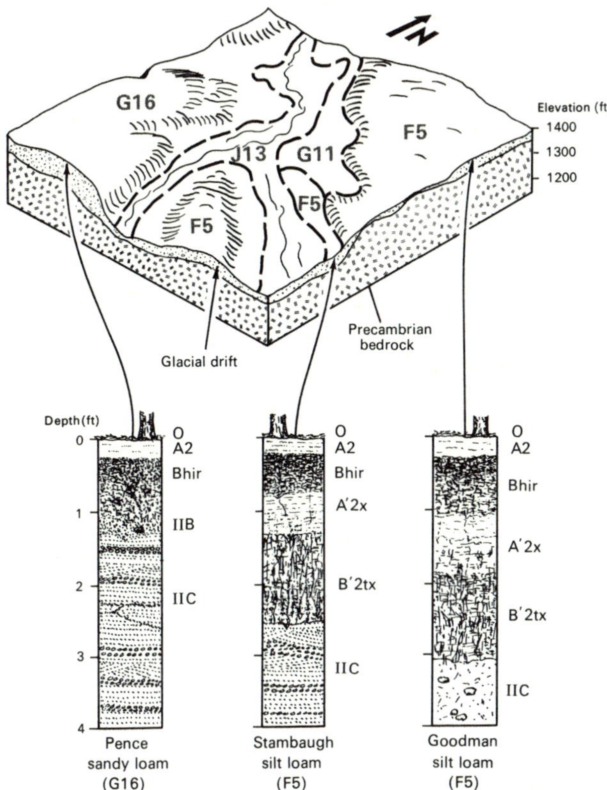

Figure 12-9. Block diagram showing landscape positions of representative soils of soil associations F5, G11, G16, and J13, Florence County.

(Typic Hapludalf) soils, and by acid glacial outwash sand and gravel in the Onamia loam (Typic Hapludalf). Associated depressions are occupied by acid mossy, sedgey, and woody peats. In the vicinity of Rib Mountain near Wausau, the Rozellville, Marathon, Rietbrock, and Mead soils are predominant (see F15 for descriptions of them). Rock outcrops and silty Lithosols are conspicuous on Rib Mountain, where loess has filled some cracks in the quartzite.

About half the area is in cropland and pasture and half in forest. Facilities for skiing, hiking, and picnicking are developed on Rib Mountain, where some steep and rocky terrain also serves as a wildlife refuge.

F5, F16, F24. *Stambaugh and associated silt loams over acid glacial outwash sand and gravel.*

- F5. The rolling to undulating Stambaugh and Goodman silt loam, and Padus and Iron River loams, association.
- F16. The undulating Stambaugh silt loam, Padus and Iron River loams, and peat, association.
- F24. The nearly level Stambaugh silt loam, and Padus, Pence, and Iron River loams, association.

Although the total acreage of the nearly level and undulating bodies of these soils is small, they provide productive, easily worked soil areas in the midst of rougher and usually stonier land. The dominant Stambaugh soil (Alfic Haplorthod) has formed in about 3 feet of silty covering over sand and gravel. It

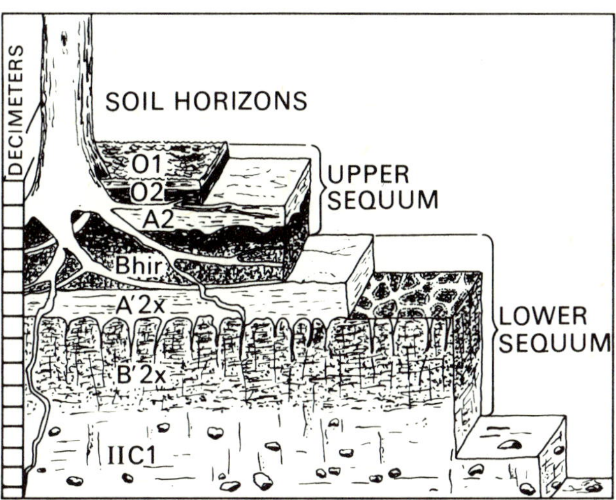

Figure 12-10. Cutaway block diagram of a bisequal soil, Goodman silt loam (Alfic Haplorthod), showing two fragipan horizons, with extensions of the upper one down a polygonal system of cracks into the lower, somewhat finer textured one.

is associated with soils that are progressively shallower and coarser (and hence more droughty) over sandy glacial drift: Padus loam (Alfic Haplorthod), Pence sandy loam (Typic and Entic Haplorthod), and Iron River sandy loams (Alfic Fragiorthod). In rolling landscapes, silt-capped acid sandy loam glacial till overlies the outwash in places. The Goodman silt loam (Alfic Haplorthod), less droughty than the Stambaugh soil, has formed in these materials (Figs. 12-9, 12-10). Both soils support excellent stands of northern hardwood forest.

Fragipans are probably most prominent in these soils in Florence County, where the pans commonly occur between the depths of 1 and 3 feet and impede water movement and root growth (Olson and Hole, 1967). In northern landscapes like this one, spodic and underlying fragipan horizons become better developed down-slope near wetlands, whereas in southern landscapes (B13; Fig. 2-52) planosolic-like A2 and Bt sequences (without fragipans) are prominent on the borders of wetlands.

F6, F17, F25, F26. *Antigo and associated silt loams over acid to neutral glacial outwash sand and gravel.*

- F6. The rolling to undulating Antigo and Norrie silt loam, and Onamia and Kennan loams, association.
- F17. The undulating Antigo silt loam, Onamia loam, and Brill and Poskin silt loam, association.
- F25. The nearly level Antigo and Brill silt loam, and Onamia loams, association.
- F26. The nearly level Poskin, Brill, and Antigo silt loam, and Onamia loams, association.

The Antigo silt loam (Typic Glossoboralf) is a productive soil that dominates these associations. The Antigo Flats (Fig. 13-3), a silt-blanketed outwash plain, has long been an outstanding farming area in the midst of less valuable, generally rolling, and hilly terrain. The availability of abundant water for irrigation enhances the productivity of the Antigo soil. By its expansion, the city of Antigo competes for the use of this soil. Associated with the well-drained Antigo soil are the excessively drained Onamia loam (Typic Hapludalf), the moderately well drained Brill silt loam,[2] also a Typic Glossoboralf, and the somewhat poorly drained Poskin silt loam (Aquic Glossoboralf). In rolling landscapes, silt-covered acid sandy loam glacial till may replace or overlie the outwash. Two soils have developed in these materials: Norrie silt loam (Typic Glossoboralf), where the silty covering is nearly 3 feet thick, and Kennan loam (Typic Hapludalf), where the silty layer is less than 20 inches thick.

Scattered from Polk County to Waushara County are associated small glacio-lacustrine plains on which the Campia (Typic Glossoboralf) catena has developed in calcareous silts (Ranney, 1966; Soil Conservation Service, 1967a). The clay bands mentioned by Weidman (1903) as being associated with Antigo soils may have been observed by him in the Campia silt loam. These soils are further discussed in Chapter 13.

Rolling Antigo soils are developed in highly pitted outwash, which is extensive in parts of Langlade and Barron counties (see Fig. 4 in Nygard and Hole, 1949). In southern Florence County, Fence silt loam (Alfic Haplorthod) has developed in deep lacustrine acid silts and very fine sands.

F7. *The rolling to undulating Cushing and associated Alstad and Bluffton loams on dolomitic loamy glacial till.*

The presence of a calcareous loam glacial till in the substratum of the Cushing silt loam (Glossic Eutroboralf) accounts in large part for the high agricultural productivity of the landscape as compared to that of the acid, more droughty Chetek sandy loam of the G8 soil association lying immediately to the east. The contrast is between an older Woodfordian glacial deposit derived from acid bedrock terrain to the east, and the young till of the Grantsburg glacial lobe that advanced over carbonate rocks in Minnesota. Soil associates in the F7 soilscape are the somewhat poorly drained Alstad silt loam and loam (Aquic Eutroboralf), also formed in 20 to 42 inches of deposit over the calcareous till, and the poorly drained Bluffton loam and silty clay loam (Mollic Haplaquept) in depressions.

F8, F13. *Jewett, Renova, and associated dark, well-drained silt loams developed under prairie cover on dolomitic glacial drift.*

F8. The rolling to undulating Jewett and Waukegan silt loam, and Dakota loams, association.

F13. The undulating Renova, Ostrander, Sargeant, and Floyd silt loam association.

In St. Croix and Pierce counties are bodies of soil developed in moderately deep (about 2 feet thick) silty material over acid loam to clay loam glacial till of early Woodfordian age, and associated glacial outwash. The dark, well-drained Waukegan (Pillot) silt loam (Typic Hapludoll) and the coarser textured Dakota loam (Typic Argiudoll) have formed under prairie vegetation over outwash. An equally dark, well-drained soil on the till uplands is the Ostrander silt loam (Typic Argiudoll), with which is associated the somewhat poorly drained Floyd silt loam (Aquic Hapludoll). A prairie-border soil (Mollic Hapludalf), intermediate in darkness between Ostrander and Renova, is the Racine soil, formerly called Jewett silt loam, which occurs in soil association F8, formed under a prairie called the Star Prairie by settlers. The Renova silt loam (Typic Hapludalf; Fig. 12-4) is well drained and is the forest soil equivalent of the Ostrander. Associated with these soils in Pierce County are the Wykoff (Typic Hapludalf), which overlies a less clayey acid till, and the Edith gravelly loam (Entic Hapludoll), shallow over coarse acid till. The poorly drained associate of the Renova is the Sargeant silt loam (Typic Glossaqualf). The soilscapes in which these soils occur are quite productive of crops, pasture, and hardwood timber. However, free lime, which occurs in abundance under the Cushing soil (F7), is scarcer. The glacial till is finer in texture than that under the Milaca, Cloquet, and Iron River soils of northwestern Wisconsin (G13).

F12, F22. *Almena and associated moderately deep, somewhat poorly drained silt loams developed over acid compact glacial till.*

F12. The undulating Spencer, Almena, Auburndale, and Adolph silt loam association.

F22. The nearly level Almena, Auburndale, and Spencer silt loam, and peat, association.

Near the Woodfordian glacial end moraine across north-central Wisconsin are the undulating and nearly level areas of Almena and associated soils that differ from the Withee soilscapes (F11, F21) in having a thicker silty soil (3 to 4 feet thick, rather than about 2 feet) and a somewhat more permeable till substratum. The agricultural productivity of these deeper silty soils is somewhat better than that of the corresponding shallower ones (F11, F21). The Almena silt loam (Aeric Glossaqualf) predominates and is associated with the moderately well drained Spencer silt loam (Typic Glossoboralf), poorly drained Auburndale silt loam (Typic Glossaqualf), very poorly drained Adolph silt loam (Typic Haplaquoll), and acid peats (Fibrists and Hemists). Landforming on farms and ditching in some wetlands and along roads has speeded up surface drainage of these soilscapes in wet seasons.

In Pierce County soils of F12 overlie sandy clay loam till like that under Renova soils (F13). Locally, as in some parts of northeastern Eau Claire County, the well-drained Otterholt (Typic Glossoboralf) and Seaton (Typic Hapludalf) soils are found together, the latter where the somewhat coarse silt and fine sand covering is thicker than 4 feet. Near the terminal moraine in Taylor County, depth of silty covering over till changes abruptly from place to place. Sand and gravel lenses at the contact between silty soil and till indicate that erosion and deposition by water occurred locally, just prior to deposition of loess. Field workers have noted that the subsoil in the Otterholt may show a variation in color which is interpreted not as an effect of impeded drainage, but as a result of tonguing of A2 horizon material into the Bt horizon. Almena soils in Dunn County are less mottled than is usual.

F14, F15, F23. *Fenwood, Rozellville, Dolph, and associated shallow silt loams over weathered Precambrian bedrock.*

F14. The undulating Fenwood, Marathon, Rozellville, and Cable silt loam association, with some stony areas.

2. Repeated irrigation of some potato fields on the Antigo Flats has resulted in mottling of the Antigo soil to a degree similar to that of natural mottling of the Brill soil.

Table 12-1. Classification of major soils of the F14, F15, and F23 soil associations of the Wausau area

Soil map number	Thickness of upper silty soil (in.)	Nature of the substratum[a] beneath the silty soil	Naturally well- to moderately well drained soils	Naturally somewhat poorly drained soils	Naturally poorly drained soils
F14	< 30	Decomposed course-grained granite[b]	Marathon silt loam, sandy loam (Typic Glossoboralf)	Mylrea silt loam, sandy loam (Aquic Glossoboralf)	Cable loam, silt loam (Typic Haplaquept)
	< 15	Fine-grained igneous and metamorphic rock	Fenwood silt loam (Typic Glossoboralf)	Rietbrock silt loam (Aquic Glossoboralf)	
F15	< 15	Micaceous residuum	Rozellville silt loam, sandy loam (Typic Glossoboralf)	Meadland silt loam, sandy loam (Aquic Glossoboralf)	
F23	≤30	Clayey residuum, with some talc and mica	Rudolph silt loam, loam (Typic Glossoboralf)	Dolph silt loam, loam (Aquic Glossoboralf)	Altdorf silt loam (Aeric Glossaqualf)

[a]The geologic map by Dutton and Bradley (1970) shows felsic rocks and undifferentiated mafic rocks in this part of Wisconsin.
[b]Syenite is extensive locally.

F15. The undulating Rozellville, Marathon, Rietbrock, and Mead stony silt loam association.

F23. The nearly level Dolph and Altdorf silt loam association, with some stony areas.

On the upland slopes leading down to the Wisconsin River, between Wausau and Wisconsin Rapids, Precambrian bedrock comes near the surface and residual materials from it have affected the soil profiles (Figs. 2-3, 2-4). Four kinds of bedrock-influenced materials are recognized in the classification of these soils (Table 12-1), the upper horizons of which formed from silty coverings. A shallow layer of till may lie between the silty layer and the decomposed bedrock material, as under bodies of deep Marathon soils. Steep slopes, stoniness, and wetness limit agricultural development of the soils in many places, but the well-drained, uneroded areas are productive under good management. The presence of well-drained to moderately well drained soils makes these areas quite different from adjacent areas of F21. These landscapes have unusual scenic qualities because of the relief in excess of 100 feet.

The rolling, well-drained Marathon soil is extensive in Marathon and Wood counties and grades westward into undulating, somewhat poorly drained Withee and associated soils developed in a silt covering over glacial till (Nelson, 1942; Hole, 1943).

The bedrock structure of gneisses and schists persists into the lower solum of some of these soils. The compact subsoil in the body of F23 of southwestern Langlade County was termed a *pan* as early as 1903 by Weidman.

CHAPTER **13**

Soil Region G: Soils of the Northern Loamy Uplands and Plains

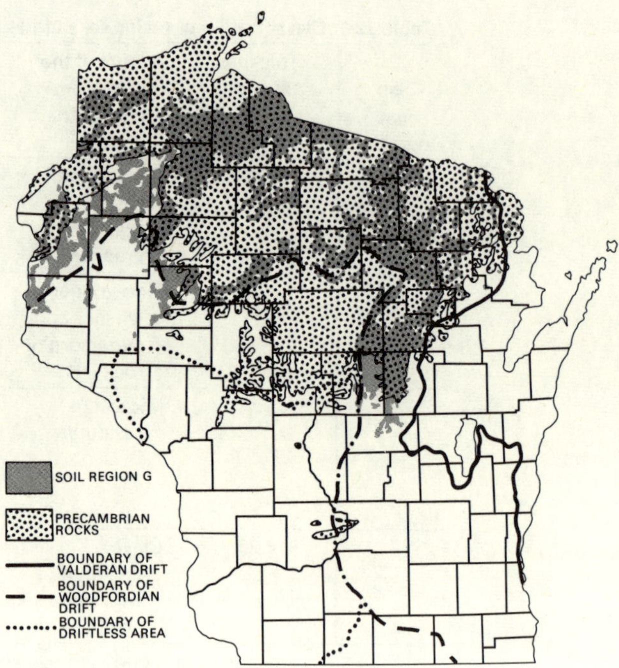

Figure 13-1. Index map showing the geographic relationship of Soil Region G to bodies of Precambrian bedrock and major glacial boundaries.

During the era of lumbering toward the end of the last century and for the first decade of the twentieth century, northern Wisconsin became a "cutover" land. Most of the virgin forest was felled. Numerous fires swept through the litter of trimmings and burned into unlogged stands in many places. The villages of Saxon and Upson in Iron County are examples of northern settlements that were once active lumber towns in this soil region. The removal of the forest that had clothed the drumlins, eskers, moraines, and pitted glacial till and outwash exposed the landforms to view. Much of the surficial rich organic layers of the upland forest soils was wasted away in the process. Menominee Indian lands, the area of the former Menominee County, still contain at the time of this writing the last extensive remnants of forests spared from clear cutting and of soils untouched by the plow. Even so, fires have taken their toll in this area during the last century, and present-day selective cutting creates some disturbance in both vegetation and soil.

Hilly morainic terrain in this region of nearly six million acres is in second- and third-growth timber and pastureland, for the most part. Complex (kame-kettle) slopes are common in rough moraines; simple slopes occur on drumlins and unpitted moraines. Rolling to level areas of farmland are commonly used for dairying and potato production programs. In the vicinity of lakes, these loamy soils are used for recreational activities and for wildlife protection. National, state, and county forests are concentrated on these soils.

The region lies almost entirely in the Northern Highland (Fig. 2-1) that is underlain by Precambrian bedrock (Fig. 13-1). As a result, the soil materials are typically acid, even to a depth of tens of feet (Fig. 2-5).[1] Although gritty silt loam is a common surface soil texture in the region (Fig. 2-28), the general absence of a distinct silty covering more than 10 inches thick distinguishes these soils from those of Soil Region F. Boulders are unevenly distributed on the land surface and in the soil throughout much of the region. Undoubtedly, significant amounts of windblown silt and clay have been trapped by the vegetation and incorporated into the loamy soil sola by physical and biological mixing processes since deglaciation. The stonelines that are occasionally apparent within or at the base of the solum (Jacobson, 1969) indicate that erosion by wind and water of fines from bare glacial drift has occurred. Parts of the landscape underwent deflation and erosion while other parts were sites of accumulation of sediment. Tree-tipping in storms has given about a fifth of the upland surface a small scale roughness (relief of about 3 feet) called *cradle-knoll* microtopography (Milfred, Olson, and Hole, 1967).[2]

The end moraine zones or belts have rolling and hilly portions (G1 through G8) amounting to about 17% of the region. Rolling (G9 through G19) and undulating upland (G20 through G26) occupy, each, a little more than a third of the region. The remaining 7% of the region is in scattered outwash flats.

About 75% of the area is occupied by naturally well- to excessively drained soils, 15% by peat, and the rest by poorly drained mineral soils.

These soils have developed under northern mesic forest and pine forest, with numerous patches of swamp conifers in kettles. At present about 65% of the area of the region is in woodland, about half of which is pastured (more in southern townships). About 20% of the land is in cropland that is harvested. This is vastly less than was anticipated in 1918 by authors of an agricultural bulletin entitled *Farm Making in Upper Wisconsin; Hints for the Settler* (Packer and Delwiche, 1918), and by Whitson, Dunnewald, and Thompson (1919), who wrote that

1. The presence of calcareous glacio-lacustrine silts in the Campia soil profile (see comment under F25) and carbonates (< 2%) in reddish till in end moraine in Taylor County (soil association G22; Hole, 1943) provides examples of local concentrations of calcareous material. Marl in bogs in northern Florence County (F5) is related to outcrops of Precambrian marble in adjacent uplands (Hole et al., 1962, p. 116).

2. The authors estimated the volume of soil and root mass tipped up by a white pine in Menominee lands to be 900 cubic feet, of which about 15% was root volume.

"one hundred thousand farms of eighty acres each are waiting the farmer in upper Wisconsin."

The dark brown Podzol B horizon (spodic horizon; Bhir) is noticeable in road cuts throughout most of the region. In northern counties (Fig. 2-40) a double soil profile is apparent, by color at least, in deep loams and sandy loams. The lower sequum may be expressed as a fragic A'2-B't horizon sequence (Figs. 12-9, 12-10). Where the fragipan is well developed, lateral seepage of water is notable, especially in wet seasons, and slump on roadbanks is a common result. Even in the absence of the fragipan, movement of percolating water through the soil to the water table is guided by stratification and irregular discontinuities in the glacial drift, as well as by joints. At the time of the spring thaw, the soils of the entire landscape are probably saturated, and perched and subsoil water tables merged.

The fresh, little-weathered condition of the feldspar and mafic minerals in the sand fractions (Hole and Schmude, 1959) in the reddish-brown till (Fig. 2-16) and associated outwash indicates that there is a vast reserve of plant nutrients in these soils. However, the supply of immediately available nutrients is not great, except where farmers have carried on a modern crop-fertilization program. A detailed study of the mineralogy of soils of the Antigo catena (Al-Rawi et al., 1969) indicated that some weathered clay (vermiculite, montmorillonite, chlorite, kaolinite, allophane) was an original, though sparse, component of the glacial drift, having been mixed, from some unspecified source, with the glacially comminuted Precambrian rock materials.

Mor (raw litter and humus mat) is characteristic of soils under hemlock stands; mull (mineral soil with incorporated humus) is found where sugar maple is prominent in the forest and where spring ephemerals like trillium and squirrel corn are abundant (Curtis, 1959; Milfred, Olson, and Hole, 1967).

The glacial till is commonly reddish brown (5YR 4/4) in color (Figs. 2-16, 2-20), but the distribution of tills of reddish (2.5YR 4/4) and brownish (10YR 4/4) colors is apparently without regular pattern. Near river valleys and lakes the glacial drift becomes extremely variable, with outwash, till, and lacustrine materials interspersed.

In northwestern Wisconsin reversal of topography occurs with respect to bodies of lacustrine sands and silts (Bevent soil of G4, G14, G16, and Campia soil of F25 and F26). Whereas Saylesville silt loam and similar soils on lacustrine deposits in landscapes of Jefferson County and adjacent southern areas (Region B) are in depressions, soils of lacustrine deposits in the Town of Birchwood in southeastern Washburn County are on local flat-topped elevations. These landforms stand like low, miniature mesas as much as 40 feet above the surrounding till-derived Milaca and Cloquet soils. The explanation is that stagnant masses of glacial ice persisted in these landscapes for centuries, serving as enclosing walls for lakes in which silty deposits accumulated. The surrounding moraine was covered with hills of ice at the time (Jacobson, 1969).

Spodosol profiles have become obscure in cultivated fields because of mixing of horizons by plowing, oxidation of organic matter in the spodic horizon (Bhir), and interruption, by removal of the forest, of the flow of dark colloids to replenish the Bhir horizon. Dark A1 horizons are being formed by farming operations, and at the same time the spodic horizons are fading. Each year more soil boundaries become recognizable at fence lines between cropland and forestland. Even without cultivation the Bhir horizon appears to have a half-life of 100 years after removal of hemlock (and other conifers) by cutting or natural blow-down and burning. Studies in the Menominee Indian lands further indicate that, with eventual establishment of maple forest on a former hemlock site, the Bhir horizon becomes undetectable by 400 years after the removal of the hemlock (Hole, 1975).

Variations in soil texture and depth of soil to water table in this region are reflected in crop yields. This is illustrated by estimates of annual yields of alfalfa-brome hay per acre (Beatty et al., 1966) that range from 2 to 4 tons (dry weight) under a high level of management, including adequate fertilization and drainage: 2.0, Cloquet, Marenisco, Vilas, and Omega loamy sands; 2.5, Cloquet, Pence, and Wyocena sandy loams; 2.75, Chetek, Cloquet, Dakota, and Onamia sandy loams; 3.0, Dakota, Gogebic, Padus, Pence, Onamia, and Stambaugh loams; 3.5, Ahmeek, Freer, Iron River, Kennan, and Milaca silt loams; 4.0, Bevent and Santiago silt loams and fine sandy loams.

In the legend of the soil map (Plate 1) the soil associations are listed in order from the hilliest to the nearly level areas. In the following discussion the units (Fig. 13-2) are regrouped with more emphasis on soil series than on topography.

G1, G9, G10, G20. *Gogebic and associated shallow silt loams over acid reddish glacial till, with some bedrock outcrops.*

- G1. The hilly to undulating Gogebic and Iron River loams association, stony with bedrock outcrops.
- G9. The rolling to undulating Gogebic sandy loam, Marenisco loamy sand, and Ahmeek loam, association, with bedrock outcrops.
- G10. The rolling to undulating Gogebic sandy loam, Marenisco loamy sand, Ahmeek and Cable loams, and peat, association.
- G20. The undulating Gogebic, Iron River, and Cloquet loams and sandy loam, Vilas sand, and Cable loams, association.

These hilly to undulating Orthods (Podzol soils) and Inceptisols (Low Humic Gley and Acid Brown Forest soils) are prominent on reddish-brown glacial drift in the upland of Douglas, Bayfield, Ashland, and Iron counties (Nygard, McMiller, and Hole, 1952). The Penokee-Gogebic ranges are characterized by the Gogebic (Alfic Fragiorthod; Fig. 14-5) and Ahmeek (Typic Fragiorthod) sandy loams, with fragipans, and by numerous rock outcrops (Fig. 2-3). The Ahmeek soils may have an unusually thick (5 inch) A1 horizon and a dense fragipan,[3] as in Florence County (Hole et al., 1962). Other Spodosols vary in surface texture from loam (Iron River soils, Alfic Fragiorthods) to loamy sand (Marenisco and Vilas soils, Typic and Entic Haplorthods; and Cloquet soils, Dystric Eutrochrepts). The wet soil associate is the Cable loam (Typic Haplaquept). The reddish color of the glacial drift (2.5-5YR 4/4, moist) derives

3. The fragipan resists excavation. It breaks out in slabs that are dark reddish brown (5YR 4/2, moist), vesicular, coarse platy, and coherent.

Figure 13-2. Sequence of maps showing distribution of soil associations in Soil Region G.

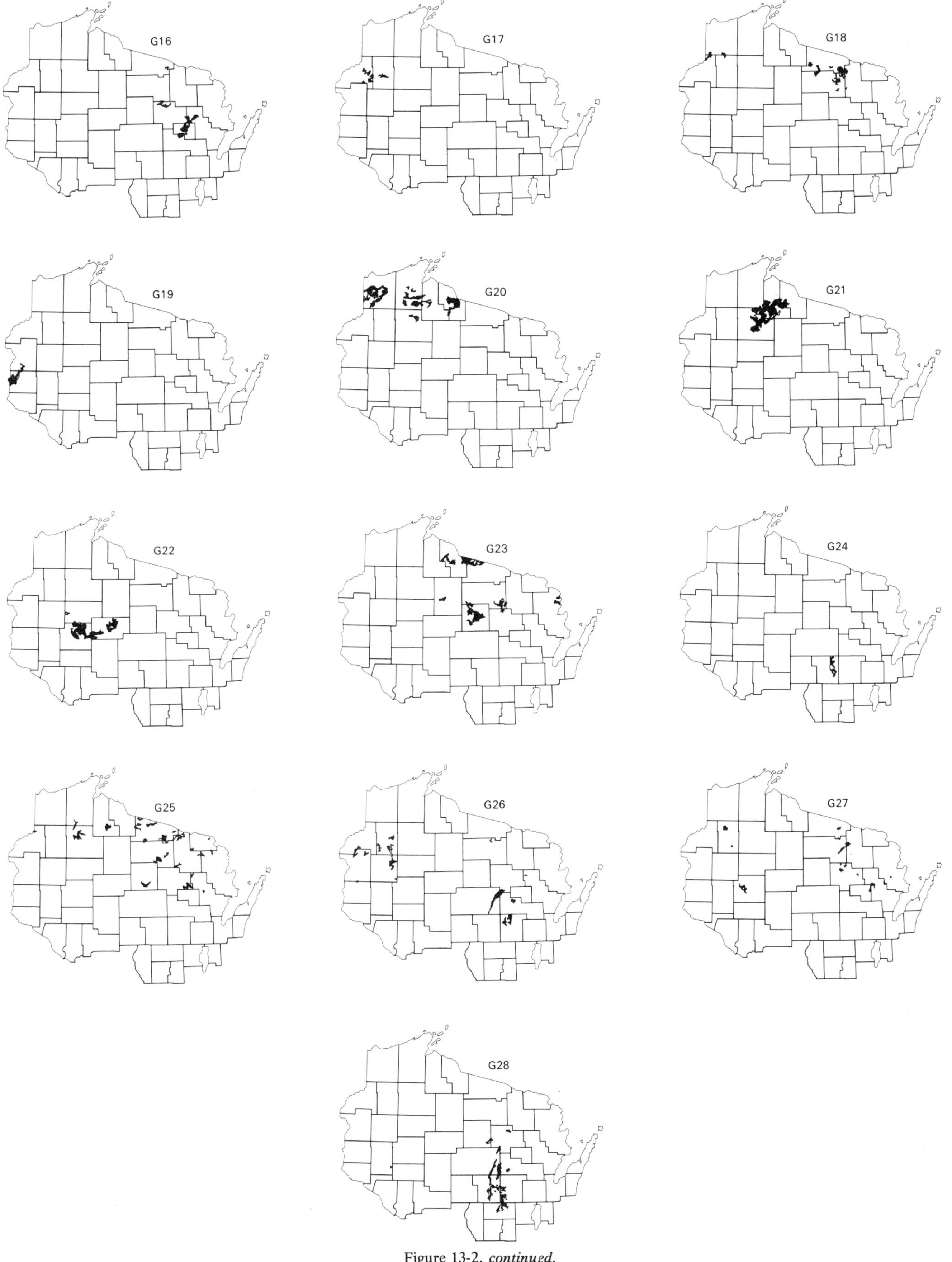

Figure 13-2, *continued.*

from the pulverized iron formations encountered by the glacier. Soils of the region are commonly stony. About 6 miles northwest of Hurley in Iron County are several small, nearly level areas of boulders and broken rock (Whitson et al., 1916). Where the underlying till has a loam texture, the soil is classified in the Wakefield series. The Munising soils are over sandy loam till and have prominent fragipans.

G2, G3, G11, G21, G23. *Iron River and associated loams over acid reddish sandy glacial till.*

- G2. The hilly to undulating Iron River and Pence loams, Goodman, Monico, and Stambaugh silt loam, and peat, association, with some stony areas.
- G3. The hilly to undulating Iron River and Pence loams, Vilas sand, and peat, association.
- G11. The rolling to undulating Iron River, Padus and Pence loams, Vilas sand, and peat, association.
- G21. The undulating Iron River, Gogebic, and Cable loams, and peat, association.
- G23. The undulating Iron River, Gogebic, and Monico loams, Marenisco loamy sand, and peat, association.

The Iron River loam (Alfic Fragiorthod; Fig. 13-3) is the prominant soil in these hilly to undulating soilscapes, with acid dark brown (7.5YR 4/4, moist), stony, coarse glacial till substratum containing dark rock material from iron formations and basalts. The Monico loam (Aquic Dystrochrept) is the somewhat poorly drained soil associate. Other common soils are the Pence (Typic and Entic Haplorthods) and Padus (Alfic Haplorthod) loams developed from loamy coverings over acid outwash sand and gravel. The loamy solum of the Pence is not thick enough over outwash sand and gravel to provide for a double profile such as is observed in the deeper Padus loam. The Gogebic, Goodman, Vilas, and Cable soils are described elsewhere in this report. Bodies of peat are numerous.

Whitson et al. (1916) reported a red clay subsoil layer exposed in railroad cuts in eastern Vilas County. Thwaites (1929) mapped the Winegar terminal moraine in northern Vilas County and noted the color changes in the upper part of the till produced by the bleaching effect of podzolization.

G4, G13, G22. *Milaca and associated loams over acid reddish glacial till.*

- G4. The hilly to undulating Milaca loam, Cloquet and Pence sandy loam, Vilas sand, and peat, association.
- G13. The rolling to undulating Milaca, Cloquet, Iron River, and Cable loams, and peat, association.
- G22. The undulating Milaca and Cloquet loams, Santiago, Freer, and Cable silt loam, and peat, association.

The Milaca (Amery) loam (Typic Fragiochrept; Figs. 13-4, 13-5)[4] and the coarser Cloquet loam (Dystric Eutrochrept) formed in glacial drift that is reddish brown (5YR 4/3, moist) in color and is usually acid, though it may be calcareous locally at a depth of 100 inches or so. The fragipan of the Milaca is usually more than a foot thick. It has been suggested that the dense, platy glacial till (the C horizon) was compacted by the weight of the glacier before deglaciation. Bedrock outcrops are

4. Formerly classified as a Typic Hapludalf (Gray-Brown Podzolic).

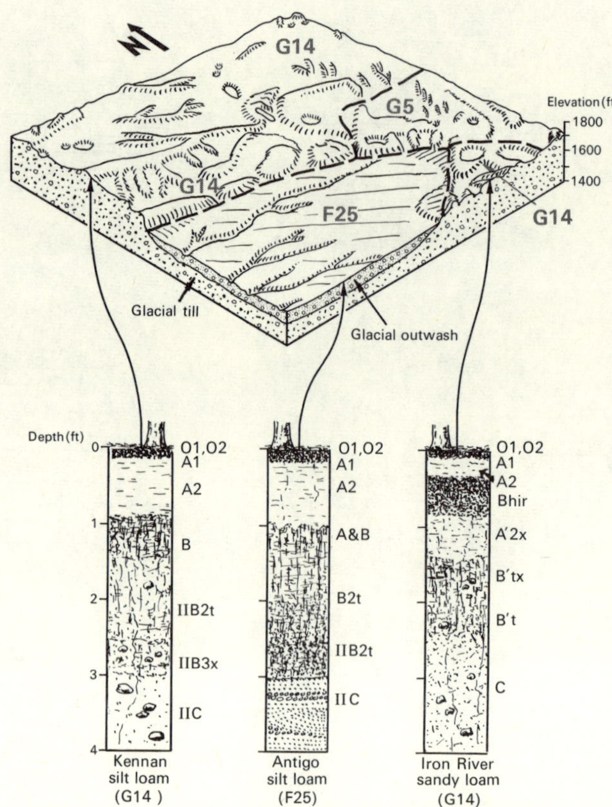

Figure 13-3. Block diagram showing landscape positions of major soils of T.32N., R.12E., Langlade County.

common among these soils in parts of Rusk and Barron counties. In Barron County the Milaca soil commonly has an 8-inch gritty silt loam A horizon. The other associated soils are described elsewhere in this publication. The landscape is characteristically spotted with kettles containing small peat bogs bordered in places by Cable loams (Typic Haplaquepts). Eskers and kames are numerous in parts of Taylor County. Cradle-knoll or tree-throw microrelief is common.

The southernmost body of soil association G13, on the Dunn-St. Croix County line, is underlain to the north by reddish-brown glacial till characteristic of Milaca soils, in the central part by thin till over sandstone, and in the south by both reddish-brown sandy loam and grayish-brown clay loam glacial tills. Recent detailed soil mapping recognizes little Milaca here, but considerable Santiago, Freer, Arland (with Hixton, Gale, and Boone), with some Otterholt and Renova. Huntsville and Lawson, Cumulic Hapludolls (Alluvial soils), occupy narrow valley bottoms (personal communication, John E. Langton, 1971).

Reddish-brown till contains as much as 2% calcium carbonate (equivalent basis) in parts of north-central Taylor County (soil association G22; Hole, 1943).

G5, G6, G14, G15, G24. *Kennan and associated loams over brown sandy glacial till.*

- G5. The hilly to undulating Kennan, Iron River, and Pence loams, Vilas sand, and peat, association.

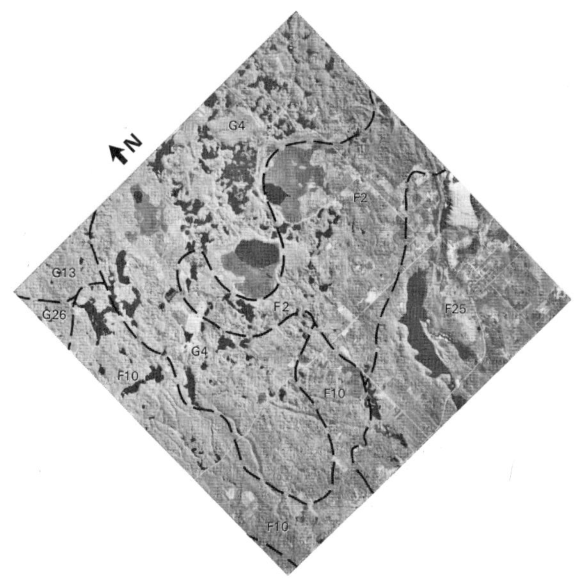

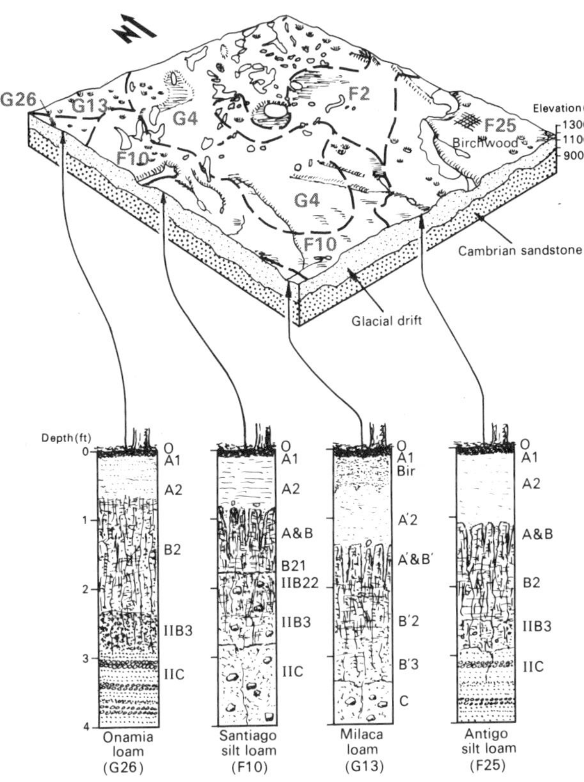

Figure 13-4. Aerial photo map of T.37N., R.10W., Washburn County. The area shown is 6 miles on a side.

Figure 13-5. Block diagram showing landscape positions of major soils of T.37N., R.10W., Washburn County.

G6. The hilly to undulating Kennan, Wyocena, and Onamia loams association, with some stony areas.

G14. The rolling to undulating Kennan, Iron River, and Pence loams, and peat, association.[5]

G15. The rolling to undulating Kennan, Wyocena, Onamia, and Bevent sandy loam, and peat, association.

G24. The undulating Kennan, Wyocena, and Onamia loams association.

Kennan loam (Typic Hapludalf) is an extensive soil in the end moraine upland of north-central Wisconsin, formed in glacial till that is brown (10YR-7.5YR 4/4) rather than reddish (see previous Gogebic and Milaca soilscape descriptions). Common associates are the coarser Iron River loams (Alfic Fragiorthods; Fig. 13-3) and farther south, the Wyocena loam (Typic Hapludalf). Pence and Onamia loams (described elsewhere) are formed in coverings over acid outwash sand and gravel. Bevent (Alfic Haplorthod) is in acid sandy and coarse silty sediments of glacial lake origin. Vilas sand (Entic Haplorthod) is inextensive. Peat bogs are numerous.

G7, G16. *Padus and associated loams over acid glacial outwash sand and gravel.*

G7. The hilly to undulating Padus and Pence loams, Vilas sand, and peat, association.

G16. The rolling to undulating Padus and Pence loams, Omega sand, Stambaugh silt loam, and peat, association.

Padus (Alfic Halporthod) and Pence (Typic and Entic Haplorthods) soils are, respectively, deep loams (about 3 feet) and shallow sandy loams (2 feet) underlain by acid sand and gravel outwash. Where the outwash has no loamy covering, Vilas (Entic Haplorthod) and Omega (Entic Haplorthod) soils occur. Patches of silty covering are sites of Stambaugh silt loam (Alfic Haplorthod). Peat bogs are numerous in kettle depressions.

G8. *The hilly to rolling Chetek and associated Scandia sandy loam and Omega sand, over acid glacial outwash, association, with peat.*

The mineral soils of this group are derived from acid glacial outwash having local silty and loamy coverings. Organic soils are found in numerous peat bogs. The moderately deep (3 feet) Scandia (Typic Hapludalf) and shallow (20 inches) Chetek (Eutric Glossoboralf) soils have weakly developed subsoil (Bt) horizons. Omega loamy sand (Entic Haplorthod) is found where the outwash sand is without a loamy covering.

G12. *The rolling to undulating Cloquet, Gogebic, and Pence loams association, with peat.*

This soilscape contains a mixture of soils over reddish acid glacial outwash and till. Included are the shallow Pence sandy loams (Typic and Entic Haplorthods), the moderately deep Cloquet loamy sand (Dystric Eutrochrept), and the moderately deep (3 feet) Gogebic sandy loam (Typic Fragiorthod). The

5. On the soil map, Plate 1, an unlabeled area of G14 is located in T.29N., R.14E.

glacial till under the Gogebic soils is high in content of upper Keweenawan sandstone. Fragipans are distinct in these till-derived soils. Peat bogs occupy the numerous depressions.

G17, G18, G25, G27. *Pence and associated shallow sandy loams over acid outwash sand and gravel.*

G17. The rolling to undulating Pence and Cloquet sandy loam, Stambaugh silt loam, and peat, association.

G18. The rolling to undulating Pence sandy loam, Vilas sand, and peat, association.

G25. The undulating Pence and Padus loams, Stambaugh silt loam, and Vilas and Omega sand, association.

G27. The nearly level Pence sandy loam, Vilas sand, Stambaugh and Padus loams, and peat, association.

The dominant soil in these associations is the shallow (20 inches) Pence sandy loam (Typic and Entic Haplorthods) formed over deep acid glacial outwash sand and gravel. The Cloquet sandy loam (Dystric Eutrochrept) is only slightly deeper. Where no loamy covering is present on the outwash sand, Vilas loamy sand (Entic Haplorthod) has formed. Padus loam and Stambaugh silt loam (both Alfic Haplorthods) are about 3 feet deep over coarse sediments. Peat bogs occupy depressions in the landscape.

G19, G26, G28. *Onamia and associated loams over neutral to acid glacial outwash sand and gravel.*

G19. The rolling to undulating Onamia, Chetek, and Dakota loams association.

G26. The undulating Onamia and Chetek loams and sandy loam, Antigo silt loam, and peat, association.

G28. The nearly level Onamia and Chetek sandy loam, Antigo silt loam, and peat, association.

The Onamia loam (Eutric Glossoboralf; Figs. 13-4, 13-5) dominates the soilscape and is about 3 feet deep over well-drained acid glacial outwash sand and gravel. Associated is the prairie equivalent, the Dakota loam (Typic Argiudoll) and the shallower (20 inches) forest soil, the Chetek loam (Eutric Glossoboralf). Where 3 feet of silty material overlies the outwash, Antigo soils (Typic Glossoboralfs) have formed. Peat bogs occupy depressions.

Associated with the Onamia and Chetek soils are Inceptisols and Entisols that differ from them in lacking the argillic horizon.

The small body of G28 near Elk Mound in Dunn County includes Onamia and related soils on some deposits resembling kames and eskers, probably of an early Woodfordian glacial advance.

CHAPTER **14**

Soil Region H: Soils of the Northern Sandy Uplands and Plains

Many of the features of Soil Region C are also present in Soil Region H. The soilscapes in both regions are characterized by extremes in temperature, quick growth of vegetation in the spring of the year, dessication during summer droughts, and an abundance of sites near lakes for summer homes and tourist facilities. However, the climate is relatively cooler and the growing season rather shorter in Region H (Figs. 2-33, 2-34, 2-35). The soils reflect these conditions by the presence of a stronger brown B horizon that, in many places, qualifies as a spodic (Podzol) horizon. The vegetation and its acid residues have favored this kind of soil formation.

The northern sandy areas are pine barrens, aspen forests, and bracken-grasslands, with unforgettable fragrances, such as that of sweet fern on a hot summer day. These kinds of vegetation are characteristic of plant successions following fire and consequent loss of litter and humus (O horizons) (Curtis, 1959; Chamberlin, 1883). Some remnants still exist of the once extensive white pine forests (Plate 6; Milfred, Olson, and Hole, 1967). The droughtiness of the soils makes the vegetative cover vulnerable to kindling by lightning. Man-induced fire was common in slash after lumbering operations. Prior to that, some burning may have been practiced by Indians to improve the berry crop and to increase suitable browse for deer. Production in the soil of growth-inhibiting compounds by herbaceous vegetation may be a factor in preventing reinvasion of "stump prairies" by forest.

These very sandy lands of northern Wisconsin occupy about two and a half million acres divided into three major areas and more than a half-dozen smaller ones (Figs. 14-1, 14-2). First is an interlobate morainic area that extends from near the end of the Bayfield Peninsula to Polk County, and is characterized by kettles, ridges, and hills, among which are interspersed pitted outwash plains (Ableiter and Hole, 1961). The second body of sandy soils is in Oneida and Vilas counties, where stagnant ice features are numerous, including crevasse fillings and eskers in both simple and complex patterns, along with kames and kettles (Hole and Schmude, 1959). On the basis of topography and concentration of boulders, Thwaites (1929) mapped some drumlins and "terminal moraines" near Trout Lake in Vilas County. The third major area is the broad band of sandy soils from Florence, through Marinette into Oconto counties. The landforms are similar to those of the other two major areas (Hole et al., 1962). Other patches of sand are scattered from Barron County to Oconto County and include soils on outwash terraces in the Wisconsin River valley in the vicinity of Wausau. Lakes and peat bogs occupy many of the kettles in these landscapes.[1] In places, calcareous lacustrine clays of Region I border on or are inclusions in the sandy soil associations. Stratification of sand with silts and clays may occur in the substratum, but over large areas the sands seem to be many tens of feet deep with only minor variations in texture and mineralogy.

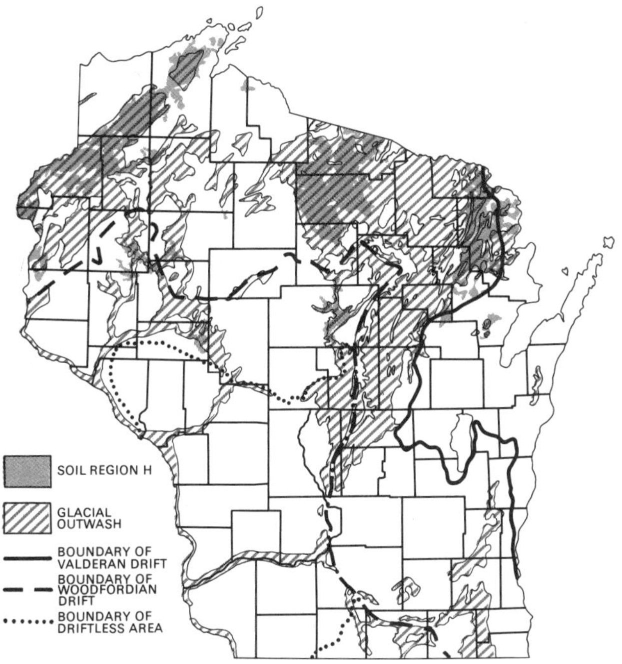

Figure 14-1. Index map showing the geographic relationship of Soil Region H to bodies of glacial outwash and major glacial boundaries.

These soils resemble those of Region G in their low content of available potassium, but differ from them in the great variability in content of available phosphorus (Fig. 14-3), due chiefly to the irregular distribution of the mineral apatite in the sands (Beatty and Corey, 1961, 1962). A mineralogical analysis of the sands and coarse silts of Hiawatha and Vilas sands from Oneida County (Hole and Schmude, 1959) showed about 89% quartz, 5% feldspar, 4% quartzite, and 2% heavy minerals (garnet, tourmaline, hornblende, kyanite, sillimanite, andalusite, staurolite, leucoxene, hematite, magnetite, ilmentite, with small amounts of zircon and apatite) by weight. Microscopic examination reveals little evidence of weathering of the sands, but shows coatings of iron oxide and clay. Madison and Lee (1965) reported as much as 25% feldspar and 2% heavy minerals in sands of northern Wisconsin.

The sandy barrens of this region are low in productivity, not because of lack of total reserves of plant nutrients, but because of limited waterholding and available nutrient exchange capacities, and periodic loss of soil humus by uncontrolled burning in times past.

Pine barrens, with scattered jack pine, Hill's oak, and prairie vegetation, and pine forest of jack pine, red pine, and white

1. R. F. Black has suggested (personal communication, 1969) that large kettles may have been formed by early Woodfordian ice blocks (about 28,000 years old) that did not melt away until post-Cary time.

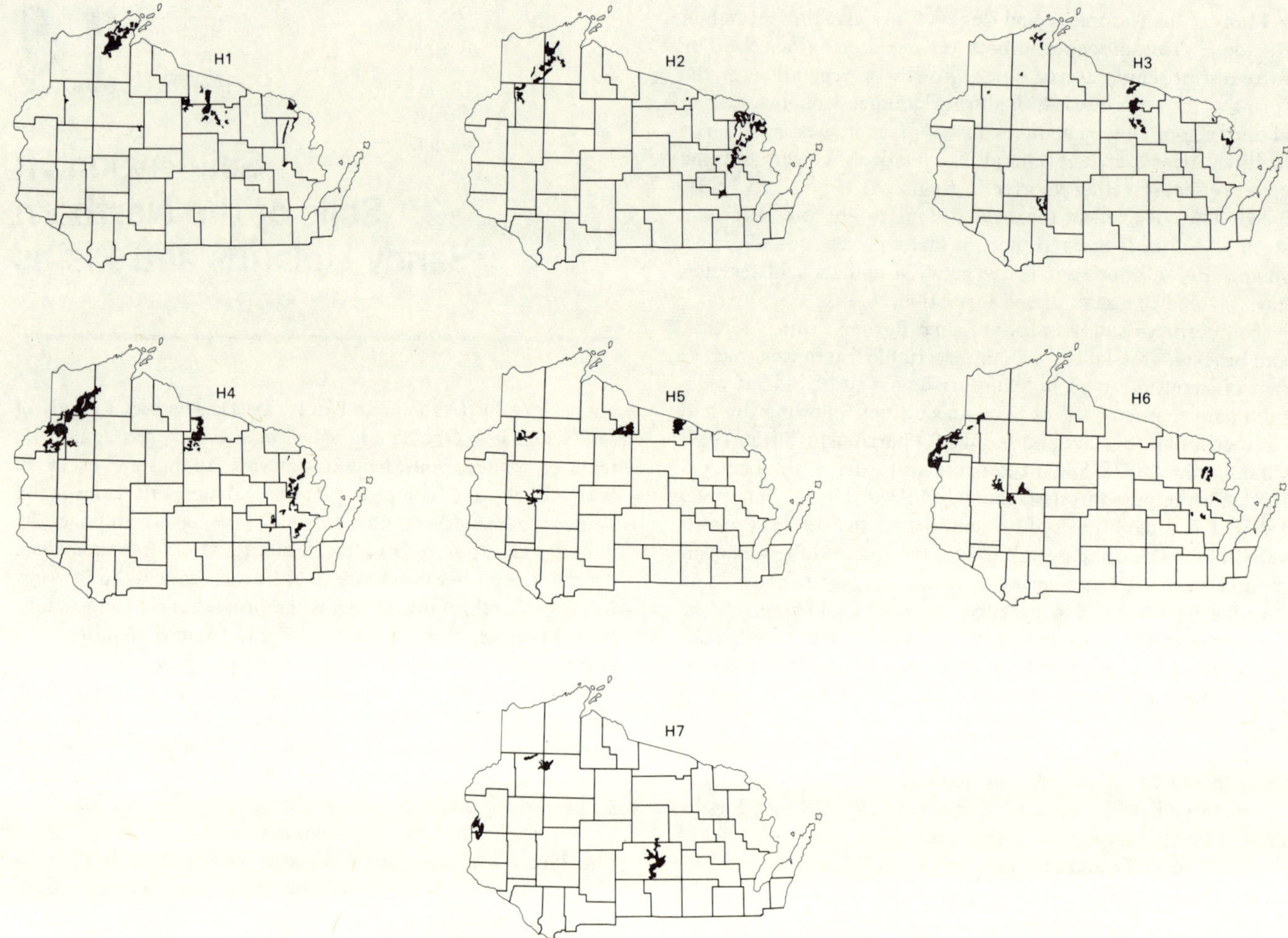

Figure 14-2. Sequence of maps showing distribution of soil associations in Soil Region H.

pine are the principal plant communities native to the upland soils. Black spruce, tamarack, and white cedar predominate in the wetlands, along with sphagnum moss, leatherleaf, and labrador tea. A vegetative sequence on upland soils from pine forest on hill slopes to sedge grassland in frost pockets has been described by Curtis (1959), but no detailed study has been reported of the corresponding soil sequence.

The region is about equally divided among hilly, rolling, and undulating landscapes. In general, the soils at lower upland positions have maximum development of Spodosol (Podzol) horizons, while soils on drier slopes and hill crests have minimal spodic horizons. This may not only be a result of differences in moisture regime and quantities of organic litter produced per acre, but may also reflect greater disturbances connected with logging and burning on drier sites. Spodosol formation is relatively rapid in sands and is favored by accumulation of acid litter and humus (O1, O2 horizons) on the soil surface, bleaching of the A2 horizon by movement of organic compounds from the organic layers into the mineral soil, and subsequent deposition of organic matter and iron compounds in the Bhir horizon.

Most of the soils of the region are used for forestry, recreation, and wildlife, but there is a little grazing and cropping. Cranberry production is developed locally in organic soils and sands. Gaikawad and Hole (1961) compared properties of a maximum Spodosol, Wallace loamy sand (Fig. 1-7; formerly called Au Train) under native forest with those of the same soil under a corn crop in Florence County. They noted an increase in fertility in the cultivated field. Production of red and white pine in board feet per acre per year is estimated to be from 400 to 500 on Vilas, Omega, and Hiawatha soils.

In those few places where plant communities have been stable, sandy soils have developed distinct horizons in relatively short periods of time, pedologically speaking. Milfred, Olson, and Hole (1967) report a well-developed Spodosol (Podzol), Vilas loamy sand (Typic Haplorthod), in hemlock-white pine stand number 35 in Menominee lands. They gave this soil a very high podzolization rating, expressed on a special pedogenetic numerical scale. Formation of major soil horizons in sandy material takes place in terms of centuries, compared with the thousands of years required for the argillic B horizon (Bt) to develop in silty soils (Franzmeier and Whiteside, 1963; Arnold and Riecken, 1964). This is not surprising since the sands, with only about one hundredth of the particle surface area of the silty soils, can more quickly develop visible coating on the sand grains in the Bhir horizon. Water movement

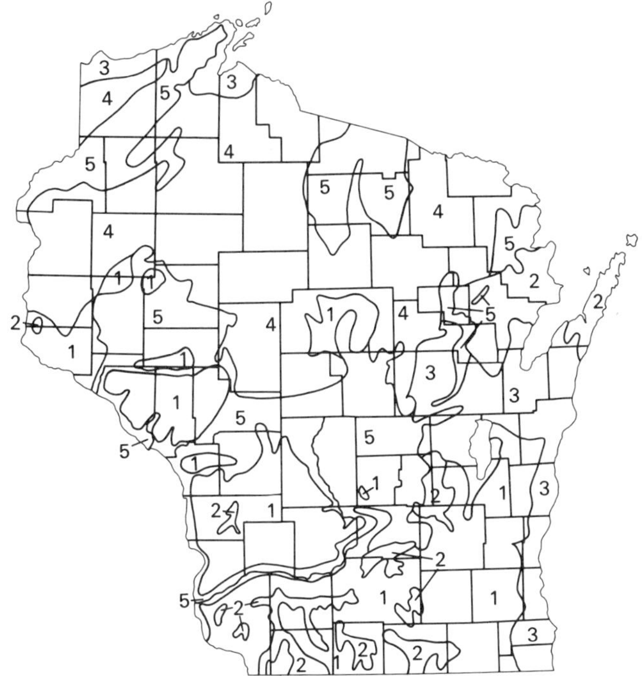

Group 1 - Forest soils; subsoil phosphorus mostly available to plants

Group 2 - Prairie soils; organic phosphorus accumulated in subsoils is relatively unavailable to plants

Group 3 - Soils low in content of available phosphorus but higher in available potassium in subsoil than most soils of Wisconsin

Group 4 - Soils with medium content in subsoil of phosphorus, but low in content of potassium, available to plants

Group 5 - Soils unusually variable in content of available phosphorus, and low in content of available potassium

Figure 14-3. Subsoil fertility provinces of Wisconsin (after Beatty and Corey, 1962).

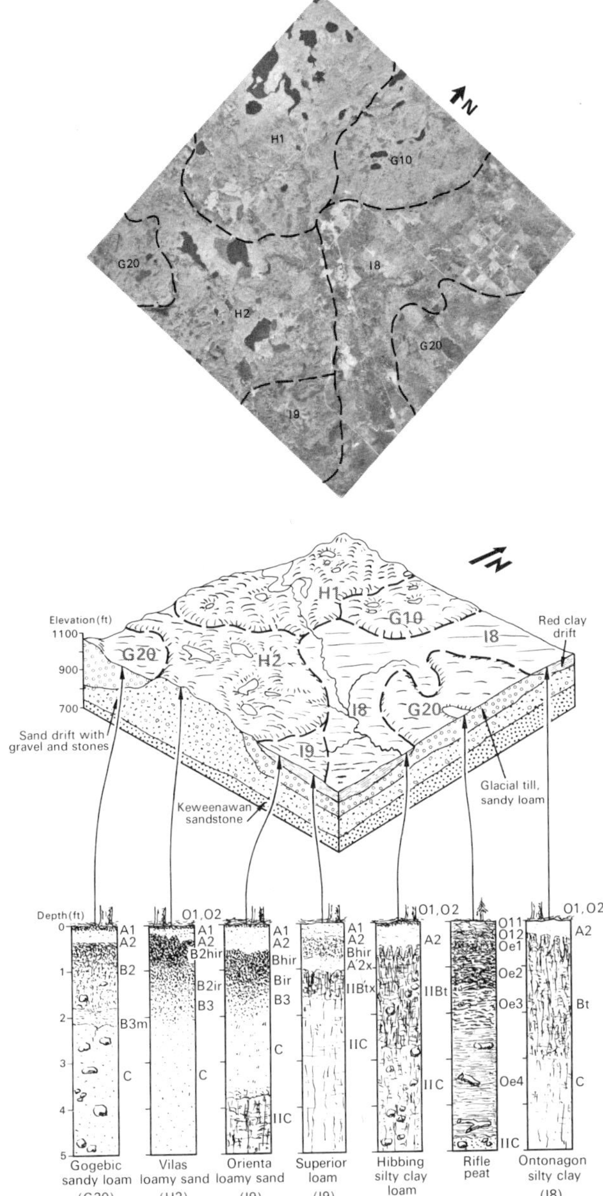

Figure 14-4. Aerial photo map of T.46N., R.7W., Bayfield County. The area shown is 6 miles on a side.

Figure 14-5. Block diagram showing landscape positions of major soils of T.46N., R.7W., Bayfield County.

during a heavy rain in summer has been observed to follow preferentially down tongues of A2 and Bhir horizons in sandy Spodosols. This suggests that, once started, irregularities in spodic horizons may be accentuated by differential movement of percolating water.

Soil scientists have devoted a great deal of attention to sandy soils, a fact upon which Aldo Leopold commented when he wrote (1949) that "soil experts . . . would have a hard life without the Sand Counties. Where else would their podzols, gleys and anaerobics find a living?"

These soil associations (Fig. 14-2) are arranged in the soil map legend (Plate 1) in order from hilly to level terrain. In the following discussion they are regrouped on the basis of soil series.

H1, H2, H3, H4, H5, H6. *Vilas, Omega, and associated sandy soils over acid glacial drift sand.*

H1. The hilly to rolling Vilas, Omega, and Hiawatha loamy sand and sand, Pence sandy loam, and peat, association.

H2. The rolling Vilas, Omega, and Hiawatha loamy sand and sand, and peat, association.

H3. The undulating Vilas, Omega, and Hiawatha loamy sand and sand, Pence sandy loam, and peat, association.

H4. The undulating Omega and Vilas loamy sand and sand, Pence sandy loam, and peat, association.

H5. The nearly level Vilas and Omega loamy sand and sand, Pence sandy loam, and peat, association.

H6. The nearly level Omega and Vilas loamy sand and sand, Chetek and Pence sandy loam, and peat, association.

The Vilas (Entic Haplorthod; Figs. 14-4, 14-5) and Omega (Entic Haplorthod; formerly called Brown Podzolic; Nygard,

McMiller, and Hole, 1952) soils dominate the more than two million acres of these soilscapes. Two thirds of the area is rolling to undulating. The remainder is about equally divided between hilly and nearly level lands. The Vilas soil represents the more stable and less droughty land surface, on which forest growth has been uninterrupted enough to yield distinct Spodosol horizons. The more extensive Omega soil records influence of scattered trees and of a variety of prairie and sedge-grassland plants, as well as disturbance by fire, erosion by wind and water, and cultivation during periods of human occupance. The presence of numerous kettles, ridges, stabilized sand dunes, and outwash and lacustrine plains makes the landscapes complex topographically, floristically, hydrologically, and pedologically. Some Hiawatha soil (Typic Haplorthod) has formed at sites where banding of the sandy substratum with small amounts of silt and clay has favored forest growth.

Pence (Entic Haplorthod) and Chetek (Eutric Glossoboralf) soils, although shallow and droughty, are more loamy than the sands already named. Peat bogs under sphagnum moss, black spruce, and tamarack are numerous. Beaver have been responsible for the creation of some small wetlands (Hole et al., 1962). Red clay deposits have been observed under sands along the St. Croix River near the mouth of Wolf Creek in Polk County (H6) (Weidman, 1914). In northeastern Eau Claire County, Roscommon loamy sand (Mollic Psammaquent) is associated with Vilas sand in soil association H3.

H7. *The nearly level Pence sandy loam and Omega and Au Gres sands association over acid glacial outwash sand and gravel.*

The dominant Pence sandy loam (Entic and Typic Haplorthod) is 10 to 20 inches deep over acid sand and gravel outwash (Fig. 2-5). The Omega loamy sand (Entic Haplorthod) is droughty and the Au Gres loamy sand (Entic Haplaquod) has a high water table. This soil association is shown in four widely separated bodies in northern counties (Fig. 14-2).

CHAPTER **15**

Soil Region I: Soils of the Northern and Eastern Clayey and Loamy Reddish Drift Uplands and Plains

A few miles northeast of the city of Green Bay is a historic site known for more than two centuries as Red Banks. The red clay materials exposed in the vicinity by wave action are by no means confined to the shorelands. The reddish clayey soils of Soil Region I cover about 2.6 million acres, or more than 7% of the state, distributed throughout fifteen counties. Soil Region I also includes some patches of gray upland clay soils in Burnett County.

The soils of this region are largely concentrated in two subregions (Figs. 15-1, 15-2), one bordering the south shore of Lake Superior and the other adjacent to Green Bay, Lake Winnebago, and Lake Michigan, north of Milwaukee. Small bodies of clayey soils are notable in other places, namely in Florence County and in an area in Adams County and western Marquette County, as well as in Burnett County. Most bodies of these soils fall within the boundaries of the Valderan glacial ice advance and of extinct glacial lakes as delineated by Thwaites (Flint, 1945; also see Black, 1970b) (Fig. 15-1).

About two thirds of the region is less sloping than 6% gradient and includes almost equal areas of nearly level undulating till plains and bottom lands of extinct glacial lakes.

Under many feet of glacial drift are Keweenawan sedimentary and volcanic formations in the Lake Superior area, and Paleozoic sedimentary formations in southern and eastern areas. The bedrock has contributed locally to the soil initial materials via the drift.

This soil region crosses a variety of major plant communities (Plate 6): boreal forest, northern mesic forest, pine forest, southern mesic forest, southern oak forest, and oak savanna. The soil spectrum ranges from the dark Mollic Hapludalfs and even some Argiudolls, under local prairies (Horn, 1957), to Alfic Haplorthods where a thin sand or silt smear is present at the surface (Fig. 14-5). Much work remains to be done to correlate variations in these soil profiles with original and relict biotic communities.

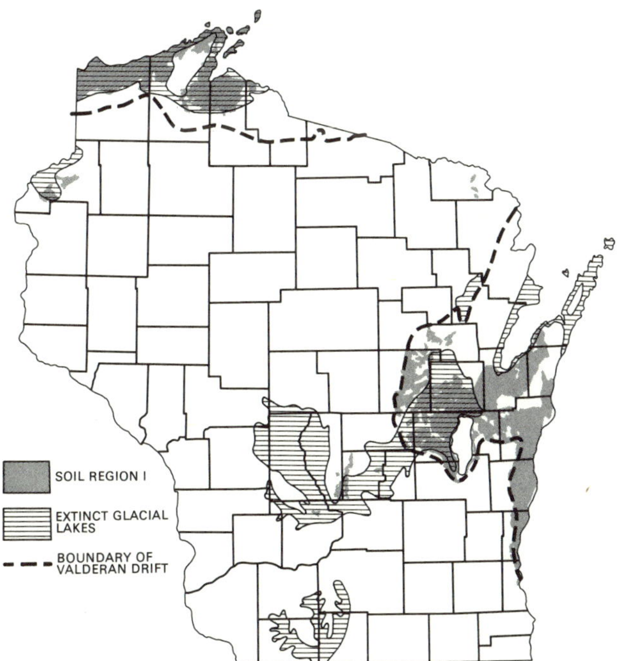

Figure 15-1. Index map showing the geographic relationship of Soil Region I (black areas) to Valderan glacial deposits and boundary.

In the eastern subregion about 60% of the landscape is cropland, 15% pastureland, and 25% in woodland and wetland (see Chapter 16, Soil Region J). Adjacent to Lake Superior, the corresponding percentages are approximately 35, 20, and 45. Cultivation entails autumn plowing, which leaves the soil exposed to erosion for considerable periods, but promotes improved structure or tilth in the plow layer.

The red silt and clay (Figs. 2-12, 2-15, 2-19) are colored by a glacially pulverized Precambrian iron formation which was distributed widely by former high-standing glacial lakes (Flint, 1945; Alden, 1918b) that bordered wasting and retreating glaciers thousands of years ago. The content of iron oxide (Fe_2O_3) is about 3% in the coarse and medium clay (Petersen, 1965; Petersen, Lee, and Chesters, 1968). Iron concretions are present in the C horizon of the Oshkosh soil. Mixed with this red sediment in the C horizons are silts consisting of quartz and feldspar and limestone-derived carbonates. Soils in the southern subregion, where dolomite bedrock is extensive, contain more carbonates than those to the north do (Fig. 15-3; Table 15-1) (Janke, 1962). The noncarbonate material is quite uniform in composition in the C horizon in both major subregions. The noncarbonate clay fraction is high in content of the mineral montmorillonite (Petersen, Lee, and Chesters, 1968). The glacial advance (Valderan) that deposited the red till reached what is now Manitowoc County about 11,800 years ago, as determined by dating of wood of trees that were overridden and buried by the ice at the site of the Two Creeks Forest Bed (Black, 1970b; Lee and Horn, 1972). Later glaciolacustrine deposits were laid down over the till locally (Fig. 15-4). Rounded fragments of the Valderan till, called clay balls, are observed in post-Valderan outwash and beach deposits (Zakrzewska, 1970).

The differences in the composition of carbonates and clay between northern (Hibbing, Ontonagon, and associated soils) and eastern (Oshkosh and associated soils) clay regions provide some explanations for soil differences. Upland soils are leached slightly deeper in the less calcareous northern landscapes. Mineral wetland soils are shallower (16-inch solum in the Pickford soil near Ashland as compared to the 24-inch solum in the Poygan soil near Oshkosh) in the more clayey northern areas,

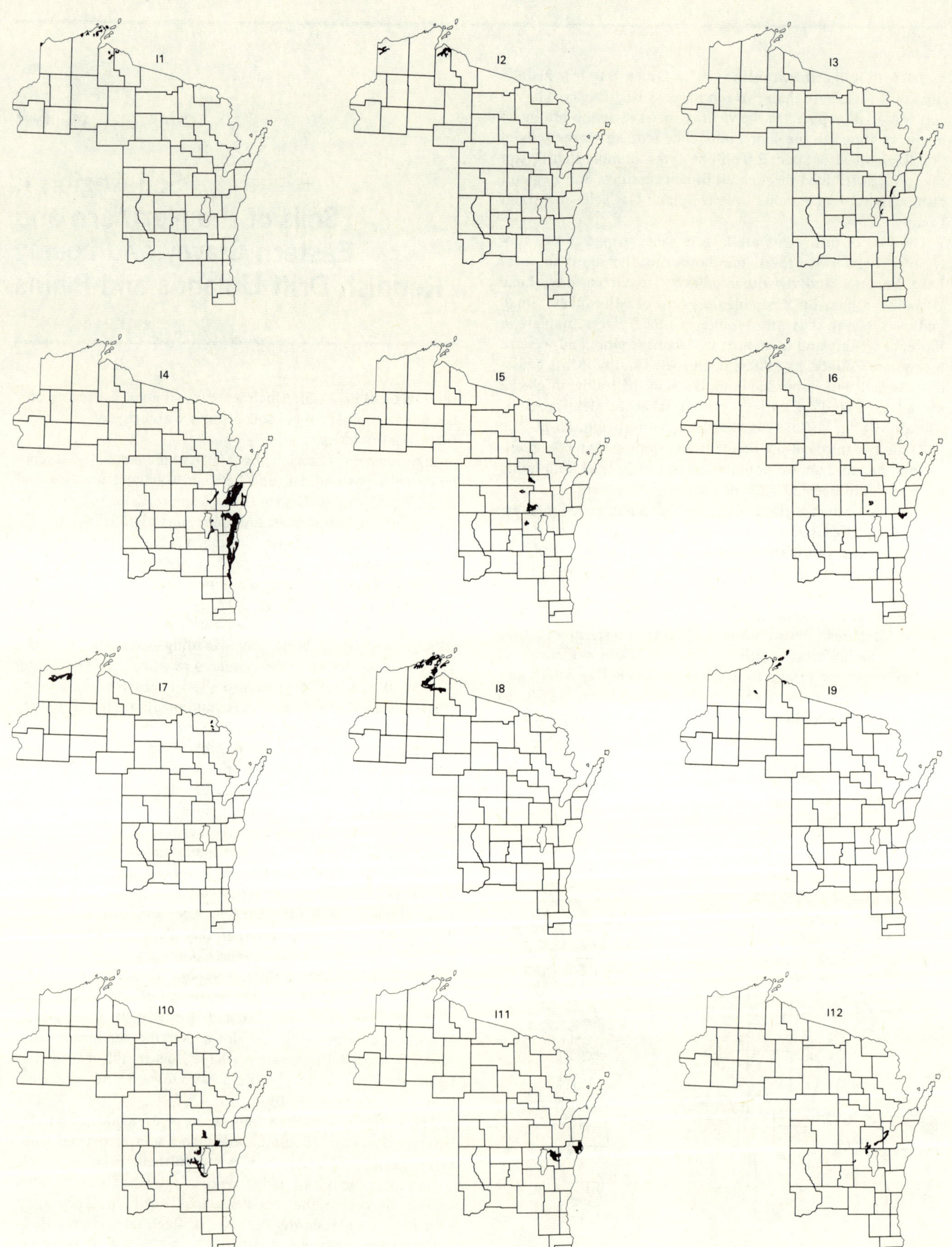

Figure 15-2. Sequence of maps showing distribution of soil associations in Soil Region I.

Figure 15-2, *continued.*

where cooler subsoil temperatures may limit root growth and related dewatering of the soil during the growing season.

Soil productivity in the region is influenced by many factors, including texture of soil profile, depth to water table, shortness of the growing season, and length of day during the growing season. In a favorable fog belt adjacent to Lake Superior, growth in height of red pine is as rapid as 48 inches per year, or nearly four times the usual rate (Wilde, 1965). Estimates for annual yield (tons, dry weight, per acre) of alfalfa-brome hay have been made (Beatty et al., 1966) for major soils, under high levels of management that include adequate fertilization and drainage: 2.75, Casco and Dalbo soil series; 3.0, Fox loam, Gogebic, Lapeer, Superior sandy loam; 3.5, Fox silt loam, Hibbing, Ontonagon, Pickford, Rudyard, Superior loam; 3.75, Manawa; 4.0, Theresa, Poygan; 4.5, Kewaunee.

Just after snow melt in May, the red clay soils are largely saturated, even on uplands. Clay holds water more tenaciously than silt does, as has been demonstrated by field and laboratory studies; hence, these red clay soils can provide an agricultural or silvicultural crop with less available water during a growing season than can the more silty soils of Soil Region F (personal communication, A. J. Wojta, 1955). Neither group of soils can be said to be characteristically droughty, however. The presence in I8 and I18 soilscapes of as much as a foot of silty material over clay substratum lowers the suitability of a site for crops, in the spring, because of a persistent perched water table condition in the silt.

In the legend of the soil map (Plate 1) the soil associations of this region are arranged in order from the steepest to the most nearly level (Fig. 15-2). In the following discussion, the units are regrouped chiefly on the basis of soil series.

I1, I2, I7, I8, I9, I18. *Hibbing, Superior, and associated soils formed on reddish clayey glacial till.*

I1. The rolling to hilly Hibbing clay loam, and Leonidas, Superior, and Ogemaw sandy loams, association.

I2. The rolling to hilly Hibbing, Pickford, and Ontonagon loams and silty clay loam, and Bibon sand, association.

I7. The undulating Hibbing silty clay loam, Leonidas and Gogebic loams, and Bibon sand, association.

I8. The undulating Hibbing, Rudyard, and Pickford silty clay loam, and Hiawatha loamy sand, association.

I9. The undulating Superior, Orienta, and Pickford loams, and Manistee and Hiawatha loamy sand, association.

I18. The nearly level Hibbing, Rudyard, Pickford, and Ontonagon silty clay loam and Superior loams, association.

The Hibbing (Typic Eutroboralf) and Superior (Alfic Haplorthod) are the most extensive soils in this association, and have formed, respectively, without and with a loam covering (Fig. 14-5) over reddish-brown (2.5YR 5/4) calcareous and somewhat stony silty clay glacial drift, containing 10 to 15% carbonates, 55% clay and 35% silt (Fig. 15-3). The Rudyard (Aquic Eutroboralf) somewhat poorly drained soil associates and the Pickford (Aeric Haplaquept) poorly drained soil associates lie in low places, including drainageways. Sandy materials are present not only as coverings 10 to 20 inches thick (Superior soils, Alfic Haplorthods), 20 to 40 inches thick (Leonidas, Manistee, Ogemaw soils), and 40 to 60 inches thick

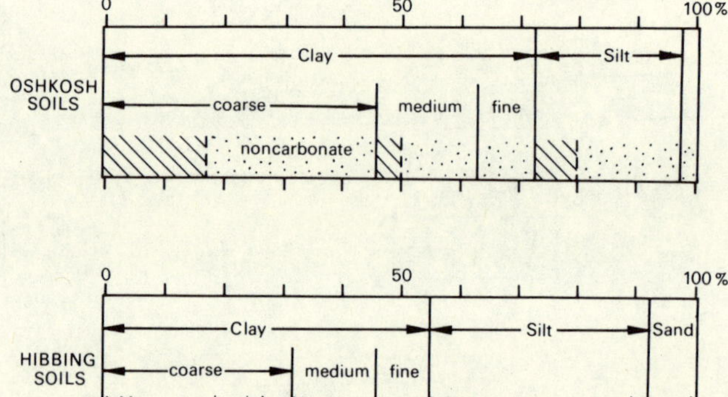

Figure 15-3. Analytical data for two eastern (Oshkosh) and two northern (Hibbing) red clay soils. Barred areas indicate material containing carbonates.

(Bibon, Orienta soils; Typic and Aquic Haplorthods, respectively), but also as very much deeper deposits (Hiawatha soils). Somewhat poorly drained conditions are characteristic of Ogemaw and Orienta sands (both Aquic Haplorthods). The landscape is thus a mosaic of fertile clayey soils, productive of such crops as hay and oats, and leached sands capable of supporting good stands of coniferous trees. Exposure of these soils in hilly terrain, along stream banks and lake shores, can lead to severe erosion of both clay and sand materials. Proximity to Lake Superior ameliorates the climate (see Chapter 2, section on climate), and permits cultivation of small fruit and tree fruit crops on these soils.

I3, I4, I5, I6, I10, I11, I12, I13, I14, I15, I16, I20. *Kewaunee, Hortonville, Manawa, Oshkosh, and associated soils formed on reddish clayey glacial till.*

I3. The rolling to hilly Kolberg, Summerville, and Kewaunee silt loam and silty clay loam association, with limestone and shale rockland.

I4. The rolling to undulating Kewaunee, Hortonville, Manawa, and Poygan silt loam and silty clay loam association.[1]

I5. The rolling to undulating Hortonville, Kewaunee, Manawa, and Poygan silt loam association.

I6. The rolling to undulating Onaway loam, and Theresa, Hortonville, Fox, and Casco loams, association.

I10. The undulating Kewaunee, Manawa, and Poygan silty clay loam association.[2]

I11. The undulating Kewaunee, Manawa, and Poygan silt loam and loams association.[3]

I12. The undulating Kewaunee and Manawa silt loam and loams association.

I13. The undulating Kewaunee, Manawa, Poygan, and Hortonville loams and silt loam, and Tustin loamy sand, association.

I14. The undulating Manawa and Poygan silty clay loam, and Rimer and Tustin sandy loam, association.

1. On the soil map, Plate 1, in T.12N., R.22E., 4 should read I4.
2. On the soil map, Plate 1, in T.21N., R.19E., 10 should read I10.
3. On the soil map, Plate 1, in T.19N., R.19E., 11 should read I11.

Table 15-1. Analytical data for two eastern (Oshkosh) and two northern (Hibbing) clay soil substrata

Carbonate material (%)	Non-carbonate material (%)				
	Total	sand (2-0.05 mm)	silt (.05-.002 mm)	clay (<.002 mm)	Free Fe_2O_3
Oshkosh soil substrata					
29	71	0.5	17	54	1.13
Hibbing soil substrata					
13	87	6.0	31	50	1.83

Source: Data from Petersen, Lee, and Chesters, 1968. Percentages are by weight.

I15. The undulating Kewaunee, Kolberg, and Manawa silt loam and loams association.
I16. The undulating Hortonville, Manawa, and Poygan loams, and Shiocton fine sandy loam, association.
I20. The nearly level Kewaunee, Oshkosh, Manawa, and Poygan silty clay loam association.

The dominant soil of this group of soil associations is the Kewaunee (Typic Hapludalf; Fig. 15-4), which is developed in reddish-brown (5YR 5/3-4/4) glacial till of silty clay loam to clay loam texture, containing in the fine earth fraction 25 to 35% carbonates by weight, 30 to 40% each silt and clay and 30% sand. Coarser fractions include 10 to 20% gravel (largely dolomitic) with a few erratic cobbles and boulders of Precambrian crystalline rocks and dolomite. Manawa soils (Aquollic Hapludalfs) are the somewhat poorly drained and Poygan (Typic Haplaquolls) the poorly drained associates found in depressions and drainageways. The Bt horizon of the Kewaunee soil contains 40 to 60% clay. Some level areas of clay till, with less than 10% gravel, overlying glacio-lacustrine sediments have been mistaken for lake-laid deposits, particularly in soil association I20. Sandy clay loam to sandy loam tills occur locally where the material is thin over sandy drift, sandstone, or granitic Precambrian bedrock (Horn, 1959).

Less extensive than Kewaunee soils are the similar Hortonville (Glossoboric Hapludalf) and Oshkosh (Typic Hapludalf) soils (Figs. 15-4 and 11-5).

Where shallow over the yellowish-brown Woodfordian till of the earlier Cary stage, the reddish-brown (Valderan) till is typically loam in texture, with more gravel and cobbles than usual, and is yellowish red in color (5YR 5/6). This condition has been observed on glacial moraines in Kewaunee, Manitowoc, Brown, and Waushara counties. The Hortonville soils have developed in this material, with Bt horizons containing 27 to 35% clay. Associated with the Hortonville in soil association I6 are the Onaway, Theresa, Fox, and Casco soils, already considered in Chapters 8 and 11 on Soil Region B and Soil Region E (where Rimer, Tustin, and Shiocton are also discussed). Oshkosh soils are found on nearly level plains of extinct glacial lakes, and have developed in sediments that have a clay content of about 70% (Fig. 15-3), carbonate content of 29%, and almost no sand or gravel. The profile is shallower and less developed than in the Kewaunee soils because the Oshkosh soils

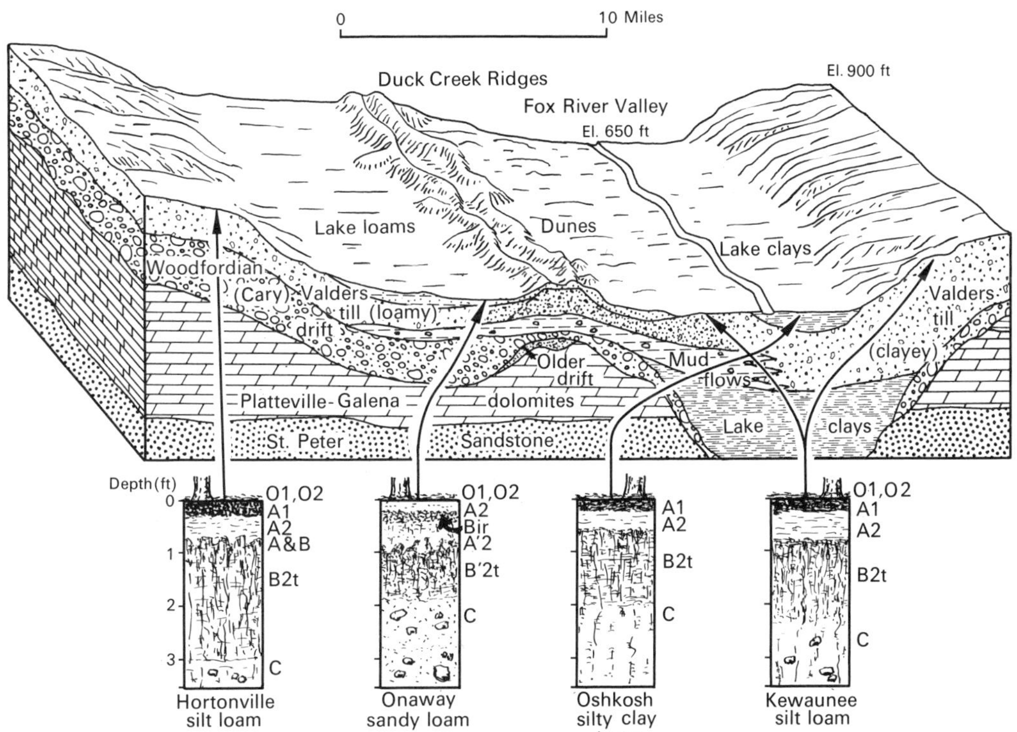

Figure 15-4. Block diagram showing landscape positions of representative soils of soil associations I3 and I20 in an east-west transect across Brown County. The soils labeled in the diagram are classified as follows: Hortonville, Glossoboric Hapludalf; Onaway, Alfic Haplorthod; Oshkosh, Typic Hapludalf; Kewaunee, Typic Hapludalf. (After Beaver, 1966.)

are younger, more clayey, and usually somewhat higher in carbonate content.

Kolberg (Glossic Eutroboralf) and Summerville (Entic Lithic) soils are shallow (20 to 40 inches) over dolomitic limestone along the Silurian ("Niagara") Escarpment in Door, Brown, and Calumet counties (Fig. 2-3). Some areas of the somewhat poorly drained Bonduel (Aquic Eutroboralf), shaly Shullsburg (Aquic Argiudoll), and the shallow, droughty Sogn (Lithic Haplustoll) silt loams are associated with the Kolberg.

Southeast of Green Bay, Kewaunee soils of soil association I4 have developed in thin red clayey glacial drift (Valderan) overlying Woodfordian outwash sand and gravel. Both geologic erosion and accelerated erosion have developed gullies in these materials, with the result that a dissected strip of Casco loams marks an escarpment between the I4 soilscape to the east and the I20 soilscape to the west.

Average depth of leaching in well-drained soils on till is about 23 inches in soilscape I4 where it borders soilscape B17 on the east, north, and west, but is about 44 inches to the south (see discussion of B17 in Chapter 8).

Under small patches of prairie, dark analogues (Argiudolls) of Kewaunee and Oshkosh soils have formed in Winnebago and Fond du Lac counties.

The Valders glacial till is in places overlain by old beach deposits on which Casco loam and related soils have been identified.

I17, I19, I21, I22. *Ontonagon, Oshkosh, Briggsville, Braham, and associated soils formed on glacio-lacustrine clay.*

- I17. The undulating Briggsville and Poygan loam and clay loam, and Tustin and Lapeer sandy loam, association.
- I19. The nearly level Ontonagon, Hibbing, and Rudyard silty clay loam association.
- I21. The nearly level Oshkosh, Manawa, and Poygan silty clay loam, and Tustin sandy loam, association.
- I22. The nearly level Braham and Blomford loams, and Dalbo silt loam, association.

These nearly level soils are dominantly silty clay loams, in both the northern (Ontonagon) and eastern (Oshkosh) subregions. The Ontonagon soils (Typic Eutroboralfs; see Fig. 14-5) are about 24 inches deep to calcareous lacustrine clay substratum as compared to about 21 inches in the more clayey Oshkosh soils (Typic Hapludalfs; see Fig. 15-4). The Oshkosh soil overlies clays containing about 30% carbonates by weight; the Ontonagon soil overlies clays containing about 12% carbonates. Locally, silty coverings about 6 inches thick overlie the clayey soil, thickening the acid solum by that much. Virgin profiles of the Ontonagon catena of soils may be observed in a state forest of two-century-old white pine trees on Madeline Island of the Apostle Islands.

Wetlands of Rudyard (Aquic Eutroboralf) and Pickford (Aeric Haplaquept) soils in the north and Manawa (Aquollic Hapludalf) and Poygan (Typic Haplaquoll) in the east occupy depressions, along with some bodies of peat. Small depressions usually contain somewhat poorly drained soils (Rudyard, Manawa) and large depressions receiving drainage from a considerable area are occupied by poorly and very poorly drained soils (Pickford, Poygan). The Pickford soil in the northern landscape is 14 inches deep to carbonates, in contrast to the 27 inches in the Poygan in the southeast. The difference may be ascribed to wetter conditions with a higher water table in the northern landscape. Considerable "landforming" has been done by grading machinery to provide surface drainage through depressions in fields.

The Briggsville (Typic Hapludalf) is somewhat less clayey (45% in the Bt horizon rather than 70% clay) than the associated Oshkosh series and, although considered to be well to moderately well drained on the basis of morphological characteristics, has about the same hydraulic conductivity as the somewhat poorly drained Almena silt loam of soilscape F22 (personal communication, J. Bouma, 1972).

Chamberlin (1883) first described the fine blocky structure of the C horizon of the Ontonagon soil. In his studies of this soil material, he noted that a magnet picked up grains of magnetite when drawn through a handful of the dry, pulverized red clay.

In the vicinity of Lake Winnebago, gravelly beach deposits of ancient glacial Lake Oshkosh transect the clayey soil continuum of soil associations I20 and I21.

The "gray" calcareous glacial drift of Burnett County (soil association I22) is actually light yellowish brown (10YR 6/4) to very dark grayish brown (10YR 3/2), with numerous mottles. It was laid down by the Grantsburg lobe of the late Wisconsinan (Mankatoan of Minnesota) glacial ice and the glacial Lake Grantsburg which lay in front of the ice in Minnesota and Wisconsin (Thwaites, 1947). In transitional landscapes between sand plains and gray till plains, the moderately well drained Braham (Arenic Eutrochrept) and somewhat poorly drained Blomford (Mollic Haplaquept) sandy loams and loams have formed in 20 to 36 inches of loamy covering over loam to clay loam glacial till, which is calcareous at a depth of about 42 inches. The Dalbo silt loam (Typic Eutroboralf) has formed in less than 20 inches of silty material over lake-laid stone-free clay that is calcareous at a depth of 24 to 30 inches. The Dalbo soils lie on the landscape as nearly level bodies about 5 to 30 acres in size separated by wetlands in finger-shaped depressions, including some peat bogs. The Braham soils are in larger undulating to rolling upland units, with Blomford soils in slight depressions. The Braham is leached of carbonates to depths of 30 to 60 inches in Wisconsin.

Even the somewhat poorly drained Dalbo and the well-drained Braham, Ontonagon, Oshkosh, and Briggsville soils have high or perched water tables at the time of the early spring thaw. During the growing season vegetation usually pumps water out of these soils to the point that they crack, sometimes to the depth of about 2 feet, because of the capacity of the abundant clays to shrink on drying.

The practice of ridging Ontonagon, Hibbing, and some associated soils by plowing pairs of furrow slices together and planting seedlings on them has notably improved the survival of white and red pine plantations on those soils, by elevating the root systems above the seasonal perched water table. The use of Hibbing and Ontonagon clay materials (I18, I19) to top-dress Vilas and Omega sands has not been beneficial to pine plantations because the clay impedes drainage (Wilde, 1965).

Soil association I17 in Adams and Columbia counties includes (1) a variant of Briggsville having a deep (20 to 40 inches) silt cover over lacustrine reddish clayey sediments, (2) Oshtemo sandy loam, and (3) Plainfield sand (McColley, 1971).

CHAPTER **16**

Soil Region J: Soils of Stream Bottoms and Major Wetlands

Gerhard B. Lee

Soil Region J includes major areas of alluvial soils, peat and muck soils (Histosols), and poorly drained mineral soils (Figs. 2-50, 16-1, 16-2). The areas shown on the soil map (Plate 1) total approximately 2,850,000 acres, or about 8.2% of Wisconsin's land area. The actual area on the land may be nearly double that; many small areas of these soils cannot be shown on the soil map because of its scale. These areas include thousands of kettle-hole wetlands and numerous colluvial deposits on footslopes, in small intermittent drainageways, and in overwash deposits on the edges of wetlands (areas of discharge of water by through-flow of water and by evaporation).

Vegetative growth is vigorous on alluvial soils of floodplains. Recurrent floods may damage plant cover, especially immediately adjacent to rivers and streams, but they also add new soil material that is usually fertile and of a medium texture that is favorable to plant growth. Plant species range from the scarce sagebrush to the abundant ragweed, willow, and silver maple. Marshland cover includes the cattail and bulrush, which together are probably more productive in terms of mass of vegetative growth per acre than any other plant community of Wisconsin. Natural vegetation of muck and peat bogs is characterized by slowly growing black spruce and shrubby leatherleaf, sedge, and sphagnum moss. Soils of Region J thus have a wide range in productivity, from about the lowest to the highest in the state. These are among the least studied of Wisconsin soils. They comprise areas in which modern zoning ordinances prohibit the building of many kinds of permanent structures, but permit recreational, wildlife, and flood-control uses. Much can be learned from future investigations of interaction of these soils with the environment.

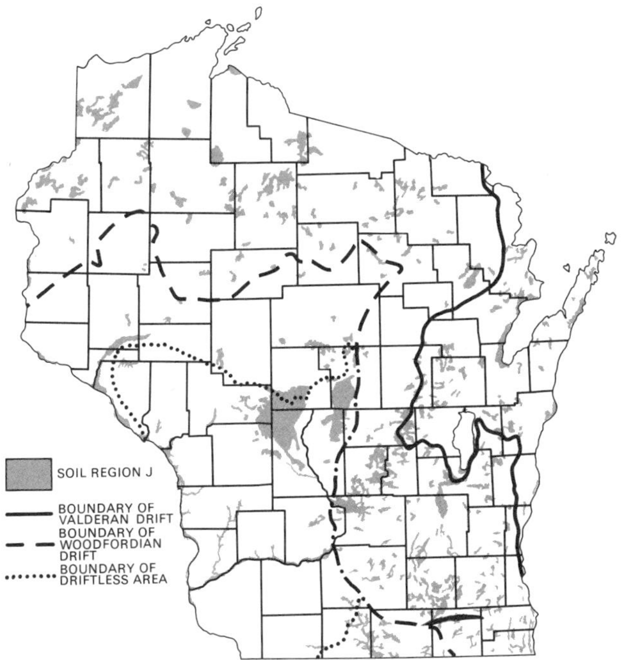

Figure 16-1. Index map showing the geographic relationship of Soil Region J to major glacial boundaries.

ALLUVIAL SOILS

All soils that have formed in alluvium (stream deposits), or colluvium (soil material transported down-slope by gravity and/or slope-wash) are commonly called alluvial soils even though they may vary widely in color, texture, and other characteristics. The chief diagnostic characteristic is irregularity in content of organic matter from horizon to horizon with depth, although there is a general vertical decrease within each A1 and A1b horizon. Alluvial soils in Wisconsin are composed of relatively young sediments and have not been stable long enough for distinctive subsoil horizons to form, with few exceptions. Alluvial soils that are fairly uniform in composition and natural drainage have been classified as soil series and given names such as Arenzville and Orion. Others, particularly those formed in heterogeneous deposits, are classified as undifferentiated land classes. The latter term indicates that the soils present are extremely variable within short distances, or that they are unstable and may change from year to year as flood waters erode them and deposit new sediments. The generally low value of such lands makes detailed classification and mapping unprofitable.

Two major associations of alluvial soils are recognized in Region J. The first is J1, dominated by soils of the Arenzville, Orion, and Ettrick series (Figs. 2-51, 7-9, 7-10). The second is J2, a land class called "wet alluvial soils, undifferentiated" (Figs. 11-6, 16-2, 16-4). The largest mappable areas of J2 are along the lower reaches of the Wisconsin and Chippewa rivers, and along the Mississippi River.

J1. *Arenzville, Orion, and Ettrick silt loam.*

Arenzville and Orion soils (Figs. 2-50, 2-51, 7-9, 7-10) consist of silty, light-colored alluvial sediments 40 inches or more in thickness over a dark buried presettlement soil that resembles the Ettrick (see below). Arenzville soils (Typic Udifluvents) are well to moderately well drained. Orion soils (Aquic Udifluvents) are somewhat poorly drained, being mottled and drab in color at depth. In both soils the A horizon is granular or subangular blocky in structure and slightly darker in color than subsoil horizons. There is, however, very little soil development below the A1 or Ap horizon. Crude stratification or lamination is common in subsurface layers; thin lenses of sand or gravel may be present but are not common. These two soils are asso-

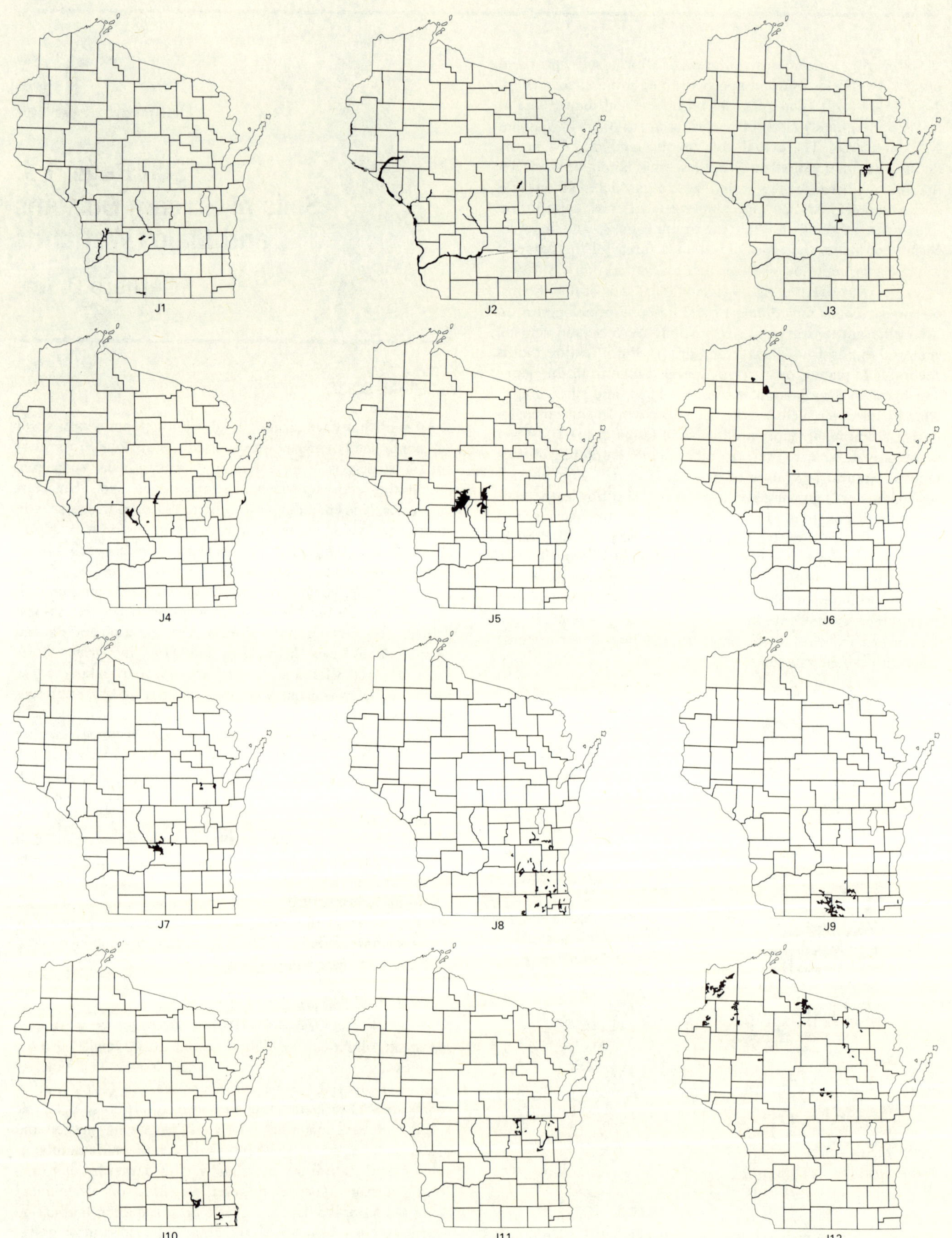

Figure 16-2. Sequence of maps showing distribution of soil associations in Soil Region J.

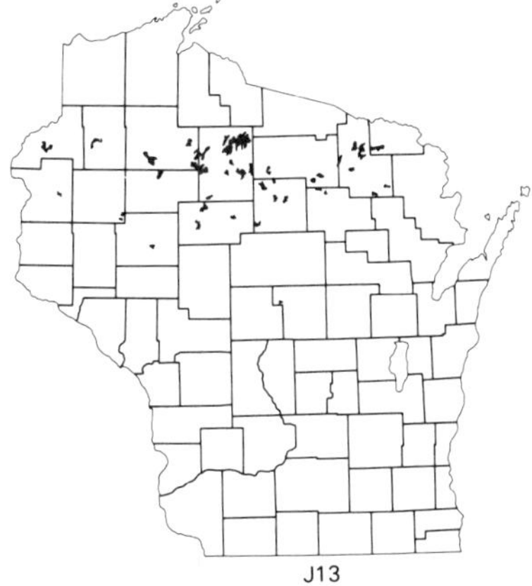

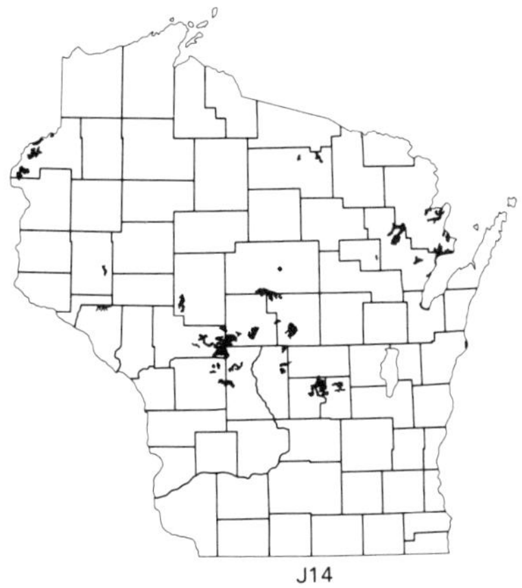

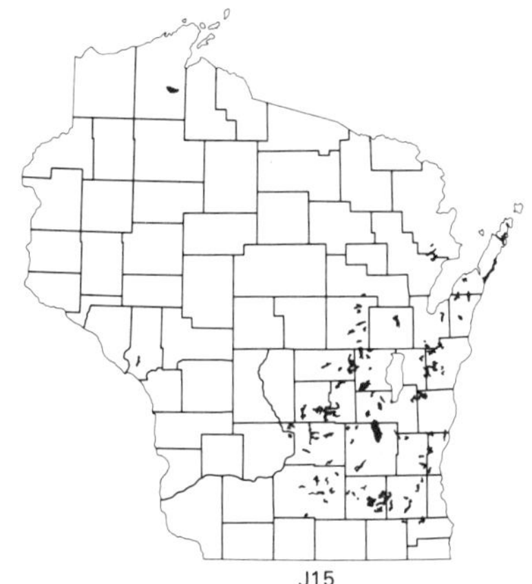

Figure 16-2, *continued.*

ciated in the Chippewa River valley in Pepin County, but are most extensive in southwestern Wisconsin. Sixty-two thousand acres are shown on the map and there are many additional areas too small to delineate.

Ettrick silt loam (Typic Argiaquoll) is a poorly drained associate of Arenzville and Orion. It is an older soil with a higher content of clay in the subsoil. A less well developed, poorly drained, alluvial soil is Otter, a Cumulic Haplaquoll. Millington, also dark colored and poorly drained (Cumulic Haplaquoll), is found in loamy alluvium.

Other common alluvial soils in southwestern counties are Huntsville (well drained) and Lawson (somewhat poorly drained). Both are deep, silty, dark-colored soils classified as Cumulic Hapludolls.

Akan silt loam, a somewhat poorly drained soil formed in silty alluvium over lacustrine silts and clays, and Boaz silt loam, also somewhat poorly drained, and formed in deep silts, are, like the Ettrick, somewhat older alluvial soils. They have been on stable surfaces for long enough periods to have developed cambic (juvenile) B horizons. These soils are classified as Mollic and Aeric Haplaquepts, respectively. Where undisturbed they support sedge meadows that include numerous composites and the bottle gentian.

Well-drained, loamy soils formed in alluvium include Terril and Ankeny (both Cumulic Hapludolls), Kickapoo (Typic Udifluvent), and Caryville (Fluventic Haploboroll). The latter is formed in 10 to 20 inches of loamy alluvium over sand and has lower soil temperatures than the other soils mentioned in this paragraph.

Accelerated erosion in some parts of southwestern Wisconsin has resulted in local washing of cherty rubble from steep stony lands onto silty bottomlands. The detailed pattern of soils of alluvial fans has not yet been mapped.

Alluvial soils formed in slope-wash on footslopes of cultivated fields include dark-colored soils such as Worthen (well drained) and Littleton (somewhat poorly drained), both Cumulic Hapludolls; the light-colored, well-drained Chaseburg (Typic Udifluvent); Orion (formerly called Osseo), somewhat poorly drained (Aquic Udifluvent); and the poorly drained Sawmill (Cumulic Haplaquoll). The Cumulic Hapludolls of the Worthen and Littleton series (formerly called Judson) are darker equivalents of Chaseburg and Orion.

In southeastern Wisconsin, silty alluvial soils, which formed mainly on footslopes from post-agricultural colluvial sediments over older soils, include dark-colored, well-drained soils such as Troxel (Typic Argiudoll), Worthen (Cumulic Hapludoll), and Radford (somewhat poorly drained, Aquic, Fluventic Hapludoll), and light-colored soils such as Juneau (well drained, Typic Udifluvent) and Pistakee (somewhat poorly drained, Aquic Udifluvent) formed in silty alluvial and colluvial deposits. Poorly drained associates include Washtenaw (Typic Haplaquent), formed in colluvial silts, and Wallkill (Thapto Histic Fluvaquent; Fig. 8-7), formed in colluvial silts over peat or muck deposits.

Some soil scientists observe that the Arenzville has no detectable B horizon but that older colluvial soils, such as Chaseburg, may have cambic or even weak argillic horizons. The difference will probably be of less interest in the future, if accelerated deposition causes Arenzville-like soils to be formed over Chaseburg and Juneau soils.

Caryville fine sandy loam, a Fluventic Haploboroll, is the principal alluvial soil in Soil Region C, on high bottoms along the Wisconsin River in Portage County and the Chippewa River in Dunn County.

In Soil Region E, DePere (Typic Udifluvent, well drained, mesic); Jump River[1] (Typic Udifluvent, well drained, frigid); and Stinson (Aquic Udifluvent, somewhat poorly drained) are formed in reddish silt loam to silty clay alluvium. In north-central and far northern Wisconsin (Soil Regions F and G mainly) Brule (Typic Udifluvent), a well-drained soil formed in loamy alluvial sediments, has been recognized, along with several other unnamed series.

Numerous small bodies of valley bench soils such as Dakota and Bertrand are included in soil association J1 on the soil map (Plate 1).

Alluvial soils of the Arenzville, Orion, Ettrick, or similar associations, along streams, are subject to frequent flooding or ponding, unless protected by structures such as dams or levees.

Less well drained members of the association have high water tables. These factors severely limit their use for many purposes.

1. The native forest on the "Jump River Bottom" was notable because of the abundance of species of the southern mesic forest. In 1973 a virgin stand of 210 acres was still present in Price County.

In the southern part of the state, well- or somewhat poorly drained alluvial soils that are not flooded too frequently and that occur in areas large enough to cultivate are commonly used for corn. Because of the great productivity of these soils, high-value crops such as tobacco are grown in some places. Poorly drained soils may also by cropped if they are artificially drained and protected from flooding. Otherwise, wet and frequently flooded alluvial soils are set aside for pasture, parks, and other nonintensive uses.

Zoning ordinances recognize that alluvial soils are not well suited for homesites or other uses where flooding would easily cause property damage or hazard to life. The use of soil maps to delineate alluvial soils and adjacent low, poorly drained soils as flood-hazard zones for regulatory purposes has been shown to be feasible in many rural areas (Lee, Parker, and Yanggen, 1972).

Alluvial soils formed from colluvium on footslopes ordinarily are not flooded. However, poorly drained members have a high water table, may be ponded at times, and usually require artificial drainage in order to be cropped. Seepage in subsurface layers often occurs in moderately well and somewhat poorly drained alluvial soils on footslopes. These well- to moderately well drained colluvial soils are among the most productive soils in the landscape, being deep and friable, and having a good moisture supply. They are usually present in small bodies, however, and, except where small acreage crops such as tobacco are grown, are included in the cropping system of larger fields.

J2. *Wet alluvial soils, undifferentiated.*

These J2 soils (Figs. 7-4, 11-6, 16-3, 16-4) are characterized by a high water table and are subject to frequent overflow. As a result they are variable in composition and wet much of the time. Texture depends on the source of the sediment and on stream characteristics and may include strata of sand, silt, and clay, and, in some places, mucky or peaty layers. Some large areas of marsh and other land that is periodically inundated are also included in this unit. Altogether, 200,000 acres are shown on the map. Many other areas are too small to be included.

Because of the hazards of frequent flooding, these soils have severe limitations for most uses except as wildlife and recreational areas or for pasture and certain forest crops.

Weidman (1914) reported 15 to 20 square miles of a light loam soil suitable for truck crops between Caryville and Meridian in Dunn County, in the midst of sandy loam alluvial soils of the Chippewa River bottom.

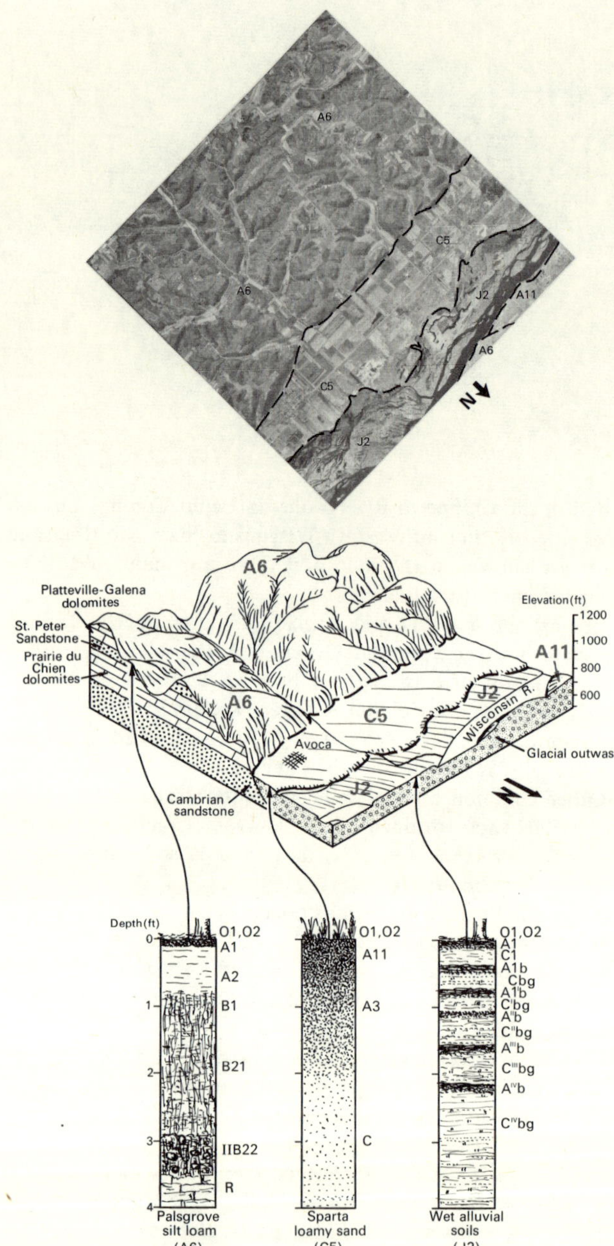

Figure 16-3. Aerial photo map of T.8N., R.1E., Iowa County. The area shown is 6 miles on a side.

Figure 16-4. Block diagram showing landscape positions of major soils of T.8N., R.1E., Iowa County.

MAJOR WETLANDS

The largest continuous areas of wetlands (Fig. 16-1; Plate 1) (Whitson and Ullsperger, 1919) are in the central part of the state, in low-lying portions of the bed of glacial Lake Wisconsin (Wood, Jackson, Monroe, and Juneau counties). In nearby Portage and Adams counties wetlands are in kettles and other depressions of the outwash plain that lies just west of the north-trending Woodfordian moraine on the west. In eastern and southeastern Wisconsin, wet soils are in enclosed glacial depressions such as kettles or in lowlands between drumlins and morainal ridges. Some of them are on footslopes below ledges, or cuestas. For example, just east of Waupun in Dodge and Fond du Lac counties, relatively soft Maquoketa Shale was deeply eroded by glacial ice and meltwaters at the foot of the Silurian Escarpment, forming the large depression containing the Horicon Marsh.

Wetlands in northern Wisconsin are also in depressions in glacial drift. Sphagnum bogs are mainly in kettles. Many bodies of other wet soils are scattered throughout the region.

In the southwestern part of the state, the well-developed dendritic drainage system, which was not disrupted by recent glaciation, allows for few wetland sites. Most are in oxbows of meandering streams on bottomlands, at seepage sites on footslopes, marshlands of flood plains, and in tributary valleys that were blocked at the confluence by massive deposits of outwash in the Wisconsin or Mississippi River valleys. A few wet mineral soils are on broad uplands.

On the soil map, wetlands are divided into two general classes. One of these, the wet mineral soils, includes soil associations J3 through J11 (Fig. 16-2). The other, consisting primarily of organic soils (peat and muck), includes soil associations J12 through J15 (Fig. 16-2).

Wet Mineral Soils

J3. *Granby, Shawano, and Emmet loamy sand and sandy loam, and shallow peat.*

The wet Granby loamy sand (Typic Haplaquoll) predominates, and is associated with Shawano loamy sand (Typic Udipsamment) and Emmet sandy loam (Alfic Haplorthod) on rises and with shallow peat in depressions.

These soils occur principally along the southwest shore of Green Bay in areas formerly occupied by Lake Michigan at higher water levels following glaciation. Water tables are generally high and the soils have drab, mottled subsoils as a result. The soils are only slightly acid in reaction in the solum and they support a good growth of coniferous and hardwood vegetation.

J4. *Newton, Plainfield, and Morocco sand and loamy sand, and shallow peat.*

Between the wet Newton (Typic Humaquept) and excessively drained Plainfield (Typic Udipsamment) lies the less extensive Morocco (Aquic Udipsamment) sand.

This group of sandy soils, which range from very wet to excessively drained and include areas of shallow peat, occurs in Portage County (near the Buena Vista Marsh) and in Adams and Juneau counties. It occurs in a landscape with only a few feet of microrelief. Small ridges are occupied by the excessively drained Plainfield (Typic Udipsamment) soils and the moderately well drained Nekoosa soils, while at slightly lower elevations somewhat poorly drained Morocco (Aquic Udipsamment) and Newton (Typic Humaquept) soils occur on nearly level slopes. Numerous shallow depressions are filled with 1 to 2 feet of acid peat formed from reed and sedge materials (Adrian peat, Terric Medisaprist). The mineral soils in this association are extremely sandy. They have no visible textural B horizon and the chief variations are in the amount of organic matter which is accumulated in the surface and the degree of gleying and mottling of the subsoil. Some are used for cranberry production.

J5. *Newton, Morocco, and Au Gres sands, and shallow peat.*

The sands are Typic Humaquepts (Newton), Aquic Udipsamments (Morocco), and Entic Haplaquods (Au Gres). There are a variety of Histosols (peats).

These soils occupy wet plains and shallow bogs in central Wisconsin. Morocco and Au Gres soils are somewhat poorly drained, Newton soils are poorly drained, and peat is saturated with water for most of each year. All of the mineral soils are very sandy. Morocco soils have a thick black A horizon on mottled sand. The Bir and A2 horizons of Au Gres are distinctive. All horizons are acid and pH values may be as low as 4.0 in some horizons of Au Gres sand and peat. The peat is formed mostly from remains of reed and sedge materials. Included in the area are small islands of sandy soils, such as Nekoosa, which are less affected by ground water than are the major soils. The poor soil drainage severely limits farming.

On the shore of Lake Michigan in Sheboygan County this soil association includes sand dunes, both stabilized and active.

J6. *Cable, Monico, Auburndale, and Freer loams and silt loam, and peat.*

These silty soils are Typic Haplaquepts (Cable, Warman), Aquic Dystrochrepts (Monico), Typic Glossaqualfs (Auburndale), and Aeric Ochraqualfs (Freer); they are associated with Hemists (peat).

These wet soils occupy depressions in glacial drift in a broad belt across northern Wisconsin. Several other loamy or sandy soils with restricted drainage are associated. Many bodies too small to show on the soil map occur in this region. All of these soils are somewhat poorly, poorly, or very poorly drained. Peats are acid and may vary considerably in depth and degree of decomposition. Most peat consists of remains of reeds and sedges with some included wood.

J7. *Wauseon, Keowns, Tustin, and Rimer loams and sandy loam.*

The wet Wauseon (Typic Haplaquoll) and Keowns (Mollic Haplaquept) soils predominate over the better drained Tustin (Arenic Hapludalf) and Rimer (Arenic Ochraqualf). These soils are in wetlands scattered from Shawano to Sauk counties.

The soils have formed in stratified lake sediments of silt to sand texture. Among associated soils are the Shiocton, Salter, and Seward. Shiocton and Keowns soils are coarse silts with weak horizonation and drab mottled colors. They liquify easily and cannot be drained with tile. Tustin, Rimer, and Wauseon soils consist of sandy loam or loam upper strata over clay loam strata of variable thickness. Tustin is well drained, Rimer somewhat poorly drained, and Wauseon poorly drained.

In Columbia County, Granby sandy loam and some alluvial lands are included in this unit (McColley, 1971).

J8. *Pella, Brookston, and Virgil silt loam and silty clay loam.*

These are the predominant black silty wetland soils of southeastern Wisconsin. The naturally poorly drained Pella (formerly called Kokomo and Elba) is a Haplaquoll and Brookston is a Typic Argiaquoll. The Virgil (Mollic Ochraqualf) and associated Kendall and Lamartine soils are somewhat poorly drained under natural conditions.

These soils occur in depressions of glacial uplands in southeastern and south-central Wisconsin (Fig. 8-13). The depressions differ from those described in J15 principally by being occupied largely by mineral soils rather than by muck and peat. The Pella and Brookston soils have thick black A1 horizons and dark gray B horizons of silty clay loam texture. Brookston soils have poorly sorted gravelly drift at depths of 2 or 3 feet, while Pella soils consist of moderately fine sediments throughout. Virgil soils are silty throughout and occur at the up-slope margins of the depressions occupied by Brookston and Pella soils. Artificial drainage of the latter two soils may favor development of mottling in the gleyed horizon.

In Waukesha County this soil association also includes Montgomery (Typic Haplaquoll), Martinton (Aquic Argiudoll), and Hebron and Saylesville (both Typic Hapludalfs) soils, formed from lacustrine deposits.

J9. *Matherton, Will, and Pella silt loam and silty clay loam.*

In wet glacial outwash areas of southeastern Wisconsin are found the somewhat poorly drained Matherton (Udollic Ochraqualf) silt loam, and the poorly drained Will and Pella (both Typic Haplaquolls) silt loams (Fig. 2-6).

Except for Pella these soils developed in loamy sediments and are underlain by coarse glacial outwash. Pella soils differ by being underlain by moderately fine lake sediments. Other associated soils are two Typic Haplaquolls, the Montgomery silty clay loam on lacustrine silts and clays and the Marshan silt loam over outwash deposits.

J10. *Navan, Hebron, Aztalan, and Pella loams and silty clay loam.*

The numerous wet places in glacial lake beds of loamy, silty, and clayey texture in southeastern Wisconsin contain the wet Navan (Typic Argiaquoll) and Pella (Typic Haplaquoll) soils, and associated better drained Aztalan (Aquic Argiudoll) and Hebron (Typic Hapludalf) soils.

Navan, Hebron, and Aztalan soils have formed in loamy deposits which are 18 to 36 inches thick over silty clay loam lake sediments. Hebron soils are moderately well drained, Aztalan and the related Mosel series soils are somewhat poorly drained, and Navan soils are very poorly drained. Pella soils are formed primarily in silty lake sediments and are very poorly drained.

J11. *Zittau, Poygan, Poy, and Borth loams and silty clay loam.*

These wet silty and clayey soils are found in ancient glacial lake beds of the red drift region of northeastern Wisconsin.

They are predominantly fine textured with sand and gravel beds at 3 feet or more in many places. Poygan soils (Typic Haplaquolls) have formed on very thick clayey deposits; Zittau (Aquollic Hapludalf), Borth (Typic Hapludalf), and Poy (Typic Haplaquoll) have formed in clayey sediments only 20 to 40 inches thick over calcareous sand and gravel. The sola are generally no more than 30 inches thick over calcareous materials. These soils are associated with areas of moderately well drained, fine-textured soils such as Oshkosh and Winneconne.

Organic Soils[2]

Histosols (peats and mucks) occur in thousands of bogs, large and small, profusely dotting Wisconsin landscapes (except in Soil Regions A and D) (Figs. 2-52, 8-7, 8-8, 9-4, 9-5, 14-5). A bog represents a stage in wetland succession from a clear lake to a eutrophied ("dead") lake, to a marsh, and finally to a bog. If allowed to proceed for many thousands of years, natural processes will convert all the remaining lakes of the state into Histosols.

2. The "humus soils" of Chamberlin (1883).

The accumulation of organic materials at wet sites is considered geogenetic rather than pedogenetic. The layers of organic matter reflect the vegetation patterns (Grittinger, 1971) and the plant succession in each bog. Drainage, cultivation, and consequent aeration of a peat deposit initiates pedogenetic processes (Isirimah, 1969) called ripening (soil formation). Ripening includes (1) physical disintegration of plant parts (the transformation from peaty material to muck), and (2) biochemical decomposition, called moulding, by which surface horizons are converted to a sawdust-like moder material (by mites, insects, and small worms) or to a crumb-like mull material (by earthworms) (Langton and Lee, 1964; Lee and Manoch, 1974). Platy structure may be either geogenetic or pedogenetic in origin, but blocky and prismatic structures result from drying and wetting cycles during ripening. The presence of muck horizons buried under peat may record ponding of water behind beaver dams.

Many bogs are wildlife refuges; some of them are officially designated as scientific areas for research on flora and fauna.

Several thousand acres of sphagnum peat in Jackson and Juneau counties are used for sustained-yield sphagnum production. The sphagnum is harvested about every fifth year for sale for horticultural uses.

Drainage and agricultural development of peats and mucks has been greatest in southeastern Wisconsin, moderate in the central sandy plain, and least in northern counties (Albert, 1945, 1951). In southeastern counties muck farms have produced head lettuce, carrots, onions, mint oil, lawn sod, and other special crops[3] (Albert and Zeasman, 1953). Cranberry culture is practiced on about 6,000 acres in central and northern counties.

Present areas of agriculturally developed organic soils are estimated by J. A. Schoenemann (personal communication, 1972) to be 24,000 acres in southeastern counties, 21,000 in central counties, 500 in southwestern counties, and 8,000 in northern counties, for a total of 53,500 acres in the state, or about 2% of the total area of peats and mucks in the state.

Burning of peat in connection with forest fires in "cutover country" had already destroyed considerable volumes of organic soils by 1903, when Weidman made a reconnaissance soil survey of north-central Wisconsin. Uncontrolled fires continue to be a problem in bogs. Some areas mapped as peat in early surveys in Price County are wet mineral soils today. This probably indicates loss of shallow peat in the interim by slow oxidation or burning after drainage.

Early in this century, some Wisconsin bogs were used as disposal basins for effluent such as that from cheese factories. Although this practice has been discontinued, the enormous absorptive capacity of peat and muck may be utilized in the future in a variety of specially designed systems for handling liquid wastes.

Bogs cause problems for highway construction in that emplacement of large fills of stable material is usually necessary.

3. Potatoes, celery, cabbage, sweet corn, peas, and lima beans are also reported by Professor J. A. Schoenemann, Horticulture Department, College of Agricultural and Life Sciences, University of Wisconsin-Madison.

J12. *Moss peat and acid sedge and woody peats, Au Gres sand, and Cable loams.*

The organic soils and associated sandy and loamy wet mineral soils of soil association J12 occur in depressions within the sandy drift region of northern Wisconsin.

Organic deposits may range in depth from 1 or 2 feet to 40 feet or more. Deep bogs may be occupied by a succession of peat deposits which reflect changes in past climates and vegetation. Substrata under these organic soils are usually loamy or sandy; pH values of the organic deposits may be as low as 3.6 to 4.0. Vegetative cover on these bogs includes open meadows of sedges, and black spruce and tamarack stands which may vary in quality from a few sparse, stunted trees to reasonably good stands.

Associated with the "Manitowish marshes" in eastern Iron County are several square miles of level wet Au Gres sand (Whitson et al., 1916).

In Bayfield County Typic and Hemic Borosaprists and Typic Borohemists are prominent in bogs.

J13. *Raw acid sedge and woody peats with thin moss covering, and Cable and Freer silt loam.*

These soils occur in landscape positions similar to those of association J12 but are bordered by silty or loamy glacial drift rather than sandy drift. Physical and chemical properties and land-use limitations and hazards of the organic deposits are similar to those described for J12.

Dawson peat (Terric Borosaprist; Fig. 1-8) has been reported from Taylor County. Houghton (Typic Medisaprist) and Couderay (Hemic Borofibrist) series are represented in Rusk County. A Rifle peat (Typic Borohemist, euic) of Bayfield County is illustrated in Fig. 14-5.

A study in Oneida County showed that partial or complete cutting of spruce and balsam fir lowland forest admits light to the ground cover and permits live sphagnum moss ("S-horizon," 0 to 7 inches thick) to expand from nearby peat bogs, causing site quality deterioration as much as fourfold with respect to tree growth (Keller and Watterston, 1962). The moss insulates the soil and keeps it cool, increases soil acidity, promotes saturation of the soil with water, and impoverishes the soil by holding plant nutrients in unavailable forms in the moss itself.

A detailed study of a Fibrist in the Bogus Marsh of Langlade County (T.33N., R.10E.; see Frazier and Lee, 1971) reported a vegetative cover of black spruce, tamarack, sphagnum moss, leatherleaf, and labrador tea. The peat was extremely acid (pH 3.2-4.2) with low ash content (2.1-6.4%, whole dry soil) and high carbon/nitrogen ratio (C/N = 37-61).

J14. *Acid sedge peat and muck, and Au Gres, Newton, and Morocco sands.*

These soils occur extensively in Adams, Portage, Juneau, and Wood counties in central Wisconsin. Peat and muck deposits are usually shallow and are interspersed with "islands" of wet sandy soils. The peat and muck may vary locally in degree of decomposition, acidity, and thickness. Au Gres, Newton, and Morocco soils are all sandy and wet. Au Gres and Morocco soils are somewhat poorly drained; Newton soils are poorly drained. Also present in the association are Dillon, Nekoosa, and occasional tracts of Plainfield soils. In Monroe County Saugatuck sand (Aeric Haplaquod) is associated with these soils.

In Pepin and adjacent counties deep peats (Hemists) commonly consist of 40 inches of black peat overlying less oxidized brown peat. Deep Houghton (Typic Medisaprist; Figs. 8-7, 9-4, 9-5) and shallow Adrian (Terric Medisaprist) mucks are found in Pepin, Buffalo, and Portage counties. Parts of the Mondovi Marsh in Buffalo County are drained and used for special crops, including horseradish.

Over considerable areas of J14 near Babcock, City Point, and other parts of the central sandy region, a layer of sphagnum peat is maintained about a foot deep by slow growth of sphagnum moss, except where harvesting, burning, and artificial drainage have interfered with the process. Under the sphagnum are layers of woody and sedge peat resting on gray siliceous sand (Catenhusen, 1950).

In eastern Jackson County there are sphagnum and cranberry bogs with borders of swamp forest. Even though these bogs are abundantly watered, the high acidity and the coldness of the water (insulated from the heat of the sun by layers of peat) so hinder water uptake that for plants the habitat is physiologically dry. Leaves of many species of Heath family—cranberry, labrador tea, and leatherleaf—bear some resemblance to leaves of certain desert plants. The carnivorous sundew and pitcher plant have evolved in these bogs because of lack of available nitrogen and other nutrients.

Greenwood and Spalding peats (Typic Borohemists, dysic) are reported from Menominee Indian lands (Milfred, Olson, and Hole, 1967); Carlisle muck (Typic Medisaprist) has been reported from Monroe County. In northern Oconto County (T.31N., R.17, 18E.) there is a vast wetland locally called the Brazean Swamp. Construction of State Highway 64 in 1926 raised the water table despite twelve culverts under the road, killing a large "deer yard" of white cedar and other swamp conifers north of the highway. In 1962 the Wisconsin Department of Natural Resources began construction of a drainage ditch, now nearly 2 miles long, to lower the water table to its original level of 1926. These changes in the water table and the vegetative cover are bound to be reflected in the horizons of the peats, as are the various natural events that have occurred during the 10,000 years since glaciation. The Tawas mucky peat (Terric Borosaprist) and Lupton mucky peat (Typic Borosaprist) have been described in the area.

J15. *Slightly acid to alkaline sedge and woody peats and mucks and Pella, Poygan, and Brookston silt loam and silty clay loam.*

These soils occur in many depressions throughout glaciated portions of eastern and southern Wisconsin. These depressions range from a few acres to several square miles in size. Only a few of the largest (such as the Horicon and Theresa marshes) can be shown on the generalized soil map. The proportion of organic soils and mineral soils varies greatly from depression to depression. Some contain little peat, while others have only a thin border of mineral soils at the edges of large areas of peat. A variety of mineral soils occur in these depressions. In addition to the soils already named, Pella, Colwood, Keowns, Saylesville, Ashkum, Abington, Angelica, Roscommon, Wauseon, and Navan soils are common. The organic soils, such as Chippeny, Tawas, Carbondale, and Lupton peats, and Edwards and Suamico mucks, consist of partially to highly decomposed reed and sedge materials, with occasional fragments of wood. The deposits are often stratified and range from 1 to 40 feet or more in thickness. The mineral soils are all characterized by thick black A horizons and gray or mottled underlying horizons. Most are medium or fine textured and are calcareous at shallow depths.

Houghton (Fig. 8-7), Adrian, Palms, and Willett peats and mucks are reported from Trempealeau, Waukesha, and Columbia counties. Some small bodies of calcareous Edwards muck over marl (not shown on the map) occur in soil association F5 in northern Florence County (Hole et al., 1962). A mastodon skeleton was found in this soil association near Deerfield in Dane County (Dallman, 1968). A detailed study of a Hemist (33 cm of sapric horizons over hemic soil) in the Middleton Marsh in Dane County (T.7N., R.8E.; see Frazier and Lee, 1971; Lee and Manoch, 1974) reports a cover of grasses and sedges, medium to slight acidity (pH 5.8-6.1) and low carbon/nitrogen ratio (11-16). A Saprist in the University Marsh in Dane County (T.7N., R.9E.) originally supported arrowhead and bulrushes, and was moderately acid, with a moderate carbon/nitrogen ratio. Organic coatings had formed on prismatic and subangular blocky peds in the subsoil. Sheboygan (Hemic Medisaprist) is present in Sheboygan County under sedges, reeds, and grasses. Ogden (Terric Medisaprist) and Houghton (Typic Medisaprist) are present in Fond du Lac County.

Part III

Properties and Occurrence of Major Soil Series in Wisconsin Landscapes

CHAPTER 17

Some Properties of the Soil Series

This section summarizes information concerning the soil series listed in the legend of the soil map (Plate 1) and mentioned in the text of preceding chapters. Detailed soil profile descriptions and interpretations of these soils are not presented here. This information is available for study in the files of the U.S. Soil Conservation Service and the Geological and Natural History Survey, University of Wisconsin-Extension in Madison. The bibliography lists published sources of data cited.

The data presented here and results of studies published elsewhere, particularly by the Soil Conservation Service (1967a) and the Geological and Natural History Survey (Milfred, Olson, and Hole, 1967; Milfred and Hole, 1970), afford exciting opportunities for study and interpretation of Wisconsin soils. Fig. 17-1 illustrates how analytical results can be graphed for easy comparison. The data curves show amounts of available (exchangeable) cations, particularly the abundant plant nutrients calcium and magnesium, that are present in profiles of representative soils, and indicate relationships between these amounts and proportions of sand, clay, carbonates, and organic matter in the soils. The most fertile of these mineral soils contain the most organic matter and the least sand, and overlie calcareous materials.

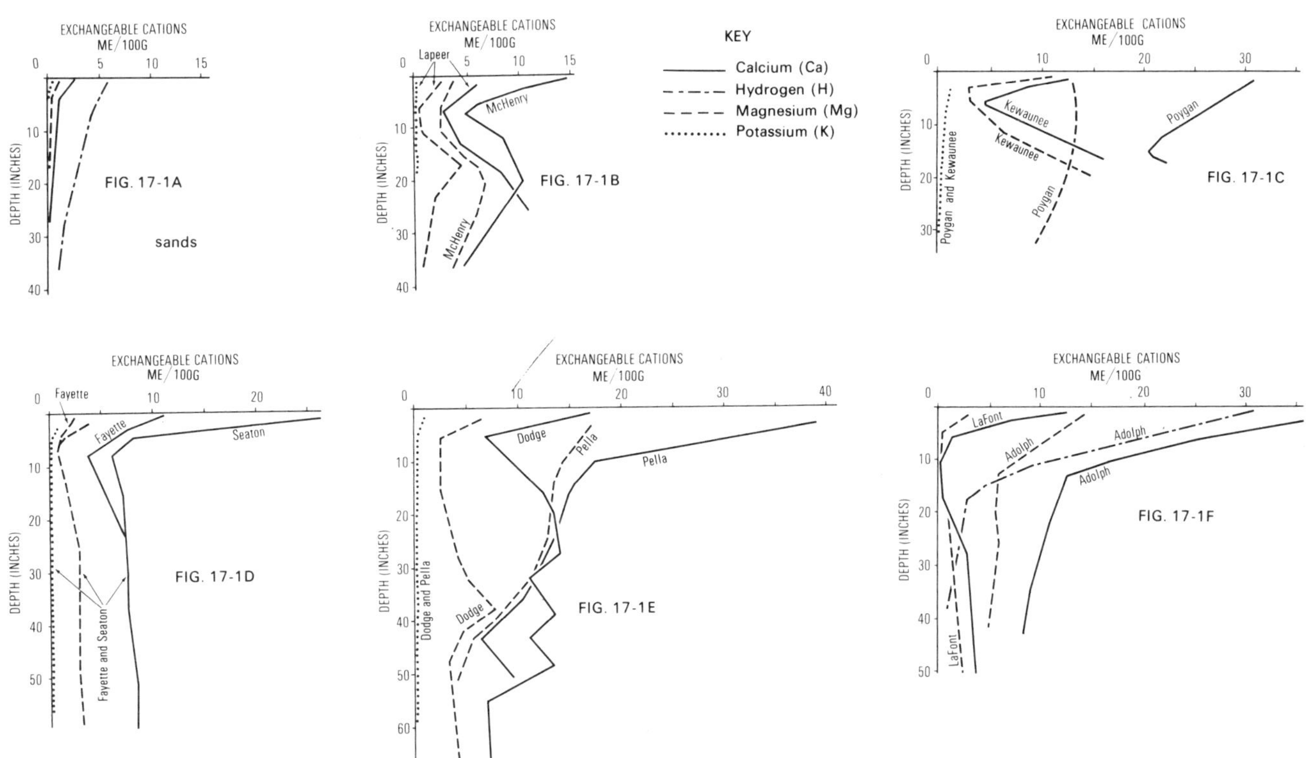

Figure 17-1. Data curves for exchangeable cation content of major horizons of several Wisconsin soils (data from Soil Survey Investigations Report No. 17, Soil Conservation Service, 1967a).

Contents of exchangeable calcium (Ca^{++}) are indicated by solid lines, exchangeable magnesium (Mg^{++}) by dashed lines, exchangeable hydrogen (H^{+}) by dash-dot lines, and exchangeable potassium (K^{+}) by dotted lines. Exchangeable sodium (Na^{+}) is not shown because quantities are minute.

Fig. 17-1A shows the characteristic low reserves of exchangeable calcium and magnesium in very sandy, excessively drained soils (Plainfield, Sparta, and the less fertile Boone). Fig. 17-1B shows data for the well-drained Lapeer sandy loam, developed from glacial till, and McHenry silt loam, formed in both a loess covering and an underlying till. The curves for the more fertile McHenry soil lie to the right of those for the Lapeer. Data for the Miami silt loam would fall between those for these two soils. Fig. 17-1C shows relationships for the fine-textured, well to moderately well drained Kewaunee soils and the poorly drained Poygan, both on clayey reddish-brown, calcareous glacial drift. The Poygan soil has a high content of available nutrients because of high contents of organic matter and clay and because it is in a lowland that receives leachate from the upland Kewaunee soils. Depletion of clay, organic matter, and exchangeable cations in the A2 horizon is evident in the Lapeer, McHenry, and Kewaunee soils. Fig. 17-1D compares two loess-derived soils, of which the Seaton silt loam is the less leached. Figs. 17-1E and 17-1F represent upland (Dodge and Lafont) and corresponding wetland (Pella and Adolph) soils. The Dodge-Pella landscape is much more fertile than the Lafont-Adolph landscape.

This chapter does not present much data on content of exchangeable cations in soils, but does report information on cation exchange capacity and a number of other properties.

One hundred and eighty-eight soil series, four miscellaneous stony or rocky land units, one wet alluvial soils (undifferentiated) unit, and peats and mucks constitute the list of units in the legend of the soil map. These units and some others mentioned in preceding chapters are listed below in alphabetical order, with selected information given in abbreviated form. Data for the Fayette silt loam are given below in italics by way of example. Laboratory data are lacking for many soil series. In such cases estimates of properties are given (see item 10, below). The information is given in the following order:

1. Name of soil series: *Fayette.*
2. Texture(s) of surface soil: *sil.* The following abbreviations are used:

s = sand	l = loam
ls = loamy sand	sil = silt loam
lfs = loamy fine sand	sicl = silty clay loam
sl = sandy loam	sic = silty clay
fsl = fine sandy loam	cl = clay loam
vfsl = very fine sandy loam	c = clay

3. Map legend units pertaining to the soil: (*A5, A6, A7, D1, D2, D3*); and relevant text figures: (*Figs. 2-51, 2-53, 7-4, 7-9, 7-10*).
4. Classification of the soil, according to both the new USDA classification (Soil Conservation Service, 1960) and the older classification (Baldwin, Kellogg, and Thorp, 1938; Thorp and Smith, 1949), and natural drainage condition: *Typic Hapludalf, fine-silty, mixed, mesic* (*Gray-Brown Podzolic*)*; well drained.*
5. Initial materials from which the soil formed: *solum* [i.e., A and B horizons] *is 40 to 55 in. thick, developed within a loess deposit at least that thick over residuum, i.e., weathered rock material, and Paleozoic bedrock. In the valley phase of Fayette silt loam, textural discontinuities occur due to incorporation of sandstone and limestone bedrock debris into the silty material during slow downhill creep over a period of thousands of years.* [Any color notations are for moist soil.]
6. Major soil horizons present: *Ap* [or *O1, O2, A1*], *A2, B1, B2t, B3, C.* [Note: The following soil horizon designations are used: Organic horizons—O1, O2; mineral surface horizons—A1, A2, A3; intertonguing surface soil and subsoil horizons—A&B; subsoil horizons—B1, B2, B3; for clay accumulations—Bt; for organic matter accumulations—Bh; for iron accumulations, Bir; for organic matter and iron accumulations, Bhir; initial material, below the true soil—C. A second (lower) sequence of A and B horizons is indicated by the prime mark, as in A′2&B′2, to differentiate these from the A2 and B2 horizons above them. Fragipans are indicated by the symbol x, cemented horizons by m, strong gleying (graying by waterlogging) by g, buried horizons by b, plowed horizons by p, as in Ap. Roman numerals II and III indicate second and third geologic (initial) materials, different from each other and from the material from which the horizons above II layers formed. See Buol, Hole, and McCracken, 1973. The list of soil horizons is not necessarily exhaustive in each instance.]
7. Representative properties of selected horizons [see table below]:

Some properties of Fayette silt loam

Horizon	Depth in.	OM[a] %	Sand[b] %	Silt[c] %	Clay[d] %	CEC[e] me/100g	BS[f] %	pH[g]	BD[h] g/cc	Fibers[i] unrub. %	Fibers[i] rub. %
A1	0-3	—	—	—	—	—	—	—	—	—	—
A2	3-10	2.7	7	78	15	12	50	5.6	1.3	0	0
B2t	15-36	0.3	4	65	31	22	60	4.7	1.6	0	0
C	42-60	0.1	7	63	30	20	66	5.2*	1.6	0	0

*The C horizon is calcareous at a depth of 6 to 10 ft. It is only near the Mississippi River valley that the loess is as thick as that.

[a]Organic matter, percent by weight (% carbon X 1.7) in the fine earth (<2mm material).

[b]Sand particles are 0.50-2.0 mm in dia. Percentages are of total fine earth, ovendried, by weight.

[c]Silt particles are .002-0.50 mm in dia. Percentages are of total fine earth, ovendried, by weight.

[d]Clay particles are <.002 mm in dia. Percentages are of total fine earth (<2 mm material), ovendried, by weight.

[e]Cation exchange capacity, determined with NH_4OAC_3, is in terms of milliequivalents per 100 grams of ovendry soil.

[f]Base saturation percentage by sum of cations indicates the percentage of CEC occupied by exchangeable calcium, magnesium, potassium, and sodium.

[g]Soil reaction is measured in terms of pH value, which is the logarithm of the reciprocal of the H ion concentration.

[h]Bulk density is determined by dividing the weight in grams of 1 cubic centimeter of ovendry soil by the volume, which is 1 cc.

[i]This property, percent of fibers by volume in unrubbed (unrub.) and rubbed (rub.) soil, applies only to organic soils (Histosols), of which Adrian muck is an example (p. 127 below).

8. Other properties: *Clay films and light gray silt coatings, the latter 0.1 mm thick, are observed on the surfaces of blocky peds in the Bt; silt coatings not visible when moist; calcareous C material has a carbonate content of 15%* $CaCO_3$ *equiv.*
9. Setting: *On convex ridges, side slopes, and some high stream terraces, with slope gradients of 3 to 25% in the "Driftless Area"; original vegetation was oak savanna; all but steep land cleared and in crops or pasture; mean ann. soil temp. 48-52°F, mean ann. ppt. 32-34 in.*
10. History and references: *Estab., Fayette Co., Ia., 1919; Muckenhirn et al., 1955; Klingelhoets, 1959; Glocker, 1966; Milfred, 1966; Soil Conservation Service, 1967a.* [Where no reference to data is listed in the following pages, soil properties are estimated on the basis of field descriptions and laboratory data for similar soils.]
11. Productivity index: *75 bu. oats; 475-550 bd. ft. red pine; 500-600 bd. ft. white pine.* [Estimate from Beatty et al., 1966. Yields per acre under a high level of management are given. Where artificial drainage is mentioned as a prerequisite, this applies to oats, not to pine.]
12. Catena number: *catena no. 3.* [See Chapter 18, "Major Soil Toposequences (Catenas) of Wisconsin."]

Adolph l-sil (F3, F12, F21); Typic Haplaquoll, coarse-loamy, mixed, frigid (Humic Gley); very poorly drained; solum ranges from 30 to 50 in. in thickness and is developed in silty material 18 to 34 in. thick over reddish-brown glacial till that may be slightly calcareous at 60 in. but is composed largely of igneous rock materials. Horizons: O1, O2, A1, A13, B2g, IIB3, IIC.

Some properties of Adolph silt loam

Hori-zon	Depth in.	OM %	Sand %	Silt %	Clay %	CEC me/100g	BS %	pH	BD g/cc
A13	8-16	1.0	23	56	21	20	81	6.9	1.1
B2g	16-32	0.3	15	65	20	19	94	7.0	1.4
IIB3	32-44	0.1	54	35	11	9	95	7.3	1.5
IIC	44-54	0.1	60	31	9	7	94	7.6	1.6

Other: A stoneline may occur at the silt-till contact. Setting: In low-lying wet flats and depressions, slope 1% gradient, or, rarely, seepage spots on slopes in ground moraine landscapes; originally forested with lowland hardwoods and now still the same except for a few fields cleared for pasture and for raising forage crops; mean ann. soil temp. 40-45°F, mean ann. ppt. 25-32 in. Estab., Mille Lacs Co., Minn., 1927; Soil Conservation Service, 1967a; where artificially drained, 60 bu. oats; not suitable for timber production; catena no. 94.

Adrian muck (J15); Terric Medisaprist, sandy or sandy-skeletal, mixed, euic, mesic (Bog); very poorly drained; solum ranges from 16 to 50 in. in thickness and is developed in herbaceous, fibrous material overlying calcareous sands and loamy sands. Horizons: O1, Oa1, Oa2, Oa3, Oa4, IIC.

Some properties of Adrian muck

Hori-zon	Depth in.	OM %	Sand %	Silt %	Clay %	CEC me/100g	BS %	pH	BD g/cc	Fibers unrub. %	Fibers rub. %
Oa1	0-16	51	5	40	4	250	30	7.0	0.2	12	<5
Oa4	27-34	68	8	22	2	190	40	5.5	0.3	12	<5
IIC	34-60	0.2	87	8	5	2	11	8.0	1.7	0	0

Other: The solum has about 55% ash content; and four organic horizons have structure as follows: (1) moderate, medium granular, (2) weak coarse subangular blocky, (3) weak thick platy, and (4) massive; most of the organic material is herbaceous in origin; a few woody fragments are present. Setting: In ancient lake basins of a few to hundreds of acres in size; slopes are less than 2% gradient; surrounding mineral soils are loamy; original vegetation was sedges, reeds, grasses, with some willow, alder, aspen; less than 10% of the area is drained and used for pasture, hay, or truck crops; mean ann. soil temp. 40-45°F; mean ann. ppt. 30 in., supplemented by runoff from surrounding slopes. Estab., Sanilac Co., Mich., 1955; Milfred, Olson, and Hole, 1967; not suitable for oats or pine; catena no. 51.

Ahmeek sl-sil (G9, G10); Typic Fragiorthod, coarse-loamy, mixed, frigid (Brown Podzolic intergrading to Acid Brown Forest soil or Sol Brun Acide); well to moderately well drained; solum, 18 to 30 in. thick, is developed in silty material less than 2 ft. thick over dark reddish-brown (5YR-2.5YR 4/4) acid stony loam glacial till less than 4 ft. thick over bedrock; outcrops of basic igneous and metamorphic rocks are common. Fragipan (x) is present. Horizons: O1, O2, A1, Bhir, Bx, C.

Some properties of Ahmeek silt loam

Hori-zon	Depth in.	OM %	Sand %	Silt %	Clay %	CEC me/100g	BS %	pH	BD g/cc
A1	0-3	14.0	27	55	18	50	40	5.2	1.2
Bhir	3-16	2.6	52	33	15	26	26	5.1	1.4
Bx	16-60	1.4	60	30	10	16	60	5.7	1.8
C	60-75	0.3	38	40	22	18	80	6.3	1.6

Other: Numerous stones from Precambrian "iron formation"; till may be a clay loam. Setting: On hilly to undulating moraines on bedrock ridges, slopes 0 to 15% gradient; originally under northern mesic forest and now mostly in pasture and woodlots; mean ann. soil temp. 41°F, mean ann. ppt. 30 in. Estab., Houghton Co., Mich. (name from Keweenaw Co., Mich.); Soil Conservation Service, 1952; Hole et al., 1962; 70 bu. oats; 300-325 bd. ft. red pine; 350-400 bd. ft. white pine; catena no. 126.

Akan sil (J1, J2); Mollic Haplaquept, fine-silty, mixed, nonacid, mesic (Low Humic Gley); somewhat poorly drained; solum is 24 to 36 in. thick, developed in bluish-gray (7.5YR 5/0) calcareous lake-laid silty clay. Horizons: O1, A1, Bg, Cg.

Some properties of Akan silt loam

Hori-zon	Depth in.	OM %	Sand %	Silt %	Clay %	CEC me/100g	BS %	pH	BD g/cc
A1	0-8	8.0	20	60	20	55	85	6.5	1.3
Bg	8-26	2.0	18	50	32	23	92	6.0	1.5
Cg	26-60	0.1	15	50	35	21	95	7.5	1.7

Other: There are variations in thickness of cambic B horizon and development of mottling. Setting: In nearly level areas in river valleys where intermittent flooding occurs; slopes 0 to 3% gradient; originally under swamp hardwoods; about half the area has been drained and used for pasture or corn; mean ann. soil temp. 46°F, mean ann. ppt. 32 in. Proposed, Town of Willow, Richland Co., Wis., 1947; Robinson and Klingelhoets, 1959; 65 bu. oats; not suitable for pine; catena no. 41.

Alban fsl-sil (E4); Typic to Spodic Glossoboralf, coarse-loamy, mixed, frigid (Gray-Brown Podzolic); well to moderately well drained; solum is 20 to 40 in. thick, developed in loamy deposits 20 to 42 in. thick over reddish-brown acid stratified silts and fine sands of glacio-lacustrine origin. Horizons: O1, O2, A2, Bhir, A′2x&B′2x, A′2x, B′2x, C.

Some properties of Alban loam

Hori-zon	Depth in.	OM %	Sand %	Silt %	Clay %	CEC me/100g	BS %	pH	BD g/cc
Bhir	1-9	2.0	40	45	15	12	44	5.7	1.3
A′2x	9-13	0.5	35	55	10	6	10	6.0	1.5
B′2x	16-24	0.3	36	52	12	7	40	6.0	1.6
C	24-50	0.1	45	44	11	5	80	5.7	1.5

Setting: On nearly level to gently rolling glacio-lacustrine plains bordered by moraines; slopes 3 to 20% gradient; originally under northern mesic forest, now about equally divided among pasture, cropland, and woodland; mean ann. soil temp. 45°F, mean ann. ppt. 30 in. Proposed, NW ¼, NE ¼, Sec. 34, T.26N., R.9E., Marathon Co., Wis., 1960; Milfred, Olson, and Hole, 1967; 65 bu. oats; 350-400 bd. ft. red pine; 300-600 bd. ft. white pine; catena no. 120.

Almena sil (F3, F12, F20, F22); Aeric Glossaqualf, fine-silty, mixed, frigid (Gray-Brown Podzolic); somewhat poorly drained; solum is 30 to 40 in. thick, developed entirely in a loess covering that is 30 to 60 in. thick over acid (calcareous in a strip of land about 20 miles long and 1 mile wide near Auburndale, Wis.; Hole, 1943), compact, reddish-brown, gravelly sandy loam to loam glacial till. Horizons: O1, O2, A1, A2, As&B2t, B2tg, B3t, IIC.

Some properties of Almena silt loam

Hori-zon	Depth in.	OM %	Sand %	Silt %	Clay %	CEC me/100g	BS %	pH	BD g/cc
A1	0-3	9.4	7	76	17	32	45	5.0	1.1
A2	3-14	1.8	5	85	10	12	25	5.0	1.5
B2tg	19-32	2.0	12	65	23	20	45	4.8	1.6
IIC	40-65	0.1	65	25	10	9	65	5.5	1.9

Other: All horizons below the A1 show mottling and some platy structure; tongues and skeletans (gritty films; see Brewer, 1964) of bleached silt and very fine sands partially coat surfaces of prisms of upper Bt; clay skin in Bt horizon is 31% clay, 64% silt; iron and manganese oxide nodules constitute more than 50% of the 0.1-2 mm sand in the A1 horizon (0-4 in.) and 25-50% in the B32 horizon (32-41 in.). Setting: On level ground moraine under lowland hardwoods and northern forest, largely cleared and cropped to hay, small grains, and corn for silage; mean ann. soil temp. 42-45°F, mean ann. ppt., 28-32 in. Estab., Barron Co., Wis., 1950; Soil Conservation Service, 1967a; where artificially drained, 70 bu. oats; 450-500 bd. ft. white pine; catena no. 94.

Alpena sl-l (E4, E6); Udorthentic Haploboroll, sandy-skeletal, carbonatic, mixed, frigid (Regosol; Rendzina-like); well drained; solum ranges from 4 to 14 in. in thickness, and is developed in sand and gravel of glacial lake beaches, kames, outwash plains, eskers, and deltas. Horizons: O1, O2, A1, B, C.

Some properties of Alpena sandy loam

Hori-zon	Depth in.	OM %	Sand %	Silt %	Clay %	CEC me/100g	BS %	pH	BD g/cc
A1	0-6	16.0	91	6	3	40	98	8.5*	1.8
C	9-20	0.3	99	1	0.3	1	100	8.5*	2.5

*calcareous

Other: Volume of fragments larger than 2 mm in diameter may be 60% in the A1 horizon and 80% in the C horizon. Setting: On glacial drift deposits, with slope gradients of 0 to 50% (mostly under 20%); the original vegetative cover of mixed northern mesic forest, with some xeric species present, is largely undisturbed; used for recreation, forestry, and some pasture; mean ann. soil temp., 44°F, mean ann. ppt., 28 in. Estab., Alpena Co., Mich., 1924; not suitable for row crops; 350-400 bd. ft. white pine; catena no. 148.

Alstad l-sil (F7): Aquic Eutroboralf, fine-loamy, mixed, mesic (Gray Wooded); somewhat poorly drained; solum is 32 to 48 in. thick, developed in calcareous loam to clay loam (brown; 10YR 5/3) glacial till with as much as a foot of loess covering in places. Horizons: O1, O2, A1, A2, Btg, C.

Some properties of Alstad silt loam

Hori-zon	Depth in.	OM %	Sand %	Silt %	Clay %	CEC me/100g	BS %	pH	BD g/cc
A2	4-18	7.0	25	55	20	26	65	7.0	1.5
Btg	18-44	0.6	30	35	35	20	80	5.3	1.6
C	44-60	0.1	40	32	28	12	100	7.0	1.7

Other: Clay skins and light-gray silt coatings on faces of blocky peds of Btg and upper C; all horizons below the A1 are mottled. Setting: In nearly level depressions in glacial ground moraine; originally under southern mesic forest cover; now mostly in pasture and fields cropped to corn, small grains, and forage crops; mean ann. soil temp. 44°F, mean ann. ppt. 30 in. Proposed, Burnett Co., Wis., Sec. 20, T.38N., R.18W.; where artificially drained, 70 bu. oats; 450-500 bd. ft. white pine; catena no. 107.

Altdorf sil (F23); Aeric Glossaqualf, fine, mixed, frigid (Low Humic Gley); poorly drained; solum ranges from 40 to 72 in. in thickness and is developed in 15 to 30 in. of silty material over clayey residuum from micaceous schist. Horizons: O1, O2, A1, A2g, (A&B)g, I-II, IIB2t, IIB3t, IIC.

Some properties of Altdorf silt loam

Hori-zon	Depth in.	OM %	Sand %	Silt %	Clay %	CEC me/100g	BS %	pH	BD g/cc
A1	0-3	12.0	10	70	20	50	50	6.5	1.0
A2g	3-12	2.0	15	75	10	15	40	5.8	1.5
IIB2t	23-49	0.3	10	20	70	27	70	7.0	1.7
IIC	62-90	0.1	50	45	5	5	80	7.0	1.8

Other: Rock structure visible in lower IIC. Setting: In level or depressional sites on bedrock-controlled upland, slopes of less than 2% gradient originally under lowland hardwoods, now mostly in woodlots and pasture but with some in drained cropland; mean ann. soil temp. 46°F, mean ann. ppt. 28-32 in. Proposed, Wood Co., Wis., 1960; 60 bu. oats; not suitable for timber production; catena no. 104.

Amery sl-sil. *See* Milaca.

Angelica l-sil (E2, E3, E5, E9); Mollic or Aeric Haplaquept, fine-loamy, mixed, nonacid, frigid (Humic Gley); poorly and very poorly drained; solum is 15 to 25 in. thick, developed in reddish-brown (5YR 5/4) calcareous loam glacial till. Horizons: O1, O2, A1, B2g, B3t, C.

Some properties of Angelica loam

Hori-zon	Depth in.	OM %	Sand %	Silt %	Clay %	CEC me/100g	BS %	pH	BD g/cc
A1	0-4	18.0	45	35	20	60	70	6.3	1.0
B2g	4-12	1.0	55	27	18	15	75	7.0	1.4
B3t	12-15	0.2	53	25	22	11	80	7.0	1.5
C	15-48	0.1	52	30	18	10	100	8.0*	1.6

*calcareous

Other: The O2 horizon is a 2-in. layer of black muck; some pedons have limestone fragments throughout. Setting: depressions in till plains, slopes less than 2% gradient; originally in lowland hardwood and conifer forest and still in woodland except where cleared for pasture; mean ann. soil temp. 44-46°F, mean ann. ppt. 30 in. Proposed, Shawano Co., Wis., 1946; where artificially drained, 65 bu. oats; 350-400 bd. ft. white pine; catena no. 135.

Ankeny ls-sl (J1); Cumulic Hapludoll, coarse-loamy, mesic (minimal Brunizem; Alluvial soil); well drained; solum is 44 to 60 in. thick, developed in somewhat sandy local alluvium that is dark yellowish-brown (10YR 4/4, moist) in color. Horizons: O1, O2, A1, B, C.

Some properties of Ankeny loamy sand

Horizon	Depth in.	OM %	Sand %	Silt %	Clay %	CEC me/100g	BS %	pH	BD g/cc
A1	0-24	12.0	60	30	10	30	85	6.3	1.2
C	50-60	0.3	80	15	5	3	90	7.0	1.6

Other: The A horizon is very dark brown (10 YR 2/2) and is 2 to 3 ft. thick. Setting: On low footslopes, alluvial fans; slope gradients are 1 to 5%; original vegetation of tall prairie grasses is now replaced by bluegrass sod or crops of corn, oats, or legume hay; mean ann. soil temp. 47°F, mean ann. ppt. 32 in. Estab., Polk Co., Ia., 1958; 70 bu. oats; 400-450 bd. ft. white pine; catena no. 165.

Antigo sil (F6, F17, F25, F26) (Figs. 2-53, 2-54, 4-7, 13-3, 13-5); Typic Glossoboralf, fine-silty over sandy or sandy-skeletal, mixed, frigid (Gray-Brown Podzolic or Gray Wooded); well drained; solum is 24 to 45 in. thick, developed in a silty deposit 20 to 36 in. thick, overlying sand and gravel that contains only occasional fragments of carbonate rock. Horizons: O1, O2, A1, A2, B&A, B2t, I&IIB3, IIC.

Some properties of Antigo silt loam

Horizon	Depth in.	OM %	Sand %	Silt %	Clay %	CEC me/100g	BS %	pH	BD g/cc	Gravel*
A2	3-15	2.5	16	75	9	15	62	4.9	1.3	1
B2t	15-30	1.1	12	63	25	20	45	4.8	1.6	5
I&IIB3	30-33	0.3	51	37	12	10	65	4.7	1.8	12
IIC	33-60	0.2	91	4	5	6	80	5.0	1.9	25

*on the basis of the whole soil

Other: The A2 tongues into the B2t. Setting: On nearly level to sloping glacial outwash plains, both unpitted and pitted, with slopes of 1 to 3% gradient common; original vegetation was northern mesic forest, but level lands are nearly all cleared and in crop production (potatoes, oats, hay, silage corn) with some irrigation; sloping areas are commonly in pasture; mean ann. soil temp. 41-45°F, mean ann. ppt. 28-32 in. Estab., Langlade Co., Wis., 1947; Milfred, Olson, and Hole, 1967; Al-Rawi et al., 1969; 80 bu. oats; 475-575 bd. ft. red pine; 300-600 bd. ft. white pine; catena no. 112.

Arenzville sil (J1) (Fig. 2-51); Typic Udifluvent, coarse-silty, mixed, nonacid, mesic (Alluvial soil); well to moderately well drained; a postsettlement stratified alluvial deposit, light in color (10YR 5/3 moist), 20 to 40 in. thick, overlies a buried black silt loam A horizon. Horizons: O1, A1, C, Ab, Cb. The thickness of the A-C profile over the Ab is proportional to the area of the contributing drainage basin.

Some properties of Arenzville silt loam

Horizon	Depth in.	OM %	Sand %	Silt %	Clay %	CEC me/100g	BS %	pH	BD g/cc
A1	0-10	7	10	75	15	25	70	7.0	1.2
C	10-25	4	8	80	12	15	65	6.3	1.3
Ab	25-40	20	8	72	20	60	88	6.1	1.2
Cb	40-60	3	6	78	16	20	93	7.0	1.6

Other: The Ab horizon is more than 10 in. thick and may be mottled; the Cb horizon may contain chert, gravel, and sand in places, amounting to 5 to 20% of the volume. Setting: On nearly level floodplains with slopes of less than 3% gradient; submergence in flood waters is frequent but of short duration; original vegetation was lowland hardwoods; present uses are for crops, pasture, and woodland; mean ann. soil temp. 47-51°F, mean ann. ppt. 29-32 in. Estab., Cass Co., Ill., 1939; Hole, 1956a; Robinson and Klingelhoets, 1961; Klingelhoets, 1962; where protected from steam overflow, 70 bu. oats; not suitable for timber production; catena no. 39.

Arland sl-sil (D7, D9); Typic Hapludalf grading to Eutric Glossoboralf, fine-loamy over sandy-skeletal, mixed, frigid (Gray-Brown Podzolic and Gray Wooded); well drained; solum is 24 to 36 in. thick, developed in brown (7.5YR 5/4) sandy loam glacial till 20 to 40 in. thick over weakly cemented Cambrian sandstone. Horizons: O1, O2, A1, A2, B1, IIB2t, B3, IIC, R.

Some properties of Arland loam

Horizon	Depth in.	OM %	Sand %	Silt %	Clay %	CEC me/100g	BS %	pH	BD g/cc
A2	2-10	3.0	30	60	10	10	10	4.4	1.2
IIB2t	16-25	0.5	20	60	20	13	40	5.0	1.5
IIC	30-36	0.1	88	6	6	4	35	5.1	1.7

Other: Bleached silt coatings may be present on ped faces in the upper Bt horizon; some pebbles throughout the solum. Setting: On nearly level to rolling uplands, slopes of 3 to 15% gradient common; original vegetation was northern mesic forest, but bodies of this soil have been cleared for the most part and used for production of corn, small grains, hay, and for pasture; mean ann. soil temp. 43-46°F, mean ann. ppt. 30-35 in. Estab., Barron Co., Wis., 1950; Robinson et al., 1958; Soil Conservation Service, 1967a; 65 bu. oats; 400-450 bd. ft. red pine; 350-400 bd. ft. white pine; catena no. 108.

Ashdale sil (A1, A2) (Fig. 7-5); Typic Argiudoll, fine-silty, mixed, mesic (Brunizem); well drained; solum is 45 to 60 in. thick, developed in 36 to 45 in. of loess resting in cherty reddish-brown clay residuum 0.5 to 2 ft. thick overlying limestone (dolomite) bedrock. Horizons: O1, O2, A1, A2, A3, B2t, IIB3t, R.

Some properties of Ashdale silt loam

Horizon	Depth in.	OM %	Sand %	Silt %	Clay %	CEC me/100g	BS %	pH	BD g/cc
A1	0-12	6.0	2	78	20	20	60	5.6	1.2
B2t	15-43	2.0	5	70	25	17	62	5.2	1.4
IIB3t	43-51	0.5	2	48	50	25	55	4.8	1.6

Other: Chert gravel may occupy 20% of the volume of the B3 horizon; the red residual clay consists chiefly of montmorillonite, vermiculite, and illite. Setting: On gently to moderately sloping (3 to 12% gradient) ridge tops in dissected uplands of "unglaciated" areas; original vegetation was tall prairie; nearly all bodies of this soil have been plowed and used for corn, oats, hay, and pasture; mean ann. soil temp. 47°F, mean ann. ppt. 28-40 in. Estab., Lafayette Co., Wis., 1964; Kaddou, 1960; Slota, 1969; Black, 1970a; 70 bu. oats; where forest soil microorganisms have been introduced, 400-450 bd. ft. red pine; catena no. 7.

Ashippun l-sicl (B1) (Fig. 8-7); Aquollic Hapludalf, fine-loamy, mixed, mesic (Gray-Brown Podzolic); somewhat poorly drained; solum is 20 to 40 in. thick, developed in 10 to 20 in. of silty material overlying highly calcareous shaly loam glacial till. Horizons: O1, O2, A1, A2, B21t, IIB22t, IIC.

Some properties of Ashippun loam

Hori-zon	Depth in.	OM %	Sand %	Silt %	Clay %	CEC me/100g	BS %	pH	BD g/cc
A2	8-11	1.7	33	45	22	17	98	7.5	1.3
B21t	11-20	0.2	2	62	36	23	100	8.0*	1.4
IIB22t	20-26	0.3	21	41	38	21	100	8.5*	1.6
IIC	26-30	0.1	49	35	16	8	100	8.5*	1.8

*calcareous

Other: Horizons below the A1 are quite mottled; the II22 horizon contains 5 to 25% shale fragments by volume, and the IIC contains more; some segregated lime is observed in the IIC. Setting: On slopes of 1 to 12% gradient on moraines of shaly glacial till; presence of seeps on slopes accounts for the occurrence of some bodies of this soil at higher positions than would be expected; the original southern mesic forest and oak savanna have been largely replaced, after drainage, by pasture, hay, oats, and corn; mean ann. soil temp. 45°F, mean ann. ppt. 30 in. Estab., Dodge Co., Wis., 1969; 65 bu. oats; not suitable for pine; catena no. 164.

Ashkum sil-sicl (B9, B19, B20) (Fig. 8-10); Typic Haplaquoll, fine, mixed, mesic (Humic Gley); poorly drained; solum is 30 to 40 in. thick, developed in less than 15 in. of loess overlying gray (5Y 5/2) calcareous silty clay loam glacial till in Woodfordian age. Horizons: O1, O2, A1, B2g, IIB2g, IIC.

Some properties of Ashkum silty clay loam

Hori-zon	Depth in.	Om %	Sand %	Silt %	Clay %	CEC me/100g	BS %	pH	BD g/cc
A1	0-12	8.0	14	48	38	40	90	7.0	1.0
B2g	12-18	1.5	18	46	36	25	94	7.5	1.4
IIB2g	18-34	0.8	4	52	44	20	95	7.8*	1.5
IIC	34-50	0.7	20	50	30	10	100	8.0*	1.8

*calcareous

Other: Gleying has produced gray colors with yellowish-brown mottles; crayfish holes and sedge-root channels are present to a depth of 4 ft.; some manganese and iron oxide concretions in sand fractions, especially in the coarse sand of the IIB horizon; some calcite concretions in the sand of the IIC horizon. Setting: In nearly level depressions on glacial moraines, with slopes of less than 2% gradient; original sedge meadow vegetation is replaced in most places by corn and soybeans on artificially drained fields; mean ann. soil temp. 47°F, mean ann. ppt. 20-40 in. Estab., Iroquois Co., Ill., 1940; Soil Conservation Service, 1967a; where artificially drained, 65 bu. oats; not suitable for timber production; catena no. 53.

Auburndale sil (F1, F9, F12, F18, F19, F20, J6) (Fig. 12-7, 12-8); Typic Glossaqualf, fine-silty, mixed, frigid (Low Humic Gley); poorly drained; solum is 30 to 40 in. thick and is formed in 30 to 48 in. of loess over reddish-brown (5YR 4/4) acid sandy loam glacial till. Horizons: O1, O2, A1, A2, B&Ag, B2tg, B3g, IICg.

Some properties of Auburndale silt loam

Hori-zon	Depth in.	OM %	Sand %	Silt %	Clay %	CEC me/100g	BS %	pH	BD g/cc
A1	0-4	12.0	11	65	24	40	50	5.3	0.9
A2	4-10	0.8	14	70	16	12	45	5.2	1.5
B2tg	14-29	0.2	20	60	20	18	70	4.9	1.6
IICg	40-50	0.1	77	18	5	5	90	6.5	1.8

Other: A2 horizon is mottled and tongues into the mottled Bt horizon; platy structure is prominent in all horizons below the A1, grading from fine above to coarse below. Setting: In level or depressional areas, with slopes of less than 2% gradient, in silt-blanketed acid glacial ground moraine landscape; original vegetation was lowland hardwoods; perhaps 15% of the area of bodies of this soil has been cleared and is used for pasture or hayland; mean ann. soil temp. 43-44°F, mean ann. ppt. 28-32 in. Estab., Langlade Co., Wis., 1947; Hole and Schmude, 1959; Soil Conservation Service, 1967a; where artificially drained, 60 bu. oats; 250-300 bd. ft. white pine; catena no. 94.

Au Gres s-ls (E12, J5, J12, J14); Entic Haplaquod, sandy, mixed, frigid (weakly developed Groundwater Podzol); somewhat poorly drained; solum is 20 to 48 in. thick, developed in Woodfordian and Valders glacio-fluvial or lacustrine sands that are acid to neutral at depth. Horizons: O1, O2, A1, A2, Bhir, B3, C.

Some properties of Au Gres sand

Hori-zon	Depth in.	OM %	Sand %	Silt %	Clay %	CEC me/100g	BS %	pH	BD g/cc
A2	2-13	1.8	90	7	3	3	24	5.8	1.3
Bhir	13-30	2.0	88	6	6	6	27	5.2	1.4
C	40-60	0.1	95	3	2	2	37	6.0	1.5

Other: The soil is mottled below the A1 horizon; chunks of weakly cemented Ortsein may be found in the upper third of the Bhir horizon; the A2 and Bhir horizons tongue downward in places; a thin peat mat may rest on the A1 horizon. Setting: On glacial drift plains (on till, outwash, or lacustrine deposits) with slopes of 0 to 3% gradient; original vegetation of swamp conifers is largely undisturbed; in places the soil has been cleared and artificially drained for pasture and for production of pickles or blueberries; mean ann. soil temp. 43-47°F, mean ann. ppt. 27-33 in. Estab., Franklin Co., N.Y., 1955; Milfred, Olson, and Hole, 1967; where artificially drained, 55 bu. oats; 325-375 bd. ft. red pine; 500.600 bd. ft. white pine; catena no. 146.

Aztalan l (J10); Aquic Argiudoll, fine-loamy, mixed, mesic (Brunizem); somewhat poorly drained; solum is 30 to 42 in. thick (the same as depth to carbonates), developed in 24 to 30 in. of loamy material (outwash) overlying glacio-lacustrine silts and clays. Horizons: O1, O2, A1, B21t, IIB22t, IIB3, IIC.

Some properties of Aztalan loam

Hori-zon	Depth in.	OM %	Sand %	Silt %	Clay %	CEC me/100g	BS %	pH	BD g/cc
A1	0-12	9.0	50	35	15	40	67	6.8	1.3
B21t	12-27	1.5	30	45	25	14	84	7.0	1.4
IIB22t	27-37	1.0	7	55	38	22	95	7.5	1.5
IIC	43-60	0.1	3	65	32	17	100	8.0*	1.6

*calcareous

Other: The thickness of the A1 horizon ranges from 10 to 16 in. Setting: Slopes are commonly less than 2% gradient; original vegetation was prairie and sedge meadow; nearly three quarters of the area has been drained and cropped to corn, small grains, and hay; mean ann. soil temp. 48°F, mean ann. ppt. 31 in. Proposed, Jefferson Co., Wis., 1966; Milfred and Hole, 1970; where artificially drained, 60 bu. oats; not suitable for timber production; catena no. 90.

Baraboo sil (B3) (Fig. 7-12); Typic Hapludalf, fine-silty, mixed, mesic (Gray-Brown Podzolic); well drained; solum is 20 to 40 in. thick, developed in that depth of loess over quartzite bedrock. Horizons: O1, O2, A1, A2, B2t, B3, R.

Some properties of Baraboo silt loam

Horizon	Depth in.	OM %	Sand %	Silt %	Clay %	CEC me/100g	BS %	pH	BD g/cc
A2	2-10	1.5	8	77	15	13	58	5.0	1.3
B2t	10-30	0.3	12	60	28	20	62	5.3	1.5

Other: Some quartzite fragments scattered throughout the pedon. Setting: On sloping to steep upland with thin glacial drift and loess covering over quartzite bedrock which outcrops in places; original vegetation was largely oak savanna; the less rocky and steep slopes have been cleared and are in crops and pasture; other areas are in forest; mean ann. soil temp. 47°F, mean ann. ppt. 27-33 in. Estab., Sauk Co., Wis., 1925; Geib et al., 1925; 70 bu. oats; 400-450 bd. ft. red pine; 400-450 bd. ft. white pine; catena no. 11.

Berrien ls-sl (C18, I17); Arenic to Typic Hapludalf, sandy over clayey to loamy, mixed, mesic (Gray-Brown Podzolic); moderately well drained; solum is 40 to 60 in. thick, developed in sandy material overlying loamy or clayey substratum, of glacial outwash and lacustrine origin. Horizons: O1, O2, A1, A2, B2t, C.

Some properties of Berrien loamy sand

Horizon	Depth in.	OM %	Sand %	Silt %	Clay %	CEC me/100g	BS %	pH	BD g/cc
A1	0-2	1.8	80	10	10	30	30	5.0	1.4
B2t	50-60	0.3	87	8	5	8	15	5.0	1.5
C	60-70	0.1	28	35	37	18	94	8.5*	1.5

*calcareous

Other: The C horizon ranges in texture from loam to clay; depth to mottling is from 16 to 36 in. Setting: On nearly level to gently sloping (0 to 6% gradient) glacial outwash, delta, or lacustrine deposits; the original vegetation, oak savanna, has largely been replaced by dairy farm crops, truck, and small fruit crops; mean ann. soil temp. 46°F, mean ann. ppt. 30 in. Estab., Van Buren Co., Mich., 1922; type location is in the NE½, NW¼, Sec. 23, T.4N., R.12W., Allegan Co., Mich.; 60 bu. oats; 400-450 bd. ft. white pine; catena no. 91.

Bertrand sil (A11, A12) (Fig. 7-4); Typic Hapludalf, fine-silty, mixed, mesic (Gray-Brown Podzolic); well drained; solum is 40 to 48 in. thick. formed in 36 to 50 in. of silty deposits over acid sand and gravel outwash. Horizons: O1, O2, A1, A2, B1, B2t, IIB3, IIC.

Some properties of Bertrand silt loam

Horizon	Depth in.	Om %	Sand %	Silt %	Clay %	CEC me/100g	BS %	pH	BD g/cc
A2	3-8	1.5	5	80	15	14	72	5.6	1.2
B2t	14-29	0.3	2	63	35	20	60	5.1	1.4
IIC	38-47	0.1	88	6	6	5	53	5.2	1.8

Other: Bt horizon seems to be slightly more developed than that of Fayette silt loam. Setting: On natural high terraces in the lower Wisconsin River valley and tributaries thereof with slopes commonly 0 to 4% gradient; original vegetation was oak savanna; bodies of this soil are almost entirely cleared and in crops or pasture; mean ann. soil temp. 47°F, mean ann. ppt. 27-33 in. Estab., Mississippi Co., Mo., 1921; Beatty, 1960; Klingelhoets, 1962; 75 bu. oats; 450-550 bd. ft. red pine; 500-600 bd. ft. white pine; catena no. 25.

Bevent ls-fsl (G14); Alfic Haplorthod, sandy, mixed, frigid (Podzol to Gray-Brown Podzolic transitional soil); well to moderately well drained; solum is 25 to 35 in. thick, developed in 20 to 42 in. of sandy deposit over lacustrine acid sands and coarse silts. Horizons: O1, O2, A1, A2, Bir, A′2x, B′2tx, B3, C.

Some properties of Bevent loamy sand

Horizon	Depth in.	OM %	Sand %	Silt %	Clay %	CEC me/100g	BS %	pH	BD g/cc
A2	1-3	1.0	78	17	5	8	60	5.8	1.5
Bir	3-10	1.3	64	25	11	8	40	5.2	1.5
A′2x	10-20	0.2	72	21	7	4	55	5.1	1.6
B′2tx	20-27	0.4	55	30	15	9	65	5.4	1.6
C	27-40	0.1	80	13	7	4	70	5.7	1.7

Other: Gravel content is variable, as is sand content; discontinuous bands of silty clay loam to clay loam may occur in the B2 horizon. Setting: Slopes are mostly 0 to 6% gradient; original vegetation was northern mesic forest; more than half has been cleared for the production of small grains and forage crops; mean ann. soil temp. 42-47°F, mean ann. ppt. 28-32 in. Proposed, Marathon Co., Wis., 1957; 65 bu. oats; 475-575 bd. ft. red pine; 500-600 bd. ft. white pine; catena no. 121.

Bibon s-ls (I2, I7); Typic Haplorthod, sandy, mixed, frigid (Podzol); well drained; solum is 12 to 24 in. thick, developed in 40 to 60 in. of acid sandy deposits over reddish-brown (5YR-2.5YR 4/4) clay that is calcareous with depth. Horizons: O1, O2, A1, A2, Bir, B3, C1, IIC2.

Some properties of Bibon loamy sand

Horizon	Depth in.	OM %	Sand %	Silt %	Clay %	CEC me/100g	BS %	pH	BD g/cc
A2	1-3	0.6	87	9	4	6	20	5.2	1.6
Bir	3-10	1.2	85	9	6	7	17	5.4	1.6
C1	15-50	0.1	96	2	2	2	25	5.5	1.7
IIC2	50-70	0.2	5	25	70	25	98	6.0	1.6

Other: Sand is mostly fine sand. Setting: On nearly level to rolling (0 to 15% gradient) glacio-lacustrine upland; original vegetation was boreal forest which was logged and replaced by northern mesic forest, except for small acreages used for pasture, small grains, and forage crops; mean ann. soil temp. 39-43°F, mean ann. ppt. 32-36 in. Proposed, Bayfield Co., Wis.,

1928; Ableiter and Hole, 1961; 45 bu. oats; 300-375 bd. ft. red pine; 450-500 bd. ft. white pine; catena no. 153.

Blomford sl-l (I22); Mollic Arenic Haplaquept, sandy over loamy, mixed, frigid (Low Humic Gley); poorly drained; solum is 36 to 52 in. thick, developed in 20 to 36 in. of sandy outwash material over calcareous loam glacial drift. Horizons: O1, O2, A1, A2, B1, IIB2tg, IIC.

Some properties of Blomford sandy loan

Hori-zon	Depth in.	OM %	Sand %	Silt %	Clay %	CEC me/ 100g	BS %	pH	BD g/cc
A1	0-4	8.0	67	25	8	36	80	5.8	1.4
A2	4-15	2.0	75	20	5	10	60	5.6	1.6
IIB2tg	24-48	0.3	26	40	34	18	60	5.9	1.7
IIC	48-66	0.1	35	35	30	15	98	7.0	1.7

Other: Substratum is calcareous at 40 to 60 in. depth; coarse fragments occupy as much as 8% of the volume of the II horizons, and a thin stone line is present at the interface between I and II materials; thin sandy coatings on surfaces of peds in upper IIB2tg. Setting: On nearly level surfaces of depressions on till plains, in places at the edges of lacustrine plains; original vegetation was lowland hardwoods of northern mesic forest landscapes but is now largely replaced by corn, soybeans, and forage crops; mean ann. soil temp. 42-44°F, mean ann. ppt. 28-32 in. Estab., Isanti Co., Minn., 1953; 55 bu. oats; 400-450 bd. ft. red pine; 400-450 bd. ft. white pine; catena no. 122.

Blount l-sicl (B9, B19) (Fig. 8-10); Aeric Ochraqualf, fine, illitic, mesic (Gray-Brown Podzolic); somewhat poorly drained; solum is 20 to 40 in. thick, developed in less than 20 in. of silty covering over calcareous light yellowish-brown (10YR 6/4), mottled light gray (10YR 6/1) silty clay loam to clay loam glacial till. Horizons: O1, O2, A1, A2, IIB2t, IIC.

Some properties of Blount silt loam

Hori-zon	Depth in.	OM %	Sand %	Silt %	Clay %	CEC me/ 100g	BS %	pH	BD g/cc
A2	5-10	3.0	25	55	20	15	70	5.4	1.6
IIB2t	10-32	1.0	15	40	45	25	75	4.8	1.7
IIC	32-45	0.5	20	55	25	12	100	8.5*	1.8

*calcareous

Other: The B2t horizon is mottled; the II horizons contain some gravel; the IIC horizon contains 33% calcium carbonate equivalent. Setting: On slopes of 1 to 3% gradient on gently rolling till plains; original vegetation was lowland hardwoods in oak savanna landscape, but is now replaced by corn, soybeans, small grains, and forage crops, except for a few hardwood woodlots; mean ann. soil temp. 47-49°F, mean ann. ppt. 30-32 in. Estab., Vermillion Co., Ill., 1931; Link and Demo, 1970; 65 bu. oats; not suitable for timber production; catena no. 54.

Bluffton sil-sicl (F7); Mollic Ochraqualf, fine-loamy, mixed, nonacid, frigid (Humic Gley); poorly drained; solum is 20 to 40 in. thick, developed in less than 20 in. of silty covering over calcareous reddish silty clay loam to silty clay glacio-lacustrine deposits. Horizons: O1, O2, A1, A2, IIB2tg, IIC.

Some properties of Bluffton silt loam

Hori-zon	Depth in.	OM %	Sand %	Silt %	Clay %	CEC me/ 100g	BS %	pH	BD g/cc
A1	0-5	10.0	15	60	25	37	78	7.0	1.5
A2	5-14	2.0	10	63	27	16	57	7.5	1.6
IIB2tg	14-30	1.0	10	45	45	27	60	7.6	1.6
IIC	30-40	0.1	45	23	32	18	97	8.5*	1.7

*calcareous

Other: Horizons mottled below the A1. Setting: On slopes of 1 to 3% gradient in undulating glacio-lacustrine plains; original vegetation was lowland hardwoods of oak savanna and southern oak forest landscapes, but is now largely replaced by corn, soybeans, and forage crops; mean ann. soil temp. 42-44°F, mean ann. ppt. 28-32 in. Estab., Wadena Co., Minn., 1926; where artificially drained, 60 bu. oats; not suitable for timber production; catena no. 118.

Boaz sil (J1) (Fig. 2-51); Aeric Haplaquept, fine-silty, mixed, nonacid, mesic (Alluvial soil intergrading toward a Low Humic Gley); somewhat poorly drained; *see* Akan, which Boaz resembles except for the darker, thicker A1 horizon of Akan. Estab., Richland Co., Wis., 1949; Robinson and Klingelhoets, 1959; catena no. 42.

Bonduel l-sil (E6); Aquic Eutroboralf, fine-loamy, mixed, frigid (Gray-Brown Podzolic and bisequal Podzol over Gray Wooded); somewhat poorly drained; solum is 20 to 40 in. thick, developed in the same thickness of glacial drift, with patches of silty covering, over dolomite bedrock. A fragipan (x) is present. Horizons: O1, O2, A1, A2, Bir, A′2x, B′2t, B3, IIR.

Some properties of Bonduel loam

Hori-zon	Depth in.	OM %	Sand %	Silt %	Clay %	CEC me/ 100g	BS %	pH	BD g/cc
A2	3-8	2.0	40	45	15	30	80	7.0	1.6
A′2x	10-13	0.5	50	40	10	10	60	7.2	1.8
B′2t	13-17	0.6	40	30	30	18	90	7.5	1.7

Other: Horizons are mottled below the A1 horizon; there is some tonguing of the A′2x into the B2t horizon; as much as 4 in. of alkaline residual-like loam caps the bedrock in places; the bedrock is massive or with widely spaced, clay-filled fissures. Setting: On slopes of 1 to 3% gradient on a bedrock-controlled till plain; original vegetation was northern mesic forest and lowland hardwoods, much of which has been cleared for the production of small grains, corn, and forage crops; mean ann. soil temp. 45°F, mean ann. ppt. 30 in. Proposed, Shawano Co., Wis., 1946; where artificially drained, 70 bu. oats; 400-450 bd. ft. white pine; catena no. 139.

Boone s-ls (D4, D5, D6, D8, D11) (Figs. 1-3, 10-4); Typic Quartzipsamment, mesic, uncoated (Lithosol); excessively drained; solum is 3 to 6 in. deep, consisting of an A1 horizon over loose sand containing sandstone fragments, and underlain at 20 to 40 in. by weakly to strongly consolidated sandstone bedrock. Horizons: O1, O2, A1, C, R.

Some properties of Boone sand

Horizon	Depth in.	OM %	Sand %	Silt %	Clay %	CEC me/ 100g	BS %	pH	BD g/cc
A1	0-4	3.0	93	5	2	7	40	5.6	1.6
C	4-60	0.1	96	2	2	2	5	5.5	1.7

Other: The sand fraction is about equally divided into coarse, medium, and fine sand; an incipient Spodosol horizon sequence may show by color, but sand grains are only slightly coated in such a case; the sandstone consists of 95% or more quartz. Setting: On ridge tops and valley slopes of as much as 40% gradient; original vegetation was oak, pine, and some hemlock; some areas are cleared and in pasture or have a second-growth cover of woods; mean ann. soil temp. 43-50°F, mean ann. ppt. 31 in. Estab., Bates Co., Mo., 1908, where it was a somewhat wet soil with argillic horizon and fragipan; in the 1920s and 1930s the Boone series included in Wisconsin what are now called Fayette, Gale, Hixton, Norden, and Boone; in the 1940s and 1950s the present concept came into use; Thomas, Carroll, and Wing, 1962; Thomas, 1964; Soil Conservation Service, 1967a; Haszel, 1968; 35 bu. oats; not suitable for timber production; catena no. 22.

Borth l-sicl (J11) (Fig. 11-5); Typic Hapludalf; clayey over sandy or sandy-skeletal, mixed, mesic (Gray-Brown Podzolic); well drained; solum is 30 to 45 in. thick, developed in 20 to 40 in. of reddish clay over calcareous sand or sand and gravel. Horizons: O1, O2, A1, A2, B2t, IIC.

Some properties of Borth silt loam

Horizon	Depth in.	OM %	Sand %	Silt %	Clay %	CEC me/ 100g	BS %	pH	BD g/cc
A2	3-7	2.0	22	60	18	14	63	6.2	1.0
B2t	7-35	0.8	15	25	60	39	80	6.7	1.5
IIC	35-50	0.1	95	4	1	2	100	8.0*	1.6

*calcareous

Setting: On level to gently rolling land where a lake stood over glacial outwash long enough to form the clayey deposits; original vegetation was southern mesic forest for the most part, with some oak savanna and some northern mesic forest; now largely cleared and cropped or pastured; mean ann. soil temp. 45°F, mean ann. ppt. 30 in. Proposed, Waushara Co., Wis., 1950; 70 bu. oats; 400-450 bd. ft. white pine; catena no. 89.

Boyer ls-sl (B15, B30); Typic Hapludalf, coarse-loamy, mixed, mesic (Gray-Brown Podzolic); well drained; solum is 24 to 40 in. thick, developed in the same thickness of loamy material overlying a stratified sand and gravel IIC horizon. Horizons: O1, O2, A1, A2, B1, B2t, IIC.

Some properties of Boyer loamy sand

Horizon	Depth in.	OM %	Sand %	Silt %	Clay %	CEC me/ 100g	BS %	pH	BD g/cc
A1	0-4	5.0	81	12	7	3	65	5.3	1.5
A2	4-12	1.0	86	10	4	2	58	5.7	1.6
B2t	18-34	0.	55	20	25	12	55	6.9	1.7
IIC	34-50	0.1	96	2	2	1	100	8.5*	1.8

*calcareous

Other: The portion of the B2t that is a sandy clay loam is less than 10 in. thick; horizons are variable; for example, the B2t may be absent and a loamy sand B3 may replace it; or a 6-in.-thick gravelly sandy clay B2t may be present; in some pedons, the lower part of the solum consists of alternating bands 2 to 4 in. thick of A2 and B2t horizons; the B2t horizon may tongue down 6 to 12 in. into the IIC horizon. Setting: On level to hilly glacial outwash deposits; original vegetation was oak savanna; about 60% of the area is now in corn, small grains, soybeans, field beans, forage crops; 25% is in pasture and the remainder in woodland; mean ann. soil temp. 46-49°F, mean ann. ppt. 33 in. Estab., Berrien Co., Mich., 1938; Milfred and Hole, 1970; Link and Demo, 1970; Haszel, 1971; 55 bu. oats; 400-450 bd. ft. white pine; catena no. 80.

Braham sl-l (I22); Arenic Eutrochrept to Arenic Hapludalf, sandy over loamy, mixed, frigid (Gray-Brown Podzolic); well drained; solum is 40 to 50 in. thick, developed in a sandy deposit 20 to 36 in. thick overlying calcareous loam glacial drift (till and glacio-lacustrine deposits). Horizons: O1, O2, A1, A2, B1, IIB2t, IIC.

Some properties of Braham sandy loam

Horizon	Depth in.	OM %	Sand %	Silt %	Clay %	CEC me/ 100g	BS %	pH	BD g/cc
A2	3-8	1.0	77	15	8	9	60	6.2	1.5
B1	8-24	0.3	78	12	10	3	45	6.5	1.6
IIB2t	24-42	0.1	57	20	23	10	70	7.5	1.7
IIC	42-50	0.1	45	35	20	7	100	8.5*	1.8

*calcareous

Other: Depth to free carbonates ranges from 40 to 60 in.; coarse fragments occupy as much as 8% of the volume of the II horizons, but are absent in the overlying sandy horizons; a thin stone line may be present at the interface between I and II materials. Setting: On slopes of 2 to 12% gradient in landscapes transitional between sand plains and till plains; original vegetation was southern mesic forest and oak savanna, but is now largely replaced by soybeans, corn, and alfalfa; mean ann. soil temp. 43-45°F, mean ann. ppt. 30 in. Estab., Isanti Co., Minn., 1953; 70 bu. oats; 350 bd. ft. white pine; catena no. 122.

Briggsville sl-sil (I17) (Fig. 9-5); Typic Hapludalf, fine, mixed, mesic (Gray-Brown Podzolic); moderately well drained; solum is 24 to 40 in. thick, developed in less than 18 in. of loamy deposits over calcareous reddish-brown glacio-lacustrine silts and clays (35-50% clay in IIBt). Horizons: O1, O2, A1, A2, B1, IIB2t, IIC.

Some properties of Briggsville silt loam

Horizon	Depth in.	OM %	Sand %	Silt %	Clay %	CEC me/ 100g	BS %	pH	BD g/cc
A2	3-11	0.5	28	52	20	20	65	5.8	1.6
B1	11-17	0.3	15	55	30	25	68	5.5	1.7
IIB2t	17-38	0.4	7	50	43	30	85	5.7	1.8
IIC	38-48	0.1	3	45	52	20	98	8.5	1.7

*calcareous

Other: Depth to free carbonates is from 24 to 40 in.; a mottled zone a fraction of an inch thick is present just below the interface between I and II materials. Setting: On slopes of 1 to 12% gradient of old glacial lake basins; original vegetation was southern mesic forest and oak savanna, now largely replaced by row crops, hay, and pasture; mean ann. soil temp. 47°F, mean ann. ppt. 30 in. Proposed, Marquette Co., Wis., 1963; Peck and Lee, 1961; 70 bu. oats; 400-450 bd. ft. white pine; catena No. 84.

Brill sil (F17, F25, F26); Typic Glossoboralf, fine-silty over sandy or sandy-skeletal, mixed, frigid (Gray-Brown Podzolic); moderately well drained; solum is 20 to 42 in, thick, developed in 20 to 40 in. of silty material overlying acid sand and gravel glacial outwash. Horizons: O1, O2, A1, Bir, A'2, B&A', B'2t, IIB'3t, IIC.

Some properties of Brill silt loam

Horizon	Depth in.	OM %	Sand %	Silt %	Clay %	CEC me/ 100g	BS %	pH	BD g/cc
A'2	10-15	0.8	15	65	20	11	40	5.1	1.2
B'2t	19-27	0.6	17	55	28	14	58	4.8	1.6
IIC	36-60	0.1	97	1	2	3	50	5.2	1.8

Other: Some stones occur in horizons in the I material, and on slopes may be slightly concentrated in the A1 horizons, probably as a result of creep from tree-tip mounds at higher positions in the landscape. Setting: In slight depressions on unpitted outwash plains, or at footslopes in pitted outwash; original vegetation was northern mesic forest, but now is largely replaced by row crops, hay, and pasture; mean ann. soil temp. 43-45°F, mean ann. ppt. 32 in. Estab., Langlade Co., Wis., 1941; Milfred, Olson, and Hole, 1967; 80 bu. oats; 500-600 bd. ft. white pine; catena no. 112.

Brookston l-sicl (B11, B14, B32, J8, J15) (Fig. 8-8); Typic Argiaquoll, fine-loamy, mixed, mesic (Humic Gley); poorly drained; solum is 30 to 45 in. thick, developed in 2 or 3 ft. of silty material over calcareous loam to silty clay loam glacial till. Horizons: O1, O2, A1, B1gt, IIB2gt, IIC.

Some properties of Brookston silty clay loam

Horizon	Depth in.	OM %	Sand %	Silt %	Clay %	CEC me/ 100g	BS %	pH	BD g/cc
A1	0-14	15.0	7	65	28	70	85	6.9	1.2
B1gt	14-20	2.0	4	58	38	40	92	7.0	1.5
IIC	46-66	0.1	28	45	27	30	100	8.5*	1.7

*calcareous

Other: Part of the B2t horizon is developed in the glacial till; some stones and sand may be present throughout the solum. Setting: On slopes of 0 to 3% gradient in depressions in glacial till plains; original vegetation was lowland hardwoods and sedge meadows; now largely cleared, drained, and used for corn, soybeans, leguminous forage crops, and pasture; mean ann. soil temp. 48-49°F, mean ann. ppt. 31 in. Estab., White Co., Ind., 1915; Milfred and Hole, 1970; Steingraeber and Reynolds, 1971; where artificially drained, 65 bu. oats; not suitable for timber production; catena no. 52.

Brule sl-sil (I1, I2, I4, I19); Typic Udifluvent, coarse-loamy, mixed, frigid (Alluvial soil); moderately well drained; solum is 18 to 30 in. thick, developed in stratified alluvium of sands and silts with some stones, in the red clay region adjacent to Lake Superior. Horizons: O1, O2, A1, C.

Some properties of Brule sandy loam

Horizon	Depth in.	OM %	Sand %	Silt %	Clay %	CEC me/ 100g	BS %	pH	BD g/cc
A1	0-2	3.0	70	22	8	20	50	5.7	1.3
C	2-60	0.1	77	18	5	12	70	6.3	1.7

Other: In places an incipient spodic B horizon is evident. Setting: On nearly level (0 to 2% gradient) floodplains of streams; original vegetation was a mixture of swamp conifers and hardwoods, with northern mesic forest; second-growth forest occupies most of the area, but some bodies have been cleared for pasture or hay production; mean ann. soil temp. 41°F, mean ann. ppt. 30 in. Estab., Iron Co., Mich., 1930; not suitable for oats; 500-575 bd. ft. red or white pine; catena no. 156.

Burkhardt sl-l (C9); Typic Hapludoll, sandy, mixed, mesic (Brunizem); well drained; solum is 16 to 20 in. thick, developed in deep acid sand and gravel glacial outwash. Horizons: O1, O2, A1, B, C.

Some properties of Burkhardt sandy loam

Horizon	Depth in.	OM %	Sand %	Silt %	Clay %	CEC me/ 100g	BS %	pH	BD g/cc
A1	0-10	3.0	68	20	12	25	60	6.4	1.3
B	10-19	1.0	78	12	10	10	50	5.8	1.7
C	19-60	0.1	95	3	2	2	20	5.8	1.8

Other: Faces of peds are commonly one Munsell notation lower in chroma than are ped interiors; gravel content of the IIC horizon is between 15 and 40%. Setting: On a wide range of slopes on level to rolling outwash terraces, plains, and moraines; original vegetation was prairie and oak savanna, but is now largely replaced by small grains, corn, and forage crops; mean ann. soil temp. 43-46°F, mean ann. ppt. 30 in. Estab., Washington Co., Minn., 1941; Thomas, 1964; Haszel, 1968; 55 bu. oats; 400-450 bd. ft. red pine; 400-450 bd. ft. white pine; catena no. 114.

Cable l-sil (F2, F10, F14, G10, G13, G20, G21, G22, J6, J12, J13) (Figs. 12-5, 12-7); Typic Haplaquept, coarse-loamy, mixed, frigid (Low Humic Gley); very poorly drained; solum is 20 to 32 in. thick, developed in 12 to 30 in. of silty materials overlying acid sandy loam to loam glacial till. Horizons: O1, O2, A1, Bg, IICg.

Some properties of Cable silt loam

Horizon	Depth in.	OM %	Sand %	Silt %	Clay %	CEC me/ 100g	BS %	pH	BD g/cc
A1	0-6	10.0	30	55	15	38	55	5.6	1.2
Bg	6-30	1.0	30	52	18	19	70	5.2	1.4
IICg	30-60	0.1	38	40	22	12	85	5.0	1.7

Other: The A1 horizon is no more than 7 in. thick over the gleyed material below, so that these soils "plow up white." Setting: On slopes of 1 to 3% gradient in depressions in moraine landscape; original vegetation was sedge meadows, lowland hardwoods, and lowland conifers, and this is still largely unchanged by man; pastures may extend onto bodies of this soil; mean ann. soil temp. 42-44°F, mean ann. ppt. 30 in. Proposed, Clark Co., Wis., 1942; Hole and Schmude, 1959; Hole et al., 1962; where artificially drained, 65 bu. oats; not suitable for timber production; catena no. 94.

Calamine sil, sicl (A4); Typic Argiaquoll, fine, illitic, mesic (Humic Gley); very poorly drained; solum is 36 to 45 in. thick, developed in 15 to 30 in. of loess over clayey residuum from shale with bedrock shale at 30 to 50 in. Horizons: O1, O2, A1, B21tg, IIB22tg, IICg, IIR.

Some properties of Calamine silt loam

Hori-zon	Depth in.	OM %	Sand %	Silt %	Clay %	CEC me/ 100g	BS %	pH	BD g/cc
A1	0-20	20.0	15	60	25	42	88	7.0	1.2
IIB22tg	27-40	2.0	7	55	38	25	95	7.5	1.4
IICg	40-60	0.5	3	45	52	23	100	8.5*	1.6

*calcareous

Other: Solum thickness is essentially the depth to shale bedrock; the bedrock is platy to massive, soft to consolidated. Setting: On upland flats, in depressions, and at sloping seepage spots; slopes are usually less than 3% gradient; original vegetation was sedge meadow and this still exists except in the few places where artificial drainage has been successful and the soil is used for crops and pasture; mean ann. soil temp. 48-50°F, mean ann. ppt. 32 in. Establ., Lafayette Co., Wis., 1964; Watson, 1966; Denning, Hole, and Bouma, 1973; where artificially drained, 65 bu. oats; not suitable for timber production; catena no. 12.

Campia sil (F26); Typic Glossoboralf, fine-silty, mixed, frigid (Gray-Brown Podzolic and Gray Wooded); well drained; solum is 32 to 45 in. thick, developed in deep glacio-lacustrine silts that are calcareous at a depth of about 6 ft. Horizons: O1, O2, A1, A2, A&B, B2t, B3t, C1, C2.

Some properties of Campia silt loam

Hori-zon	Depth in.	OM %	Sand %	Silt %	Clay %	CEC me/ 100g	BS %	pH	BD g/cc
A2	3-10	1.5	10	80	10	10	60	6.3	1.0
B2t	22-29	0.5	8	66	26	20	56	5.1	1.4
C1	40-72	0.2	15	67	18	15	70	5.4	1.5
C2	72-80	0.1	15	70	15	12	100	7.8*	1.6

*calcareous

Other: The C horizon may contain silty clay loam layers, and the color of these horizons ranges from the typical 10YR 5/4 to 5YR 5/2, as in Price Co. Setting: On level to undulating glacio-lacustrine plains surrounded by moraines that may or may not have silt coverings; original vegetation was northern mesic forest, now largely removed so that the soil may be cropped to small grains, corn, and hay, or may be pastured; mean ann. soil temp. 42°F, mean ann. ppt. 31 in. Estab., Barron Co., Wis., 1949; Ranney, 1966; Soil Conservation Service, 1967a; Ranney and Beatty, 1969; 80 bu. oats; 450-550 bd. ft. red pine; 500-600 bd. ft. white pine; catena no. 119.

Carbondale muck (J15); Hemic Borosaprist, euic, frigid (Bog); very poorly drained; the solum ranges in thickness from 51 in. to about 15 ft. and is developed from herbaceous organic materials overlying calcareous glacial drift. Horizons: O1, Oa1, Oa2, Oa3, Oe4, IIC.

Some properties of Carbondale muck

Hori-zon	Depth in.	OM %	Sand %	Silt %	Clay %	CEC me/ 100g	BS %	pH	BD g/cc	Fibers unrub. %	Fibers rub. %
Oa1	0-5	53	0	27	20	175	30	6.3	0.1	35†	10
Oe4	39-60	68	0	17	15	200	35	6.3	0.2	70‡	35
IIC	60-80	0.2	40	40	20	15	95	8.5*	1.6	0	0

*calcareous †half woody ‡mostly herbaceous

Other: The solum has about 20% ash content; the four organic horizons below the O1 have the following structures—weak, medium granular in Oa1 and Oa2, massive in Oa3 and Oa4; a dark brown hemic material overlies a nearly black sapric material. Setting: In small (10 acres) to large (about 1,000 acres) depressions in glacial till, outwash, and lacustrine landscapes; slopes are less than 2% gradient except on occasional peat mounds over upwelling water; surrounding mineral soils are loamy over calcareous drift; the original vegetation, largely still undisturbed, is swamp conifer and hardwood forest; less than 15% is drained and used for pasture, hay, or truck crops, and this mostly in southeastern counties; mean ann. soil temp. 46°F, mean ann. ppt. 32 in., supplemented by surface runoff and seepage from adjacent slopes. Estab., Chippewa Co., Mich., 1927; not suitable for oats or pine; catena no. 135.

Carlisle muck (J15) (Fig. 2-52); Typic Medisaprist, euic, mesic (Bog); very poorly drained; the solum ranges from 51 in. to about 15 ft. in thickness and is developed from woody and herbaceous deposits overlying calcareous glacial drift. Horizons: O1, Oa1, Oa2, Oa3, Oa4, Oa5, Oa6, IIC.

Some properties of Carlisle muck

Hori-zon	Depth in.	OM %	Sand %	Silt %	Clay %	CEC me/ 100g	BS %	pH	BD g/cc	Fibers unrub. %	Fibers rub. %
Oa1	0-4	55	0	25	20	190	20	6.8	0.1	10†	5
Oa5	37-41	70	0	18	12	200	35	6.3	0.2	40‡	10
IIC	80-100	0.3	40	40	20	15	95	8.5*	1.6	0	0

*calcareous †primarily grass root fibers
‡herbaceous fibers with 0.5 to 6-in. fragments of woody material

Other: The woody fragments throughout the solum are twigs, branches, logs, and stumps that occupy 15 to 30% by volume; soil structure grades from granular above, through coarse subangular blocky, to massive below. Setting: In depressions of one to hundreds of acres in extent in glacial drift landscapes; slopes are less than 2% gradient except on occasional peat mounds over artesian springs; surrounding mineral soils are loamy over calcareous drift; original vegetation was swamp hardwood and conifer forest; much of the area in large bogs has been drained and the soils used to produce truck crops and sod, particularly in southeastern counties; mean ann. soil temp. 47°F, mean ann. ppt. 32 in., supplemented by surface runoff and seepage from adjacent slopes. Estab., Livingston Co., Mich., 1923; Hole, 1956b; Isirimah, 1969; Frazier and Lee, 1971; not suitable for oats or pine; catena no. 52.

Caryville l-sl (J1, J2); Fluventic Haploboroll, sandy, mixed, mesic or frigid (Alluvial soil); moderately well drained; solum is 15 to 20 in. thick, developed in about that much loamy material overlying sand containing reddish-brown (5YR 4/4) layers. Horizons: O1, A1, AC, IIC.

Some properties of Caryville sandy loam

Hori-zon	Depth in.	OM %	Sand %	Silt %	Clay %	CEC me/100g	BS %	pH	BD g/cc
A1	0-15	4.0	73	15	12	12	92	7.0	1.2
IIC	18-48	0.1	87	10	3	1	95	7.0	1.4

Other: A few thin laminae of fine gravel and of silt occur in the "Driftless Area." Setting: On nearly level alluvial flood plains; slopes less than 3% gradient; original vegetation was brush; mostly cleared for production of corn, soybeans, alfalfa; mean ann. soil temp. 45°F, mean ann ppt. 32 in. Proposed, Eau Claire, Wis., 1954; 70 bu. oats; 500 bd. ft. red or white pine; catena no. 157.

Casco sl-sil (B16, B18, B30, B31, I6) (Figs. 8-4, 8-9); Typic Hapludalf, fine-loamy over sandy or sandy-skeletal, mixed, mesic (Gray-Brown Podzolic); well drained; solum is 12 to 20 in. thick, developed in a loamy mantle 10 to 20 in. thick overlying calcareous, coarse sand and gravel glacial outwash. Horizons: O1, O2, A1, A2, B2t, IIC.

Some properties of Casco loam

Hori-zon	Depth in.	OM %	Sand %	Silt %	Clay %	CEC me/100g	BS %	pH	BD g/cc
A1	0-3	8.0	33	44	23	25	88	7.0	1.3
B2t	8-17	1.0	51	15	34	37	95	7.3	1.6
IIC	17-60	0.1	88	7	5	3	100	8.0*	1.8

*calcareous

Other: A silt mantle less than 12 in. thick may be present; the maximum development of the B2t horizon is commonly at the bottom of the solum; clay films (argillans) are prominent in the B2t. Setting: Slope gradients range from 1% on outwash plains to 45% in kettle moraines and on eskers; original vegetation was somewhat open oak-hickory forest, and hilly land is still this way; most of the land has been cleared, however, and is in pasture or cropped to corn, small grains, alfalfa; mean ann. soil temp. 45°F, mean ann. ppt. 31 in. Estab., Fairfield Co., O., 1956; Milfred and Hole, 1970; Link and Demo, 1970; Haszel, 1971; Steingraeber and Reynolds, 1971; 55 bu. oats; 300-375 bd. ft. white pine; catena no. 77.

Cassel silt (F21); Aeric Glossaqualfs, fine-loamy, mixed, frigid (Gray-Brown Podzolic); somewhat poorly drained; solum is 32 to 38 in. thick, developed in 15 to 30 in. of acid loess over acid clay loam glacial till. Horizons: O1, O2, A1, A2, A&B, IIB2t, IIC.

Some properties of Cassel silt loam

Hori-zon	Depth in.	OM %	Sand %	Silt %	Clay %	CEC me/100g	BS %	pH	BD g/cc
A1	0-4	5.0	28	58	14	12	25	5.0	1.2
A2	4-18	0.5	35	55	10	8	12	4.6	1.5
IIB2t	21-32	0.3	29	41	30	20	40	4.1	1.7
IIC	32-40	0.1	33	40	27	15	60	4.6	1.7

Other: Occasional lenses of sand and gravel occur in the profile. Setting: On undulating till plain in west-central Marathon Co.; slopes are less than 3% gradient; original vegetation was nothern mesic forest and is nearly all replaced by pasture and cultivated fields in a rotation of forage crops, small grains, and corn for silage; mean ann. soil temp. 43°F, mean ann. ppt. 32 in. Proposed, Marathon Co., Wis., 1942; Soil Conservation Service, 1967a; 70 bu. oats; 450-500 bd. ft. white pine; catena no. 98.

Chaseburg sil (J1); Typic Udifluvent, coarse-silty, mixed, nonacid, mesic (Alluvial soil); well drained; solum is 3 to 4 in. thick, developed in silty and very fine sandy alluvium. Horizons: O1, A1, C.

Some properties of Chaseburg silt loam

Hori-zon	Depth in.	OM %	Sand %	Silt %	Clay %	CEC me/100g	BS %	pH	BD g/cc
A1	0-28	7.0	30	55	15	18	90	7.0	1.2
C	28-60	0.1	30	55	15	9	87	6.3	1.6

Other: Stratification consists of fine lenses and layers of slightly differing proportions of coarse silt and fine sand; the C horizon is platy as a result. Setting: In the upper ends of valleys, alluvial fans, and footslopes in the "Driftless Area"; slopes commonly 2 to 6% gradient; original vegetation was oak savanna, but this has been largely cleared to allow use of these soils for pasture or a rotation of hay, oats, and corn; mean ann. soil temp. 47°F, mean ann. ppt. 32 in., supplemented by surface runoff and seepage from slopes above. Estab., Vernon Co., Wis., 1938; Hole, 1956a; Beatty, 1960; Slota and Garvey, 1961; Robinson and Klingelhoets, 1961; Slota, 1969; 70 bu. oats; not suitable for pine; catena no. 37.

Chelsea ls (C5); Alfic Udipsamment, sandy, mixed, mesic (Regosol); excessively drained; solum is 4 to 12 ft., developed in aeolian sand. Horizons: O1, O2, A1, A2, A2&B2t, C.

Some properties of Chelsea sand

Hori-zon	Depth in.	OM %	Sand %	Silt %	Clay %	CEC me/100g	BS %	pH	BD g/cc
A1	0-4	2.0	91	4	5	4	8	4.2	1.2
B2t*	36-70	0.2	85	5	10	3	3	4.9	1.5
C	70-90	0.1	94	3	3	1	7	5.1	1.7

*The B2t is repetitive, occurring as bands.

Other: Carbonates may be present at a depth of 60 in.; the B2t horizon occurs as bands containing slightly more clay and iron oxide than the interlayers; the aggregate thickness of the bands is about 6 in. in the 5-ft. profile. Setting: On crests of escarpments, and as footslope and tributary valley fill deposits made by wind in the "Driftless Area"; slopes range from 5 to 30% in gradient; original vegetation was oak savanna, a small portion of which has been removed to permit use of the soils for pasture and for a rotation of hay, oats, and corn; mean ann. soil temp. 47°F, mean ann. ppt. 32 in. Estab., Tama Co., Ia., 1938; Hole, 1956a; Robinson and Rich, 1960; Robinson and Klingelhoets, 1961; 35 bu. oats; 350-400 bd. ft. red pine; 450-500 bd. ft. white pine; catena no. 23.

Chetek sl-l (D10, G8, G19, G26, G28, H6); Eutric Glossoboralf, coarse-loamy, mixed, frigid (Gray-Brown Podzolic); well drained; solum is 16 to 20 in. thick, developed in a loamy covering of that thickness over acid coarse sand and gravel glacial outwash. Horizons: O1, O2, A1, A2, B2t, IIB3t, IIC.

Some properties of Chetek sandy loam

Horizon	Depth in.	OM %	Sand %	Silt %	Clay %	CEC me/100g	BS %	pH	BD g/cc
A1	0-3	5.0	58	30	12	22	45	5.4	1.3
B2t	13-18	1.0	58	28	14	15	50	5.8	1.6
IIC	20-36	0.1	92	5	3	3	60	6.0	1.8

Other: Gravel contents range from 15 to 35% in the profile. Setting: Slopes are commonly less than 5% gradient but on escarpments are as much as 30%; this soil is on glacial outwash plains and kettle moraines; original vegetation was northern mesic forest, and about half of this has been cleared to use for potatoes, corn, oats, hay, and pasture; mean ann. soil temp. 40-45°F, mean ann. ppt. 30 in. Estab., Langlade Co., Wis., 1947; Robinson, et al., 1958; Haszel, 1968; 55 bu. oats; 400-450 bd. ft. red or white pine; catena no. 115.

Clifford l-sil (F9, F18, F19) (Fig. 12-7); Aqualfic Haplorthod, coarse-loamy, mixed, frigid (Podzol); somewhat poorly drained; solum is 15 to 28 in. thick, developed in 12 to 24 in. of acid coarse silt loam overlying acid, reddish-brown sandy loam to loam glacial tiil. Horizons: O1, O2, A2, Bhir, IIBx, IIC.

Some properties of Clifford silt loam

Horizon	Depth in.	OM %	Sand %	Silt %	Clay %	CEC me/100g	BS %	pH	BD g/cc
A2	0-5	2.0	22	64	14	7	26	4.7	1.4
Bhir	5-12	3.0	22	66	12	6	50	4.7	1.5
IIC	22-30	0.1	68	24	8	5	53	5.0	1.8

Other: All mineral horizons are mottled; the fragipan (x) is incipient. Setting: On slopes of less than 2% gradient in depressions and on lower slopes on till plains; original vegetation was northern mesic forest that was largely taken for lumber and has been replaced by aspen, yellow birch, and young conifers; some land is used for pasture; mean ann. soil temp. 41-42°F, mean ann. ppt. 30 in. Proposed, Oneida Co., Wis., 1958; Hole and Schmude, 1959; where artificially drained, 70 bu. oats; 400-450 bd. ft. white pine; catena no. 127.

Cloquet ls-sl (G12, G13, G17, G20, G22) (Fig. 12-5); Dystric Eutrochrept, sandy, mixed, frigid (Podzol); well drained; solum is 15 to 30 in. thick, developed in reddish-brown acid stony glacial drift. Horizons: O1, O2, A1, A2, B, C.

Some properties of Cloquet sandy loam

Horizon	Depth in.	OM %	Sand %	Silt %	Clay %	CEC me/100g	BS %	pH	BD g/cc
A1	0-1	8.0	77	12	11	8	25	5.0	1.4
A2	1-10	2.0	77	14	9	9	28	5.2	1.3
B	10-36	0.6	92	5	3	3	60	5.4	1.6
C	36-42	0.1	96	2	2	2	70	5.6	1.8

Other: This is a weakly developed Podzol and hence falls into the Inceptisols in the new classification; an incipient fragipan may be present at the bottom of the solum. Setting: On undulating to hilly moraines with slopes of 3 to 45% gradient; original vegetation was northern mesic forest and pine forest, mostly cut for lumber and now replaced by forest succession except in the few places where the soil is used for pasture, hay, oats, or potatoes; mean ann. soil temp. 43-44°F, mean ann. ppt. 39 in. Estab., Pine Co., Minn., 1935; Robinson, et al., 1958; Ableiter and Hole, 1961; 50 bu. oats; 400-450 bd. ft. red pine; 450-500 bd. ft. white pine; catena no. 132.

Conover sl-sil; Udollic Ochraqualf, fine-loamy, mixed, mesic (Gray-Brown Podzolic); somewhat poorly drained member of the Octagon catena, a prairie-border associate of the Miami series.

Crosby sl-sil: *See* Conover and Lamartine.

Crown sl-l (D10, G8, G19, G26, G28, H6); Aquic Eutroboralf, coarse-loamy, mixed, frigid (Gray-Brown Podzolic); somewhat poorly drained; solum is 16 to 20 in. thick, developed in a loamy covering of that thickness over acid coarse sand and gravel glacial outwash. Horizons: O1, O2, A1, A2, B2t, IIB3, IIC.

Some properties of Crown loam

Horizon	Depth in.	OM %	Sand %	Silt %	Clay %	CEC me/100g	BS %	pH	BD g/cc
A1	0-4	3.0	51	34	15	34	60	6.3	1.2
B2t	10-15	0.2	52	30	18	20	65	5.7	1.5
IIC	18-30	0.1	92	5	3	3	70	5.8	1.8

Other: There are variations in mottling, and in proportion of gravel and cobbles in substratum. Setting: In slight depressions and flats on outwash plains, with slopes of less than 3% gradient; original vegetation was northern mesic forest, much of which has been removed to allow production of potatoes, corn, oats, and hay and to make room for pasture; mean ann. soil temp. 43°F, mean ann. ppt. 30 in. Proposed, Isanti Co., Minn., 1953; 50 bu. oats; 400-450 bd. ft. white pine; catena no. 115.

Curran sil (A12, A13); Udollic Ochraqualf, fine-silty, mixed, mesic (Gray-Brown Podzolic); somewhat poorly drained; solum is 40 to 55 in. thick, developed in 36 to 50 in. of silty material overlying acid fluvial or glacio-fluvial sand. Horizons: O1, O2, A1, A2, B2tg, IIB3tg, IICg.

Some properties of Curran silt loam

Horizon	Depth in.	OM %	Sand %	Silt %	Clay %	CEC me/100g	BS %	pH	BD g/cc
A1	0-3	7.0	12	70	18	19	70	6.5	1.2
A2	3-16	2.0	12	73	15	14	50	5.8	1.3
IIB3tg	44-48	0.5	54	30	16	23	60	5.4	1.5
IICg	48-60	0.1	92	5	3	3	90	5.8	1.7

Other: Soil horizons below the A1 are mottled. Setting: On level stream terraces on slopes of less than 3% gradient; original vegetation was oak savanna and lowland hardwoods, but is now almost entirely replaced by corn, small grains, hay, and pasture; mean ann. soil temp. 47-49°F, mean ann. ppt. 33 in. Estab., Henderson Co., Ill., 1947; Robinson and Klingelhoets, 1959, 1961; Beatty, 1960; where artificially drained, 65 bu. oats; 450-500 bd. ft. white pine; catena no. 25.

Cushing l-sil (F7); Glossic Eutroboralf, fine-loamy, mixed, frigid (Gray-Brown Podzolic); well drained; solum is 32 to 46 in. thick, developed in 20 to 42 in. of loamy covering over calcareous loam glacial till. Horizons: O1, O2, A1, A2, B&A, B2t, C.

Some properties of Cushing loam

Horizon	Depth in.	OM %	Sand %	Silt %	Clay %	CEC me/100g	BS %	pH	BD g/cc
A2	3-16	1.2	48	34	18	18	62	5.8	1.2
B2t	20-42	0.5	36	30	34	20	65	5.6	1.5
C	42-60	0.1	47	30	23	12	100	8.5*	1.7

*calcareous

Other: All horizons may be somewhat stony; free lime may occur at depths of 3 to 6 ft., the shallower depths being on the steeper slopes; the lower B horizon may be slightly mottled. Setting: On slopes of 2 to 12% gradient in late Wisconsin glacial moraines of the Grantsburg Lobe that moved east from Minnesota; original vegetation was southern mesic forest, which is now almost entirely replaced by corn, small grains, forage crops, and pasture; mean ann. soil temp. 43-44°F, mean ann. ppt. 30 in. Proposed, Polk Co., Wis., 1950; 75 bu. oats; 450-500 bd. ft. red pine; 500-600 bd. ft. white pine; catena no. 107.

Dakota sl-l (A12, A14, C8, C9, C15, C16) (Figs. 2-53, 2-54, 7-10, 9-4, 10-6, 12-4); Typic Argiudoll, fine-loamy over sandy or sandy-skeletal, mixed, mesic (Brunizem); well drained; solum is 40 to 48 in. thick, developed in 24 to 40 in. of loamy material overlying acid sand and gravel glacial outwash. Horizons: O1, O2, A1, B2t, IIB3t, IIC.

Some properties of Dakota sandy loam

Horizon	Depth in.	OM %	Sand %	Silt %	Clay %	CEC me/100g	BS %	pH	BD g/cc
A1	0-14	8.0	53	32	15	17	52	5.8	1.3
B2t	14-35	1.5	37	40	23	14	40	5.4	1.5
IIC	43-65	0.1	95	3	2	3	50	5.5	1.7

Other: Usually less than 10% by volume of gravel is present in the IIC. Setting: On broad outwash plains and narrow stream terraces, on slopes ranging from 1 to 15% in gradient; original vegetation was tall grass prairie with scattered oaks, but this has been replaced by corn, soybeans, small grains, forage crops, and pasture; mean ann. soil temp. 47-49°F, mean ann. ppt. 32 in. Estab., Dakota Co., Minn., 1941; Robinson and Klingelhoets, 1959, 1961; Slota and Garvey, 1961; Thomas, Carroll, and Wing, 1962; Thomas, 1964; Haszel, 1968; 65 bu. oats; 400-450 bd. ft. red pine; 500 bd. ft. white pine; catena no. 27.

Dalbo sil (I22); Aquic Eutroboralf, fine, mixed, frigid (Gray-Brown Podzolic); somewhat poorly drained; solum is 24 to 40 in. thick, the same as the depth to carbonates in the calcareous silty gray glacio-lacustrine deposits from which the soil formed. Horizons: O1, O2, A1, A2, B&A, B2t, C.

Some properties of Dalbo silt loam

Horizon	Depth in.	OM %	Sand %	Silt %	Clay %	CEC me/100g	BS %	pH	BD g/cc
A2	5-9	2.0	20	60	20	15	58	5.7	1.3
B2t	14-28	0.6	5	52	43	27	80	5.2	1.5
C	28-60	0.1	10	65	25	12	98	8.5*	1.7

*calcareous

Other: The A2 horizon tongues into the B2t horizon; the soil is free of stones and gravel. Setting: On nearly level to level glacio-lacustrine plains; original vegetation was lowland hardwoods and northern mesic forest, replaced now over three fourths of the area by corn, soybeans, oats, legume hay, and pastureland; mean ann. soil temp. 43-45°F, mean ann. ppt. 29 in. Estab., Isanti Co., Minn., 1956; where artificially drained, 75 bu. oats; 375 bd. ft. white pine; catena no. 118.

Dancy sl-l (C17); Aeric Glossaqualf, fine-loamy, mixed, frigid (Low Humic Gley); poorly drained; solum is 24 to 50 in. thick, developed in 20 to 40 in. of sandy deposits over acid clayey residuum from micaceous granitic rocks. Horizons: O1, O2, A1, A2, I-IIB&A, IIB2tg, IIC.

Some properties of Dancy sandy loam

Horizon	Depth in.	OM %	Sand %	Silt %	Clay %	CEC me/100g	BS %	pH	BD g/cc
A2	7-13	2.0	80	10	10	20	30	6.5	1.6
IIB2tg	27-32	1.0	32	33	35	20	60	7.2	1.7
IIC	41-60	0.1	65	25	10	5	90	6.8	1.8

Other: Fine fragments of igneous bedrock commonly occupy 1 to 10% of the volume of the lower solum but locally take up 10 to 30% of the volume; mottles may be many and prominent in A&B and B&A horizons. Setting: In level and depressional sites in rock-controlled landscapes, with slopes on Dancy soils of less than 2% gradient; original vegetation was lowland hardwoods, now replaced in a few places by cleared pasture or by alders, sedges, and grasses; mean ann. soil temp. 43°F, mean ann. ppt. 32 in. Proposed, Wood Co., Wis., 1945; where artificially drained, 65 bu. oats; 350-400 bd. ft. white pine; catena no. 149.

Dawson peat (J14) (Fig. 1-8); Terric Borosaprist, sandy, mixed, dysic, frigid (Bog); very poorly drained; solum is 16 to 50 in. thick, developed in herbaceous organic materials over acid glacial drift. Horizons: O1, O2, Oi1, Oa1, IIAb, IIICg.

Some properties of Dawson peat

Horizon	Depth in.	OM %	Sand %	Silt %	Clay %	CEC me/100g	BS %	pH	BD g/cc	Fibers unrub. %	Fibers rub. %
Oa1	8-38	54	0	26	20	200	5	3.5	0.1	40	10
IIICg	40-60	0.2	85	10	5	8	15	4.5	1.7	0	0

Setting: In kettles and other depressions in till and outwash plains and moraines; original vegetation, still largely undisturbed, was leatherleaf, bog rosemary, blueberries, laurel, cranberries, sphagnum moss, with occasional black spruce and tamarack trees; mean ann. soil temp. 44°F, mean ann. ppt. 30 in. Estab., Luce Co., Mich., 1929; catena no. 34.

Del Rey l-sil (B33); Aeric Ochraqualf, fine, illitic, mesic (Gray-Brown Podzolic); somewhat poorly drained; solum is 25 to 35 in. thick, developed in as much as 18 in. of sandy loam deposits (although these may be absent) overlying calcareous glacio-lacustrine silts and clays. Horizons: O1, O2, A1, A2, IIB2t, IIB3, IIC.

Some properties of Del Rey loam

Hori-zon	Depth in.	OM %	Sand %	Silt %	Clay %	CEC me/ 100g	BS %	pH	BD g/cc
A2	4-9	2.0	45	40	15	18	65	5.8	1.0
IIB2t	9-33	0.7	7	50	43	25	88	6.5	1.5
IIC	41-61	0.1	5	60	35	20	100	8.0*	1.7

*calcareous

Setting: In slight depressions and gentle slopes of less than 6% gradient bordering wetlands; original vegetation was lowland hardwood forest, now largely drained and cropped or pastured; mean ann. soil temp. 46°F, mean ann. ppt. 32 in. Estab., Iroquois Co., Ill., 1940; Milfred and Hole, 1970; where artificially drained, 65 bu. oats; 400-450 bd. ft. white pine; catena no. 85.

Delton ls-l (C18, E10) (Figs. 9-6, 11-6); Arenic Hapludalf, coarse-loamy over clayey, mixed, mesic (Gray-Brown Podzolic); well drained; solum is 30 to 50 in. thick, developed in a sandy deposit 20 to 40 in. thick overlying acid lacustrine stratified silts and clays. Horizons: O1, O2, A1, A2, IIB2t, IIC.

Some properties of Delton loam

Hori-zon	Depth in.	OM %	OM %	Sand %	Silt %	Clay %	CEC me/ 100g	BS %	pH	BD g/cc
A2	3-30	0.3	2.0	45	40	15	18	65	5.8	1.0
IIB2t	30-40	0.5	0.6	7	50	43	25	78	6.5	1.5
IIC	40-50	0.1	0.4	5	60	35	20	80	6.6	1.7

Other: The I-IIB21t horizon is present and has a loamy intermixture of clayey materials (below) and sandy materials (above); the lower B and C horizons may be slightly mottled. Setting: On nearly level to sloping (less than 5% gradient) portions of glacial lake basins; original vegetation was southern oak forest; now largely cleared and cropped or in pasture; mean ann. soil temp. 44°F, mean ann. ppt. 32 in. Estab., Marquette Co., Wis., 1969; 450-500 bd. ft. red pine; 450-550 bd. ft. white pine; catena no. 123.

Denrock sil-sicl (A11); Aquic Argiudoll, fine, montmorillonitic, mesic (Brunizem); somewhat poorly drained; solum is 3 to 5 ft. thick, developed in lacustrine clays underlain at 5 to 4 ft. by sand. Horizons: O1, O2, A1, IIB2t, IIIC.

Some properties of Denrock silt loam

Hori-zon	Depth in.	OM %	Sand %	Silt %	Clay %	CEC me/ 100g	BS %	pH	BD g/cc
A1	0-13	10.0	20	60	20	30	82	5.7	1.2
IIB2t	19-25	2.0	22	40	38	38	92	5.3	1.6
IIIC	50-60	0.2	80	10	10	3	90	6.3	1.8

Other: The IIB2t horizon is typically 2.5YR 4/6 to 5YR 3/3 (moist) in color. Setting: On nearly level slopes on stream terraces and floodplains in the "Driftless Area" (slope gradient, 0-2%); original vegetation was prairie, now replaced by corn, soybeans, oats, alfalfa, and some pasture; mean ann. soil temp. 47°F, mean ann. ppt. 32 in. Estab., Henderson Co., Ill., 1947; 70 bu. oats; 450-500 bd. ft. red or white pine; catena no. 35.

DePere sil-sicl (I4, I12); Typic Udifluvent, fine, mixed, nonacid, mesic (Alluvial soil); well to moderately well drained; solum is 6 to 12 in. thick, developed in clayey alluvium. Horizons: O1, A1, C.

Some properties of DePere silty clay loam

Hori-zon	Depth in.	OM %	Sand %	Silt %	Clay %	CEC me/ 100g	BS %	pH	BD g/cc
A1	0-8	12	7	55	38	35	95	7.5	1.2
C	8-40	1	8	50	42	28	98	8.5*	1.5

*calcareous

Other: The color of the C horizon ranges from 5YR 3/4 to 2.5YR 4/4 (moist); coarse material may be encountered below a depth of 40 in.; light-colored patches and veins of lime ($CaCO_3$) commonly appear in the C horizon. Setting: On slopes of 1 to 5% gradient on stream floodplains; original vegetation was northern mesic forest, but much of this has been cleared to establish pasture; mean ann. soil temp. 45°F, mean ann. ppt. 28 in. Proposed, Fond du Lac Co., Wis., 1968; not suitable for row crops because of small size of areas; 400-450 bd. ft. white pine; catena no. 158.

Derinda sil (A4) (Fig. 7-8); Typic Hapludalf, fine, mixed, mesic (Gray-Brown Podzolic); well to moderately well drained; the solum ranges in thickness from 20 to 40 in. and is developed from 15 to 30 in. of loess over thick neutral to calcareous shale into which some of the B2 horizon has extended. Horizons: O1, O2, A1, A2, B21t, IIB22t, IIC, R.

Some properties of Derinda silt loam

Hori-zon	Depth in.	OM %	Sand %	Silt %	Clay %	CEC me/ 100g	BS %	pH	BD g/cc
A1	0-3	12.0	15	65	20	25	70	6.4	1.2
B21t	7-12	0.5	10	60	30	21	60	6.3	1.4
IIB22t	18-23	0.7	7	50	43	27	75	7.0	1.5
IIC	25-30	0.2	7	53	40	15	98	8.5*	1.7

*calcareous

Other: G. B. Lee and coworkers report from unpublished data that the Maquoketa Shale IIC horizon material contains about 40% $CaCO_3$ equivalent (the Ca:Mg ratio is about 1.8) and that when this is removed, the particle size distribution changes from the above-quoted figures to these—62%c, 37%si, 1%s; the IIC horizon is greenish gray to olive in color (5GY 6/1 to 5Y 6/3, moist) and contains lenses of limestone and chert; montmorillonite is abundant in the loess-derived horizons and illite in the clay of the shale-derived horizons. Setting: On bedrock-controlled hills in the "Driftless Area"; slopes of 4 to 12% gradient are common; original vegetation was oak savanna, most of which has been replaced by pasture or by fields of corn, oats, and hay; mean ann. soil temp. 47°F, mean ann. ppt. 32 in. Estab., Jo Daviess Co., Ill., 1941; Watson, 1966; 70 bu. oats; not suitable for pine; catena no. 13.

Detour sl-sil (E7); Aquic or Aquic Lithic Eutrochrept, fine-loamy, mixed, frigid (Low Humic Gley); (in places, is an Entic Haplaquod or weakly developed Podzol); poorly drained; solum is 15 to 40 in. thick, developed from 42 to 60 in. of dolomitic glacial till over fissured dolomite. Horizons: O1, O2, A1, A2, Bg, C, R.

Some properties of Detour loam

Horizon	Depth in.	OM %	Sand %	Silt %	Clay %	CEC me/100g	BS %	pH	BD g/cc
A1	0-6	12.0	42	40	18	21	75	6.8	1.3
Bg	10-20	1.0	56	22	22	14	75	7.0	1.5
C	20-40	0.3	65	20	15	12	85	8.2*	1.7

*calcareous

Other: The volume of stones, cobbles, and gravel in the solum ranges from 5 to 35%; in some pedons the B horizon is transitional from Bt to Bir. Setting: On till plains of rock-controlled uplands in the Door Peninsula, on slopes of 0 to 5% gradient; original vegetation was swamp conifer and hardwood forest and this still remains except in a few areas that have been cleared for permanent pasture; mean ann. soil temp. 44°F, mean ann. ppt. 28 in. Estab., Chippewa Co., Mich., 1927; 60 bu. oats; not suitable for pine; catena no. 140.

Dodge sil (B7, B13, B25, B29) (Figs. 2-52, 2-53, 2-54, 8-8, 8-13, 11-1); Typic Hapludalf, fine-silty, mixed, mesic (Gray-Brown Podzolic); well drained; solum is 30 to 42 in. thick, the same as the depth to carbonates, and is developed in 20 to 36 in. of loess overlying dolomitic (about 42% $CaCO_3$ equiv.) gravelly loam glacial till. Horizons: O1, O2, A1, A2, B1, B21t, IIB22t, IIC.

Some properties of Dodge silt loam

Horizon	Depth in.	OM %	Sand %	Silt %	Clay %	CEC me/100g	BS %	pH	BD g/cc
A1	0-3	9.0	5	78	17	12	80	6.5	1.2
A2	3-11	1.5	10	75	15	17	70	6.0	1.3
B21t	17-24	0.5	6	64	30	24	70	5.3	1.5
IIB22t	29-38	0.3	50	24	26	19	90	6.3	1.6
IIC	38-60	0.2	65	30	5	3	100	8.5*	1.8

*calcareous

Other: A few stones are present in the horizons developed from glacial till; about a third of the B2 horizon is developed in till-dominated material. Setting: On upland slopes of 2 to 20% gradient on moraines and drumlins in southeastern Wisconsin; original vegetation was oak savanna and southern mesic forest, now largely replaced by pasture or fields of corn, oats, and hay; mean ann. soil temp. 47°F, mean ann. ppt. 32 in. Proposed, Dodge Co., Wis., 1951; Hole, 1956b; Soil Conservation Service, 1967a; Steingraeber and Reynolds, 1971; 75 bu. oats; 400-450 bd. ft. red pine; 475-550 bd. ft. white pine; catena no. 51.

Dodgeville sil (A2, A3) (Figs. 2-53, 2-54, 7-5); Typic Argiudoll, fine-silty over clayey, mixed, mesic (Brunizem); well drained; solum is 24 to 40 in. thick, developed from 15 to 30 in. of loess over clayey residuum resting on limestone at a depth of 24 to 40 in. Horizons: O1, A1, B1t, B21t, IIB22t, IIR.

Some properties of Dodgeville silt loam

Horizon	Depth in.	OM %	Sand %	Silt %	Clay %	CEC me/100g	BS %	pH	BD g/cc
A1	0-13	6.0	9	65	26	30	61	6.1	1.1
B21t	16-23	0.6	13	55	32	26	58	5.5	1.4
IIB22t	23-36	0.3	5	30	65	40	60	5.7	1.5

Other: Angular chert fragments occupy about 10% of the volume of the IIB2 horizon, in which clay content ranges from 55 to 75%. Setting: On ridges in rock-controlled landscapes of the "Driftless Area"; slope gradients range from 1 to 20%; the original prairie vegetation has been replaced, except along railroad tracks, edges of cemeteries, and some fence rows and roadsides, by pasture, corn, oats, or hay; mean ann. soil temp. 47°F, mean ann. ppt. 32 in. Estab., Iowa Co., Wis., 1912; Whitson et al., 1917; Hole, 1956a; Robinson and Klingelhoets, 1961; Glocker, 1966; Watson, 1966; Salem and Hole, 1968; 60 bu. oats; not suitable for pine unless the soil is inoculated with forest soil, after which production may be 400-450 bd. ft. red pine, 450-500 ft. white pine; catena no. 10.

Dolph l-sil (F23); Aquic Glossoboralf, fine, mixed, frigid (Gray-Brown Podzolic); somewhat poorly drained; solum is 24 to 40 in. thick, developed in less than 30 in. of loess over weathered schist. Horizons: O1, O2, A1, A2, A&B, IIB2t, IIC, IIR.

Some properties of Dolph silt loam

Horizon	Depth in.	OM %	Sand %	Silt %	Clay %	CEC me/100g	BS %	pH	BD g/cc
A1	0-3	10.0	17	65	18	20	45	5.8	1.2
IIB2t	19-26	0.2	22	30	48	25	40	5.4	1.5
IIC	36-57	0.1	54	27	19	12	60	6.2	1.7

Other: Soil tongues down from the A2 into cracks in the IIB2, and from the IIB2 into the IIC; the IIB2 horizon is dark reddish brown (2.5YR 3/4) with mottles of 5YR 5/6; this soil is sometimes classified as an Aeric Ochraqualf. Setting: In nearly level and depressional sites on bedrock-controlled uplands; slope gradients are 1 to 3%; original vegetation was swamp hardwoods with some swamp conifers, but somewhat more than half of the area is in cleared pasture or has been drained to produce hay, oats, and silage corn; mean ann. soil temp. 44°F, mean ann. ppt. 30 in. Proposed, Marathon Co., Wis., 1943; 75 bu. oats; 400-450 bd. ft. white pine; catena no. 104.

Downs silt (A1) (Fig. 7-5); Mollic Hapludalf, fine-silty, mixed, mesic (intergrade between Brunizem and Gray-Brown Podzolic); well drained; solum is 36 to 45 in. thick, developed in more than 50 in. of loess over residuum on bedrock (usually dolomite). Horizons: O1, O2, A1, A2, B1, B2t, B3, C.

Some properties of Downs silt loam

Horizon	Depth in.	OM %	Sand %	Silt %	Clay %	CEC me/100g	BS %	pH	BD g/cc
A1	0-7	7.0	7	73	20	30	55	5.7	1.2
B2t	16-32	0.6	6	65	29	27	65	5.1	1.5
C	38-50	0.2	16	69	15	20	70	5.8	1.6

Other: The A2 horizon is brown (10YR 5/3, moist) with at least some light gray coatings of silt on the surfaces of the platy peds, and continuing down the surfaces of cracks in the B2 horizon. Setting: On broad ridge tops with slope gradients of 2 to 12%; original vegetation was oak savanna, which has almost everywhere been replaced by cropland in hay, oats, or corn; mean ann. soil temp. 47°F, mean ann. ppt. 32 in. Estab., Stephenson Co., Ill., 1938; Robinson and Klingelhoets, 1959, 1961; Watson, 1966; 75 bu. oats; 400-450 bd. ft. red pine; 450-500 bd. ft. white pine; catena no. 2.

Dubuque sil (A3, A5, A6, A8, A9, B6) (Figs. 2-51, 2-53, 7-9); Typic Hapludalf, fine-silty, mixed, mesic (Gray-Brown

Podzolic); well drained; solum is 24 to 40 in. thick, developed in 20 to 30 in. of loess over clayey, cherty residuum less than 6 in. thick resting on dolomite at a depth of about 36 in. Horizons: O1, O2, A1, A2, B1, B2t, IIB2t, IIR.

Some properties of Dubuque silt loam

Horizon	Depth in.	OM %	Sand %	Silt %	Clay %	CEC me/ 100g	BS %	pH	BD g/cc
A1	0-3	5.0	6	76	18	16	61	6.3	1.1
B2t	16-22	1.0	10	60	30	27	50	5.7	1.4
IIB2t	22-27	0.3	5	40	55	38	60	5.8	1.5

Other: Clay films are common on surfaces of peds in the B2 horizon, as well as some light gray silty coatings; angular chert fragments are present in the IIB2 horizon. Setting: On ridges and valley walls at slope gradients of 3 to 30% in the "Driftless Area"; original vegetation was oak savanna except in Richland County, where southern mesic forest prevailed; about two thirds of the area has been cleared for use as pasture (on steeper slopes) and for fields of hay, oats, and corn; mean ann. temp. 47°F, mean ann. ppt. 32 in. Estab., Dubuque Co., Ia., 1920; Whitson et al., 1928; Kaddou, 1960; Beatty, 1960; Slota and Garvey, 1961; Robinson and Klingelhoets, 1961; Watson, 1966; Glocker, 1966; Slota, 1969; 60 bu. oats; 400-450 bd. ft. red pine; 450-500 bd. ft. white pine; catena no. 9.

Durand sil (B5, B21); Typic Argiudoll, fine-loamy, mixed, mesic (Brunizem); well drained; solum is 4 to 8 ft. thick, developed in 15 to 25 in. of loess overlying over calcareous sandy loam to loam glacial till. Horizons: O1, O2, A1, A3, B2t, IIB2t, IIC.

Some properties of Durand silt loam

Horizon	Depth in.	OM %	Sand %	Silt %	Clay %	CEC me/ 100g	BS %	pH	BD g/cc
A1	0-10	7	7	65	28	28	70	5.7	1.1
IIB2t	22-64	0.6	25	40	35	18	65	5.2	1.4
IIC	64-85	0.1	57	25	18	11	97	8.5*	1.6

*calcareous

Other: The subsoil is reddish brown (5YR 4/4, moist) as contrasted with the yellowish-brown (10YR 4/5) color of the glacial till; this is considered to be a paleosol. Setting: On gently rolling to rolling moraines with slope gradients from 1 to 12%; the original prairie vegetation has been replaced by corn, soybeans, small grains, and pasture; mean ann. soil temp. 44°F, mean ann. ppt. 30 in. Estab., Carroll Co., Ill., 1967; 70 bu. oats; not suitable for pine without inoculation with forest soil; catena no. 45.

Edith ls-sl (G8, G19); Entic Hapludoll, sandy-skeletal, mixed, frigid (Regosol); excessively drained; solum is 3 to 7 in. thick, developed in deep, slightly calcareous to acid glacial outwash and coarse till. Horizons: O1, O2, A1, C.

Some properties of Edith sand

Horizon	Depth in.	OM %	Sand %	Silt %	Clay %	CEC me/ 100g	BS %	pH	BD g/cc
A1	0-5	12.0	89	7	4	30	72	6.5	1.6
C	5-20	0.2	97	2	1	1	65	5.7	1.8

Other: One fifth to one half of the soil volume may be occupied by gravel, cobbles, and stones, particularly below the depth of 15 in. Setting: On rolling to steep moraines, eskers, and kames, with slopes of 10 to 40% gradient; the original vegetation of oak savanna is still present, for the most part, although the prairie component has been considerably replaced by bluegrass and other invaders, where the soil has been used for pasture; mean ann. soil temp. 44°F, mean ann. ppt. 30 in. Estab., Washington Co., Minn., 1941; not suitable for agricultural crops; 400-450 bd. ft. red or white pine; catena no. 159.

Elburn sil (B22); Aquic Argiudoll, fine-silty, mixed, mesic (Brunizem); somewhat poorly drained; solum is 45 to 60 in. thick, developed in 36 to 50 in. of loess over calcareous sandy loam or loam glacial till or outwash. Horizons: O1, O2, A1, B1, B21t, IIB32, IIC.

Some properties of Elburn silt loam

Horizon	Depth in.	OM %	Sand %	Silt %	Clay %	CEC me/ 100g	BS %	pH	BD g/cc
A1	0-11	12.0	6	70	24	30	70	6.3	1.1
B21t	19-26	1.0	4	66	30	22	75	6.1	1.3
IIB32	42-50	0.6	50	35	15	10	90	7.4	1.7
IIC	50-60	0.1	72	23	5	2	98	8.5*	1.8

*calcareous

Other: A concentration of clay may occur at the interface between loess and glacial drift. Setting: On slopes of 0 to 7% gradient on glacial drift plains; original vegetation was prairie, now replaced by pasture, hay, small grains, corn, and soybeans; mean ann. soil temp. 47°F, mean ann. ppt. 30 in. Estab., Kendall Co., Ill., 1941; 65 bu. oats; not suitable for pine; catena no. 47.

Elkmound sl-l (D6); Lithic Dystrochrept, loamy, mixed, mesic (Lithosol); excessively drained; solum is 12 to 20 in. thick, developed in platy, somewhat cemented sandstone. Horizons: O1, O2, A1, A2, B, R.

Some properties of Elkmound sandy loam

Horizon	Depth in.	OM %	Sand %	Silt %	Clay %	CEC me/ 100g	BS %	pH	BD g/cc
A1	0-2	5.0	67	22	11	12	65	6.3	1.2
B	4-17	0.3	68	20	12	8	60	5.7	1.4

Setting: On slopes of 1 to 30% gradient in rolling and hilly landscapes; the original oak savannah vegetation is still intact except for suppression of prairie plants where grazing has been practiced, and except in a few nearly level areas where this soil has been included in cropped fields; mean ann. soil temp. 47°F, mean ann. ppt. 31 in. Proposed, Green Co., Wis., 1968; not suitable for oats; 450-500 bd. ft. red or white pine; catena no. 160.

Elliott l-sicl (B20) (Fig. 8-10); Aquic Argiudoll, fine, illitic, mesic (Brunizem); somewhat poorly drained; solum is 20 to 45 in. thick, developed in less than 20 in. of loess over dolomitic olive-brown silty clay loam to clay loam glacial till. Horizons: O1, O2, A1, A3, IIB2t, IIB3, IIC.

Some properties of Elliott silty clay loam

Horizon	Depth in.	OM %	Sand %	Silt %	Clay %	CEC me/ 100g	BS %	pH	BD g/cc
A1	0-10	14.0	19	53	28	32	69	5.5	1.0
IIB2t	14-30	2.0	11	37	52	28	65	5.8	1.5
IIC	36-48	0.4	23	47	30	17	98	8.5*	1.7

*calcareous

Other: The IIB2 and IIC horizons are mottled. Setting: On slopes of 1 to 3% gradient on till plains; the original prairie vegetation has been replaced by hay, small grains, soybeans, and corn; mean ann. soil temp. 47°F, mean ann. ppt. 32 in. Estab., Ford Co., Ill., 1929; Pedersen, 1954; Link and Demo, 1970; Steingraeber and Reynolds, 1971; 70 bu. oats; not suitable for pine; catena no 53. (See Chapter 8, above.)

Elm Lake ls-sl (D8, D11, D12) (Fig. 10-4); Typic Haplaquent, sandy over loamy, mixed, acid, frigid (Low Humic Gley); poorly drained; solum is 1 to 3 in. thick, developed in 20 to 40 in. of sandy sediment over interbedded acid sandstone and shale. Horizons: O1, O2, A1, C1, IIC2, IIR.

Some properties of Elm Lake loamy sand

Horizon	Depth in.	OM %	Sand %	Silt %	Clay %	CEC me/ 100g	BS %	pH	BD g/cc
A1	0-1	12.0	80	12	8	10	60	5.8	1.3
C1	1-23	0.2	91	5	4	5	55	5.5	1.5
IIC2	23-36	0.1	25	45	30	27	52	4.5	1.6

Other: The interstratified sandstone and shale are typically fragmented or even unconsolidated to a depth of about 50 in. Setting: On level areas and in slight depressions, on slope gradients of 0 to 2%; original vegetation was swamp hardwoods and conifers, second-growth stands of which occupy most areas; some land has been cleared for pasture or drained and heavily fertilized for crop production; mean ann. soil temp. 45°F, mean ann. ppt., 32 in. Proposed, Wood Co., Wis., 1958; not suitable for oats or pine; catena no. 15.

Emmert sl-l (G2, G3, G11); Typic Udorthent, sandy-skeletal, mixed, frigid (Regosol). *See* Kiva.

Emmet sl-l (E1, E2, E3, E4, J3); Alfic Haplorthod, coarse-loamy, mixed, frigid (Podzol); well to excessively drained; solum is 24 to 50 in. thick, developed in sandy material overlying "pink" (7.5YR 5/4) calcareous sandy loam to loam glacial till. Horizons: O1, O2, A1, A2, Bir, A′2, IIB′2t, IIC.

Some properties of Emmet loamy fine sand

Horizon	Depth in.	OM %	Sand %	Silt %	Clay %	CEC me/ 100g	BS %	pH	BD g/cc
A2	1-12	3.0	68	28	4	6	62	6.2	1.3
Bir	12-24	0.7	76	20	4	3	40	6.0	1.3
A′2	24-30	0.1	73	22	5	2	54	6.5	1.6
IIB′2t	30-38	0.5	59	22	19	10	63	7.4	1.7
IIC	38-60	0.1	68	25	7	3	90	8.5*	1.8

*calcareous

Other: Thickness of solum equals depth to carbonates; coarse fragments may occupy as much as 15% of the solum by volume; vesicular structure may be common in the solum; the A′2 horizon is a fragipan (A2m) in many pedons. Setting: On moraines and drumlins with slope gradients of 0 to 50%; original vegetation was northern mesic forest, and this has been replaced by second- and third-growth woodland on steep slopes, by pasture and cherry and apple orchards on rolling land, and by hay, oats, and corn for silage on less sloping land; mean ann. soil temp. 44°F, mean ann. ppt. 28 in. Estab., Antrim Co., Mich., 1923; Soil Conservation Service, 1967a; 45 bu. oats; 400-450 bd. ft. red pine; 450-500 bd. ft. white pine; catena no. 137.

Ettrick sil-sicl (A13, J1); Typic Argiaquoll, fine-silty, mixed, mesic (Humic Gley); poorly drained; solum is 24 to 48 in. thick, developed in silty alluvial deposits 24 to 40 in. thick overlying "bluish-gray" (5Y 5/1) stratified silt and fine sand. Horizons: O1, O2, A1, ABg, B2tg, IIB3g, IICg.

Some properties of Ettrick silt loam

Horizon	Depth in.	OM %	Sand %	Silt %	Clay %	CEC me/ 100g	BS %	pH	BD g/cc
A1	0-10	14.0	12	70	18	50	97	7.5	1.1
B2tg	14-26	2.0	10	55	35	20	100	7.5	1.4
IIB3g	26-35	0.5	12	60	28	15	96	7.5	1.6
IICg	35-60	0.1	20*	70†	10	7	100	7.5	1.7

*in silty layers; 78% in sandy layers †in silt, and 12% in sandy layers

Other: Dark organic stains and even clay films coat many ped faces in the IIB3g horizon. Setting: On nearly level high stream bottoms with slope gradients of 1 to 2% in valleys of the "Driftless Area"; original vegetation was swamp hardwoods and sedge meadows, and areas subject to frequent flooding are still in the same condition, except where pastured; about half of the area is drained and utilized for field crops; mean ann. soil temp. 47°F, mean ann. ppt. 32 in. Estab., Trempealeau Co., Wis., 1939; Robinson and Klingelhoets, 1959; Slota and Garvey, 1961; 60 bu. oats; not suitable for pine; catena no. 41.

Ewen sl-sicl (I4, I10, I11, I12, I18, I19, I21); Typic Udifluvent, fine-silty, mixed, mesic (Alluvial soil); well to moderately well drained; solum is 2 to 8 in. thick, developed in reddish-brown alluvium of silts and clays. Horizons: O1, O2, A1, C.

Some properties of Ewen silty clay loam

Horizon	Depth in.	OM %	Sand %	Silt %	Clay %	CEC me/ 100g	BS %	pH	BD g/cc
A1	0-8	10.0	12	60	28	22	95	7.5	1.2
C*	8-42	0.8	8	57	35	18	98	8.0†	1.5

*buried A1 horizons commonly present †calcareous

Other: Degree and intensity of mottling in the C horizon varies; below a depth of 42 in. various kinds of glacial or glacio-lacustrine drift may be present. Setting: On stream bottoms in regions of reddish-brown clayey glacial drift (Regions E and I of the soil map) on gradients of 0 to 4%; original vegetation was northern mesic forest, which is reproducing after logging in about half the area, and is replaced by pasture in other areas; mean ann. soil temp. 40°F near Lake Superior and 44°F near Lake Michigan, mean ann. ppt. 32 in. Estab., Ontonagon Co., Mich., 1941; 75 bu. oats; 450-500 bd. ft. white pine; catena no. 161.

Fabius sl-sil (B31); Aquic Argiudoll, fine-loamy over sandy or sandy-skeletal, mixed, mesic (Brunizem); somewhat poorly drained; solum is 12 to 20 in. thick, the same as the depth to coarse outwash substratum and to calcareous material underlying loamy deposits. Horizons: O1, O2, A1, A2, B2t, IIC.

Some properties of Fabius loam

Hori-zon	Depth in.	OM %	Sand %	Silt %	Clay %	CEC me/ 100g	BS %	pH	BD g/cc
A1	0-7	9.0	50	30	20	17	75	6.4	1.2
B2t	10-18	0.6	48	22	30	19	60	6.1	1.6
IIC	18-40	0.1	90	8	2	2	100	8.5*	2.0

*calcareous

Other: Depth to mottling ranges from 6 to 16 in.; the B2 horizon ranges in texture from heavy sandy loam to light clay loam. Setting: On glacial outwash plains and beach ridges with slopes of 0 to 2% gradient; original vegetation was lowland hardwood forest, only a little of which has been allowed to reproduce itself; some land is used for pasture, but most of the area has been put in cultivated crops; mean ann. soil temp. 45°F, mean ann. ppt. 32 in. Estab., Lapeer Co., Mich., 1967; 65 bu. oats; 450-500 bd. ft. white pine; catena no. 77.

Fayette sil: *See* p. 126.

Fence sl-sil (F25); Alfic Haplorthod, coarse-loamy, mixed, frigid (Podzol with double profile: Podzol over weakly developed Gray-Brown Podzolic); well drained; solum is 32 to 45 in. thick, developed in more than 2 ft. of silty deposits overlying stratified silts and fine sands of glacio-lacustrine beds. A fragipan (x) may be present. Horizons: O1, O2, A1, A2, Bhir, A′2x, A′2&B′2t, B′2t, C.

Some properties of Fence silt loam

Hori-zon	Depth in.	OM %	Sand %	Silt %	Clay %	CEC me/ 100g	BS %	pH	BD g/cc
A2	2-4	4.0	13	75	12	12	28	6.1	1.2
Bhir	4-14	6.0	19	71	10	9	35	5.8	1.3
A′2x	14-18	1.0	26	67	7	11	40	5.7	1.5
B′2t	26-35	0.6	20	65	15	9	52	5.4	1.6
C	35-50	0.1	50	40	10	7	60	5.8	1.7

Other: The A′2 horizon tongues into the B′2 horizon. Stratification in the C horizon may be associated with mottling; occasional gravel lenses are present below 42 in. Setting: On nearly level to gently sloping glacio-lacustrine plains and in lowlands between esker ridges on slopes of 0 to 15% gradient; original vegetation was northern mesic forest, most of which has been removed to allow use of the land for pasture and crops of hay, oats, and corn for silage; mean ann. soil temp. 42°F, mean ann. ppt. 32 in. Proposed, Florence Co., Wis., 1958; Hole et al., 1962; 75 bu. oats; 450-500 bd. ft. red pine; 500-575 bd. ft. white pine; catena no. 154.

Fenwood sil (F14); Typic Glossoboralf, fine-loamy, mixed, frigid (Gray-Brown Podzolic); well to moderately well drained; solum is 36 to 48 in. thick, developed in about 15 in. of silt over loamy residuum on fine-grained Precambrian crystalline bedrock. Horizons: O1, O2, A1, A2, A&B, B21t, IIB22t, R.

Some properties of Fenwood loam

Hori-zon	Depth in.	OM %	Sand %	Silt %	Clay %	CEC me/ 100g	BS %	pH	BD g/cc
A2	3-12	1.0	42	40	18	17	50	4.8	1.3
B21t	24-31	0.4	20	45	35	15	70	5.0	1.6
IIB22t	31-43	0.6	28	40	32	12	78	5.8	1.7

Other: Coarse fragments occupy 10% by volume of the B21 horizon and 20% by volume of the IIB22 horizon. Setting: On slopes of 3 to 12% gradient on rock-controlled uplands in the Wisconsin River valley of central Wisconsin in a portion of the "Driftless Area"; original vegetation was northern mesic forest, somewhat more than half of which has been replaced by pasture, hay, oats, and silage corn; mean ann. soil temp. 44°F, mean ann. ppt. 31 in. Proposed, Marathon Co., Wis., 1970; 75 bu. oats; 450-500 bd. ft. red pine; 500-575 bd. ft. white pine; catena no. 102.

Flagg sil (B10); Typic Hapludalf, fine-silty, mixed, mesic (Gray-Brown Podzolic); well drained; solum, including the paleosol, is 5 to 8 ft. thick, developed from 36 to 50 in. of loess over weathered sandy loam to loam glacial till that is calcareous below the paleosol. Horizons: O1, O2, A1, A2, B21t, IIB22t, IIC.

Some properties of Flagg silt loam

Hori-zon	Depth in.	OM %	Sand %	Silt %	Clay %	CEC me/ 100g	BS %	pH	BD g/cc
A1	0-3	10.0	12	70	18	18	70	5.7	1.2
B21t	18-42	0.5	8	60	32	17	65	5.3	1.5
IIB22t	42-75	0.6	9	55	36	18	78	5.5	1.7
IIC	75-90	0.1	52	30	18	10	98	8.5*	1.7

*calcareous

Other: Colors in the paleosol (IIB2 horizon) are reddish brown (5YR 4/4), and the texture ranges from silty clay loam to sandy clay loam or clay loam. Setting: On rolling moraines with slope gradients of 2 to 18%; original vegetation of oak savanna has largely been replaced by fields of hay, small grains, and corn; mean ann. soil temp. 48°F, mean ann. ppt. 32 in. Proposed, Carroll Co., Ill., 1960, Haszel, 1971; 75 bu. oats; 450-500 bd. ft. red pine; 500-575 bd. ft. white pine; catena no. 44.

Floyd l-sil (F13); Aquic Hapludoll, fine-loamy, mixed, mesic (Brunizem); somewhat poorly drained; solum 40 to 60 in. thick, developed in less than 24 in. of silty sediment over firm loam to clay loam glacial till that is calcareous within a foot below the sediment cover. Horizons O1, O2, A1, A3, B1, B21, IIB22, IIC1, IIC2.

Some properties of Floyd silt loam

Hori-zon	Depth in.	OM %	Sand %	Silt %	Clay %	CEC me/ 100g	BS %	pH	BD g/cc
A1	0-16	12.0	20	54	26	28	53	5.8	1.2
IIB22	38-45	1.0	22	40	38	18	50	5.6	1.5
IIC1	45-50	0.2	35	30	35	14	60	5.8	1.7
IIC2	50-60	0.1	45	25	30	12	98	8.5*	1.7

*slightly calcareous

Other: Depth to carbonates ranges from 4 to 6 ft.; stones are present in the till-derived horizons. Setting: On gentle lower slopes of 0 to 5% gradient in rolling moraine landscapes; original vegetation was prairie, now replaced by pasture and, where drained, fields of hay, small grains, and corn; mean ann. soil temp. 43°F, mean ann. ppt. 30 in. Estab., Floyd Co., Ia., 1922; Haszel, 1968; 60 bu. oats; not suitable for pine unless the soil is inoculated with forest soil; catena no. 105.

Fox sl-sil (B4, B17, B18, B30, B31, B32, B33, B34, I6) (Figs. 1-6, 2-53, 2-54, 8-4, 8-9, 11-1); Typic Hapludalf, fine-loamy over sandy or sandy-skeletal, mixed, mesic (Gray-Brown Podzolic); well drained; solum is 24 to 40 in. thick, developed in less than 20 in. of silty material overlying calcareous glacial outwash sand and gravel. Horizons: O1, O2, A1, B1t, B21t, IIB22t, IIC.

Some properties of Fox silt loam

Horizon	Depth in.	OM %	Sand %	Silt %	Clay %	CEC me/100g	BS %	pH	BD g/cc
A1	0-2	7.0	18	60	22	40	65	6.5	1.2
B21t	16-22	0.5	18	50	32	22	70	5.7	1.5
IIB22t	22-31	0.7	50	15	35	25	88	6.8	1.7
IIC	34-60	0.1	89	6	5	5	100	8.5*	2.0

*calcareous

Other: The B2 horizon adjacent to the underlying sand and gravel is commonly highest in clay content and called the "beta" B; a horizontal banding has been observed in the solum as high as the uppermost subhorizon of the B horizon. Setting: On level and hilly glacial outwash deposits of glacio-fluvial plains and kettle moraines, with slope gradients ranging from 0 to 40%; original vegetation was oak savanna; much has been altered in hilly areas by grazing, and on less sloping land by clearing and cultivation for production of hay, small grains, and corn; mean ann. soil temp. 47°F, mean ann. ppt. 32 in. Estab., Columbia Co., Wis., 1911; Milfred and Hole, 1970; Haszel, 1971; 60 bu. oats; 325-400 bd. ft. red pine; 400-450 bd. ft. white pine; catena no. 76.

Freeon silt (F3, F10, F20); Typic Glossoboralf, fine-loamy, mixed, frigid (Gray-Brown Podzolic); moderately well drained; solum is 20 to 40 in. thick, developed in 15 to 36 in. of loess overlying acid, reddish-brown sandy loam to loam glacial till. Horizons: O1, O2, A1, A2, B1&A2, B21t, IIB22t, IIB3g, IIC.

Some properties of Freeon silt loam

Horizon	Depth in.	OM %	Sand %	Silt %	Clay %	CEC me/100g	BS %	pH	BD g/cc
A2	2-11	1.4	17	75	8	10	13	5.0	1.2
B21t	14-20	0.5	22	60	18	13	40	5.0	1.4
IIB22t	20-26	0.6	45	35	20	13	50	5.0	1.6
IIC	28-38	0.1	71	17	12	7	50	5.3	1.7

Other: The till-derived horizons are stony; the till ranges in color from reddish brown (5YR 4/4) to dark brown (7.5YR 4/4). Setting: On glacial moraine slopes of 0 to 5% gradient; original vegetation was northern mesic forest, which has been replaced over more than half the area by pasture, hay, small grains, corn, and potatoes; mean ann. soil temp. 44°F, mean ann. ppt. 30 in. Estab., Barron Co., Wis., 1949; Robinson et al., 1958; Haszel, 1968; 75 bu. corn; 450-500 bd. ft. white pine; catena no. 96.

Freer sil (F2, F3, F10, F20, G22, J6, J13); Aeric Ochraqualf, fine-loamy, mixed, frigid (Gray-Brown Podzolic); somewhat poorly drained. Horizons: O1, O2, A1, A2g, B&A, B21tg, IIB22t, IIB3t, IIC.

Some properties of Freer silt loam

Horizon	Depth in.	OM %	Sand %	Silt %	Clay %	CEC me/100g	BS %	pH	BD g/cc
A2g	4-13	1.6	13	75	12	15	15	5.3	1.2
B21tg	18-24	0.5	15	60	25	18	40	5.2	1.5
IIB22tg	24-32	0.6	42	38	20	14	50	5.2	1.6
IIC	42-55	0.1	68	20	12	10	52	6.2	1.7

Other: The A2 horizon tongues down cracks between peds in the B horzon; coarse fragments of crystalline rocks occupy about 1% of the volume of the loess-derived horizons and 10% of the till-derived horizons. Setting: On gentle slopes and in slight depressions with slope gradients of 0 to 2%; original vegetation was a mixture of northern mesic forest and swamp hardwoods and conifers, which has been allowed to reproduce itself on about half of the area, but the remainder of the area has been cleared for use as pasture or for production of hay, small grains, and corn; mean ann. soil temp. 43°F, mean ann. ppt. 30 in. Estab., Mille Lacs Co., Minn., 1952; Robinson et al., 1958; Haszel, 1968; 70 bu. oats; 350-400 bd. ft. white pine; catena no. 96.

Gale sil (D1, D2, D3, D5, D7, D9) (Figs. 2-53, 10-6); Typic Hapludalf, fine-silty over sandy or sandy skeletal, mixed, mesic (Gray-Brown Podzolic); well drained; solum is 20 to 40 in. thick, developed in 15 to 34 in. of loess overlying unconsolidated sandstone. Horizons: O1, O2, A1, A2, B1, B2t, IIB3, IIC, IIR.

Some properties of Gale silt loam

Horizon	Depth in.	OM %	Sand %	Silt %	Clay %	CEC me/100g	BS %	pH	BD g/cc
A2	4-8	1.5	15	70	15	10	18	5.7	1.2
B2t	13-25	0.5	19	50	31	18	20	5.5	1.6
IIC	31-39	0.1	90	5	5	5	50	5.8	1.8

Other: The bedrock is partially indurated at a depth of about 40 in. Setting: On slope gradients of 2 to 30% on valley slopes and narrow ridge tops in southwestern and west-central Wisconsin; original vegetation was oak savanna except in Richland County and vicinity where southern mesic forest prevailed; only on steepest slopes has the original vegetation been allowed to reproduce itself; on areas of less steep slopes, pasture and cropped fields (hay, small grains, oats) have been developed; mean ann. soil temp. 46°F, mean ann. ppt. 30 in. Estab., Trempealeau Co., Wis., 1940; Beatty, 1960; Robinson and Klingelhoets, 1961; Thomas, Carroll, and Wing, 1962; Haszel, 1968; Slota, 1969; 60 bu. oats; 400-450 bd. ft. red pine; 425-500 bd. ft. white pine; catena no. 16.

Garwin sicl (A1); Typic Haplaquoll, fine-silty, mixed, mesic (Humic Gley). An inextensive associate of Tama in Grant County.

Gogebic sil-l (G1, G9, G10, G12, G20, G21, G23, I7) (Figs. 2-53, 2-54, 14-5); Alfic Fragiorthod, coarse-loamy, mixed, frigid (Podzol); well drained; solum is 20 to 40 in. thick, developed in a loamy covering of that thickness over reddish-brown (2.5YR 4/4) acid sandy loam glacial till. Horizons: O1, O2, A1, A2, Bhir, B3x, C.

Some properties of Gogebic loam

Horizon	Depth in.	OM %	Sand %	Silt %	Clay %	CEC me/100g	BS %	pH	BD g/cc
A2	0-5	2.0	52	40	8	15	35	4.4	1.3
Bhir	5-21	3.0	60	33	7	19	27	5.5	1.3
B3x	21-26	1.5	69	26	5	10	35	5.8	1.7
C	26-40	0.7	72	19	9	5	60	5.3	1.7

Other: The A2 horizon is a reddish gray (5YR 5/2) to pinkish gray (5YR 6/2) in color; the platy, vesicular fragipan (B3m) is variable in degree of development; the profile is usually stony throughout; the till is commonly alkaline at a depth of about 6 ft. Setting: On glacial moraines with slope gradients of 0 to 25%; original vegetation was northern mesic forest, which is in most areas allowed to reproduce itself; some land is in pasture, and where stones have been cleared, in hay, potatoes, or small grains; mean ann. soil temp. 41°F, mean ann. ppt. 30 in. Estab., Gogebic Co., Mich., 1949; Ableiter and Hole, 1961; 60 bu. oats; 450-500 bd. ft. red pine; 475-550 bd. ft. white pine; catena no. 128.

Goodman sil (F5, G2) (Figs. 12-9, 12-10); Alfic Haplorthod, coarse-silty, mixed, frigid (Podzol); well drained; solum is 32 to 45 in. thick, developed in 24 to 40 in. of silty covering over reddish-brown (5YR 4/4) acid loam to sandy loam glacial till. Horizons: O1, O2, A1, A2, Bhir, Bir, A′2x, A&B, IIB′2tx, IIC.

Some properties of Goodman silt loam

Horizon	Depth in.	OM %	Sand %	Silt %	Clay %	CEC me/100g	BS %	pH	BD g/cc
A2	0-4	3.0	11	74	15	8	50	6.0	1.1
Bhir	4-11	5.0	12	72	16	10	40	5.7	1.1
A′2x	14-20	1.5	25	63	12	8	60	5.5	1.5
IIB′2t	25-38	1.0	73	12	15	11	70	5.7	1.5
IIC	38-45	0.1	75	14	11	7	75	5.8	1.7

Other: Platy structure is dominant throughout this profile below the A1 horizon; weak subangular blocky structure is present in the IIB′2tx, the A′2x material tongues down into cracks in the IIB′2tx. Setting: On glacial moraines with slope gradients of 2 to 12%; original vegetation was northern mesic forest, which is allowed to reproduce itself over considerably more than half of the area; where cleared, the land is used for pasture and for crops that include hay, small grains, and corn for silage; mean ann. soil temp. 42°F, mean ann. ppt. 29 in. Proposed, Marinette Co., Wis., 1954; Hole et al., 1962; 75 bu. oats; 450-500 bd. ft. red pine; 475-550 bd. ft. white pine; catena no. 124.

Gotham ls-lfs (C1, C2, C3, C5, C6, C11); Psammentic Hapludalf, sandy, mixed, mesic (Gray-Brown Podzolic); excessively drained; solum is 20 to 40 in. thick, developed in acid glacial outwash sand. Horizons: O1, A1, A2, B1, B2t, B3, C.

Some properties of Gotham loamy sand

Horizon	Depth in.	OM %	Sand %	Silt %	Clay %	CEC me/100g	BS %	pH	BD g/cc
A1	0-7	1.5	88	5	7	9	45	5.1	1.2
B2t	20-32	0.2	82	8	10	6	22	5.5	1.5
C	38-60	0.1	95	3	2	1	20	6.0	1.6

Other: The argillic B horizon is at least 6 in. thick; at depths of 40 to 72 in., in the C horizon, dark brown bands of sandy loam or loamy fine sand may be present. Setting: On nearly level to undulating glacial outwash plains on slopes of 1 to 5% gradient; original vegetation was open savanna; most of the area is now in crops (small grains, corn, soybeans); mean ann. soil temp. 47°F, mean ann. ppt. 32 in. Estab., Richland Co., Wis., 1956; Robinson and Klingelhoets, 1959; Slota and Garvey, 1961; Watson, 1966; 50 bu. oats; 450-500 bd. ft. red pine; 500-575 bd. ft. white pine; catena no. 162.

Granby ls-sl (E12, E13, J3) (Fig. 11-7); Typic Haplaquoll, sandy, mixed, mesic (Humic Gley); poorly drained; solum is 30 to 40 in. thick, developed in glacial outwash and glacio-lacustrine sand that is calcareous at a depth of 30 to 60 in. Horizons: O1, O2, A1, B2g, B3g, Cg.

Some properties of Granby loamy sand

Horizon	Depth in.	OM %	Sand %	Silt %	Clay %	CEC me/100g	BS %	pH	BD g/cc
A1	0-10	10.0	82	8	10	15	90	6.5	1.2
B2g	10-16	1.0	91	4	5	4	70	6.3	1.5
Cg	32-42	0.1	95	2	3	2	95	7.3	1.7

Other: The mollic epipedon (surface horizon) ranges in thickness from 10 to 16 in. Setting: On level or depressional areas of outwash and lake plains on slopes of 0 to 2% gradient; original vegetation was swamp hardwoods and conifers; most of the area is now in pasture or crops (small grains, hay, corn, small fruits, vegetables); mean ann. soil temp. 44°F, mean ann. ppt. 32 in. Estab., Oswego Co., N.Y., 1917; Parker, Kurer, and Steingraeber, 1970; 50 bu. oats; not suitable for pine; catena no. 81.

Greenwood mucky peat (J13); Typic Borohemist, dysic, frigid (Bog); very poorly drained; solum 50 in. to 20 ft. deep, developed in herbaceous (mossy and fibrous) debris overlying glacial drift, Horizons: O1, Oal, Oel, Oe2, Oe3, IIC.

Some properties of Greenwood mucky peat

Horizon	Depth in.	OM %	Sand %	Silt %	Clay %	CEC me/100g	BS %	pH	BD g/cc	Fibers unrub. %	Fibers rub. %
Oal	0-4	60	0	24	16	120	40	4.5	0.1	15	10
Oe3	19-28	80	0	13	7	130	10	4.8	0.3	40	30
IIC	60-70	1	78	12	10	8	5	5.6	1.7	0	0

Other: Ground water is stagnant in this soil. Setting: In kettles and other depressions in glacial landscapes of northern Wisconsin; original vegetation of sphagnum, leatherleaf, and labrador tea, with scattered black spruce, remains undisturbed in most areas left in the natural state; mean ann. temp. 41°F, mean ann. ppt. 32 in. Estab., Ogemaw Co., Mich., 1923; Hole et al., 1962; Milfred, Olson, and Hole, 1967; not suitable for oats or pine; catena no. 100.

Guenther ls-sl (C17); Alfic Haplorthod, sandy over loamy, mixed, frigid (Podzol); well drained; solum is 24 to 50 in. thick, developed in 20 to 40 in. of sandy deposits over acid clayey to loamy residuum over Precambrian crystalline bedrock. Horizons: O1, O2, A1, A2, Bir, IIA′2, IIA&B, IIB′2t, IIC.

Some properties of Guenther loamy sand

Horizon	Depth in.	OM %	Sand %	Silt %	Clay %	CEC me/100g	BS %	pH	BD g/cc
A2	3-9	2.0	86	6	8	6	14	5.8	1.3
Bir	9-25	3.0	80	10	10	12	5	5.4	1.3
IIB′2t	30-37	0.5	27	45	28	18	35	5.5	1.5
IIC	37-60	0.1	36	40	24	13	40	6.0	1.7

Other: Small fragments of igneous rock occupy 10 to 30% by volume of the IIC horizon, and decrease in number upward in the solum; IIA′2 material tongues down cracks in the IIB′2 horizon; the horizons in the second initial material are fragic. Setting: On rock-controlled uplands of north-central Wisconsin, with slope gradients of 3 to 15%; original vegetation was northern mesic forest and red and white pine; most of the area has been cleared for pasture and for production of hay, small grains, and corn for silage; mean ann. soil temp. 43°F, mean ann. ppt. 30 in. Proposed, Portage Co., Wis., 1944; 50 bu. oats; 400-450 bd. ft. red pine; 450-500 bd. ft. white pine; catena no. 149.

Hebron sl-sil (B33, J10); Typic Hapludalf, fine-loamy, mixed, mesic (Gray-Brown Podzolic); well drained; solum is 20 to 40 in. thick, developed in loamy deposits 20 to 36 in. thick over glacio-lacustrine silty clay loam and stratified silt and silty clay. Horizons: O1, O2, A1, A2, B1t, B21t, IIB22t, IIC.

Some properties of Hebron loam

Hori-zon	Depth in.	OM %	Sand %	Silt %	Clay %	CEC me/ 100g	BS %	pH	BD g/cc
A2	3-11	1.5	40	40	20	18	70	6.6	1.3
B21t	16-24	0.5	33	35	32	17	65	6.3	1.5
IIB22t	24-29	0.7	7	55	38	20	80	7.4	1.6
IIC	29-40	0.1	8	60	32	16	95	8.5*	1.7

*calcareous

Other: Depth to carbonates is 20 to 40 in. Setting: On glacio-lacustrine deposits on benches or plains with slope gradients of 1 to 20%; original vegetation was largely oak savanna; except for steep woodland or pasture, now in cropland (hay, small grains, corn); mean ann. soil temp. 47°F, mean ann. ppt. 32 in. Estab., Jefferson Co., Wis., 1912; Milfred and Hole, 1970; 70 bu. oats; 400-450 bd. ft. white pine; catena no. 90.

Hesch sl-l (D5, D7, D8, D9, D10); Typic Argiudoll, coarse-loamy, mixed, mesic (Brunizem); well drained; solum is 20 to 40 in. thick, developed in loamy material of that thickness overlying firm sandstone bedrock. Horizons: O1, A1, A3, B1, B2t, B3t, C, R.

Some properties of Hesch sandy loam

Hori-zon	Depth in.	OM %	Sand %	Silt %	Clay %	CEC me/ 100g	BS %	pH	BD g/cc
A1	0-8	10.0	57	25	18	20	75	6.5	1.2
B2t	19-29	1.0	52	28	20	14	80	6.7	1.6

Other: Small fragments of sandstone are commonly present in the solum. Setting: On valley slopes with gradients of 5 to 25%; the original prairie vegetation has been replaced by pasture or fields of hay, small grains, and corn; mean ann. soil temp. 47°F, mean ann. ppt. 31 in., supplemented by runoff from upper slopes. Estab., La Crosse Co., Wis., 1957; Beatty, 1960; Slota and Garvey, 1961; Robinson and Klingelhoets, 1961; 55 bu. oats; not suitable for pine unless forest soil is introduced as an inoculant; catena no. 20.

Hiawatha s-ls (H1, H2, H3, I8, I9); Typic Haplorthod, sandy, mixed, frigid (Podzol); excessively drained; solum is 30 to 60 in. thick, developed in acid sandy glacial drift of pinkish hue (7.5YR-5YR). Horizons: O1, O2, A2, Bhir, B3, C.

Some properties of Hiawatha loamy sand

Hori-zon	Depth in.	OM %	Sand %	Silt %	Clay %	CEC me/ 100g	BS %	pH	BD g/cc
A2	0-7	1.0	83	15	2	3	30	4.8	1.3
Bhir	7-16	2.5	82	12	6	12	10	5.4	1.3
C	33-50	0.1	93	5	2	2	30	6.0	1.7

Other: Degree of ortstein development in the Bhir ranges from incipient to nearly continuous; a few stones and cobbles may be present in the solum and C horizon; at places a finer-textured material underlies the sandy C horizon at a depth of 9 ft. or so; an incipient fragipan may be present below the Bhir horizon, extending into the C horizon as deep as 60 in. Setting: On sandy glacial drift hills and plains, with slope gradients from 2 to 25%; original vegetation was pine, now replaced (after logging) by second-growth hardwoods and conifers, except where limited pasture is maintained or still more limited cropland (hay, small grains, silage corn); mean ann. soil temp. 41°F, mean ann. ppt. 28 in. Estab., Alger Co., Mich., 1929; Soil Conservation Service, 1952; Ableiter and Hole, 1961; Hole et al., 1962; 35 bu. oats; 400-450 bd. ft. red or white pine; catena no. 147.

Hibbing l-sicl (I1, I2, I7, I8, I18, I19) (Figs. 2-53, 14-5, 15-3); Typic Eutroboralf, fine, mixed, frigid (Gray Wooded); well to moderately well drained; solum is 30 to 34 in. thick, developed in less than 8 in. of loess over calcareous reddish-brown (5YR 5/4) glacial till. Horizons: O1, O2, A1, A2, A&B, IIB2t, IIC.

Some properties of Hibbing silt loam

Hori-zon	Depth in.	OM %	Sand %	Silt %	Clay %	CEC me/ 100g	BS %	pH	BD g/cc
A2	2-8	1.0	14	60	26	16	55	5.7	1.3
IIB2t	11-34	0.6	5	45	50	28	70	6.0	1.5
IIC	34-60	0.1	10	45	45	15	100	8.5*	1.6

*calcareous

Other: The A2 tongues down cracks in the IIB2 horizon; where more than 6 in. of silty material overlies the clayey till, a small Spodosol sequum may be present; thickness of solum is the same as depth to carbonates. Setting: On glacial moraines with slope gradients of 0 to 20%; the original northern mesic forest was logged off and regrowth has covered about half of the area, the rest of which is in pasture or fields of hay, small grains, and silage corn; mean ann. soil temp. 40°F, mean ann. ppt. 28 in. Estab., Pine Co., Minn., 1935; Hole et al., 1962; 65 bu. oats; 350-400 bd. ft. red pine; 400-450 bd. ft. white pine; catena no. 133.

Hixton sl-l (D1, D2, D4, D5, D7, D9, D10) (Fig. 2-53); Typic Hapludalf, fine-loamy over sandy or sandy-skeletal, mixed, mesic (Gray-Brown Podzolic); well drained; solum is 20 to 40 in. thick, usually also the depth to bedrock, developed in a loamy covering on sandstone. Horizons: O1, O2, A1, A2, B1, B2t, IIB3, IIIR.

Some properties of Hixton loam

Hori-zon	Depth in.	OM %	Sand %	Silt %	Clay %	CEC me/ 100g	BS %	pH	BD g/cc
A2	3-12	1.5	40	45	15	10	25	6.4	1.3
B2t	14-21	0.5	35	40	25	14	50	5.7	1.5
IIB3	25-35	0.6	87	8	5	2	60	5.8	1.7

Other: Fragments of sandstone may be present in the lower B horizon; the bedrock is not hard enough to provide a lithic contact, hence a paralithic contact is recognized. Setting: On undulating to rolling uplands with slope gradients of 2 to 20%; original vegetation ranged from oak savanna in many southwestern counties to southern mesic forest in Richland County, to northern mesic forest in portions of northwestern Wisconsin;

pasture occupies more area than woodland, and on the less sloping areas, fields of hay, small grains, and corn are usual; mean ann. soil temp. 47°F, mean ann. ppt. 30 in. Estab., Jackson Co., Wis., Hole, 1956a; Beatty, 1960; Robinson and Klingelhoets, 1961; Slota and Garvey, 1961; Watson, 1966; 52 bu. oats; 400-450 bd. ft. red pine; 450-500 bd. ft. white pine; catena no. 21.

Hochheim sl-sil (B2, B12, B24) (Fig. 8-8); Typic Argiudoll, fine-loamy, mixed, mesic (Brunizem); well drained; solum is 12 to 24 in. thick, developed in less than 20 in. of loess overlying highly calcareous sandy loam to loam glacial till. Horizons: O1, O2, A1, B2t, C.

Some properties of Hochheim silt loam

Horizon	Depth in.	OM %	Sand %	Silt %	Clay %	CEC me/100g	BS %	pH	BD g/cc
A1	0-7	10.0	23	55	22	24	85	7.0	1.2
B2t	7-16	1.0	25	40	35	17	92	8.0	1.6
C	16-24	0.1	47	35	18	8	100	8.5*	1.7

*calcareous

Other: At unplowed sites a 3-in. A2 horizon is commonly present; the gravelly loam to sandy loam glacial till has a $CaCo_3$ equivalent of 40 to 60%. Setting: On slopes of 2 to 30% gradient on drumlins and moraines; the original vegetation of southern mesic forest has nearly all been replaced by fields of hay, small grains, and corn, and by pasture; mean ann. soil temp. 47°F, mean ann. ppt. 32 in. Proposed, Dodge Co., Wis., 1955; Milfred and Hole, 1970; Schmude, 1971; 65 bu. oats; 400-450 bd. ft. white pine; catena no. 63.

Hortonville sl-sicl (E5, I4, I6, I13, I16) (Figs. 11-1, 15-4); Glossoboric Hapludalf, fine-loamy, mixed, mesic (Gray-Brown Podzolic); well to moderately well drained: solum is 24 to 40 in. thick, developed in less than 20 in. of silty deposits over reddish-brown calcareous till of clay loam, silty clay loam, or loam texture. Horizons: O1, O2, A1, A2, A&B, IIB2t, IIB3t, IIC.

Some properties of Hortonville silt loam

Horizon	Depth in.	OM %	Sand %	Silt %	Clay %	CEC me/100g	BS %	pH	BD g/cc
A2	4-8	1.5	10	65	25	20	65	5.8	1.2
IIB2t	12-30	0.7	13	55	32	19	60	5.4	1.6
IIC	35-50	0.1	10	60	30	16	98	8.5*	1.7

*slightly calcareous

Other: Depth to carbonates is the same as thickness of solum; the A2 tongues down into cracks in the B2. Setting: On moraines with slope gradients of 0 to 15%; original vegetation was southern mesic forest to oak savanna; about 10% of the area is in woodland, and the rest is in hay, small grains, corn, or pasture; mean ann. soil temp. 45°F, mean ann. ppt. 30 in. Proposed, Waupaca Co., Wis., 1966; 75 bu. oats; 400-450 bd. ft. white pine; catena no. 57.

Houghton muck (J14, J15); (Figs. 8-7, 9-4, 9-5); Typic Medisaprist, euic, mesic (Bog); very poorly drained; solum is 51 in. to 20 ft. thick, developed in herbaceous debris in which some wood fragments are included, overlying calcareous glacial drift. Horizons: O1, Oa1, Oa2, Oa3, Oa4, IIC.

Some properties of Houghton muck

Horizon	Depth in.	OM %	Sand %	Silt %	Clay %	CEC me/100g	BS %	pH	BD g/cc	Fibers unrub. %	Fibers rub. %
Oa1	0-9	55	0	25	20	190	19	7.0	0.1	5	tr
Oa2	9-13	60	0	24	16	120	25	7.0	0.2	20	10
IIC	66-75	1	73	12	15	9	90	8.5*	1.7	0	0

*calcareous

Other: Structure of upper solum is massive, breaking to coarse granular, and of lower solum is massive, breaking to thick platy. Setting: In bogs in depressions in glacial landscapes; slope gradients are 0 to 5%; original vegetation of marsh grasses, sedges, reeds, buttonbrush, and cattails has been replaced over about a third of the area by pasture or truck crops —onions, lettuce, potatoes, celery, carrots, mint, lawn grass, sweet corn; mean ann. temp. 46°F, mean ann. ppt. 30 in. Estab., Roscommon Co., Mich., 1924; not suitable for oats or pine; catena no. 51.

Hubbard ls-fsl (C8, C16); Udorthentic Haplorboroll, sandy, mixed, frigid (Brunizem); well drained; solum is 20 to 40 in. thick over acid medium sand glacial outwash. Horizons: O1, A1, A3, B2, B3, C.

Some properties of Hubbard loamy sand

Horizon	Depth in.	OM %	Sand %	Silt %	Clay %	CEC me/100g	BS %	pH	BD g/cc
A1	0-8	3.0	84	8	8	12	55	5.8	1.3
B2	14-30	0.8	78	12	10	10	45	5.4	1.6
IIC	40-72	0.1	92	5	3	3	65	6.0	1.7

Other: The A horizon ranges from 10 to 24 in. thick; the profile is pebble-free or nearly so. Setting: On glacial outwash plains and terraces, with slopes of 0 to 5% gradient; the original prairie vegetation has been almost entirely replaced by crops (corn, soybeans, small grains, hay) and pasture; mean ann. soil temp. 44°F, mean ann. ppt. 30 in. Estab., Wadena Co., Minn., 1926; Thomas, 1964; 55 bu. oats; not suitable for pine unless this soil is inoculated with forest soil, after which yields of both red and white pine are about 400-450 bd. ft.; catena no. 32.

Humbird ls-sl (D11, D12); Alfic Haplorthod, coarse-loamy over clayey, mixed, frigid (Gray-Brown Podzolic); well to moderately well drained; solum is 24 to 50 in. thick, developed in weathered Cambrian sandstone and shale over indurated sandstone. Horizons: O1, O2, A1, A2, Bir, IIA′2, IIB′2, IIIB′3, IVC, R.

Some properties of Humbird sandy loam

Horizon	Depth in.	OM %	Sand %	Silt %	Clay %	CEC me/100g	BS %	pH	BD g/cc
A2	1-8	2.0	70	20	10	7	30	5.8	1.3
Bir	8-16	3.0	61	25	14	10	45	6.1	1.3
IIA′2	16-24	1.0	90	2	8	5	50	6.4	1.5
IIB′2t	24-30	0.7	82	8	10	6	30	5.4	1.6
IIIB′3	30-36	0.2	5	50	45	25	20	4.3	1.6
IVC	36-60	0.1	91	4	5	2	65	6.4	1.8

Other: The thickness and arrangement of sandy, loamy, and clayey horizons are variable. Setting: On uplands of slope gradients of 2 to 12%; much of the original white pine forest has been cleared and replaced by pasture and cropland; mean ann. soil temp. 44°F, mean ann. ppt. 32 in. Proposed, Jackson Co., Wis., 1954; 35 bu. oats; 400-450 bd. ft. white pine; catena no. 15.

Huntsville sil (J1); Cumulic Hapludoll, fine-silty, mixed, mesic (Alluvial soil); well to moderately well drained; solum is 10 to 40 in. thick, developed from silty and fine sandy alluvium derived largely from Mollisols of uplands. Horizons: O1, A1, C.

Some properties of Huntsville silt loam

Horizon	Depth in.	OM %	Sand %	Silt %	Clay %	CEC me/100g	BS %	pH	BD g/cc
A1	0-54	14	6	70	24	24	90	6.5	1.2
C	54-60	1	7	65	28	19	88	6.5	1.4

Other: Sandy layers may be present below a depth of 3 ft. Setting: On nearly level floodplains in valleys of the "Driftless Area" and in narrower bodies in southeastern Wisconsin; the original prairie vegetation with scattered hardwoods has been nearly entirely replaced with pasture and crops (corn, soybeans); mean ann. soil temp. 47°F, mean ann. ppt. 32 in. Estab., Schuyler Co., Ill., 1930; Beatty, 1960; 70 bu. oats; not suitable for pine; catena no. 38.

Institute sl (E6, E7). Similar to Onaway soils, but shallower.

Iron River sl-sil (F5, F16, G1, G2, G3, G5, G11, G13, G14, G20, G21, G23) (Figs. 2-53, 2-54, 12-7, 13-3); Alfic Fragiorthod, coarse-loamy, mixed, frigid (Podzol); well drained; solum is 28 to 42 in. thick, developed from less than 24 in. of silty sediment over acid sandy loam till. Horizons: O1, O2, A1, A2, Bhir, Bir, IIA′2x, IIB′tx, IIC.

Some properties of Iron River silt loam

Horizon	Depth in.	OM %	Sand %	Silt %	Clay %	CEC me/100g	BS %	pH	BD g/cc
A2	0-3	2.0	23	70	7	10	50	5.0	1.3
Bhir	3-7	3.0	30	60	10	18	20	4.5	1.3
IIA′2x	16-24	1.0	62	28	10	10	10	5.4	1.5
IIB′tx	24-36	0.7	55	30	15	14	15	5.0	1.6
IIC	36-50	0.1	72	20	8	5	25	5.0	1.8

Other: The Orterde (Bhir horizon) and fragipan (IIA′2x and IIB′tx) together may range in thickness from 6 to 20 in.; color of the IIC ranges from 10YR to 5YR hues; these soils are usually stony. Setting: On glacial moraines with slope gradients of 0 to 25%; the original northern mesic forest was logged off, and has been replaced in most areas by hardwood-coniferous forest with sugar maple dominant; some pasture and cropland are maintained; mean ann. soil temp. 41°F, mean ann. ppt. 32 in. Estab., Iron Co., Mich., 1930; Soil Conservation Service, 1952; Hole et al., 1962; 70 bu. oats; 400-450 bd. ft. red pine; 450-500 bd. ft. white pine; catena no. 125.

Jackson sil (A11, A12) (Fig. 7-12); Typic Hapludalf, fine-silty, mixed, mesic (Gray-Brown Podzolic); moderately well drained; solum is 40 to 48 in. thick, formed in 36 to 50 in. of silty deposits over acid sand and gravel outwash. Horizons: O1, O2, A1, A2, B1, B2t, IIB3, C, IIC.

Some properties of Jackson silt loam

Horizon	Depth in.	OM %	Sand %	Silt %	Clay %	CEC me/100g	BS %	pH	BD g/cc
A2	3-10	1.2	5	78	17	18	75	5.6	1.3
B2t	16-30	0.6	1	65	34	22	62	5.0	1.5
IIC	42-50	0.1	82	10	8	10	68	5.3	1.8

Other: The B2 and C horizons are mottled. Setting: In slight swales in high terraces in the lower Wisconsin River valley and tributaries, and on low terraces in the tributary valleys; slope gradients are 0 to 3%; original vegetation was southern mesic forest or oak savanna, now replaced by row crops, hay, and pasture; mean ann. soil temp. 47°F, mean ann. ppt. 31 in. Estab., Shelby Co., Mo., 1903; Robinson and Klingelhoets, 1961; 70 bu. oats; 450-500 bd. ft. white pine; catena no. 25.

Jewett sil (F8) (Fig. 2-53); Eutric Glossorboralf, fine-loamy, mixed, frigid (Gray Wooded); well drained; solum is 20 to 40 in. thick, developed in 36 to 50 in. of silty covering over acid reddish-brown glacial till. Horizons: O1, O2, A1, A2, A&B, B21t, IIB22t, IIC.

Some properties of Jewett silt loam

Horizon	Depth in.	OM %	Sand %	Silt %	Clay %	CEC me/100g	BS %	pH	BD g/cc
A1	0-3	10.0	12	70	18	22	40	5.8	1.2
B21t	12-17	1.0	13	60	27	18	46	5.4	1.4
IIB22t	17-32	0.5	46	22	32	18	60	5.4	1.5
IIC	32-40	0.1	41	32	27	14	90	5.5	1.7

Other: The till ranges in texture from sandy clay loam to heavy loam and in color from 10YR to 7.5YR hues. Setting: On glacial moraines with slope gradients of 0 to 20%; the original vegetation of prairie cover and oak savanna has been replaced by pasture and fields of hay, small grains, and corn; mean ann. soil temp. 42°F, mean ann. ppt. 30 in. Proposed, St. Croix Co., Wis., 1942; 75 bu. oats; not suitable for pine unless soil is inoculated with forest soil; catena no. 95.

Judson sil (J1). *See* Worthen.

Juneau sil (J1); Typic Udifluvent, coarse-silty, mixed, nonacid, mesic (Alluvial soil); well drained; solum is 18 to 30 in. thick over a buried solum about 3 ft. thick, the first solum developed in recent coarse silty alluvial-colluvial sediments and the buried solum in presettlement heavier silty sediments over glacial till. Horizons: O1, O2, A1, C, A1b, B2b, IICb.

Some properties of Juneau silt loam

Horizon	Depth in.	OM %	Sand %	Silt %	Clay %	CEC me/100g	BS %	pH	BD g/cc
A1	0-3	8.0	12	70	18	20	90	6.6	1.2
A1b	24-30	12.0	15	65	20	25	93	6.8	1.4
B2b	40-54	1.0	12	60	28	16	95	7.0	1.6
IICb	60-70	0.1	40	40	20	12	98	8.5*	1.7

*calcareous

Other: The IIC horizon ranges in texture from sandy loam to silty clay loam. Setting: On footslopes of gradients of 0 to 6% on drumlins and moraines; original vegetation was oak savanna or southern mesic forest, now replaced by pasture or fields of hay, small grains or corn; mean ann. soil temp. 46°F, mean ann. ppt. 32 in. Proposed, Dodge Co., Wis., 1951; Steingraeber and Reynolds, 1971; 75 bu. oats; not suitable for pine; catena no. 93.

Kellner s-ls (C10, C15); Typic Udipsamment, mixed, mesic (Regosol); well drained; solum is 18 to 30 in. thick, developed in acid glacial outwash sand. Horizons: O1, O2, A1, Bhir, C.

Some properties of Kellner loamy sand

Hori-zon	Depth in.	OM %	Sand %	Silt %	Clay %	CEC me/ 100g	BS %	pH	BD g/cc
A1	0-8	3.0	91	5	4	6	28	5.1	1.2
Bhir	8-24	0.7	93	4	3	3	42	5.6	1.5
C	24-35	0.1	98	1	1	1	5	6.6	1.6

Other: Weak mollic and spodic horizons commonly present; severe wind erosion has in places truncated the soil. Setting: On outwash plains with slope gradients of 0 to 5%; the original vegetation of oak savanna has been replaced in some areas by fields of hay, oats, and corn, productive only under irrigation; mean ann. soil temp. 44°F, mean ann. ppt. 30 in. Proposed, Wood and Portage Cos., Wis., 1958; Soil Conservation Service, 1967a; 35 bu. oats; 350-400 bd. ft. red pine; 450-500 bd. ft. white pine; catena no. 117.

Keltner sil (A4) (Fig. 7-8); Typic Argiudoll, fine-silty, mixed, mesic (Brunizem); moderately well drained; solum is 40 to 50 in. thick, developed in loess 30 to 50 in. thick overlying calcareous shale. Horizons: O1, O2, A1, A3, B21t, B22t, B23t, IIB3t, IIC.

Some properties of Keltner silt loam

Hori-zon	Depth in.	OM %	Sand %	Silt %	Clay %	CEC me/ 100g	BS %	pH	BD g/cc
A1	0-8	6.0	12	63	25	27	75	5.7	1.3
B22t	20-27	0.5	4	59	37	21	80	5.8	1.4
IIB3t	38-41	0.3	9	46	55	27	92	6.5	1.7
IIC	41-60	0.1	0	40	60	24	100	8.5*	1.8

*calcareous

Other: Hard limestone is interbedded with the shale in horizon IIC; montmorillonite clay is abundant in clay of the loess-derived horizons and illite is abundant in the Maquoketa shale-derived horizons. Setting: On bedrock-controlled ridges and hillside benches, with gradients of 2 to 50%; the original prairie vegetation is replaced by pasture, hay, and row crops; mean ann. soil temp. 47°F, mean ann. ppt. 31 in. Estab. Lafayette Co., Wis., 1964; Watson, 1963; 70 bu. oats; not suitable for pine; catena no. 166.

Kendall sil (B18, B25, B32, B34) (Fig. 8-8); Aeric Ochraqualf, fine-silty, mixed, mesic (Gray-Brown Podzolic); somewhat poorly drained; solum is 48 to 60 in. thick, developed in 36 to 50 in. of loess over dolomitic sandy loam or loam glacial till; a stratified substratum phase is recognized where glacial outwash replaced the till. Horizons: O1, O2, A1, A2, B1, B2t, B31, IIB32, IIC.

Some properties of Kendall silt loam

Hori-zon	Depth in.	OM %	Sand %	Silt %	Clay %	CEC me/ 100g	BS %	pH	BD g/cc
A2	4-11	3.0	8	77	15	20	85	6.0	1.3
B2t	14-39	0.3	4	62	34	24	90	5.8	1.4
IIB32	39-54	0.2	38	40	22	12	95	7.1	1.7
IIC	54-60	0.1	66	28	6	3	100	8.5*	1.9

*calcareous

Other: The B and C horizons are mottled; iron-manganese concretions are present in the B. Setting: On footslopes and in slight depressions on glacial moraines; slope gradients are 0 to 5%; the original oak savanna and southern mesic forest have been replaced by hay, small grains, corn, soybeans, and pasture; mean ann. soil temp. 45°F, mean ann. ppt. 30 in. Estab., Kendall Co., Ill., 1941; Milfred and Hole, 1970; Steingraeber and Reynolds, 1971; 75 bu. oats; 500-600 bd. ft. white pine; catena no. 49.

Kennan sl-sil (F4, F6, G5, G14, G15, G24) (Figs. 2-53, 13-3); Typic Hapludalf, fine-loamy, mixed, mesic (Gray-Brown Podzolic); well drained; solum is 20 to 40 in. thick, developed in less than 20 in. of silty sediment over acid sandy loam glacial till of brown (10YR-7.5YR hue) color. Horizons: O1, O2, A1, A2, B1, IIB2t, IIC.

Some properties of Kennan silt loam

Hori-zon	Depth in.	OM %	Sand %	Silt %	Clay %	CEC me/ 100g	BS %	pH	BD g/cc
A2	3-13	3	25	60	15	16	40	5.2	1.3
IIB2t	18-35	0.8	33	40	27	15	50	5.4	1.5
IIC	35-50	0.1	70	20	10	8	70	5.8	1.7

Other: An incipient fragipan may be present between the B and C horizons; degree of stoniness varies. Setting: On moraines of 3 to 20% slope gradient; the original northern mesic forest was logged off and nearly half of the area is in regrowth of the forest; considerable areas are in pasture or woodland pasture; cropland includes fields of hay, small grains, and silage corn; mean ann. soil temp. 43°F, mean ann. ppt. 30 in. Estab., Langlade Co., Wis., 1947; Hole, Dahlstrand, and Muckenhirn, 1947; 70 bu. oats; 475-550 bd. ft. red pine; 500-575 bd. ft. white pine; catena no. 100.

Keowns, sl-sil (E12, J7) (Fig. 11-7); Mollic Haplaquept, coarse-loamy, mixed, nonacid, mesic (Low Humic Gley); poorly drained; solum is 15 to 30 in. thick, developed in calcareous stratified fine sand and silt. Horizons: O1, O2, A1, A3, B2g, Cg.

Some properties of Keowns silt loam

Hori-zon	Depth in.	OM %	Sand %	Silt %	Clay %	CEC me/ 100g	BS %	pH	BD g/cc
A1	0-7	10	27	55	18	24	90	7.5	1.1
B2g	12-24	2	28	52	20	17	98	8.0	1.4
Cg	24-48	0.1	35	50	15	11	100	8.5*	1.7

*calcareous

Other: Free carbonates are commonly encountered at a depth of 10 to 20 in.; the C horizon contains strata of silt and fine sand with some clay. Setting: On nearly level glacial lake plains with slope gradients of 0 to 2%; the original vegetation of swamp hardwoods and wet prairie has been replaced in about two thirds of the area by pasture and by drained fields of hay, oats, and corn; mean ann. soil temp. 47°F, mean ann. ppt. 30 in. Proposed, Washington Co., Wis., 1952; Hole, 1956b; Milfred and Hole, 1970; Schmude, 1971; 60 bu. oats; not suitable for pine; catena no. 88.

Kert sil (D13) (Fig. 10-8); Aquic Glossoboralf, fine-loamy, mixed, frigid (Gray-Brown Podzolic); somewhat poorly drained; solum is 24 to 50 in. thick, developed in 20 to 36 in. of silty covering over acid shaly sandstone. Horizons: O1, O2, A1, A2, IIA&B, IIBt&A, IVC.

Some properties of Kert silt loam

Hori-zon	Depth in.	OM %	Sand %	Silt %	Clay %	CEC me/ 100g	BS %	pH	BD g/cc
A2	5-14	0.6	20	65	15	12	25	5.8	1.3
IIA&Bt	14-22	0.3	78	14	8	6	40	5.4	1.4
IIIBt&A	22-33	0.2	25	30	45	22	35	5.2	1.5
IVC	33-40	0.1	13	45	42	20	50	4.8	1.6

Other: The soils is mottled; the A2 horizon tongues into the B horizon; platy structure dominates. Setting: On nearly level to undulating uplands of 1 to 6% gradient; the original vegetation of mixed swamp hardwood and conifer forest is allowed to reproduce itself over a considerable area; other land use includes wooded pasture, cleared pasture, and some fields of hay, small grains, and silage corn; mean ann. soil temp. 43°F, mean ann. ppt. 31 in. Proposed, Wood Co., Wis., 1943; 60 bu. oats; 450-500 bd. ft. white pine; catena no. 14.

Kewaunee sl-sicl (I3, I4, I5, I10, I11, I12, I13, I15, I20) (Figs. 2-53, 2-54, 11-1, 15-4); Typic Hapludalf, fine, mixed, mesic (Gray-Brown Podzolic); well to moderately well drained; solum is 20 to 42 in. thick, developed in less than 20 in. of silty material over calcareous reddish-brown (5YR 4/4) clay to silty clay glacial till. Horizons: O1, O2, A1, A2, B1, IIB2t, IIC.

Some properties of Kewaunee silt loam

Hori-zon	Depth in.	OM %	Sand %	Silt %	Clay %	CEC me/ 100g	BS %	pH	BD g/cc
A2	4-10	3.0	25	60	15	10	75	5.8	1.3
IIB2t	13-29	1.0	18	30	52	24	83	6.0	1.6
IIC	29-45	0.1	20	35	45	15	100	8.0*	1.8

*calcareous

Other: Thin lenses of pebbles and sand may be present in the profile. Setting: On glacial moraines with slope gradients of 2 to 30%; the original vegetation of southern mesic forest has been replaced by woodland and pasture on steeper slopes, and by fields of hay, small grains, and corn on gentler slopes; mean ann. soil temp. 45°F, mean ann. ppt. 28 in. Estab., Fond du Lac Co., Wis., 1911; Soil Conservation Service, 1967a; Parker, Kurer, and Steingraeber, 1970; 75 bu. oats; 400-450 bd. ft. white pine; catena no. 56.

Kibbie sl-sil (E11); Aquollic Hapludalf, fine-loamy, mixed, mesic (Gray-Brown Podzolic); somewhat poorly drained; solum is 20 to 40 in. thick, developed in calcareous fine sand and silt. Horizons: O1, O2, A1, A2, B2t, C.

Some properties of Kibbie loam

Hori-zon	Depth in.	OM %	Sand %	Silt %	Clay %	CEC me/ 100g	BS %	pH	BD g/cc
A2	5-11	2.5	40	40	20	18	80	6.4	1.3
B2t	11-34	0.8	18	52	30	18	90	7.0	1.6
C	34-42	0.1	35	50	15	11	100	8.5*	1.7

*calcareous

Other: Strata of fine sand, silt, and occasionally clay, in the C horizon are variable in thickness and horizontal continuity. Setting: On glacial lake plains on slopes of 0 to 6% gradient; the original vegetation of swamp hardwoods and southern mesic forest has largely been replaced by pasture and fields of hay, small grains, soybeans, and corn; mean ann. soil temp. 47°F, mean ann. ppt. 32 in. Estab., Newton Co., Ind., 1943; Milfred and Hole, 1970; 65 bu. oats; 450-500 bd. ft. white pine; catena no. 87.

Kickapoo fsl (J1); Typic Udifluvent, coarse-loamy, mixed, nonacid, mesic (Alluvial soil); well to moderately well drained; solum is 4 to 6 in. thick, developed in sandy and silty alluvium that is neutral in reaction, over a buried similar soil at 3 ft. Horizons: A1, C, A1b, Cb.

Some properties of Kickapoo sandy loam

Hori-zon	Depth in.	OM %	Sand %	Silt %	Clay %	CEC me/ 100g	BS %	pH	BD g/cc
A1	0-5	4.0	53	30	17	18	90	6.8	1.2
C	5-36	0.5	53	30	17	9	95	7.0	1.5

Other: The buried A1 horizon is very dark brown (10YR 2/2) and about 5 in. thick. Setting: On first and second bottomlands of valleys of the "Driftless Area" on slopes of 0 to 3% gradient; the original vegetation of southern mesic forest and swamp hardwoods has largely been replaced by pasture and, where flooding hazard is slight, by fields of hay, small grains, and corn; mean ann. soil temp. 47°F, mean ann. ppt. 32 in. Estab., Vernon Co., Wis., 1965; Slota, 1969; 70 bu. oats; not suitable for pine; associated with catena no. 40.

Kinross s-ls (J12); Typic Haplaquod, sandy, mixed, frigid (Groundwater Podzol intergrading to Humic Gley); poorly drained; solum is 28 to 32 in. thick, developed in acid sands and fine gravel of glacial outwash plains and lake beds. Horizons: O1, O2, A1, A2g, Bhir, Bg, Cg.

Some properties of Kinross loamy sand

Hori-zon	Depth in.	OM %	Sand %	Silt %	Clay %	CEC me/ 100g	BS %	pH	BD g/cc
A1	0-1	8	90	5	5	14	60	5.0	1.2
A2g	2-5	1	93	4	3	5	22	5.3	1.4
Bg	5-30	0.5	86	8	6	5	25	5.2	1.7
Cg	30-40	0.1	95	3	2	2	10	5.3	1.8

Other: The A1 horizon may be absent and the peaty O1-O2 mat may rest directly on the A2g; the Bhir may be absent. Setting: On flats and depressions on glacial outwash and lake plains; the original vegetation of swamp conifers and sedge meadow is undisturbed in many areas; some land has been cleared for use as permanent pasture; mean ann. soil temp. 40°F, mean ann. ppt. 31 in. Proposed, Mackinac Co., Mich., 1950; not suitable for oats or pine; catena no. 145.

Kiva sl-sil (E1, E2, E3, E4); Entic Haplorthod, sandy, mixed, frigid (Regosol); excessively drained; solum is 10 to 20 in. thick, developed in a loamy covering of that thickness over calcareous gravel and sand. Horizons: O1, O2, A1, A2, Bhir, IIC.

Some properties of Kiva sandy loam

Hori-zon	Depth in.	OM %	Sand %	Silt %	Clay %	CEC me/ 100g	BS %	pH	BD g/cc
A2	1-3	2.0	75	15	10	9	60	6.4	1.3
Bhir	3-14	1.0	80	13	7	8	70	7.5	1.4
IIC	14-20	0.1	93	5	2	1	100	8.5*	2.0

*calcareous

Other: Coarse fragments, mostly of limestone, occupy 5 to 35% of the solum and about 85% in the IIC horizon. Setting: On glacial outwash plains, lake beaches, eskers, kames, moraines with slopes of 0 to 45% gradient; the original vegetation of northern mesic forest was logged and regrowth allowed on more than half of the area; pasture and cropland (hay and small grains) occupy the remainder; mean ann. soil temp. 44°F, mean ann. ppt. 28 in. Proposed, Mackinac Co., Mich., 1950; not suitable for oats (25 bu. acre) or pine; catena no. 144.

Knowles sil (B1, B16, B27, B29) (Fig. 8-7); Typic Hapludalf, fine-silty, mixed, mesic (Gray-Brown Podzolic); well drained; solum is 24 to 36 in., developed in a loess covering nearly that thick over a few inches of glacial till over limestone. Horizons: O1, O2, A1, A2, B21t, IIB22t, IIIR.

Some properties of Knowles silt loam

Horizon	Depth in.	OM %	Sand %	Silt %	Clay %	CEC me/100g	BS %	pH	BD g/cc
A2	3-12	1.5	12	70	18	15	65	6.1	1.3
B21t	12-20	0.5	8	60	32	16	55	6.4	1.5
IIB22t	20-25	0.5	16	50	34	16	90	6.5	1.6

Other: The IIB2 horizon is a gritty silty clay loam. Setting: On rock-controlled moraine with slope gradients of 0 to 12%; the original vegetation of oak savanna and southern mesic forest is reproducing as maple-basswood forest in woodlands on steep slopes; other areas are in pasture, and, on gentler slopes, in fields of hay, small grains, and corn; mean ann. soil temp. 45°F, mean ann. ppt. 29 in. Proposed, Dodge Co., Wis., 1959; Milfred and Hole, 1970; Steingraeber and Reynolds, 1971; 60 bu. oats; 325-400 bd. ft. red pine; 400-450 bd. ft. white pine; catena no. 67.

Kolberg l-sil (I3, I15); Glossic and Typic Eutroboralf, fine, mixed, frigid (Gray-Brown Podzolic); well drained; solum 20 to 40 in. thick, developed in reddish-brown (5YR 4/4) loam glacial till of that depth, over limestone. Horizons: O1, O2, A1, A2, A&B, B&A, B2t, B3t, C, IIR.

Some properties of Kolberg loam

Horizon	Depth in.	OM %	Sand %	Silt %	Clay %	CEC me/100g	BS %	pH	BD g/cc
A2	4-11	1	40	40	20	15	85	6.8	1.3
B2t	16-23	0.6	35	35	30	19	90	7.4	1.6

Other: The A2 tongues down into the B2 horizon. Setting: On bedrock-controlled moraines with slope gradients of 1 to 12%; the original northern mesic forest has been allowed to reproduce itself on steep lands; much of the area has been cleared and is used for pasture and crops (hay, small grains, corn) and orchards; mean ann. soil temp. 44°F, mean ann. ppt. 28 in. Estab., Brown Co., Wis., 1970; 50 bu. oats; 325-400 bd. ft. red pine; 400-450 bd. ft. white pine; catena no. 138.

Lafont l-sil (F1, F19); Alfic Haplorthod, coarse-loamy, mixed, frigid (Podzol); well drained; solum is 18 to 30 in. thick, developed in less than 24 in. of silty covering over acid reddish-brown (5YR 4/4) loam glacial till. Horizons: O1, O2, A1, A2, Bhir, Bir, IIC.

Some properties of Lafont silt loam

Horizon	Depth in.	OM %	Sand %	Silt %	Clay %	CEC me/100g	BS %	pH	BD g/cc
A2	1-5	2.0	10	80	10	16	36	5.0	1.2
Bhir	5-14	3.5	13	80	7	10	11	4.9	1.2
IIC	22-30	0.1	91	5	4	4	62	5.7	1.9

Other: A weak fragipan (A′2 tonguing into a weak IIB′2) usually lies between the Bhir and the till below. Setting: On undulating glacial moraines with slope gradients of 2 to 5%; the original vegetation of northern mesic forest was logged off, and forest regrowth has taken place on most of the area; some land is in pasture and in crops; mean ann. soil temp. 42°F, mean ann. ppt. 32 in. Proposed, Price Co., Wis., 1959; Hole and Schmude, 1959; Soil Conservation Service, 1967a; 70 bu. oats; 475-550 bd. ft. red pine; 500-575 bd. ft. white pine; catena no. 127.

Lamartine sil (J8) (Fig. 2-52); Aquollic Hapludalf, fine-silty, mixed, mesic (Gray-Brown Podzolic); somewhat poorly drained; solum is 24 to 40 in. thick, developed in 20 to 36 in. of loess over calcareous loam glacial till. Horizons: O1, O2, A1, A2, B1t, B21t, IIB22t, IIC.

Some properties of Lamartine silt loam

Horizon	Depth in.	OM %	Sand %	Silt %	Clay %	CEC me/100g	BS %	pH	BD g/cc
A1	0-3	6.5	8	70	22	34	67	6.5	1.2
A2	3-11	1.6	8	71	21	22	83	6.8	1.3
B21t	17-28	1.2	3	56	41	39	87	7.3	1.5
IIB22t	28-35	0.1	62	28	10	8	100	8.0	1.6
IIC	35-60	0.1	55	34	11	4	100	8.3*	1.8

*calcareous

Other: The soil is mottled. Setting: On nearly level to gently sloping areas on footslopes or on broad ridges in glacial moraine landscapes, on slopes of 1 to 12% gradient; the original vegetation of southern mesic forest and oak savanna is almost entirely replaced by fields of hay, small grains, and corn; mean ann. soil temp. 46°F, mean ann. ppt. 30 in. Proposed, Dodge Co., Wis., 1953; Milfred and Hole, 1970; Steingraeber and Reynolds, 1971; 65 bu. oats; not suitable for pine; catena no. 51.

Lamont sl-l (A5, A7); Typic Hapludalf, coarse-loamy, mixed, mesic (Gray-Brown Podzolic); well to excessively drained; solum is 24 to 43 in. thick, developed in aeolian fine sands more than 50 in. thick. Horizons: O1, O2, A1, A2, B1, B2t, B3t, C.

Some properties of Lamont sandy loam

Horizon	Depth in.	OM %	Sand %	Silt %	Clay %	CEC me/100g	BS %	pH	BD g/cc
A2	3-7	1.5	63	25	12	11	30	5.4	1.3
B2t	13-24	0.1	55	27	18	15	8	5.7	1.6
C	32-50	0.1	89	5	6	4	65	5.2	1.8

Other: Carbonates are present at a depth of about 12 ft. Setting: On ridge tops and side slopes near the Mississippi River valley on slopes of 3 to 20% gradient; the original vegetation of oak savanna has been replaced by woodlots, pasture, and some crops (hay, small grains, corn); mean ann. soil temp. 48°F, mean ann. ppt. 32 in. Estab., Jefferson Co., Ia., 1957; Hole, 1956a; Robinson and Klingelhoets, 1961; 45 bu. oats; 450-500 bd. ft. red pine; 475-550 bd. ft. white pine; catena no. 6.

Lapeer sl-sil (B4, B8, B14, B15, B26, B28, I17) (Figs. 4-8, 8-9, 11-1); Typic Hapludalf, coarse-loamy, mixed, mesic (Gray-Brown Podzolic); well drained; solum is 20 to 40 in. thick, developed in less than 20 in. of silty covering over sandy loam dolomitic glacial till. Horizons: O1, O2, A1, A2, B1, B2t, C.

Some properties of Lapeer sandy loam

Horizon	Depth in.	OM %	Sand %	Silt %	Clay %	CEC me/100g	BS %	pH	BD g/cc
A1	0-3	3.8	56	34	10	9	88	5.1	1.2
A2	3-12	1.5	58	33	9	5	61	4.6	1.3
B2t	15-34	0.5	58	18	24	11	73	4.1	1.5
C	34-40	0.2	84	6	10	2	100	7.6*	1.8

*calcareous

Other: Gravel may occupy as much as 10% of the volume of the solum, and 25% of the volume of the C horizon. Setting: On moraines with slope gradients of 2 to 12%; the original vegetation of oak savanna is largely replaced by fields of hay, small grains, and corn, and also by pasture and woodlots; mean ann. soil temp. 46°F, mean ann. ppt. 32 in. Estab., McHenry Co., Ill., 1960; Hole, 1956b; Milfred and Hole, 1970; 60 bu. oats; 450-500 bd. ft. red or white pine; catena no. 61.

Lawson sil (J1); Cumulic Hapludoll, fine-silty, mixed, mesic (Alluvial soil); somewhat poorly drained; solum is 30 to 50 in. deep, developed in dark silty alluvium derived from Mollisols of adjacent uplands. Horizons: O1, A1, Cg.

Some properties of Lawson silt loam

Horizon	Depth in.	OM %	Sand %	Silt %	Clay %	CEC me/100g	BS %	pH	BD g/cc
A1	0-40	12	2	71	27	50	90	6.8	1.1
Cg	40-60	1	12	68	20	20	98	7.2	1.5

Other: The C horizon contains lenses of silt and very fine sand; a buried A1b horizon may be present within 5 ft. of depth; the soil is mottled. Setting: On first and second bottoms of valleys in the "Driftless Area"; the original vegetation has been replaced by pasture and fields of hay, and, where drained, corn and small grains; mean ann. soil temp. 47°F, mean ann. ppt. 31 in. Estab., Lafayette Co., Wis., 1964; Hole, 1956a; Watson, 1966; 65 bu. oats; not suitable for pine; catena no. 38.

Leeman fs-lfs (E12, E13) (Fig. 11-5); Typic Udipsamment, sandy, mixed, frigid (Regosol); excessively drained; solum is 10 to 18 in. thick, developed in calcareous dune sand. Horizons: O1, O2, A1, C.

Some properties of Leeman sand

Horizon	Depth in.	OM %	Sand %	Silt %	Clay %	CEC me/100g	BS %	pH	BD g/cc
A1	0-12	3.0	91	5	4	5	40	5.8	1.3
C	12-40	0.1	94	3	3	1	60	5.8*	1.6

*calcareous at 6 feet

Other: Incipient spodic horizon is sometimes visible; discontinuous color bands may be present in the C horizon. Setting: On dunes, beach ridges, outwash plains on slopes of 2 to 35% gradient; oak savanna vegetation has changed to more solid oak forest since fire protection was instituted; some areas are used for pasture; mean ann. soil temp. 44°F, mean ann. ppt. 28 in. Proposed, Outagamie Co., Wis., about 1953; not suitable for oats; 300-360 bd. ft. red pine; 400-465 bd. ft. white pine; catena no. 82.

Lena muck (J15); Typic Medisaprist, euic, mesic (Bog); very poorly drained; solum is 50 in. to 15 ft. thick, developed from herbaceous debris. Horizons: O1, Oa1, Oa2, Oa3, Oa4, IIC.

Some properties of Lena muck

Horizon	Depth in.	OM %	Sand %	Silt %	Clay %	CEC me/100g	BS %	pH	BD g/cc	Fibers unrub. %	Fibers rub. %
Oa1	0-10	54	0	26	20	130	27	7.0	0.1	15	4
Oa4	82-104	60	0	24	16	200	43	6.0	0.2	12	3
IIC	104-110	1	65	20	15	8	100	8.5*	1.8	0	0

*calcareous

Other: A few woody fragments are present. Setting: In kettles and other depressions in glacial drift landscapes in southeastern Wisconsin; the original vegetation of marsh grasses, sedges, reeds, with some willow and alder, is undisturbed in most areas; some bog has been drained and used for pasture or for truck crops; mean ann. soil temp. 46°F, mean ann. ppt. 32 in. Estab., Stephenson Co., Ill., 1938; not suitable for oats or pine; catena no. 49.

Leonidas sl-l (I1, I7); Dystric Eutrochrept, coarse-loamy over clayey, mixed, frigid (Gray Wooded); well drained; solum is 20 to 30 in. thick, developed in 24 to 42 in. of loamy material overlying calcareous reddish-brown loam to silty clay loam glacial till. Horizons: O1, O2, A1, A2, B2, C1, IIC2.

Some properties of Leonidas sandy loam

Horizon	Depth in.	OM %	Sand %	Silt %	Clay %	CEC me/100g	BS %	pH	BD g/cc
A2	2-12	1.0	55	27	18	20	20	4.8	1.3
B2	12-24	0.5	45	35	20	18	35	5.5	1.4
C1	24-54	0.1	75	15	10	9	60	5.8	1.6
IIC2	54-60	0.1	20	30	50	28	100	8.5*	1.7

*calcareous

Setting: On gentle slopes on glacial moraine and old lake plains, with slope gradients of 0 to 8%; the original vegetation of northern mesic forest and boreal forest has been logged off and in much of the area has been replaced by new-growth forest; some land is cleared for use as pasture and for production of hay and small grains; mean ann. soil temp. 40°F, mean ann. ppt. 30 in. Proposed, St. Louis Co., Minn.; 60 bu. oats; 400-450 bd. ft. red pine; 450-500 bd. ft. white pine; catena no. 134.

LeRoy l-sil (B27) (Fig. 11-1); Typic Hapludalf, fine-loamy, mixed, mesic (Gray-Brown Podzolic); well to moderately well drained; solum is 12 to 24 in. thick, developed in less than 20 in. of silty material over highly calcareous (about 74% $CaCO_3$ equiv.), channery loam glacial till. Horizons: O1, O2, A1, A2, B21t, IIB22t, IIC.

Some properties of LeRoy silt loam

Horizon	Depth in.	OM %	Sand %	Silt %	Clay %	CEC me/100g	BS %	pH	BD g/cc
A2	2-9	1.3	15	70	15	25	60	5.8	1.3
B21t	9-14	0.6	18	50	32	20	65	5.4	1.4
IIB22t	14-22	0.3	31	35	34	30	70	5.5	1.6
IIC	22-60	0.1	84	10	6	3	100	8.5*	1.8

*calcareous

Other: Clay films on ped surfaces in the B2 horizons may be black from organic matter; till has a $CaCO_3$ equiv. of 60 to 90%. Setting: On moraines with slope gradients of 0 to 12%; the original vegetation of oak savanna is now replaced by pasture on more sloping land and by fields of hay, small grains, and corn on gentler slopes; mean ann. soil temp. 45°F, mean ann. ppt. 32 in. Proposed, Dodge Co., Wis., 1955; 65 bu. oats; 300-350 bd. ft. red pine; 350-400 bd. ft. white pine; catena no. 65.

Littleton sil (J1); Aquic Cumulic Hapludoll, fine-silty, mixed, mesic (Alluvial soil); somewhat poorly drained; solum is 28 to 40 in. thick, developed in silty deposits more than 40 in. thick in glacial drift landscapes. Horizons: O1, O2, A1, A3, B, C.

Some properties of Littleton silt loam

Horizon	Depth in.	OM %	Sand %	Silt %	Clay %	CEC me/100g	BS %	pH	BD g/cc
A1	0-18	12	10	70	20	38	80	6.2	1.2
B	26-35	2	5	67	28	24	75	6.4	1.4
C	35-50	1	10	70	20	20	92	6.5	1.6

Other: The soil is mottled. Setting: On footslopes, bottoms of narrow drainageways in southeastern Wisconsin on slopes of 0 to 2% gradient; the original prairie vegetation has been replaced by pasture and fields of hay, small grains, soybeans, and corn; mean ann. soil temp. 46°F, mean ann. ppt. 31 in. Estab., Schuyler Co., Ill., 1930; 65 bu. oats; not suitable for pine; catena no. 36.

Lomira sil (B27, B29) (Fig. 11-1); Typic Hapludalf, fine-silty, mixed, mesic (Gray-Brown Podzolic); well to moderately well drained; solum is 24 to 42 in. thick, developed in 20 to 36 in. of loess over highly calcareous (about 65% $CaCO_3$ equiv.), channery loam glacial till. Horizons: O1, O2, A1, A2, B1, B2t, IIB3t, IIC.

Some properties of Lomira silt loam

Horizon	Depth in.	OM %	Sand %	Silt %	Clay %	CEC me/100g	BS %	pH	BD g/cc
A2	2-10	1.5	13	70	17	28	65	5.8	1.2
B2t	14-31	0.5	18	50	32	22	67	5.4	1.5
IIB3t	31-34	0.8	47	15	38	18	70	6.7	1.6
IIC	34-60	0.1	45	35	20	12	100	8.5*	1.8

*calcareous

Other: Clay films are abundant on ped faces in the B horizons; the $CaCO_3$ equiv. of the till ranges from 60 to 90%. Setting: On moraines with slope gradients of 2 to 18%; the original oak savanna cover is replaced by fields of hay, small grains, corn, and peas; mean ann. soil temp. 46°F, mean ann. ppt. 31 in. Estab., Dodge Co., Wis., 1969; 65 bu. corn; 300-350 bd. ft. red pine; 350-400 bd. ft. white pine; catena no. 64.

Longrie l-sil (E6, E7) (Fig. 11-4); Entic to Alfic Haplorthod, coarse-loamy, mixed, frigid (Gray Wooded); well to moderately well drained; solum is 20 to 40 in. thick, developed in glacial till (with local covering of loess) over dolomite. Horizons: O1, O2, A1, A2, Bhir, A′2x&B′x, B2t, C, R.

Some properties of Longrie sandy loam

Horizon	Depth in.	OM %	Sand %	Silt %	Clay %	CEC me/100g	BS %	pH	BD g/cc
A2	1-8	2	60	25	15	12	80	6.4	1.2
Bhir	8-16	3	63	20	17	11	68	5.6	1.5
B′2t	21-28	0.5	42	28	30	15	82	5.8	1.6
C	28-35	0.1	40	40	20	12	100	8.5*	1.7

*calcareous

Other: A weak fragipan may be present between the two B horizons; mottling may be present in the lower solum. Setting: On nearly level to gently sloping rock-controlled till plains; the original vegetation of northern mesic forest is largely cleared for purposes of pasturage and production of tree fruits, hay, oats, and silage corn; mean ann. soil temp. 44°F, mean ann. ppt. 28 in. Estab., Menominee Co., Mich., 1925; 65 bu. oats; 475-500 bd. ft. white pine; catena no. 139.

Loyal l-sil (F11) (Fig. 10-8); Typic Glossoboralf, fine-loamy, mixed, frigid (Gray-Brown Podzolic); well to moderately well drained; solum is 24 to 48 in. thick, developed in 15 to 30 in. of loess overlying acid, reddish-brown (5YR 4/4) sandy loam glacial till. Horizons: O1, O2, A1, A2, A&B, IIB2t, IIB3t, IIC.

Some properties of Loyal silt loam

Horizon	Depth in.	OM %	Sand %	Silt %	Clay %	CEC me/100g	BS %	pH	BD g/cc
A2	2-19	0.8	14	70	16	10	20	5.4	1.3
IIB2t	27-34	0.1	49	28	23	16	36	4.8	1.7
IIC	40-60	0.1	56	27	17	14	60	4.9	1.8

Other: The A2 tongues into the B2 horizon; some stones have worked up from the till-derived horizons into the loess-derived horizons. Setting: On gently rolling moraines with slope gradients of 6 to 15%; the original vegetation of northern mesic forest has been logged off, and allowed to reproduce itself only in a restricted area; most of the land is kept cleared for pasture and fields of hay, small grains, and silage corn; mean ann. soil temp. 43°F, mean ann. ppt. 32 in. Proposed, Clark Co., Wis., 1954; 75 bu. oats; 475-550 bd. ft. red pine; 500-575 bd. ft. white pine; catena no. 97.

Lupton muck (J13); Typic Borosaprist, euic, frigid (Bog); very poorly drained; solum is 50 in. to 15 ft. thick, developed in mixed herbaceous and woody debris. Horizons: O1, Oa1, Oa2, Oe1, Oe2, IIC.

Some properties of Lupton muck

Horizon	Depth in.	OM %	Sand %	Silt %	Clay %	CEC me/100g	BS %	pH	BD g/cc	Fibers unrub. %	Fibers rub. %
Oa1	0-10	58	0	24	18	160	32	6.3	0.1	20	5
Oa2	10-20	65	0	20	15	200	43	6.4	0.2	22	8
IIC	65-70	1	50	35	15	8	60	6.5	1.7	0	0

Other: Hemic material accounts for more than 10 in. of the subsoil; wood fragments several inches in diameter may occur throughout the solum, but woody fibers comprise less than

25% of recognizable fiber. Setting: In kettles and other wet sites in glacial drift landscapes; the original vegetation of swamp conifers is largely undisturbed; mean ann. soil temp. 41°F, mean ann. ppt. 32 in. Site of proposal unknown; not suitable for oats or red or white pine; catena no. 136.

McHenry sil (B6, B7, B8, B11, B14, B15, B23, B28) (Fig. 8-9); Typic Hapludalf, fine-loamy, mixed, mesic (Gray-Brown Podzolic); well drained; solum is 20 to 40 in. thick, developed in 20 to 36 in. of loess over dolomitic (about 26% $CaCO_3$ equiv.) gravelly sandy loam glacial till. Horizons: O1, O2, A1, A2, B1, B2t, B3t, IIC.

Some properties of McHenry silt loam

Hori-zon	Depth in.	OM %	Sand %	Silt %	Clay %	CEC me/ 100g	BS %	pH	BD g/cc
A1	0-4	10.0	11	74	15	38	65	6.4	1.2
A2	4-13	1.5	12	72	16	18	62	6.2	1.3
B2t	17-33	0.5	36	35	29	24	75	5.3	1.5
IIC	37-60	0.1	66	22	12	5	100	8.5*	1.8

*calcareous

Other: Along joints in the glacial till deposits of secondary calcite form a white coating. Setting: On glacial moraines with slope gradients of 2 to 25%; the original vegetation of oak savanna is almost entirely replaced by pasture and fields of hay, small grains, and corn; mean ann. soil temp. 47°F, mean ann. ppt. 31 in. Estab., McHenry Co., Ill., 1953; Soil Conservation Service, 1967a; Milfred and Hole, 1970; Haszel, 1971; 65 bu. oats; 300-375 bd. ft. red pine; 350-400 bd. ft. white pine; catena no. 59.

Manawa l-sicl (I4, I5, I10, I11, I12, I13, I14, I15, I16, I20, I21); Aquollic Hapludalf, and Aquic Argiudoll, fine, mixed, mesic (Gray-Brown Podzolic); somewhat poorly drained; solum is 20 to 42 in. thick, developed in less than 20 in. of silty material over calcareous reddish-brown (5YR 4/4) silty clay glacial till. Horizons: O1, O2, A1, A2, IIB2t, IIC.

Some properties of Manawa silt loam

Hori-zon	Depth in.	OM %	Sand %	Silt %	Clay %	CEC me/ 100g	BS %	pH	BD g/cc
A2	2-12	4.0	17	64	19	17	85	6.5	1.3
IIB2t	12-30	2.0	14	40	46	27	75	7.1	1.7
IIC	30-40	0.2	23	40	37	12	100	8.0*	1.8

*calcareous

Other: The pedon is mottled; stones may occur throughout the profile. Setting: On glacial moraines with slope gradients of 1 to 12%; the original vegetation of southern and northern mesic forests has been replaced by fields of hay, small grains, and corn, and by pasture and restricted areas of woodlots; mean ann. soil temp. 45°F, mean ann. ppt. 28 in. Proposed, Waupaca Co., Wis., 1945; Soil Conservation Service, 1967a; Parker, Kurer, and Steingraeber, 1970; 75 bu. oats; not favorable for growth of pine; catena no. 56.

Manistee s-ls (I2, I8); Alfic Haplorthod, sandy over clayey, mixed, frigid (Podzol), well to moderately well drained; solum is 26 to 50 in. thick, developed in 20 to 40 in. of sandy material overlying calcareous reddish-brown (5YR 5/4) clayey substratum. Horizons: O1, O2, A1, A2, Bhir, Bir, A′2, A′2& IIB′2t, IIB′2t, IIC.

Some properties of Manistee sand

Hori-zon	Depth in.	OM %	Sand %	Silt %	Clay %	CEC me/ 100g	BS %	pH	BD g/cc
A2	1-11	0.7	92	4	4	7	60	5.4	1.3
Bhir	11-16	2.0	89	5	6	10	30	5.3	1.3
A′2	26-28	0.5	94	3	3	3	20	5.7	1.6
IIB′2t	30-38	0.8	15	30	55	26	65	5.8	1.7
IIC	38-45	0.1	16	32	52	25	100	8.5*	1.7

*calcareous

Other: Fragments of Ortstein 1 to 6 in. in dia. may occur in the Bir horizon. Setting: On glacial till and lake plains with slope gradients of 0 to 25%; the original boreal and northern mesic forest cover was logged off, and most of the area is in forest regrowth, some in pasture, and a little in cropland; mean ann. soil temp. 41°F, mean ann. ppt. 30 in. Estab., Manistee Co., Mich., 1922; 50 bu. oats; 350-400 bd. ft. red pine; 475-550 bd. ft. white pine; catena no. 152.

Mann sil (F21) (Fig. 12-8); similar to Adolph except that Mann is in the fine-loamy family.

Marathon sl-sil (F14, F15); Typic Glossoboralf, coarse-loamy, mixed, frigid (Gray-Brown Podzolic); well to moderately well drained; solum is 20 to 60 in. thick, developed in 15 to 30 in. of silty covering over weathered Precambrian granite or gneiss, over solid bedrock of the same kind at a depth of 5 ft. or so. Horizons: O1, O2, A1, A2, A&B, B&A, IIB2t, IIB3t, IIC, R.

Some properties of Marathon silt loam

Hori-zon	Depth in.	OM %	Sand %	Silt %	Clay %	CEC me/ 100g	BS %	pH	BD g/cc
A2	3-10	2.0	12	75	13	10	50	5.4	1.3
IIB2t	29-38	0.6	54	28	18	7	70	4.8	1.6
IIC	72-92	0.1	94	5	1	1	50	5.2	1.9

Other: The IIC horizon includes angular residual gravel and sand that occupy about 12% of the horizon by volume; the A2 horizon tongues into the B horizon. Setting: On rock-controlled uplands with slope gradients of 3 to 12% in the Wisconsin River valley near Wausau; the original northern mesic forest has been logged off, and regrowth forest stands in many areas; other areas are in pasture and fields of hay, small grains, and silage corn; mean ann. soil temp. 43°F, mean ann. ppt. 30 in. Proposed, Marathon Co., Wis., 1927; 70 bu. oats; 400-450 bd. ft. red pine or white pine; catena no. 101.

Marenisco ls-sl (G9, G10, G23); Typic Haplorthod, sandy, mixed, frigid (Podzol); excessively drained; solum is 26 to 42 in. thick, developed in acid, sandy, reddish-brown (5YR 4/3) glacial drift. Horizons: O1, O2, A1, A2, Bhir, Bir, C.

Some properties of Marenisco loamy sand

Hori-zon	Depth in.	OM %	Sand %	Silt %	Clay %	CEC me/ 100g	BS %	pH	BD g/cc
A2	1-5	2.0	85	6	9	7	50	5.4	1.3
Bhir	5-9	3.0	81	8	11	9	40	5.4	1.4
C	30-40	0.1	91	4	5	3	70	5.6	1.8

Other: A weak fragipan may be present below the Bhir horizon; stones are not usually present in this soil. Setting: On glacial moraines with slope gradients of 0 to 25%; the original northern mesic forest was logged off and has been replaced by new-growth forest, except where land has been kept cleared for pasture or crop production (potatoes, oats, hay); mean ann. soil temp. 41°F, mean ann. ppt. 28 in. Proposed, Keweenaw Co., Mich., 1952; 40 bu. oats; 400-450 bd. ft. red pine; 450-500 bd. ft. white pine; catena no. 131.

Marshan sil (J9); Typic Haplaquoll, fine-loamy over sandy or sandy-skeletal, mixed, mesic (Humic Gley); poorly drained; solum is 18 to 36 in. thick, developed in loamy sediments of that depth overlying calcareous glacial outwash sand and gravel. Horizons: O1, O2, A1, A3g, B2g, Cg.

Some properties of Marshan silt loam

Horizon	Depth in.	OM %	Sand %	Silt %	Clay %	CEC me/100g	BS %	pH	BD g/cc
A1	0-15	20.0	15	65	20	25	80	6.4	1.2
B2g	18-33	1.0	20	57	23	18	85	6.0	1.5
Cg	33-39	0.6	93	5	2	1	95	8.5*	1.9

*calcareous

Other: The surface texture may be silty clay loam or clay loam; a calcareous variant is recognized in places where the upper A1 horizon is calcareous. Setting: In shallow drainageways in glacial outwash terraces on slopes of 0 to 2% gradient; the original vegetation of swamp hardwoods and sedge meadows is now replaced by pasture and fields of hay, oats, soybeans, and corn; mean ann. soil temp. 47°F, mean ann. ppt. 32 in. Estab., Dakota Co., Minn., 1942; where artificially drained, 65 bu. oats; not suitable for pine; catena no. 24.

Marshfield sil (F11, F21); Typic Ochraqualf, fine-loamy, mixed, frigid (Low Humic Gley); poorly drained; solum is 24 to 50 in. thick, developed in 15 to 30 in. of loess overlying acid loam glacial till. Horizons: O1, O2, A1, A2g, B1g, B21tg, IIB22tg, IIB3tg, IICg.

Some properties of Marshfield silt loam

Horizon	Depth in.	OM %	Sand %	Silt %	Clay %	CEC me/100g	BS %	pH	BD g/cc
A2g	5-16	3.0	12	70	18	13	50	5.0	1.5
B21g	19-24	0.3	7	67	26	20	80	5.1	1.5
IIB22tg	24-34	0.2	34	40	26	21	95	6.6	1.7
IICg	44-66	0.1	36	38	26	21	95	6.9	1.8

Other: The profile is mottled. Setting: On silt-covered glacial moraines on slopes of 0 to 2% gradient; the original swamp hardwood and conifer and northern mesic forest cover has been logged off and regrowth forest has replaced it except for some areas of pasture and drained fields of hay and small grains; mean ann. soil temp. 43°F, mean ann. ppt. 30 in. Proposed, Wood Co., Wis., 1970; Soil Conservation Service, 1967a; 60 bu. oats; 400-450 bd. ft. white pine; catena no. 97.

Matherton sl-sil (J9); Udollic Ochraqualf, fine-loamy over sandy or sandy-skeletal, mixed, mesic (Gray-Brown Podzolic grading to Brunizem); somewhat poorly drained; solum is 24 to 40 in. thick, developed in less than 20 in. of silty material over calcareous glacial outwash sand and gravel. Horizons: O1, O2, A1, A2g, B2tg, IIC.

Some properties of Matherton sandy loam

Horizon	Depth in.	OM %	Sand %	Silt %	Clay %	CEC me/100g	BS %	pH	BD g/cc
A2g	6-11	2.0	57	25	18	16	70	6.4	1.3
B2tg	11-35	1.0	50	20	30	17	60	6.5	1.6
IIC	35-42	0.1	94	4	2	1	100	8.5*	2.0

*calcareous

Other: Tongues of B2 horizon extend as much as 2 ft. down into the IIC horizon. Setting: On glacial outwash plains and terraces with slope gradients of 0 to 6%; the original cover of swamp hardwoods and oak savanna has been largely replaced by fields of hay, small grains, soybeans, and corn, and some pasture; mean ann. soil temp. 47°F, mean ann. ppt. 32 in. Estab., McHenry Co., Ill., 1960; Hole, 1956b; Milfred and Hole, 1970; Haszel, 1971; Steingraeber and Reynolds, 1971; 65 bu. oats; 325-400 bd. ft. red pine; 400-450 bd. ft. white pine; catena no. 76.

Mead (Meadland) sl-sil (F15); Aquic Glossoboralf, coarse-loamy, mixed, frigid (Gray-Brown Podzolic); somewhat poorly drained; solum is 20 to 40 in. thick, developed in less than 15 in. of silty material over loamy weathered micaceous Precambrian bedrock which overlies solid bedrock at a depth of 9 to 15 ft. Horizons: O1, O2, A1, A2, A&B, IIB2t, IIC.

Some properties of Mead loam

Horizon	Depth in.	OM %	Sand %	Silt %	Clay %	CEC me/100g	BS %	pH	BD g/cc
A2	3-10	1.0	40	40	20	18	55	5.6	1.3
IIB2t	14-29	0.5	39	37	24	16	35	5.5	1.6
IIC	29-36	0.1	49	36	15	8	60	4.8	1.7

Other: The A2 horizon tongues down into the B2. Setting: On rock-controlled uplands with slope gradients of 1 to 3%; the original northern mesic forest has been logged off and most areas are now in pasture, or fields of hay, small grains, and silage corn; mean ann. soil temp. 42°F, mean ann. ppt. 30 in. Proposed, Portage Co., Wis., 1970; now called Meadland because the name Mead is being used in Nevada; 70 bu. oats; not suitable for pine; catena no. 103.

Mecan ls-l (C2) (Fig. 9-5); Typic Hapludalf, coarse-loamy, mixed, mesic (Gray-Brown Podzolic); well drained; solum is 40 to 60 in. thick, developed in calcareous sandy glacial till. Horizons: O1, O2, A1, A2, B1t, B2t, B3t, C.

Some properties of Mecan loamy sand

Horizon	Depth in.	OM %	Sand %	Silt %	Clay %	CEC me/100g	BS %	pH	BD g/cc
A2	4-12	0.8	80	15	5	2	60	5.7	1.4
B2t	18-32	0.4	69	14	17	7	52	5.3	1.6
C	47-50	0.1	72	20	8	2	100	8.5*	1.8

*calcareous

Other: The B2 horizon is reddish brown (5YR 4/4). Setting: On glacial moraines with slopes of 2 to 20% gradient; the original vegetation of oak savanna has been largely cleared for production of hay, small grains, and corn, or for pasture; mean ann. soil temp. 45°F, mean ann. ppt. 31 in. Proposed, Marquette Co., Wis., 1953; Soil Conservation Service, 1967a; 50 bu. oats; 450-500 bd. ft. red or white pine; catena no. 72.

Medary sil (A11, A12); Typic Hapludalf, fine, mixed, mesic (Gray-Brown Podzolic); well drained; solum is 28 to 42 in. thick, developed in 15 to 30 in. of silty deposits over dolomitic, reddish-brown lacustrine clays and silts. Horizons: O1, O2, A1, A2, B1, IIB2t, IIC.

Some properties of Medary silt loam

Horizon	Depth in.	OM %	Sand %	Silt %	Clay %	CEC me/100g	BS %	pH	BD g/cc
A2	3-8	1.5	13	65	22	13	75	6.4	1.3
IIB2t	8-30	0.8	8	50	42	20	60	5.3	1.6
IIC	30-36	0.1	1	52	37	14	100	8.5*	1.7

*calcareous

Other: Calcium carbonate concretions are commonly present in the upper IIC horizon. Setting: On natural terraces in the Wisconsin and Mississippi River valleys, on slopes of 0 to 25% gradient; the original vegetation of oak savanna has been replaced by fields of hay, small grains, and corn, and by pasture; mean ann. soil temp. 47°F, mean ann. ppt. 31 in. Proposed, Richland Co., Wis., 1948; Robinson and Klingelhoets, 1959, 1961; Beatty, 1960; 70 bu. oats; not suitable for pine; catena no. 35.

Mequon sil-sicl (B19); Udollic Ochraqualf, fine, mixed, mesic (Gray-Brown Podzolic); somewhat poorly drained; solum is 20 to 42 in. thick, developed in less than 20 in. of silty material over calcareous silty clay loam glacial till. Horizons: O1, O2, A1, A2, B1t, IIB2t, B3, IIC.

Some properties of Mequon silt loam

Horizon	Depth in.	OM %	Sand %	Silt %	Clay %	CEC me/100g	BS %	pH	BD g/cc
A2	7-11	2.5	15	65	20	10	85	6.2	1.3
IIB2t	16-27	1.2	8	55	37	19	90	7.2	1.6
IIC	30-60	0.1	9	57	34	13	100	8.1*	1.7

*calcareous

Other: Depth to carbonates is the same as thickness of solum. Setting: On glacial moraines with slope gradients of 1 to 8%; the original vegetation of southern mesic forest has largely been replaced by fields of hay, small grains, and corn, and by pasture; mean ann. soil temp. 45°F, mean ann. ppt. 32 in. Proposed, Ozaukee Co., Wis., 1965; Parker, Kurer, and Steingraeber, 1970; Haszel, 1971; 65 bu. oats; not suitable for pine; catena no. 55.

Meridian sl-l (A12, C7, C8, C9); Mollic Hapludalf, fine-loamy over sandy or sandy-skeletal, mixed, mesic (Gray-Brown Podzolic grading toward Brunizem); well drained; solum is 20 to 40 in. thick, developed from this thickness of loamy material over acid glacial outwash sand. Horizons: O1, O2, A1, A2, B1, B2t, B3, IIC.

Some properties of Meridian sandy loam

Horizon	Depth in.	OM %	Sand %	Silt %	Clay %	CEC me/100g	BS %	pH	BD g/cc
A2	3-10	0.5	61	31	8	7	62	6.5	1.3
B2t	16-30	0.2	56	32	12	8	50	5.3	1.6
IIC	35-50	0.1	93	4	3	2	36	5.1	1.8

Other: Solum is probably a mixture of loess and outwash sand; bands of darker brown finer-textured material may be present in the II horizon. Setting: On glacial outwash plains and terraces with slopes of 2 to 20% gradient; the original vegetation of oak savanna, southern or northern mesic forest was logged off and has been replaced by fields of hay, small grains, and corn; mean ann. soil temp. 43°F, mean ann. ppt. 30 in. Estab., Richland Co., Wis., 1956; Robinson and Klingelhoets, 1959, 1961; Beatty, 1960; Soil Conservation Service, 1967a; 60 bu. oats; 450-500 bd. ft. red or white pine; catena no. 28.

Merrillan ls-sl (D8, D11, D12) (Fig. 10-4); Aqualfic Haplorthod, coarse-loamy over clayey, mixed, frigid (Podzol); somewhat poorly drained; solum is 20 to 40 in. thick, developed in the same thickness of weathered material over Cambrian sandstone and shale. Horizons: O1, O2, A1, A2, Bir, A′2, IIB′2t, IIIR.

Some properties of Merrillan sandy loam

Horizon	Depth in.	OM %	Sand %	Silt %	Clay %	CEC me/100g	BS %	pH	BD g/cc
A2	4-5	1.5	65	25	10	9	65	5.4	1.3
Bir	5-12	2.0	62	27	11	12	50	5.2	1.4
A′2	12-18	0.5	74	18	8	7	30	5.1	1.7
IIB′2t	18-23	0.7	37	25	38	15	45	4.8	1.7

Other: The A′2 horizon is absent in some pedons. Setting: On uplands with slope gradients of 1 to 4%; most of the white pine forest was logged off; land is now in woodland and woodland pasture, and in cleared pasture and forage, crops, small grains, and corn for silage; mean ann. soil temp. 44°F, mean ann. ppt. 32 in. Proposed, Clark Co., Wis., 1954; 30 bu. oats; 250-325 bd. ft. white pine; catena no. 15.

Miami sl-sil (B8, B11, B13, B14, B23, B25, B26) (Figs. 8-4, 8-7, 8-13, 11-1);Typic Hapludalf, fine-loamy, mixed, mesic (Gray-Brown Podzolic); well drained; solum is 24 to 42 in. thick, developed in less than 20 in. of silty material over calcareous gravelly loam till (about 40% $CaCO_3$ equiv. in till). Horizons: O1, O2, A1, A2, IIB2t, IIB3, IIC.

Some properties of Miami silt loam

Horizon	Depth in.	OM %	Sand %	Silt %	Clay %	CEC me/100g	BS %	pH	BD g/cc
A1	0-3	6.0	12	77	11	22	69	6.2	1.2
A2	3-12	2.0	9	80	11	15	47	4.6	1.3
IIB2t	12-29	0.5	37	27	36	26	74	5.4	1.6
IIC	36-46	0.1	50	38	12	10	100	8.2*	1.8

*calcareous

Other: Content by volume of coarse fragments (mostly igneous) in the solum ranges from 5% in the A to 20% in the lower IIB horizon. Setting: On glacial moraines and drumlins with slope gradients of 0 to 25%; the original vegetation of oak savanna or southern mesic forest is now largely replaced by fields of hay, small grains, and corn, and by pasture; mean ann. soil temp. 47°F, mean ann. ppt. 32 in. Estab., Montgomery Co., O., 1910; the original definition was much more inclusive as indicated by Whitson et al., 1917; Hole, 1956b; Soil Conservation Service, 1967a; Steingraeber and Reynolds, 1971; Haszel, 1971; 70 bu. oats; 300-375 bd. ft. red pine; 350-400 bd. ft. white pine; catena no. 52.

Milaca (Amery) sl-sil (D9, F2, F10, G4, G13, G22) (Figs. 12-5, 13-5); Typic Fragiochrept, coarse-loamy, mixed, frigid (Gray-Brown Podzolic); well drained; solum is 34 to 48 in. thick, developed in less than 12 in. of loess over acid reddish-brown (5YR 4/3) sandy loam glacial till. Horizons: O1, O2, A1, A2, B1, IIB2x, B3x, IICx.

Some properties of Milaca silt loam

Horizon	Depth in.	OM %	Sand %	Silt %	Clay %	CEC me/100g	BS %	pH	BD g/cc
A1	0-1	9.0	34	54	12	22	32	5.5	1.3
A2	1-10	2.0	38	53	9	15	40	5.3	1.4
IIB2x	19-28	0.5	49	37	14	10	50	5.7	1.6
IICx	40-55	0.1	76	14	10	8	60	5.8	1.7

Other: Depth to the fragipan (horizons marked x) is about 20 in.; coarse fragments are present throughout the profile. Setting: On glacial moraines with slope gradients of 2 to 25%; the original northern mesic forest was logged off, and regrowth forest occupies about one fourth of the area, fields of hay, oats, and corn about half, and pasture and woodland pasture the remainder; mean ann. soil temp. 42°F, mean ann. ppt. 29 in. Estab., Mille Lacs Co., Minn., 1927; Robinson et al., 1958; Soil Conservation Service, 1967a; 65 bu. oats; 500-575 bd. ft. red or white pine; catena no. 163. Note: Recent correlations recognize Amery sl-sil (Glossic Eutroboralf) and Washburn sl-sil (Alfic Haplorthod) as being more extensive than Milaca in Wisconsin.

Millington l (J1); Cumulic Haplaquoll, fine-loamy, mixed, calcareous, mesic (Alluvial soil); poorly drained; solum is about 35 in. thick, developed in dark alluvium derived from upland Udolls. Horizons: A1, B, C.

Some properties of Millington loam

Horizon	Depth in.	OM %	Sand %	Silt %	Clay %	CEC me/100g	BS %	pH	BD g/cc
A1	0-15	20.0	48	30	22	20	95	8.5*	1.2
C	35-50	1.0	50	25	25	14	100	8.5*	1.6

*calcareous

Other: Snail shells present throughout profile. Setting: On nearly level floodplains, on slopes of 0 to 2% gradient; the original prairie and marsh vegetation is now largely replaced by pasture; mean ann. soil temp. 47°F, mean ann. ppt. 31 in. Estab., Kendall Co., Ill., 1941; not suitable for oats or pine; catena no. 36.

Mingo sil (B10); Aeric Ochraqualf, fine-loamy, mixed, mesic (Gray-Brown Podzolic); somewhat poorly drained; solum is 45 to 60 in. thick, developed in 20 to 30 in. of loess over calcareous sandy loam or loam till. Horizons: O1, O2, A1, A2, B21t, IIB22t, IIC.

Some properties of Mingo silt loam

Horizon	Depth in.	OM %	Sand %	Silt %	Clay %	CEC me/100g	BS %	pH	BD g/cc
A2	7-11	7.0	27	55	18	10	50	5.6	1.3
B21t	15-25	0.6	24	52	24	15	40	5.4	1.5
IIB22t	25-40	0.2	44	24	32	17	30	5.1	1.7
IIC	50-60	0.1	50	22	28	14	80	5.8	1.7

Other: This profile is mottled. Setting: On glacial moraines on slopes of 0 to 2% gradient; the original vegetation of oak savanna is largely replaced by fields of hay, oats, and corn, and by pasture; mean ann. soil temp. 47°F, mean ann. ppt. 32 in. Proposed, Jasper Co., Ia., 1942; in Wisconsin this soil is now combined with Kendall; 65 bu. oats; not suitable for pine; catena no. 46.

Monico sil (G2, G23, J6); Aquic Dystrochrept, coarse-loamy, mixed, frigid (Gray-Brown Podzolic); somewhat poorly drained; solum is 20 to 40 in. thick, developed in less than 24 in. of silty sediment over acid sandy loam glacial till. Horizons: O1, O2, A1, A2, Bir, C.

Some properties of Monico silt loam

Horizon	Depth in.	OM %	Sand %	Silt %	Clay %	CEC me/100g	BS %	pH	BD g/cc
A2	1-2	2.0	33	52	15	8	60	4.8	1.3
Bir	2-6	3.0	31	51	18	11	50	4.5	1.4
C	24-36	0.1	58	30	12	5	80	5.3	1.8

Other: The profile is mottled; the till is dark brown (7.5YR 4/4). Setting: In slight depressions on glacial moraines, on slopes of 0 to 2% gradient; the original northern mesic forest cover was logged off and most of the land is now in regrowth forest, although some is in pasture and cropland (hay and oats). Proposed, Oneida Co., Wis., 1958; Hole and Schmude, 1959; 70 bu. oats; not suitable for pine; catena no. 125.

Montgomery l-sicl (J9); Typic Haplaquoll, fine, mixed or montmorillonitic, noncalcareous, mesic (Humic Gley); poorly drained; solum is 20 to 40 in. thick, developed in less than 18 in. of loamy covering over calcareous brown silts and clays. Horizons: O1, O2, A1, B2g, Cg.

Some properties of Montgomery loam

Horizon	Depth in.	OM %	Sand %	Silt %	Clay %	CEC me/100g	BS %	pH	BD g/cc
A1	0-15	20.0	12	50	38	22	80	6.5	1.3
B2g	15-29	1.0	12	50	38	16	90	8.1*	1.6
Cg	38-48	0.1	13	55	32	14	100	8.5*	1.7

*calcareous

Other: Crayfish have caused formation of krotovinas. Setting: In depressions in glacial lake plains on slope gradients of 0 to 1%; the original vegetation of swamp hardwoods and sedge meadows has been replaced, after drainage, by fields of small grains, soybeans, and corn; mean ann. soil temp. 47°F, mean ann. ppt. 32 in. Estab., Monroe Co., Ind., 1922; Steingraeber and Reynolds, 1971; Schmude, 1971; where artificially drained, 65 bu. oats; not suitable for pine; catena no. 85.

Morley l-sil (B1, B9, B19) (Fig. 8-10); Typic Hapludalf, fine, illitic, mesic (Gray-Brown Podzolic); well to moderately well drained; solum is 20 to 42 in. thick, developed in less than 20 in. of silty material over calcareous silty clay loam to clay loam till. Horizons: O1, O2, A1, A2, IIB2t, IIB3, IIC.

Some properties of Morley silt loam

Horizon	Depth in.	OM %	Sand %	Silt %	Clay %	CEC me/100g	BS %	pH	BD g/cc
A1	0-4	5.0	26	54	20	12	70	6.1	1.2
A2	4-9	1.5	26	53	21	12	66	6.2	1.3
IIB2t	9-28	1.0	17	38	45	19	75	7.1	1.6
IIC	42-48	0.3	13	54	33	8	100	8.5*	1.7

*calcareous

Other: The profile contains some smooth dark concretions (Fe-Mn oxides); calcite concretions are also present in the C horizon. Setting: On glacial moraines with slope gradients of 2 to 7%; the original cover of southern mesic forest has been largely replaced by fields of hay, oats, corn, by pasture, and by urban land uses; mean ann. soil temp. 47°F, mean ann. ppt. 32 in. Estab., Will Co., Ill., 1952; Hole, 1956b; Soil Conservation Service, 1967a; Link and Demo, 1970; Steingraeber and Reynolds, 1971; 70 bu. oats; not suitable for pine; catena no. 54.

Morocco sl-lfs (C13, J4, J5, J14) (Fig. 9-6); Aquic Udipsamment, sandy, mixed, mesic (Low Humic Gley); somewhat poorly drained; solum is 8 to 15 in. deep, developed in acid glacial outwash sand. Horizons: O1, O2, A1, A2, B, Cg.

Some properties of Morocco sand

Horizon	Depth in.	OM %	Sand %	Silt %	Clay %	CEC me/100g	BS %	pH	BD g/cc
A1	0-4	4.0	89	7	4	5	88	4.8	1.4
Cg	30-48	0.1	93	4	3	2	92	5.4	1.6

Setting: On level glacial outwash and lake plains; the original oak savanna is partially replaced by fields of small grains, soybeans, and corn, and by pasture; mean ann. soil temp. 42°F, mean ann. ppt. 30 in. Estab., Fulton Co., Ind., 1943; Peck and Lee, 1961; 45 bu. oats; 250-325 bd. ft. white pine; catena no. 34.

Munising sl-l (I2, I18); Alfic Fragiorthod, coarse-loamy, mixed, frigid (Podzol); well drained; solum is 34 to 80 in. thick, developed in acid, reddish-brown (2.5YR 4/4) sandy loam glacial till. Horizons: O1, O2, A1, A2, Bhir, Bir, A′2x, B′2x, B′2t, C.

Some properties of Munising loamy sand

Horizon	Depth in.	OM %	Sand %	Silt %	Clay %	CEC me/100g	BS %	pH	BD g/cc
A2	1-9	1.0	86	7	7	6	60	4.7	1.3
Bhir	9-13	2.0	71	18	11	12	40	4.6	1.4
A′2x	29-40	0.2	80	15	5	3	30	5.2	1.8
B′2x	40-48	0.4	70	20	10	7	65	4.8	1.7
C	62-82	0.1	74	18	8	4	70	5.7	1.7

Other: The fragipan (horizons marked x) is very hard and brittle. Setting: On glacial moraines with slope gradients of 2 to 12%; the original northern mesic forest was logged off, and large areas are in forest regrowth; some land is in pasture and crops (hay, potatoes, strawberries); mean ann. soil temp. 40°F, mean ann. ppt. 28 in. Estab., Chippewa Co., Mich., 1927; Ableiter and Hole, 1961; 60 bu. oats; 400-450 bd. ft. red pine; 425-500 bd. ft. white pine; catena no. 129.

Muscatine sil (A1) (Fig. 7-8); Aquic Argiudoll, fine-silty, mixed, mesic (Brunizem); somewhat poorly drained; solum is 40 to 60 in. thick, developed in deep loess over residuum on dolomite. Horizons: O1, A1, B2t, B31t, B32, C.

Some properties of Muscatine silt loam

Horizon	Depth in.	OM %	Sand %	Silt %	Clay %	CEC me/100g	BS %	pH	BD g/cc
A1	0-16	7.0	3	65	32	28	70	5.4	1.2
B2t	20-32	0.6	2	63	35	16	62	5.7	1.5
C	48-64	0.1	5	65	30	13	70	7.7	1.7

Other: The profile is mottled. Setting: On broad uplands and some seepage spots on side slopes in the "Driftless Area"; the original prairie vegetation is now replaced by fields of hay, small grains, soybeans, and corn; mean ann. soil temp. 47°F, mean ann. ppt. 31 in. Estab., Muscatine Co., Ia., 1914; Hole, 1956a; Robinson and Klingelhoets, 1961; Slota, 1969; 70 bu. oats; not suitable for pine; catena no. 1.

Navan sl-sil (J10); Typic Argiaquoll, fine-loamy, mixed, noncalcareous, mesic (Humic Gley); very poorly drained; solum is 24 to 40 in. thick, developed in 18 to 36 in. of loamy material over calcareous silts and clays. Horizons: O1, A1, A3g, B2tg, IIB3g, IICg.

Some properties of Navan loam

Horizon	Depth in.	OM %	Sand %	Silt %	Clay %	CEC me/100g	BS %	pH	BD g/cc
A1	0-14	20.0	37	40	23	18	85	7.1	1.3
B2tg	18-33	1.0	33	35	32	16	90	7.6	1.5
IIB3g	33-38	0.5	8	48	44	25	95	8.5*	1.6
IICg	38-60	0.1	8	50	42	20	100	8.5*	1.7

*calcareous

Other: The clayey IIC horizon contains lenses of silt and fine sand. Setting: On level glacial lake beds with slopes of 0 to 2% gradient; the original cover of swamp hardwoods and sedge meadows has been replaced by pasture, and, after drainage, by fields of hay, small grains, and corn; mean ann. soil temp. 47°F, mean ann. ppt. 32 in. Proposed, Jefferson Co., Wis., 1957; Milfred and Hole, 1970; Link and Demo, 1970; Steingraeber and Reynolds, 1971; 65 bu. oats; not suitable for pine; catena no. 90.

Neda l-sicl (B1) (Fig. 8-7); Mollic Hapludalf, fine-loamy, mixed, mesic (Gray-Brown Podzolic intergrading toward Brunizem); well to moderately well drained; solum is 20 to 40 in. thick, developed in 10 to 20 in. of loess over highly calcareous channery, shaly loam glacial till. Horizons: O1, O2, A1, A2, B21t, IIB22t, IIB3t, IIC.

Some properties of Neda loam

Horizon	Depth in.	OM %	Sand %	Silt %	Clay %	CEC me/100g	BS %	pH	BD g/cc
A2	7-11	1.8	37	43	20	15	88	6.5	1.3
B21t	11-16	0.5	6	60	34	21	95	6.9	1.5
IIB22t	16-21	0.6	22	40	38	20	98	8.3	1.6
IIC	25-40	0.1	52	33	15	7	100	8.5	1.8

Other: Dark brown clay films are abundant on ped surfaces of the B21 horizon; shale fragments occupy 5 to 20% by volume of the IIB22 horizon and more of the IIC horizon; some mottling is present in both horizons. Setting: On sloping ground moraine near shale outcrops; slope gradients range from 8 to 20%; the original oak savanna and southern mesic forest have largely been replaced by fields of hay, oats, and corn, and by pasture; mean ann. soil temp. 45°F, mean ann. ppt. 30 in. Proposed, Dodge Co., Wis., 1955; 70 bu. oats; not suitable for pine; catena no. 164.

Nekoosa s-lfs (C4, C12, C13, C14, C17); Typic Udipsamment, sandy, mixed, mesic (Regosol); moderately well drained; solum is 18 to 36 in. thick, developed in acid glacial outwash sand. Horizons: O1, O2, A1, B1, B2, B3, C.

Some properties of Nekoosa sand

Horizon	Depth in.	OM %	Sand %	Silt %	Clay %	CEC me/ 100g	BS %	pH	BD g/cc
A1	0-5	4.0	90	6	4	10	40	5.8	1.3
B2	10-20	0.4	93	4	3	3	25	5.4	1.6
C	27-35	0.1	97	2	1	1	20	5.4	1.8

Other: The profile is mottled. Setting: On level outwash plains with slopes of 0 to 2% gradient; the original cover of swamp hardwood and conifer forest was cleared over considerable areas, but regrowth has taken place, except where pasture or cropland has been maintained; mean ann. soil temp. 43°F, mean ann. ppt. 30 in. Estab., Waushara Co., Wis., 1909; in frigid parts of Adams Co. the name Friendship is used instead; Peck and Lee, 1961; 40 bu. oats; 275-325 bd. ft. red pine; 400-450 bd. ft. white pine; catena no. 34.

Nenno l-sil (B12, B24); Aquic Argiudoll, fine-loamy, mixed, mesic (Brunizem); somewhat poorly drained; solum is 12 to 20 in. thick, developed in less than 20 in. of silty or loamy covering over highly dolomitic sandy loam to loam glacial till. Horizons: O1, O2, A1, A3, IIB2t, IIC.

Some properties of Nenno loam

Horizon	Depth in.	OM %	Sand %	Silt %	Clay %	CEC me/ 100g	BS %	pH	BD g/cc
A1	0-8	8.0	38	40	22	19	90	8.0	1.3
IIB2t	10-18	0.8	31	35	34	17	95	8.0	1.6
IIC	18-60	0.1	42	38	20	8	100	8.5*	1.8

*calcareous

Other: The till contains 40 to 60% $CaCO_3$ equiv. Setting: On glacial moraines with slope gradients of 0 to 5%; the original cover of oak savanna is replaced by pasture and fields of hay, small grains, and corn; mean ann. soil temp. 46°F, mean ann. ppt. 32 in. Proposed, Dodge Co., Wis., 1955; Schmude, 1971; 65 bu. oats; not suitable for pine; catena no. 63.

Newton ls-sl (C10, C12, C14, C17, J4, J5, J14) (Fig. 9-4); Typic Humaquept, sandy, mixed, mesic (Humic Gley); poorly drained; solum is 24 to 36 in. thick, developed in acid glacial outwash sand. Horizons: O1, A1, Bg, Cg.

Some properties of Newton sand

Horizon	Depth in.	OM %	Sand %	Silt %	Clay %	CEC me/ 100g	BS %	pH	BD g/cc
A1	0-12	7.0	90	6	4	13	50	5.3	1.2
Cg	30-40	0.1	97	2	1	1	20	5.4	1.7

Other: The A1 horizon is 8 to 15 in. thick; "bog iron ore" concretions may be present in the solum. Setting: On nearly level glacial outwash and lake plains, with slope gradients of 0 to 1%; the original cover of sedge meadows and swamp hardwoods has been replaced in some areas by permanent pasture and, with drainage, by fields of oats, soybeans, and corn; mean ann. soil temp. 44°F, mean ann. ppt. 30 in. Estab., Newton Co., Ind., 1905; 45 bu. oats; not suitable for pine; catena no. 34.

Norden fsl-sil (D1, D2, D3, D4, D7) (Figs. 10-4, 10-6); Typic Hapludalf, fine-loamy, mixed, mesic (Gray-Brown Podzolic); well drained; solum is 24 to 35 in. thick, developed in the same thickness of loamy material, a mixture of residuum and loess, over weakly cemented glauconitic sandstone. Horizons: O1, O2, A1, A2, B1, B2t, B3, R.

Some properties of Norden loam

Horizon	Depth in.	OM %	Sand %	Silt %	Clay %	CEC me/ 100g	BS %	pH	BD g/cc
A1	0-4	9.1	49	40	11	30	70	6.6	1.2
A2	4-11	1.5	51	42	7	10	50	6.2	1.4
B2t	15-23	0.4	57	23	20	11	52	5.4	1.6
R	29-60	0.1	79	11	10	6	80	7.2	1.7

Other: Coarse fragments present in lower solum and C; the greensand is usually weathered to a brown color in the B2 horizon. Setting: On rock-controlled upland ridges and valley slopes (2 to 30%gradient) in the "Driftless Area"; the original southern mesic forest and oak savanna have largely been replaced by second-growth forest, pasture, or cropped fields; mean ann. soil temp. 47°F, mean ann. ppt. 30 in. Estab., Richland Co., Wis., 1956; Robinson and Klingelhoets, 1959; Slota and Garvey, 1961; Thomas, Carroll, and Wing, 1962; Soil Conservation Service, 1967a; 65 bu. oats; 450-500 bd. ft. red or white pine; catena no. 18.

Norrie sil (F4, F6) (Fig. 2-54); Typic Glossoboralf, fine-loamy, mixed, frigid (Gray-Brown Podzolic); well drained; solum is 24 to 40 in. thick, developed in 20 to 36 in. of loamy covering on acid sandy loam dark brown (7.5YR 4/4) glacial till. Horizons: O1, O2, A1, A2, B21t, IIB22t, IIC.

Some properties of Norrie silt loam

Horizon	Depth in.	OM %	Sand %	Silt %	Clay %	CEC me/ 100g	BS %	pH	BD g/cc
A2	3-11	2.6	25	60	15	12	60	5.7	1.3
B21t	11-26	0.5	16	55	29	15	50	5.3	1.5
IIB22t	26-35	0.2	38	42	20	11	54	5.4	1.7
IIC	35-45	0.1	58	25	17	9	70	5.4	1.8

Other: A weak fragipan may be present between the two B horizons. Setting: Some areas are stony; the original northern mesic forest cover was logged off; regrowth forest has replaced it in some areas; much land is in fields of hay, oats, and silage corn, or pasture; mean ann. soil temp. 42°F, mean ann. ppt. 30 in. Proposed, Marathon Co., Wis., 1946; Ciolkosz, 1964; Milfred, Olson, and Hole, 1967; 75 bu. oats; 450-500 bd. ft. red pine; 500-600 bd. ft. white pine; catena no. 99.

Northfield sl-sil (D4, D5, D6, D8, D9, D11); Lithic Hapludalf, loamy, mixed, mesic (Lithosol); well drained; solum is 12 to 20 in. thick, developed in less than 10 in. of loess plus sandstone residuum over indurated sandstone at the bottom of the solum. Horizons: O1, O2, A1, A2, B2t, IIB3, IIR.

Some properties of Northfield loam

Horizon	Depth in.	OM %	Sand %	Silt %	Clay %	CEC me/ 100g	BS %	pH	BD g/cc
A1	0-3	4.0	40	40	20	14	60	5.7	1.3
B2t	12-18	0.4	40	35	25	12	55	5.4	1.6

Other: Sandstone fragments are present in the solum. Setting: On sandstone ridges with slopes of 5 to 20% gradient; the original oak savanna cover has on gentler slopes been replaced by fields of hay, small grains, and corn, and by pasture; mean ann. soil temp. 45°F, mean ann. ppt. 30 in. Estab., Jackson Co., Wis., 1954; Watson, 1966; 50 bu. oats; 400-450 bd. ft. red pine; catena no. 17.

Oakville s-ls (B4); Typic Udipsamment, sandy, mixed, nonacid, mesic (Regosol); excessively drained; solum is 24 to 36 in. deep, developed in neutral to calcareous aeolian sand. Horizons: O1, O2, A1, B2, B3, C.

Some properties of Oakville sand

Horizon	Depth in.	OM %	Sand %	Silt %	Clay %	CEC me/100g	BS %	pH	BD g/cc
A1	0-5	2.0	91	5	4	5	90	6.4	1.3
B2	5-22	0.3	89	6	5	3	85	5.5	1.6
C	34-66	0.1	94	3	3	1	98	7.2	1.8

Other: Discontinuous dark color bands may be present in the C horizon. Setting: On dunes, beach ridges, outwash plains, on slopes of 0 to 20% gradient; oak savanna vegetation has been largely untouched, but in places is replaced by pasture and fields of hay and small grains; mean ann. soil temp. 47°F, mean ann. ppt. 32 in. Proposed, Ionia Co., Mich., 1959; Milfred and Hole, 1970; 30 bu. oats; 300-360 bd. ft. red pine; 400-465 bd. ft. white pine; catena no. 81.

Ogemaw s-ls (I1); Aeric Haplaquod, sandy over loamy, mixed, ortstein, frigid (Groundwater Podzol); somewhat poorly drained; solum is 14 to 28 in. thick, developed in 20 to 36 in. of sandy covering over calcareous reddish-brown clay. Horizons: O1, O2, A2, Bhir, C1, IIC2.

Some properties of Ogemaw loamy sand

Horizon	Depth in.	OM %	Sand %	Silt %	Clay %	CEC me/100g	BS %	pH	BD g/cc
A2	2-13	2.0	89	3	8	4	50	4.1	1.4
Bhir	13-24	3.0	78	10	12	8	30	4.5	1.7
C1	24-30	0.1	92	3	5	2	20	5.0	1.6
IIC2	30-40	0.1	25	30	45	20	85	8.5*	1.7

*calcareous

Other: The B horizon contains some ortstein fragments; the C1 horizon may be a fragipan. Setting: In depressions on glacial lake and till plains on slopes of 0 to 2% gradient; swamp conifers cover much of the area; drained land produces potatoes, small grains, corn, and hay; mean ann. soil temp. 40°F, mean ann. ppt. 29 in. Estab., Antrim Co., Mich., 1923; Ableiter and Hole, 1961; 65 bu. oats; 400-450 bd. ft. red pine; 450-500 bd. ft. white pine; catena no. 152.

Ogle sil (B21); Typic Argiudoll, fine-silty, mixed, mesic (Brunizem); well drained; solum is 5 to 8 ft. thick, developed in 36 to 50 in. of loess overlying calcareous sandy loam to loam glacial till. Horizons: O1, O2, A1, A3, B21t, B22t, IIB23t, IIC.

Some properties of Ogle silt loam

Horizon	Depth in.	OM %	Sand %	Silt %	Clay %	CEC me/100g	BS %	pH	BD g/cc
A1	0-11	7.0	17	65	18	11	65	6.4	1.2
B21t	15-22	1.0	8	60	32	15	50	5.4	1.5
B22t	22-40	0.5	20	50	30	14	60	5.6	1.6
IIC	70-100	0.1	36	40	24	12	70	8.5*	1.8

*calcareous

Other: The solum consists of a modern soil developed in loess overlying a truncated paleosol. Setting: On rolling glacial moraines on slopes of 2 to 12% gradient; the original prairie vegetation has been replaced by fields of hay, small grains, and corn, and by pasture; mean ann. soil temp. 47°F, mean ann. ppt. 32 in. Proposed, Carroll Co., Ill., 1960; 75 bu. oats; not suitable for pine; catena no. 43.

Omega s-ls (E1, E3, G25, H1, H2, H3, H4, H5, H6, H7) (Figs. 2-53, 2-54); Spodic Udipsamment to Entic Haplorthod, sandy, mixed, frigid (Podzol); excessively drained; solum is 12 to 28 in. thick, developed in acid glacial outwash sand. Horizons: O1, O2, A1, Bhir, Bir, C.

Some properties of Omega loamy sand

Horizon	Depth in.	OM %	Sand %	Silt %	Clay %	CEC me/100g	BS %	pH	BD g/cc
A1	0-3	7.0	85	11	4	14	25	5.0	1.3
Bhir	3-17	1.4	86	9	5	5	17	5.6	1.5
C	24-36	0.1	98	1	1	1	25	5.8	1.8

Other: The C horizon has a reddish cast (7.5-5YR 5/6). Setting: On glacial sandy drift uplands with slopes of 0 to 30% gradient; the original jack pine savanna has been replaced locally by plantations of red and jack pine; mean ann. soil temp. 40°F, mean ann. ppt. 28 in. Estab., Iron Co., Mich., 1930; Soil Conservation Service, 1952, 1967a; Ableiter and Hole, 1961; Milfred, Olson, and Hole, 1967; 35 bu. oats; 400-450 bd. ft. red or white pine; catena no. 145.

Onamia sl-l (A14, C9, C16, D10, F4, F6, F17, F25, F26, G6, G15, G19, G24, G26, G27, G28) (Figs. 2-53, 2-54, 12-4, 13-5); Typic Eutroboralf to Eutric Glossoboralf, fine-loamy over sandy or sandy-skeletal, mixed, frigid (Gray-Brown Podzolic); well drained; solum is 20 to 40 in. thick, developed in this same thickness of loamy covering over acid sand and gravel glacial outwash. Horizons: O1, O2, A1, A2, B1, B2t, IIC.

Some properties of Onamia loam

Horizon	Depth in.	OM %	Sand %	Silt %	Clay %	CEC me/100g	BS %	pH	BD g/cc
A1	1-4	8.0	44	48	8	14	60	5.9	1.3
A2	4-10	2.0	46	47	7	6	30	5.3	1.5
B2t	14-24	0.5	45	35	20	15	47	5.0	1.6
IIC	24-30	0.1	92	5	3	4	53	5.2	1.8

Other: Solum is probably developed in a mixture of loess and outwash sand. Setting: On glacial outwash plains and terraces, with slope gradients of 0 to 30%; the original northern mesic forest has been largely replaced by fields of hay, small grains, silage corn, and potatoes, and by pasture; mean ann. soil temp. 43°F, mean ann. ppt. 30 in. Estab., Mille Lacs Co., Minn., 1927; Milfred, Olson, and Hole, 1967; Soil Conservation Service, 1967a; Haszel, 1968; 60 bu. oats; 500-575 bd. ft. red or white pine; catena no. 113.

Onaway sl-l (B17, E1, E2, E3, E4, E5, E6, E7, E9, I6) (Figs. 2-53, 11-1, 11-4, 11-5, 15-4); Alfic Haplorthod, fine-loamy, mixed, frigid (Podzol); well drained; solum is 15 to 30 in. thick, developed in reddish-brown (7.5YR 5/4) calcareous sandy loam glacial till. Horizons: O1, O2, A2, Bhir, A′2, A′2&B′2t, B′2t, C.

Some properties of Onaway silt loam

Hori-zon	Depth in.	OM %	Sand %	Silt %	Clay %	CEC me/ 100g	BS %	pH	BD g/cc
A2	0-3	3.0	61	31	8	12	80	5.5	1.3
Bhir	3-9	1.0	56	36	8	9	47	5.2	1.5
A′2	9-10	0.3	57	34	9	3	65	5.7	1.8
B′2t	13-21	0.5	55	27	18	12	70	7.6	1.7
C	21-40	0.2	51	32	17	7	100	8.5*	1.7

*calcareous

Other: The A′2 and B′2 are fragic. Setting: On glacial moraines on slopes of 2 to 8% gradient; the original northern mesic forest is largely replaced by fields of potatoes, hay, small grains, and silage corn, and by pasture; mean ann. soil temp. 44°F, mean ann. ppt. 30 in. Estab., Antrim Co., Mich., 1923; Soil Conservation Service, 1967a; 70 bu. oats; 450-500 bd. ft. white pine; catena no. 136.

Ontonagon l-sic (I2, I18, I19) (Figs. 2-54, 14-5); Typic Eutroboralf, very fine, mixed, frigid (Gray Wooded); moderately well drained; solum is 15 to 32 in. thick, developed in reddish-brown (2.5YR 5/4) calcareous clay. Horizons: O1, O2, A1, A2, B&A, B2t, C.

Some properties of Ontonagon silt loam

Hori-zon	Depth in.	OM %	Sand %	Silt %	Clay %	CEC me/ 100g	BS %	pH	BD g/cc
A2	2-8	1.7	33	54	13	11	36	4.1	1.4
B2t	15-21	0.5	15	33	52	22	53	4.2	1.6
C	21-40	0.1	12	21	67	20	100	7.8*	1.7

*calcareous

Other: The A2 tongues into the B2 horizon. Setting: On glacial lake plains on slope gradients of 2 to 30%; the original boreal forest was logged off, and has been largely replaced by second-growth forest and permanent pasture, with some cropland (flax, small grains, hay); mean ann. soil temp. 40°F, mean ann. ppt. 28 in. Estab., Ontonagon Co., Mich., 1921; Soil Conservation Service, 1952; Ableiter and Hole, 1961; 65 bu. oats; 300-375 bd. ft. red pine; 350-400 bd. ft. white pine; catena no. 150.

Orienta s-ls (I19) (Fig. 14-5); Aquic Haplorthod, sandy, mixed, frigid (Podzol); somewhat poorly drained; solum is 26 to 36 in. thick, developed in 40 to 60 in. of sandy material overlying calcareous clay. Horizons: O1, O2, A1, A2, Bhir, C1, IIC2.

Some properties of Orienta loamy sand

Hori-zon	Depth in.	OM %	Sand %	Silt %	Clay %	CEC me/ 100g	BS %	pH	BD g/cc
A2	1-7	2.0	86	6	8	10	17	4.5	1.3
Bhir	7-19	3.0	73	15	12	7	15	5.4	1.4
C1	19-50	0.1	90	5	5	2	6	5.4	1.8
IIC2	50-60	0.1	10	35	55	22	100	5.8*	1.7

*at the top, and 8.5 at 5 ft., calcareous

Other: The sandy and clay materials interfinger somewhat at the interface. Setting: On level portions (0 to 2% slope gradient) of glacial lake plains; the original boreal forest is largely replaced with new-growth timber, but some areas are cleared for pasture, hay, small grains, potatoes, and small fruits; mean ann. soil temp. 40°F, mean ann. ppt. 28 in. Proposed, Bayfield Co., Wis., 1928; Ableiter and Hole, 1961; 60 bu. oats; not suitable for pine; catena no. 153.

Orion sil (J1) (Figs. 7-9, 7-10); Aquic Udifluvent, coarse-silty, mixed, nonacid, mesic (Alluvial soil); somewhat poorly drained; solum is 4 to 8 in. thick, developed in 20 to 40 in. of light-colored alluvium over a buried A1b horizon. Horizons: O1, A1, C, A1b, Cb.

Some properties of Orion silt loam

Hori-zon	Depth in.	OM %	Sand %	Silt %	Clay %	CEC me/ 100g	BS %	pH	BD g/cc
A1	0-5	5.0	17	65	18	15	90	6.8	1.2
C	5-21	0.8	10	70	20	10	95	7.2	1.5

Other: The C horizon is finely stratified; the profile is mottled. Setting: On floodplains on slopes of 0 to 2% gradient; the original oak savanna and swamp hardwoods have been replaced by new-growth forest, pasture, and some fields of hay, oats, and corn; mean ann. soil temp. 47°F, mean ann. ppt. 30 in. Estab., Richland Co., Wis., 1956; Robinson and Klingelhoets, 1959, 1961; Thomas, Carroll, and Wing, 1962; Watson 1966; 65 bu. oats; not suitable for pine; catena no. 39.

Oshkosh sl-sicl (E8, I20, I21) (Figs. 11-1, 11-5, 11-6, 15-3, 15-4); Typic Hapludalf, very fine, mixed, mesic (Gray-Brown Podzolic); moderately well drained; solum is 20 to 30 in. thick, developed in less than 15 in. of silty covering over calcareous reddish-brown (5YR 5/3) clay. Horizons: O1, O2, A1, A2, B1t, B2t, B3, C.

Some properties of Oshkosh silty clay loam

Hori-zon	Depth in.	OM %	Sand %	Silt %	Clay %	CEC me/ 100g	BS %	pH	BD g/cc
A1	0-3	12.0	5	63	32	30	78	6.5	1.4
A2	3-6	2.3	6	63	31	16	65	6.3	1.6
B2t	8-19	1.1	1	33	66	20	68	7.8	1.7
C	24-35	0.2	0	25	75	9	100	8.2*	1.7

*calcareous

Other: Soft lime segregations may be present in the lower solum and upper C horizon. Setting: On glacial lake plains with slope gradients of 0 to 3%; the original northern and southern mesic forests have been replaced by fields of hay, oats, corn, and by pasture; mean ann. soil temp. 44°F, mean ann. ppt. 30 in. Proposed, Fox River valley, Winnebago and Brown cos., Wis., 1954; Soil Conservation Service, 1967a; 75 bu. oats; 450-500 bd. ft. white pine; catena no. 83.

Oshtemo ls-sl (C1); Typic Hapludalf, coarse-loamy, mixed, mesic (Gray-Brown Podzolic); well drained; solum is 40 to 60 in. thick, developed in loamy material of that depth over calcareous sand and gravel glacial outwash. Horizons: O1, O2, A1, A2, B2t, B3, IIC.

Some properties of Oshtemo sandy loam

Hori-zon	Depth in.	OM %	Sand %	Silt %	Clay %	CEC me/ 100g	BS %	pH	BD g/cc
A2	4-14	1.5	74	18	8	12	70	5.6	1.4
B2t	14-35	0.4	61	25	14	8	60	5.4	1.6
IIC	55-65	0.1	90	6	4	2	100	8.5*	1.8

*calcareous

Other: Depth to calcareous material is about 50 to 60 in. Setting: On glacial outwash deposits with slope gradients of 1 to 40%; the original oak savanna or southern mesic forest has been largely replaced by fields of hay, small grains, and corn, and by pasture; mean ann. soil temp. 47°F, mean ann. ppt. 31 in. Estab., Kalamazoo Co., Mich., 1922; Milfred and Hole, 1970; Steingraeber and Reynolds, 1971; 50 bu. oats; 500-600 bd. ft. red or white pine; catena no. 79.

Osseo sil (J1); Aquic Udifluvent, coarse-silty, mixed, nonacid, mesic (Alluvial soil); somewhat poorly drained; solum is 15 to 24 in. thick, developed in colluvial-alluvial material. Horizons: O1, O2, A1, C.

Some properties of Osseo silt loam

Horizon	Depth in.	OM %	Sand %	Silt %	Clay %	CEC me/100g	BS %	pH	BD g/cc
A1	0-20	6.5	15	65	20	17	85	6.5	1.2
C	20-30	0.3	22	60	18	10	92	6.3	1.6

Other: The profile is mottled. Setting: In small valleys and on footslopes of gradients of 2 to 20% in the "Driftless Area"; oak savanna and southern mesic forest have been replaced by pasture and by some fields of hay, small grains, and corn; mean ann. soil temp. 47°F, mean ann. ppt. 32 in. Proposed, Trempealeau Co., Wis., 1940; combined now with Orion; Beatty, 1960; 65 bu. oats; not suitable for pine; catena no. 37.

Ostrander ls-sil (F13); Typic Hapludoll, fine-loamy, mixed, mesic (Brunizem); well drained; solum is 20 to 42 in. thick, developed in less than 30 in. of silty material over calcareous yellowish-brown (10YR 5/6) loam glacial till. Horizons: O1, O2, A1, B1, IIB2, B3, IIC.

Some properties of Ostrander loam

Horizon	Depth in.	OM %	Sand %	Silt %	Clay %	CEC me/100g	BS %	pH	BD g/cc
A1	0-12	5.0	35	42	23	18	60	6.3	1.3
IIB2	20-30	0.6	50	20	30	15	50	5.3	1.6
IIC	35-50	0.1	45	35	20	9	95	8.5*	1.8

*calcareous

Other: Depth to carbonates is 44 to 76 in.; the IIB horizon contains one or more stone lines. Setting: On glacial moraines with slope gradients of 4 to 18%; the original prairie vegetation has been replaced by fields of corn, soybeans, small grains, and hay, and by pasture; mean ann. soil temp. 47°F, mean ann. ppt. 31 in. Estab., Filmore Co., Minn., 1955; Haszel, 1968; 75 bu. oats; not suitable for pine; catena no. 105.

Otterholt sil (F10); Typic Glossoboralf, fine-silty, mixed, frigid (Gray-Brown Podzolic); well drained; solum is 30 to 40 in. thick, developed in 36 to 50 in. of loess overlying acid, reddish-brown (5YR 4/4) sandy loam glacial till. Horizons: O1, O2, A1, A2, A&B, B2t, B3t, C1, IIC2.

Some properties of Otterholt silt loam

Horizon	Depth in.	OM %	Sand %	Silt %	Clay %	CEC me/100g	BS %	pH	BD g/cc
A1	0-3	5.0	8	79	13	14	72	5.6	1.2
A2	3-9	0.5	7	77	16	9	46	4.9	1.4
B2t	19-27	0.3	8	69	23	14	43	4.8	1.6
C1	30-40	0.2	15	68	17	13	42	4.8	1.7
IIC2	40-50	0.1	63	23	14	7	54	4.7	1.8

Other: The A2 tongues down into the B2 horizon. Setting: On moraines with slope gradients of 3 to 15%; the northern mesic forest has been largely replaced by fields of hay, small grains, and corn, and by pasture; second-growth forest is dominated by sugar maple; mean ann. soil temp. 43°F, mean ann. ppt. 30 in. Estab., Barron Co., Wis., 1950; Robinson et al., 1958; Soil Conservation Service, 1967a; Haszel, 1968; 80 bu. oats; 450-500 bd. ft. red pine; 500-550 bd ft. white pine; catena no. 94.

Ozaukee l-sil (B19); Typic Hapludalf, fine, mixed, mesic (Gray-Brown Podzolic); moderately well drained; solum is 20 to 42 in. thick, developed in less than 20 in. of silty sediments over calcareous, dark brown (7.5YR 5/4) silty clay loam glacial till. Horizons: O1, O2, A1, A2, B1t, IIB2t, IIB3, IIC.

Some properties of Ozaukee silt loam

Horizon	Depth in.	OM %	Sand %	Silt %	Clay %	CEC me/100g	BS %	pH	BD g/cc
A2	4-10	2.0	9	71	20	14	83	7.4	1.3
IIB2t	13-23	1.0	9	39	52	28	86	7.3	1.5
IIB3	23-29	0.7	16	47	37	16	87	7.7	1.7
IIC	29-60	0.1	14	54	32	9	100	8.0*	1.8

*calcareous

Other: The till may be a heavy silt loam in texture. Setting: On glacial moraines with slopes of 0 to 20% gradient; the original southern mesic forest cover has been replaced by fields of hay, small grains, and corn, and by pasture; mean ann. soil temp. 45°F, mean ann. ppt. 30 in. Proposed, Ozaukee Co., Wis., 1965; Soil Conservation Service, 1967a; Parker, Kurer, and Steingraeber, 1970; 70 bu. oats; 400-450 bd. ft. white pine; catena no. 55.

Padus sl-sil (F5, F16, F24, G7, G16, G25, G27); Alfic Haplorthod, coarse-loamy, mixed, frigid (Podzol); well drained; solum is 20 to 40 in thick, developed in less than 20 in. of silt sediment over acid sand and gravel glacial outwash. Horizons: O1, O2, A2, Bhir, Bir, A′2x, B′2x, IIC.

Some properties of Padus loam

Horizon	Depth in.	OM %	Sand %	Silt %	Clay %	CEC me/100g	BS %	pH	BD g/cc
A2	0-5	1.5	40	48	12	9	56	5.0	1.2
Bhir	5-13	1.6	40	50	10	8	45	5.1	1.4
A′2x	18-25	0.2	45	47	8	4	54	5.4	1.6
B′2x	25-32	0.2	52	33	15	9	73	5.4	1.7
IIC	32-50	0.1	97	1	2	3	80	6.0	1.8

Other: The horizons marked x constitute a fragipan. Setting: On glacial outwash, with slope gradients of 1 to 40%; the northern mesic forest has, except in Menominee lands, been replaced over considerable areas where slopes are gentle by fields of hay, small grains, silage corn, potatoes, and by pasture; mean ann. soil temp. 43°F, mean ann. ppt. 30 in. Proposed, Forest Co., Wis., 1960; Milfred, Olson, and Hole, 1967; 80 bu. oats; 450-500 bd. ft. red or white pine; catena no. 142.

Palms muck (J15); Terric Medisaprist, loamy, euic, mesic (Bog); very poorly drained; solum is 16 to 50 in. thick, developed in herbaceous material over loamy substratum. Horizons: Oa1, Oa2, Oa3, IICg.

Some properties of Palms muck

Hori-zon	Depth in.	OM %	Sand %	Silt %	Clay %	CEC me/ 100g	BS %	pH	BD g/cc	Fibers unrub. %	Fibers rub. %
Oa1	0-14	90	0	6	4	160	32	6.5	0.1	5	2
Oa3	28-35	98	0	1	1	190	40	6.0	0.2	5	2
IICg	35-60	0.5	40	40	20	8	70	8.5*	1.8	0	0

*calcareous

Other; Woody fragments may be present. Setting: In depressions in glacial drift landscapes; the original vegetation of sedges, reeds, and grasses, with some swamp hardwoods, is largely unaltered, but some areas have been drained and are used for pasture or production of truck crops; mean ann. soil temp. 46°F, mean ann. ppt. 31 in. Estab., Sanilac Co., Mich., 1955; Haszel, 1971; not suitable for oats or pine; catena no. 50.

Palsgrove sil (A3, A5, A6, A8, A9) (Figs. 7-9, 16-4); Typic Hapludalf, fine-silty, mixed, mesic (Gray-Brown Podzolic); well drained; solum is 40 to 55 in. thick, developed in 36 to 45 in. of loess over clayey residuum on dolomite. Horizons: O1, O2, A1, A2, B1, B2t, IIB3, R.

Some properties of Palsgrove silt loam

Hori-zon	Depth in.	OM %	Sand %	Silt %	Clay %	CEC me/ 100g	BS %	pH	BD g/cc
A1	0-4	6.0	7	80	13	27	59	5.8	1.2
A2	4-8	1.2	5	82	13	10	50	5.6	1.3
B2t	11-37	0.2	4	70	26	17	62	5.2	1.6
IIB3	37-42	0.1	23	28	49	33	54	5.2	1.7

Other: Chert fragments are present in the IIB3 horizon. Setting: On broad ridges and some valley slopes in rock-controlled landscapes of the "Driftless Area," on slope gradients of 2 to 30%; the original oak savanna cover is largely replaced by fields of hay, small grains, and corn, and by pasture; mean ann. soil temp. 45°F, mean ann. ppt. 30 in. Estab., Lafayette Co., Wis., 1964; Watson, 1966; Soil Conservation Service, 1967a; 65 bu. oats; 400-450 bd. ft. red pine; 450-500 bd. ft. white pine; catena no. 8.

Pardeeville sl-sil (B15, B28); Mollic Hapludalf, coarse-loamy, mixed, mesic (Gray-Brown Podzolic grading toward Brunizem); well drained; solum is 20 to 40 in. thick, developed in less than 20 in. of silty or loamy deposits over dolomitic dark brown (7.5YR 4/4) sandy loam glacial till. Horizons: O1, O2, A1, A2, B1, B2t, IIB3t, IIC.

Some properties of Pardeeville sandy loam

Hori-zon	Depth in.	OM %	Sand %	Silt %	Clay %	CEC me/ 100g	BS %	pH	BD g/cc
A1	0-5	4.0	65	25	10	9	67	6.1	1.3
A2	5-9	0.9	65	26	9	7	64	5.8	1.4
B2t	12-18	0.4	70	14	16	6	63	5.6	1.5
IIB3t	18-30	0.2	76	16	8	4	70	7.0	1.6
IIC	30-40	0.1	75	19	6	2	100	8.2*	1.8

*calcareous

Other: Coarse fragments occupy less than 10% of the volume of upper horizons, and more than 20% in the lower B and in the C. Setting: On glacial moraines with slopes of 2 to 20% gradient; the original oak savanna has been largely replaced by fields of hay, oats, and corn, and by pasture; mean ann. soil temp. 45°F, mean ann. ppt. 30 in. Estab., Marquette Co., Wis., 1969; Peck and Lee, 1961; Soil Conservation Service, 1967a; 60 bu. oats; 450-500 bd. ft. red or white pine; catena no. 60.

Pecatonica sil (B6, B7, B10) (Fig. 7-12); Typic Hapludalf, fine-loamy, mixed, mesic (Gray-Brown Podzolic); well drained; solum is 4 to 8 ft. thick, developed in 20 to 30 in. of loess over calcareous (about 20% $CaCO_3$ equiv.), gravelly, sandy loam or loam glacial till. Horizons: O1, O2, A1, A2, B1, IIB21t, IIB22t, IIB23t, IIB3, IIC.

Some properties of Pecatonica silt loam

Hori-zon	Depth in.	OM %	Sand %	Silt %	Clay %	CEC me/ 100g	BS %	pH	BD g/cc
A2	3-12	1.5	15	70	15	10	70	6.3	1.3
IIB21t	19-24	0.6	37	43	20	11	55	5.3	1.5
IIB22t	24-34	0.3	47	23	30	16	65	5.7	1.7
IIC	65-85	0.1	68	20	12	5	100	8.5*	1.8

*calcareous

Other: The silty material overlies a truncated paleosol. Setting: On glacial moraines on slopes of 2 to 18% gradient; the original oak savanna has largely been replaced by fields of hay, small grains, soybeans, and corn, and by pasture; mean ann. soil temp. 47°F, mean ann. ppt. 30 in. Estab., Carroll Co., Ill., 1967; Haszel, 1971; 75 bu. oats; 400-450 bd. ft. red pine; 450-500 bd. ft. white pine; catena no. 46.

Pella sil-sicl (B13, B21, B22, B25, J8, J9, J10, J15) (Figs. 2-52, 8-8, 8-13); Typic Haplaquoll, fine-silty, mixed, mesic (Humic Gley); poorly drained; solum is 40 to 54 in. thick, developed in 36 to 50 in. of loess over dolomitic sandy loam or loam till. Horizons: O1, O2, A1, B2g, IIB2g, IIC.

Some properties of Pella silty clay loam

Hori-zon	Depth in.	OM %	Sand %	Silt %	Clay %	CEC me/ 100g	BS %	pH	BD g/cc
A1	0-14	19.0	3	66	31	77	85	6.9	1.2
B2g	18-42	0.3	3	67	30	28	92	7.1	1.5
IIB2g	42-48	0.3	52	37	11	10	100	7.8*	1.7
IIC	48-60	0.1	55	37	8	8	100	8.5*	1.7

*calcareous at 20 to 40 in. depth

Other: A calcareous variant has snail shells in the A1 horizon; white seams of calcite coat surfaces of prismatic peds in the IIB2g and IIC. Setting: On level glacial lake plains with slopes of 0 to 2% gradient; the original vegetative cover of swamp hardwoods and wet prairie has been largely replaced, after drainage, by fields of hay, small grains, and corn, and by pasture; mean ann. soil temp. 47°F, mean ann. ppt. 32 in. Estab., Ford Co., Ill., 1929; see Elba in Soil Conservation Service, 1967a; Milfred and Hole, 1970; Steingraeber and Reynolds, 1971; Haszel, 1971; 65 bu. oats; not suitable for pine; catena no. 47.

Pence sl-l (F24, G3, G4, G5, G7, G11, G12, G14, G16, G17, G18, G25, G27, H1, H3, H4, H5, H6, H7) (Figs. 2-53, 2-54, 12-9);Typic Haplorthod, coarse-loamy, mixed, frigid (Podzol); well drained; solum is 10 to 24 in thick, developed in a loamy material of that thickness overlying acid glacial outwash sand and gravel. Horizons: O1, O2, A2, Bhir, I&II Bir, IIB3, IIC.

Some properties of Pence loam

Horizon	Depth in.	OM %	Sand %	Silt %	Clay %	CEC me/100g	BS %	ph	BD g/cc
A2	0-3	2.2	36	49	15	11	24	5.0	1.3
Bhir	3-11	1.8	39	53	8	10	27	5.1	1.4
IIB3	18-23	0.3	70	18	12	5	91	5.4	1.7
IIC	23-40	0.1	96	2	2	2	60	5.4	1.8

Other: A weak fragipan may be present below the Bhir horizon. Setting: On glacial outwash with slope gradients of 1 to 40%; the original northern mesic forest cover is largely replaced by fields of hay, small grains, and silage corn, and by pasture; new-growth forest covers some areas; mean ann. soil temp. 43°F, mean ann. ppt. 30 in. Estab., Bayfield Co., Wis., 1958; Ableiter and Hole, 1961; Milfred, Olson, and Hole, 1967; 50 bu. oats; 500-575 bd. ft. red or white pine; catena no. 143.

Perrot sil-sicl (A11, A12); Mollic Haplaquept, fine, mixed, nonacid, mesic (Low Humic Gley); poorly drained; solum is 36 to 45 in. thick, developed in calcareous stratified silts and clays. Horizons: O1, O2, A1, B1g, B2g, B3g, Cg.

Some properties of Perrot clay loam

Horizon	Depth in.	OM %	Sand %	Silt %	Clay %	CEC me/100g	BS %	pH	BD g/cc
A1	0-11	14.0	6	55	39	20	80	5.3	1.3
B2g	18-33	0.3	12	47	41	24	50	6.3	1.6
Cg	42-50	0.1	25	30	45	20	100	8.5*	1.7

*calcareous

Other: Depth to carbonates is usually greater than 50 in.; seams of sand are present in the C horizon. Setting: On natural terraces in the Wisconsin and Mississippi River valleys, on slopes of 0 to 2% gradient; the original vegetation of swamp hardwoods and wet prairie is largely replaced fields of hay, small grains, and corn, and by pasture; mean ann. soil temp. 47°F, mean ann. ppt. 30 in. Proposed, Trempealeau Co., Wis., 1943; 65 bu. oats; not suitable for pine; catena no. 35.

Pickford sil-sicl (I2, I8, I9, I18, I19); Aeric Haplaquept, fine, mixed, nonacid, frigid (Low Humic Gley); poorly drained; solum is 12 to 20 in. thick, developed in reddish-brown, calcareous clay. Horizons: O1, O2, A1, A3, B2g, Cg.

Some properties of Pickford silty clay loam

Horizon	Depth in.	OM %	Sand %	Silt %	Clay %	CEC me/100g	BS %	pH	BD g/cc
A1	0-5	15.0	15	46	39	32	83	6.8	1.2
B2g	7-14	0.3	8	49	43	20	58	8.5*	1.6
Cg	14-18	0.1	5	30	65	19	100	8.5*	1.7

*calcareous

Other: The profile is mottled. Setting: On glacial lake plains; the original boreal and northern mesic forests have been replaced in a limited drained area by hay, small grains, and pasture; mean ann. soil temp. 40°F, mean ann. ppt. 28 in. Proposed, Mackinac Co., Mich., 1950; Ableiter and Hole, 1961; 55 bu. oats; not suitable for pine; catena no. 150.

Pillot sil. *See* Waukegan.

Plainfield s-ls (C1, C2, C3, C4, C5, C6, C7, C10, C11, C12, C13, C14, C15, J4) (Figs. 2-53, 2-54, 9-4, 9-6, 11-1); Typic Udipsamment, sandy, mixed, mesic (Regosol); excessively drained; solum is 18 to 34 in. thick, developed from acid glacial outwash sand. Horizons: O1, O2, A1, A2, B, C.

Some properties of Plainfield sand

Horizon	Depth in.	OM %	Sand %	Silt %	Clay %	CEC me/100g	BS %	pH	BD g/cc
A1	0-5	2.2	90	6	4	7	20	5.2	1.3
C	20-32	0.1	96	2	2	1	6	5.9	1.7

Other: About 25% weatherable mineral grains are present in the dominantly quartz sand mass. Setting: On acid glacial outwash uplands with slope gradients of 0 to 20%; the original jack pine and scrub oak savanna has over considerable areas of Adams and adjacent counties been replaced by large-scale irrigation farming for production of truck crops and corn; mean ann. soil temp. 44°F, mean ann. ppt. 30 in. Estab., Waushara Co., Wis., 1909; Robinson et al., 1958; Beatty, 1960; Soil Conservation Service, 1967a; 35 bu. oats; 350-400 bd. ft. red pine; 450-500 bd. ft. white pine; catena no. 34.

Plano sil (B22, B32) (Figs. 2-53, 2-54, 8-13); Typic Argiudoll, fine-silty, mixed, mesic (Brunizem); well drained; solum is 48 to 65 in. thick, developed in 36 to 50 in. of loess over calcareous loam glacial till. Horizons: O1, A1, A3, B1, B2t, IIB3, IIC.

Some properties of Plano silt loam

Horizon	Depth in.	OM %	Sand %	Silt %	Clay %	CEC me/100g	BS %	pH	BD g/cc
A1	0-9	6.2	4	72	24	22	67	6.1	1.3
B2t	16-45	0.5	3	66	31	19	69	5.2	1.5
IIB3	45-53	0.2	50	35	15	10	83	6.0	1.6
IIC	53-60	0.1	72	23	5	2	100	8.0*	1.8

*calcareous

Other: The IIC horizon may be glacial till or stratified glacial outwash. Setting: On glacial moraines with slope gradients of 0 to 12%; the original prairie vegetation has been replaced by hay, small grains, corn, and soybeans; mean ann. soil temp. 45°F, mean ann. ppt. 31 in. Estab., Kendall Co., Ill., 1941; Soil Conservation Service, 1967a; Milfred and Hole, 1970; Haszel, 1971; 75 bu. oats; not suitable for pine; catena no. 47.

Poskin sil (F17, F26); Aquic Glossoboralf, fine-silty over sandy or sandy-skeletal, mixed, frigid (Gray-Brown Podzolic); somewhat poorly drained; solum is 20 to 36 in. thick, developed in 20 to 40 in. of sediment overlying acid glacial outwash sand and gravel. Horizons: O1, O2, A1, A2, A&B, B2t, IIB3t, IIC.

Some properties of Poskin silt loam

Horizon	Depth in.	OM %	Sand %	Silt %	Clay %	CEC me/100g	BS %	pH	BD g/cc
A1	0-3	12.0	16	73	11	25	40	5.7	1.1
A2	3-7	2.0	15	69	16	11	23	5.3	1.2
B2t	17-25	0.4	18	56	26	13	62	5.1	1.6
IIC	29-40	0.1	95	1	4	5	45	5.6	1.8

Other: The A2 tongues into the B2 horizon. Setting: On glacial outwash plains at slope gradients of 0 to 2%; the swamp hardwood and conifer forest has been largely replaced by pasture and by hay fields; mean ann. soil temp. 44°F, mean ann. ppt. 30 in. Estab., Langlade Co., Wis., 1947; Robinson et al., 1958; 70 bu. oats; not suitable for pine; catena no. 112.

Poy l-sicl (J11); Typic Haplaquoll, clayey over sandy or sandy-skeletal, mixed, mesic (Humic Gley); very poorly drained; solum is 24 to 42 in. thick, developed in 20 to 40 in. of reddish-brown clay over calcareous sand and gravel. Horizons: O1, O2, A1, B1g, B2g, B3, IIC.

Some properties of Poy silty clay loam

Horizon	Depth in.	OM %	Sand %	Silt %	Clay %	CEC me/100g	BS %	pH	BD g/cc
A1	0-8	18.0	21	40	39	54	80	6.2	1.2
B2g	12-28	0.3	14	44	42	25	93	7.2	1.7
IIC	35-45	0.1	80	11	9	5	100	8.1*	1.8

*calcareous

Other: The A1 horizon may be a silt loam. Setting: On glacial outwash and lake plains with slope gradients of 0 to 2%; the swamp hardwood and conifer forest and sedge meadows have to a considerable extent been replaced by pasture and, where drained, by fields of hay, oats, and corn; mean ann. soil temp. 43°F, mean ann. ppt. 30 in. Proposed, Waushara Co., Wis., 1942; Soil Conservation Service, 1967a; 65 bu. oats; not suitable for pine; catena no. 89.

Poygan l-sicl (E8, E10, I4, I5, I10, I11, I13, I14, I16, I17, I20, I21, J11, J15); Typic Haplaquoll, fine, mixed, mesic (Humic Gley); poorly drained; solum is 20 to 27 in. thick, developed in less than 20 in. of silty material over calcareous, reddish-brown glacial till. Horizons: O1, O2, A1, B1g, B2g, B3, C.

Some properties of Poygan silty clay loam

Horizon	Depth in.	OM %	Sand %	Silt %	Clay %	CEC me/100g	BS %	pH	BD g/cc
A1	0-8	10.0	3	44	53	45	71	5.8	1.3
B2g	13-19	0.5	3	37	60	36	88	7.8	1.6
C	27-60	0.3	3	33	64	26	100	8.2*	1.8

*calcareous

Other: The profile is mottled. Setting: On glacial lake plains with slope gradients of 0 to 2%; the original vegetation of swamp hardwoods and sedge meadows has to a considerable extent been replaced by pasture, and, where drained, by fields of corn, small grains, and hay; mean ann. soil temp. 45°F, mean ann. ppt. 30 in. Estab., Waushara Co., Wis., 1909; Soil Conservation Service, 1967a; 65 bu. oats; not suitable for pine; catena no. 56.

Puchyan ls-sl (B26); Arenic Hapludalf, coarse-loamy, mixed, mesic (Gray-Brown Podzolic); well drained; solum is 36 to 60 in. thick, developed in 20 to 40 in. of sandy deposits over silty aeolian materials over calcareous sandy loam or loam glacial till. Horizons: O1, O2, A1, A2, IIB1, IIB21t, IIIB22t, IIIC1, IVC2.

Some properties of Puchyan loamy sand

Horizon	Depth in.	OM %	Sand %	Silt %	Clay %	CEC me/100g	BS %	pH	BD g/cc
A2	4-27	0.5	82	10	8	7	80	5.7	1.3
IIB21t	30-36	0.2	47	35	18	9	60	6.3	1.7
IIIC1	46-54	0.1	25	60	15	8	90	7.5	1.9
IVC2	54-60	0.1	55	35	10	5	100	8.5*	2.1

*calcareous

Other: A layer of silty aeolian sediment commonly lies between the sandy overburden and the glacial till. Setting: On glacial drift uplands with slope gradients of 1 to 18%; the oak savanna is largely replaced by fields of canning and dairy crops; mean ann. soil temp. 44°F, mean ann. ppt. 30 in. Proposed, Green Lake Co., Wis., 1960; Peck and Lee, 1961; 60 bu. oats; 400-475 bd. ft. red or white pine; catena no. 70.

Radford sil (J1); Fluventic Hapludoll, fine-silty, mixed, mesic (Alluvial soil); somewhat poorly drained; solum is 15 to 22 in. thick, developed in silty alluvium 20 to 40 in. thick over a buried soil in glacial drift landscapes of southeastern Wisconsin. Horizons: O1, O2, A1, C, IIA1b, IICb.

Some properties of Radford silt loam

Horizon	Depth in.	OM %	Sand %	Silt %	Clay %	CEC me/100g	BS %	pH	BD g/cc
A1	0-19	8.0	11	70	19	24	95	7.7	1.2
C	19-28	0.2	8	72	20	11	98	7.8	1.6

Other: The buried soil is black for 16 in., and is a silty clay loam. Setting: On floodplains and footslopes with slope gradients of 1 to 5%; the swamp hardwood cover is largely replaced by corn, soybeans, and pasture; mean ann. soil temp. 47°F, mean ann. ppt. 32 in. Estab., Christian Co., Ill., 1947; 65 bu. oats; not suitable for pine; catena no. 92.

Renova l-sil (F13) (Fig. 12-4); Typic Hapludalf, fine-loamy, mixed, mesic (Gray-Brown Podzolic); well drained; solum is 20 to 42 in. thick, developed in 12 to 24 in. of silty covering over calcareous loam glacial till. Horizons: O1, O2, A1, A2, B21t, IIB22t, IIB3t, IIC.

Some properties of Renova silt loam

Horizon	Depth in.	OM %	Sand %	Silt %	Clay %	CEC me/100g	BS %	pH	BD g/cc
A2	3-10	1.5	12	70	18	16	60	5.3	1.3
IIB22t	19-42	0.5	46	20	34	15	40	5.1	1.6
IIC	52-60	0.1	30	38	32	10	100	8.5*	1.8

*calcareous

Other: Depth to calcareous till may range from 4 to 6 ft. Setting: On glacial moraines with slope gradients of 2 to 18%; the original oak savanna has been largely replaced by fields of corn, soybeans, small grains, and hay, and by pasture; mean ann. soil temp. 44°F, mean ann. ppt. 30 in. Estab., Fillmore Co., Minn., 1955; Haszel, 1968; 75 bu. oats; 400-450 bd. ft. red pine; 450-500 bd. ft. white pine; catena no. 106.

Richford ls-sl (C15); Arenic Hapludalf, sandy, mixed, mesic (Gray-Brown Podzolic); well drained; solum is 30 to 50 in. thick, developed in the same thickness of loamy sand to sandy loam material over acid loamy sand stratified glacial drift. Horizons: O1, O2, A1, A2, Bir, A′2, B1, B2t, B3t, C.

Some properties of Richford loamy sand

Horizon	Depth in.	OM %	Sand %	Silt %	Clay %	CEC me/100g	BS %	pH	BD g/cc
A2	3-7	0.6	86	10	4	5	80	6.4	1.3
B2t	27-34	0.3	70	20	10	6	75	5.8	1.7
C	41-60	0.1	82	10	8	4	90	5.6	1.8

Other: Coarse fragments may occupy as much as 15% by volume of the solum. Setting: On outwash terraces and plains, with slope gradients of 1 to 5%; the oak and jack pine savanna has been largely replaced by cropped fields; mean ann. soil temp. 45°F, mean ann. ppt. 30 in. Proposed, Waushara Co., Wis., 1957; Soil Conservatin Service, 1967a; 45 bu. oats; 350-400 bd. ft. red pine; 450-500 bd. ft. white pine, catena no. 116.

Richwood sil (A11) (Fig. 7-10); Typic Argiudoll, fine-silty, mixed, mesic (Brunizem); well drained; solum is 36 to 48 in. thick, developed in a silty deposit 40 to 60 in. thick over acid glacial outwash sand. Horizons: O1, O2, A1, A3, B1, B2t, B3, C1, IIC2.

Some properties of Richwood silt loam

Horizon	Depth in.	OM %	Sand %	Silt %	Clay %	CEC me/ 100g	BS %	pH	BD g/cc
A1	0-13	4.0	6	78	16	17	87	7.2	1.2
B2t	26-34	1.0	5	70	25	17	80	6.7	1.6
C1	42-48	0.1	7	70	23	17	75	5.3	1.7

Other: The A1 horizon ranges in thickness from 9 to 24 in. Setting: On glacial outwash terraces with slope gradients of 0 to 5% in the Wisconsin and Mississippi River valleys; the original prairie vegetation has been replaced by corn, small grains, and hay; mean ann. soil temp. 47°F, mean ann. ppt. 32 in. Proposed, Grant Co., Wis., 1949; Hole, 1956a; Beatty, 1960; Robinson and Klingelhoets, 1961; Soil Conservation Service, 1967a; 75 bu. oats; not suitable for pine unless some forest soil is used as an inoculant; catena no. 24.

Rietbrock sil (F15); Aquic Glossoboralf, fine-loamy, mixed, frigid (Gray-Brown Podzolic); somewhat poorly drained; solum is 36 to 50 in. thick, developed in about 15 in. of silty material over weathered fine-grained granite rock. Horizons: O1, O2, A1, A2, IIB&A, IIB2t, IIB3t, IIC.

Some properties of Rietbrock silt loam

Horizon	Depth in.	OM %	Sand %	Silt %	Clay %	CEC me/ 100g	BS %	pH	BD g/cc
A2	4-13	1.7	15	70	15	10	32	6.3	1.2
IIB2t	18-38	0.4	25	40	35	18	60	6.4	1.5
IIC	47-60	0.1	62	20	18	7	80	6.5	1.8

Other: Most of the IIB horizon is gravelly. Setting: On rock-controlled uplands with slope gradients of 2 to 4%; the original northern mesic forest has been largely replaced by fields of hay, small grains, and silage corn, and by pasture; mean ann. soil temp. 45°F, mean ann. ppt. 30 in. Proposed, Marathon Co., Wis., 1952; 70 bu. oats; 450-500 bd. ft. red or white pine; catena no. 102.

Rifle mucky peat (J13) (Fig. 14-5); Typic Borohemist, euic, frigid (Bog); very poorly drained; solum is 50 in. to 15 ft. thick, developed in herbaceous material overlying glacial drift. Horizons: Oi1, Oi2, Oe1, Oe2, Oe3, Oe4, IIC.

Some properties of Rifle mucky peat

Horizon	Depth in.	OM %	Sand %	Silt %	Clay %	CEC me/ 100g	BS %	pH	BD g/cc	Fibers unrub. %	Fibers rub. %
Oi1	0-2	95	0	3	2	12	15	4.0	0.1	90	60
Oe1	4-8	85	0	9	6	192	19	6.5	0.2	67	15
IIC	60-65	0.5	84	11	5	4	65	6.6	1.8	0	0

Other: Woody fragments occupy less than 15% of the volume of any horizon in the solum. Setting: In bogs of various sizes in glacial drift landscapes; slope gradients, 0 to 2%; the swamp conifer forest cover, with ground cover of sphagnum moss, leatherleaf blueberry, and labrador tea, is undisturbed in most bogs; mean ann. soil temp. 40°F, mean ann. ppt. 29 in. Estab., Ogemaw Co., Mich., 1923; Ableiter and Hole, 1961; not suitable for oats or pine; catena no. 150.

Rimer ls-sl (I14, J7); Arenic Ochraqualf, coarse-loamy, mixed, mesic (Gray-Brown Podzolic); somewhat poorly drained; solum is 28 to 46 in. thick, developed in 20 to 40 in. of sandy material over calcareous silty clay loam and clay. Horizons: O1, O2, A1, A2, IIB2t, IIC.

Some properties of Rimer loamy sand

Horizon	Depth in.	OM %	Sand %	Silt %	Clay %	CEC me/ 100g	BS %	pH	BD g/cc
A2	3-22	1.0	83	10	7	12	70	6.7	1.3
IIB2t	22-30	0.2	30	40	30	14	40	7.0	1.5
IIC	30-60	0.4	7	50	43	21	100	8.5*	1.8

*calcareous

Setting: On glacial lake plains, on slope gradients of 1 to 3%; the original swamp conifer and hardwood forest has been replaced largely by farm crops and pasture; mean ann. soil temp. 45°F, mean ann. ppt. 30 in. Proposed, Waushara Co., Wis., 1942; 45 bu. oats; not suitable for pine; catena no. 91.

Ringwood sil (B5, B22); Typic Argiudoll, fine-loamy, mixed, mesic (Brunizem); well drained; solum is 20 to 40 in. thick, developed in 20 to 36 in. of loess over dolomitic sandy loam glacial till. Horizons: O1, O2, A1, B1, B21t, IIB22t, B3, IIC.

Some properties of Ringwood silt loam

Horizon	Depth in.	OM %	Sand %	Silt %	Clay %	CEC me/ 100g	BS %	pH	BD g/cc
A1	0-9	4.0	14	65	21	17	69	6.6	1.2
B21t	14-23	1.1	6	62	32	21	68	5.6	1.5
IIB2t	23-30	0.3	70	18	12	7	66	6.8	1.5
IIC	35-41	0.1	74	21	5	2	100	8.0*	1.9

*calcareous

Setting: On glacial moraines, with slope gradients of 0 to 6%; the original prairie vegetation has been replaced by fields of corn, small grains, soybeans, and hay, and by pasture; mean ann. soil temp. 47°F, mean ann. ppt. 32 in. Proposed, McHenry Co., Ill., 1953; Soil Conservation Service, 1967a; 70 bu. oats; not suitable for pine unless forest soil is used as an inoculant; catena no. 58.

Ripon sil (B5, B16); Typic Argiudoll, fine-silty, mixed, mesic (Brunizem); well drained; solum is 20 to 42 in. thick, developed in 20 to 36 in of loess over calcareous glacial drift over limestone (dolomite). Horizons: O1, O2, A1, B21t, IIB22t, IIIR.

Some properties of Ripon silt loam

Horizon	Depth in.	OM %	Sand %	Silt %	Clay %	CEC me/ 100g	BS %	pH	BD g/cc
A1	0-11	12.0	12	70	18	11	80	5.7	1.2
B21t	11-32	1.1	8	60	32	15	65	5.4	1.5
IIB22t	32-34	1.0	24	40	36	17	98	7.2	1.7

Setting: On rock-controlled glacial moraines with slope gradients of 1 to 21%; the original prairie vegetation is replaced by fields of hay, corn, soybeans, and small grains, and by pasture; mean ann. soil temp. 47°F, mean ann. ppt. 32 in. Estab., Dodge Co., Wis., 1969; 70 bu. oats; not suitable for pine unless forest soil is used as an inoculant; catena no. 66.

Rockbridge sil (A11, A12); Typic Hapludalf, fine-silty over fragmental, mixed, mesic (Gray-Brown Podzolic); well drained; solum is 24 to 40 in. thick, developed in silty material 15 to 30 in. thick over stratified and mixed cherty gravel and reddish-brown sandy clay loam. Horizons: O1, O2, A1, A2, B1, B21t, IIB22t, IIB3t, IIC.

Some properties of Rockbridge silt loam

Hori-zon	Depth in.	OM %	Sand %	Silt %	Clay %	CEC me/ 100g	BS %	pH	BD g/cc
A2	6-8	1.5	11	70	19	10	75	6.8	1.3
B21t	13-20	0.6	18	50	32	9	55	5.5	1.6
IIC	35-60	0.4	45	10	45	21	80	5.5	1.7

Setting: On high terrace remnants on valley slopes of the Wisconsin River valley and its tributaries; IIC material is apparently ancient colluvium; the original oak savanna and southern mesic forest cover is replaced by fields of hay, small grains, and corn, and by pasture; mean ann. soil temp. 45°F, mean ann. ppt. 30 in. Proposed, Richland Co., Wis., 1947; Robinson and Klingelhoets, 1959; 60 bu. oats; 400-450 bd. ft. red pine; 450-500 bd. ft. white pine; catena no. 31.

Rodman sl-sil (B4) (Figs. 2-53, 2-54, 8-4, 8-9, 11-1);Typic Hapludoll, sandy, skeletal, mixed, mesic (Brunizem); excessively drained; solum is 8 to 15 in thick, developed in highly dolomitic in gravelly sandy loam glacial outwash. Horizons: O1, A1, C.

Some properties of Rodman gravelly sandy loam

Hori-zon	Depth in.	OM %	Sand %	Silt %	Clay %	CEC me/ 100g	BS %	pH	BD g/cc	Coarse fragments† (>2mm) %
A1	0-8	15.0	61	25	14	12	100	7.6*	1.8	58
C	8-30	0.2	95	4	1	1	100	8.2*	2.2	82

*calcareous †on the whole soil basis

Setting: On eskers, kames, and other glacial outwash deposits with slope gradients of 0 to 45%; the original oak savanna remains undisturbed in many areas, but is pastured in others; mean ann. soil temp. 45°F, mean ann. ppt. 31 in. Estab., Jefferson Co., N.Y., 1911; Whitson et al., 1917; Gaikawad and Hole, 1965; Milfred and Hole, 1970; Schmude, 1971; not suitable for oats or pine; catena no. 78.

Roscommon s-ls (J15); Mollic Psammaquent, sandy, mixed, frigid (Low Humic Gley); poorly drained; solum is 4 to 8 in. thick, developed in acid glacial outwash sand. Horizons: O1, O2, A1, Cg.

Some properties of Roscommon loamy sand

Hori-zon	Depth in.	OM %	Sand %	Silt %	Clay %	CEC me/ 100g	BS %	pH	BD g/cc
A1	0-4	12.0	84	10	6	12	80	6.8	1.3
Cg	4-40	0.1	94	4	2	2	90	7.2	1.8

Setting: On glacial outwash plains, on slopes of 0 to 2% gradient; swamp hardwood and conifer cover occupies most of the area; some land has been cleared and used for pasture or for production of hay and small grains; mean ann. soil temp. 40°F, mean ann. ppt. 29 in. Estab., Sanilac Co., Mich., 1955; Milfred, Olson, and Hole, 1967; not suitable for oats or pine; catena no. 147.

Rousseau s-lfs (E6); Entic Haplorthod, sandy, mixed, frigid (Podzol); well drained. Horizons: O1, O2, A1, A2, Bir, B3, C.

Some properties of Rousseau sand

Hori-zon	Depth in.	OM %	Sand %	Silt %	Clay %	CEC me/ 100g	BS %	pH	BD g/cc
A2	1-8	2.0	91	6	3	5	60	5.3	1.3
Bir	8-17	1.8	90	5	5	3	50	5.2	1.5
C	25-60	0.1	95	3	2	1	40	5.7	1.8

Setting: On glacial lake and outwash plains and dunes, with slope gradients of 0 to 30%; the original northern mesic forest was logged off and regrowth forest now replaces it, except for small areas cleared for pasture and hay production; mean ann. soil temp. 42°F, mean ann. ppt. 30 in. Estab., Montcalm Co., Mich., 1956; 50 bu. oats; 450-500 bd. ft. red or white pine; catena no. 155.

Rozellville sl-sil (F14, F15); Typic Glossoboralf, fine-loamy, mixed, frigid (Gray-Brown Podzolic); well drained; solum is 20 to 40 in. thick, developed in less than 15 in. of silty material over weathered micaceous Precambrian igneous rock. Horizons: O1, O2, A1, A2, B&A, IIB2t, IIB3t, IIC.

Some properties of Rozellville silt loam

Hori-zon	Depth in.	OM %	Sand %	Silt %	Clay %	CEC me/ 100g	BS %	pH	BD g/cc
A2	4-6	1.7	17	65	18	14	70	5.7	1.3
IIB2t	9-20	0.8	53	22	25	15	60	5.6	1.7
IIC	24-60	0.1	50	30	20	8	80	5.4	1.9

Other: The A2 tongues into the IIB2 horizon; the percentage by volume of coarse fragments is usually less than 15% in the solum. Setting: On rock-controlled uplands in the Wisconsin River valley near Wausau, on slope gradients of 3 to 6%; the original northern mesic forest has been largely cleared for production of hay, small grains, and silage corn; where bedrock outcrops or stoniness is great, the land is in pasture or woodland; mean ann. soil temp. 44°F, mean ann. ppt. 32 in. Proposed, Marathon Co., Wis., 1955; 75 bu. oats; 475-500 bd. ft. red or white pine; catena no. 103.

Rozetta sil (A5); Typic Hapludalf, fine-silty, mixed, mesic (Gray-Brown Podzolic) moderately well drained; solum is 40 to 55 in. thick, developed in loess more than 50 in. thick over residuum on dolomite. Horizons: O1, O2, A1, A2, B1, B2t, B3, C.

Some properties of Rozetta silt loam

Hori-zon	Depth in.	OM %	Sand %	Silt %	Clay %	CEC me/ 100g	BS %	pH	BD g/cc
A2	4-11	3.0	7	78	15	18	72	5.7	1.3
B2t	14-39	0.5	3	67	30	27	65	5.0	1.5
C	50-80	0.1	10	65	25	20	70	5.4*	1.7

*calcareous at 6 to 10 ft.

Setting: On nearly level ridge tops of rock-controlled uplands in the "Driftless Area," with slope gradients of 0 to 4%; the original oak savanna vegetation has been replaced by fields of hay, small grains, corn, and soybeans, and by pasture; mean ann. soil temp. 47°F, mean ann. ppt. 32 in. Estab., Henderson Co., Ill., 1947; 75 bu. oats; 475-500 bd. ft. red or white pine; catena no. 3.

Rudyard sil-sic (I8, I18); Aquic Eutroboralf, very fine, illitic, frigid (Gray Wooded); somewhat poorly drained; solum is 15 to 32 in. thick, developed in reddish-brown calcareous clay. Horizons: O1, O2, A1, A2, B2t, C.

Some properties of Rudyard silt clay loam

Horizon	Depth in.	OM %	Sand %	Silt %	Clay %	CEC me/100g	BS %	pH	BD g/cc
A1	0-2	10.1	10	50	40	32	85	5.5	1.3
A2	2-5	2.0	0	40	60	22	70	5.4	1.5
B2t	5-10	0.3	5	20	75	35	90	5.6	1.6
C	10-20	0.1	0	20	80	30	100	8.5*	1.7

*calcareous

Other: The A2 tongues into the B2 horizon; the profile is mottled. Setting: On glacial lake plains on slope gradients of 0 to 5%; the original boreal and swamp conifer forest was logged off, and has been largely replaced by new-growth forest; some areas are in pasture; mean ann. soil temp. 40°F, mean ann. ppt. 31 in. Proposed, Michigan, about 1955; Hole et al., 1962; 60 bu. oats; not suitable for pine; catena no. 150.

Ruse sl-sil (E6, E7) (Fig. 11-4); Lithic Haplaquept and Haplaquoll, loamy, mixed, nonacid, frigid (Low Humic Gley); poorly to very poorly drained; solum 10 to 20 in. thick, developed in loamy material of that thickness over limestone. Horizons: O1, O2, A1, A3, B2g, R.

Some properties of Ruse loamy sand

Horizon	Depth in.	OM %	Sand %	Silt %	Clay %	CEC me/100g	BS %	pH	BD g/cc
A1	0-7	18.0	82	10	8	11	92	7.8	1.3
B2g	12-20	1.0	63	25	12	15	98	8.0	1.7

Other: A peat or muck layer 1 to 6 in. thick may be present on the surface. Setting: On nearly level areas of limestone-controlled uplands of the Door Peninsula; swamp conifers still cover most of the area; permanent pasture, in a small acreage; mean ann. soil temp. 44°F, mean ann. ppt. 28 in. Estab., Schoolcraft Co., Mich., 1933; not suitable for oats or pine; catena no. 140.

St. Charles sil (B18, B25, B32, B34) (Figs. 2-53, 2-54, 7-12, 8-13); Typic Hapludalf, fine-silty, mixed, mesic (Gray-Brown Podzolic); well drained; solum is 48 to 55 in. thick, developed in 36 to 50 in. of loess over calcareous loam glacial till; a stratified substratum phase is recognized where glacial outwash replaces glacial till. Horizons: O1, O2, A1, A2, B1, B2t, B31, IIB32, IIC.

Some properties of St. Charles silt loam

Horizon	Depth in.	OM %	Sand %	Silt %	Clay %	CEC me/100g	BS %	pH	BD g/cc
A2	5-13	1.5	5	78	17	12	65	5.8	1.3
B2t	17-35	0.6	4	64	32	21	80	5.6	1.5
IIB32	52-62	0.2	35	42	23	17	92	6.9	1.7
IIC	62-70	0.1	65	30	5	2	100	8.5*	1.8

*calcareous

Other: Depth to carbonates ranges from 45 to more than 70 in. Setting: On glacial moraines with slope gradients of 0 to 12%; the original oak savanna and southern mesic forest have been largely replaced by fields of hay, small grains, corn, and soybeans, and by pasture; mean ann. soil temp. 45°F, mean ann. ppt. 30 in. Estab., Kendall Co., Ill., 1941; see Calamus in Soil Conservation Service, 1967a; Milfred and Hole, 1970; Haszel, 1971; Schmude, 1971; Steingraeber and Reynolds, 1971; 75 bu. oats; 450-500 bd. ft. red pine; 500-600 bd. ft. white pine; catena no. 49.

Salter l-sil (B17) (Figs. 1-4, 11-7); Typic Eutrochrept, coarse-loamy, mixed, mesic (Brown Prodzolic); well drained; solum is 24 to 36 in. thick, developed in calcareous stratified fine sand and silt. Horizons: O1, O2, A1, A2, B2, B3, C.

Some properties of Salter silt loam

Horizon	Depth in.	OM %	Sand %	Silt %	Clay %	CEC me/100g	BS %	pH	BD g/cc
A2	4-7	1.5	15	70	15	12	98	7.2	1.3
B2	7-15	0.5	12	70	18	18	80	8.0	1.5
C	24-40	0.1	58	30	12	5	100	8.5*	1.7

*calcareous

Other: The B is cambic. Setting: On glacial lake and outwash plains with slope gradients of 0 to 20%; the original southern and northern forest covers were logged off; new-growth forest covers considerable areas; other acreages are in hay, corn, canning crops (cabbage, peas, green beans); mean ann. soil temp. 45°F, mean ann. ppt. 30 in. Estab., Washington Co., Wis., 1944; Milfred and Hole, 1970; 65 bu. oats; 450-500 bd. ft. white pine; catena no. 88.

Santiago sil (F2, F10, G22) (Figs. 2-53, 12-4, 12-5, 13-5); Typic Glossoboralf, fine-loamy, mixed, frigid (Gray-Brown Podzolic); well drained; solum is 20 to 40 in. thick, developed in 15 to 25 in. of loess over acid, sandy loam reddish-brown (5YR 4/4) glacial till. Horizons: O1, O2, A1, A2, A&B, B21t, IIB22t, IIB3, IIC.

Some properties of Santiago silt loam

Horizon	Depth in.	OM %	Sand %	Silt %	Clay %	CEC me/100g	BS %	pH	BD g/cc
A1	0-3	5.0	17	75	8	40	54	5.8	1.2
A2	3-11	1.2	17	75	8	9	13	5.4	1.3
B21t	17-21	0.5	23	59	18	14	41	5.0	1.5
IIB22t	21-28	0.2	45	38	17	13	53	5.1	1.6
IIC	32-48	0.1	73	17	10	7	51	5.3	1.8

Other: A fragipan may be present in the lower IIB and IIC horizons; the A2 tongues down into the B2 horizon. Setting: On glacial moraines with slope gradients of 0 to 20%; the original northern mesic forest has been largely replaced by hay, small grains, silage corn; mean ann. soil temp. 43°F, mean ann. ppt. 32 in. Estab., Mille Lacs Co., Minn., 1927; Robinson et al., 1958; Carroll, 1964; Soil Conservation Service, 1967a; Haszel, 1968; 75 bu. oats; 500-575 bd. ft. white pine; catena no. 96.

Sargeant l-sil (F13); Typic Glossaqualf, fine-loamy, mixed, mesic (Humic Gley); poorly drained; solum is 20 to 42 in. thick, developed in less than 30 in. of silty material over a calcareous clay loam glacial till. Horizons: O1, O2, A1, A2, B1, IIB2t, IIB3, IIC.

Some properties of Sargeant silt loam

Hori-zon	Depth in.	OM %	Sand %	Silt %	Clay %	CEC me/100g	BS %	pH	BD g/cc
A1	0-5	14.0	23	65	12	10	60	5.3	1.3
IIB2t	16-28	1.0	25	40	35	15	50	5.2	1.6
IIC	35-42	0.1	35	35	30	12	65*	5.7†	1.8

*at 6 ft., BS = 100 †calcareous at about 6 ft.

Other: A pebble band may be present at the interface between silty horizons and till-derived horizons. Setting: On glacial moraines with slope gradients of 0 to 3%; the original northern mesic forest is largely replaced by woodland pasture, hay, small grains, soybeans, and corn; mean ann. soil temp. 43°F, mean ann. ppt. 30 in. Estab., Mower Co., Minn., 1949; Haszel, 1968; 60 bu. oats; 400-450 bd. ft. white pine; catena no. 106.

Saugatuck ls (H1, H2, H3, H4, H5, H6); Aeric Haplaquod, sandy, mixed, frigid, ortstein (Groundwater Podzol); somewhat poorly drained; solum is 20 to 36 in. thick, developed in acid sandy glacial drift. Horizons: O1, O2, A1, A2, Bhir, Bir, C.

Some properties of Saugatuck sand

Hori-zon	Depth in.	OM %	Sand %	Silt %	Clay %	CEC me/100g	BS %	pH	BD g/cc
A2	8-15	0.8	95	3	2	4	25	5.0	1.3
Bhir	15-23	2.0	89	6	5	9	30	4.8	1.6
C	30-40	0.2	95	3	2	1	20	5.4	1.6

Other: An ortstein present in the Bhir. Setting: On outwash plains and in depressions in sand glacial drift landscapes, on slopes of 0 to 3% gradient; swamp conifer forest cover is widespread; some areas are in cropland; mean ann. soil temp. 41°F, mean ann. ppt. 32 in. Estab., Allegan Co., Mich., 1901; Hole and Schmude, 1959; Ableiter and Hole, 1961; Hole et al., 1962; where artificially drained, 40 bu. oats; 500-600 bd. ft. white pine; catena no. 145.

Sawmill sil (J1); Cumulic Haplaquoll, fine-silty, mixed, mesic (Alluvial soil); poorly drained; solum is 36 to 70 in. thick, developed in calcareous alluvium. Horizons: O1, A1, A3, B2g, B3g, Cg.

Some properties of Sawmill silty clay loam

Hori-zon	Depth in.	OM %	Sand %	Silt %	Clay %	CEC me/100g	BS %	pH	BD g/cc
A1	0-25	12.0	12	58	30	22	80	6.3	1.2
B2g	31-54	2.0	8	60	32	17	92	7.2	1.4
Cg	60-70	1.0	8	62	30	14	100	8.5*	1.6

*calcareous

Other: The profile is mottled. Setting: On valley floors in the "Driftless Area"; swamp hardwoods and sedge meadows occupy considerable area; some land has been drained and is used to produce hay, corn, and soybeans; mean ann. soil temp. 47°F, mean ann. ppt. 32 in. Estab., DeWitt Co., Ill., 1937; where artificially drained, 60 bu. oats; not suitable for pine; catena no. 37.

Saybrook sil (B22); Typic Argiudoll, fine-silty, mixed, mesic (Brunizem); well drained; solum is 24 to 40 in. thick, developed in 20 to 36 in. of loess over dolomitic loam glacial till. Horizons: O1, O2, A1, A3, B2t, IIB3, IIC.

Some properties of Saybrook silt loam

Hori-zon	Depth in.	OM %	Sand %	Silt %	Clay %	CEC me/100g	BS %	pH	BD g/cc
A1	0-11	6.0	3	70	27	25	80	5.8	1.2
B2t	17-28	0.5	3	65	32	19	75	5.6	1.4
IIB3	28-35	0.2	32	40	28	15	80	6.4	1.6
IIC	35-50	0.1	43	35	22	10	100	8.5*	1.8

*calcareous

Setting: On glacial moraines with slope gradients of 2 to 7%; the original prairie vegetation is replaced by hay, small grains, corn, and soybeans; mean ann. soil temp. 47°F, mean ann. ppt. 32 in. Estab., Ford Co., Ill., 1929; 70 bu. oats; not suitable for pine unless inoculated with forest soil; catena no. 50.

Saylesville l-sil (J10) (Figs. 4-9, 11-1); Typic Hapludalf, fine, illitic, mesic (Gray-Brown Podzolic); well to moderately well drained; solum is 20 to 40 in. thick, developed in less than 18 in. of loamy sediment over calcareous brown silts and clays. Horizons: O1, O2, A1, A2, B1t, IIB2t, B3, IIC.

Some properties of Saylesville loam

Hori-zon	Depth in.	OM %	Sand %	Silt %	Clay %	CEC me/100g	BS %	pH	BD g/cc
A1	0-3	5.0	42	36	22	28	98	6.1	1.2
A2	3-12	0.5	64	23	13	16	90	6.2	1.4
IIB2t	16-24	0.8	15	30	55	31	100	7.0	1.6
IIC	29-36	0.4	2	65	33	10	100	8.1*	1.7

*calcareous

Setting: On glacial lake plains with slope gradients of 0 to 5%; the original oak savanna and southern mesic forest cover has been replaced by hay, small grains, corn, and some truck crops; mean ann. soil temp. 47°F, mean ann. ppt. 32 in. Proposed, Jefferson Co., Wis., 1956; Milfred and Hole, 1970; 75 bu. oats; 450-500 bd. ft. white pine; catena no. 85.

Scandia ls-sl (G8); Typic Hapludalf, coarse-loamy, mixed, mesic (Gray-Brown Podzolic); well drained; solum is 20 to 40 in. thick, developed in acid sandy glacial drift. Horizons: O1, O2, A1, A2, B1, B2t, C.

Some properties of Scandia sandy loam

Hori-zon	Depth in.	OM %	Sand %	Silt %	Clay %	CEC me/100g	BS %	pH	BD g/cc
A2	3-9	1.2	61	25	14	12	75	5.8	1.3
B2t	21-27	1.0	52	30	18	11	40	5.4	1.7
C	27-35	0.2	63	25	12	7	75	6.1	1.9

Setting: On glacial moraines with slope gradients of 0 to 30%; the northern mesic forest has been replaced on steep slopes by new-growth mixed forest and on gentler slopes by fields of hay, small grains, and corn; mean ann. soil temp. 42°F, mean ann. ppt. 30 in. Estab., Washington Co., Minn., 1924; 45 bu. oats; 400-450 bd. ft. red or white pine; catena no. 110.

Schapville sil (A4) (Fig. 7-8); Typic Argiudoll, fine, mixed, mesic (Brunizem); moderately well drained; solum is 20 to 35 in. thick, developed in 15 to 30 in. of loess over neutral to calcareous shale. Horizons: O1, O2, A1, A3, B21t, IIB22t, IIC-R.

Some properties of Schapville silt loam

Horizon	Depth in.	OM %	Sand %	Silt %	Clay %	CEC me/100g	BS %	pH	BD g/cc
A1	0-8	6.0	9	65	26	25	80	5.6	1.3
B21t	12-22	0.6	2	60	38	21	90	5.8	1.5
IIB22t	22-25	0.3	0	45	55	26	98	7.2	1.6
IIC-R	25-60	0.1	0	42	58	27	100	8.2*	1.7

*calcareous

Other: Hard limestone is interbedded with the shale in horizon IIC. Setting: On bedrock-controlled ridges with slope gradients of 4 to 12%; the original prairie vegetation is replaced by pasture, hay, small grains, and corn; mean ann. soil temp. 47°F, mean ann. ppt. 32 in. Estab., Jo Daviess Co., Ill., 1941; Haszel, 1968; 70 bu. oats; not suitable for pine; catena no. 12.

Seaton sil (A7, A8, D1) (Figs. 7-4, 7-10); Typic Hapludalf, fine-silty, mixed, mesic (Gray-Brown Podzolic); well drained; solum is 4 to 6 ft. thick, developed in coarse silty loess overlying residuum on dolomite. Horizons: O1, O2, A1, A2, B1, B2t, B3, C.

Some properties of Seaton silt loam

Horizon	Depth in.	OM %	Sand %	Silt %	Clay %	CEC me/100g	BS %	pH	BD g/cc
A1	0-4	4.0	8	81	11	13	57	5.6	1.2
A2	4-9	1.0	7	83	10	7	33	4.8	1.4
B2t	15-34	0.3	7	70	23	16	62	5.0	1.6
C	70-95	0.1	7	73	20	14	58	5.3*	1.5

*calcareous at a depth of 10 or 12 ft.

Setting: On rock-controlled uplands in or near the Mississippi River valley on slope gradients of 2 to 45%; the original oak savanna or southern mesic forest has continued by succession on steep lands, or has been converted to woodland or cleared pasture; gentler slopes are in hay, small grains, corn, and alfalfa; mean ann. soil temp. 47°F, mean ann. ppt. 32 in. Estab., Henderson Co., Ill., 1947; Hole, 1956a; Beatty, 1960; Slota and Garvey, 1961; Robinson and Klingelhoets, 1961; Soil Conservation Service, 1967a; Haszel, 1968; 70 bu. oats; 475-500 bd. ft. red pine; 500-600 bd. ft. white pine; catena no. 5.

Seckler sl-sil (C8); Aquic Argiudoll, fine-loamy over sandy or sandy-skeletal, mixed, mesic (Gray-Brown Podzolic); somewhat poorly drained; solum is 24 to 45 in. thick, developed in 20 to 40 in. of loamy deposits over acid glacial outwash sands. Horizons: O1, O2, A1, A2, B2t, C.

Some properties of Seckler silt loam

Horizon	Depth in.	OM %	Sand %	Silt %	Clay %	CEC me/100g	BS %	pH	BD g/cc
A1	0-7	12.0	20	65	15	13	80	5.7	1.2
B2t	8-23	1.0	14	60	26	14	50	5.4	1.5
C	23-36	0.1	89	7	4	1	90	6.5	1.9

Other: Reddish iron concretions (2.5YR 2/2) are scattered through the solum; the B2 horizon is brown (7.5YR 4/4) in color. Setting: On glacial outwash plains with slope gradients of 0 to 5% in the "Driftless Area"; oak savanna is largely replaced by pasture, hay, small grains, and corn; mean ann. soil temp. 47°F, mean ann. ppt. 32 in. Proposed, Trempealeau Co., Wis., 1942; 45 bu. oats; 450-500 bd. ft. white pine; catena no. 29.

Shawano s-ls (E8, E12, E13, J3) (Figs. 11-5, 11-7); Typic Udipsamment, sandy, mixed, frigid (Brown Podzolic); excessively drained; solum is 14 to 20 in. thick, developed in fine and medium neutral glacial lacustrine sands. Horizons: O1, O2, A1, B, C.

Some properties of Shawano sand

Horizon	Depth in.	OM %	Sand %	Silt %	Clay %	CEC me/100g	BS %	pH	BD g/cc
A1	0-4	3.0	88	7	5	5	90	6.5	1.3
C	24-36	0.1	94	4	2	1	98	8.1*	1.7

*slightly calcareous below 3 ft.

Other: In places a weak Spodosol (Podzol) profile is apparent in the A1-C1 transitional horizon. Setting: On glacial lake plains with slope gradients of 1 to 5%; the northern mesic forest is replaced by new-growth forest, by pine plantations, and by hay; mean ann. soil temp. 44°F, mean ann. ppt. 30 in. Proposed, Shawano Co., Wis., about 1954; 40 bu. oats; 400-450 bd. ft. red pine; 450-500 bd. ft. white pine; catena no. 82.

Sheboygan muck (J15); Hemic Medisaprist, euic, mesic (Bog); very poorly drained; solum is 50 in. to 15 ft. deep over calcareous glacial drift. Horizons: Oa1, Oa2, Oa3, Oa4, IIC.

Some properties of Sheboygan muck

Horizon	Depth in.	OM %	Sand %	Silt %	Clay %	CEC me/100g	BS %	pH	BD g/cc	Fibers unrub. %	Fibers rub. %
Oa1	0-15	53.0	0	27	20	200	30	6.0	0.1	10	5
Oa2	15-24	68.0	0	17	15	150	20	6.5	0.3	32	12
IIC	50-60	0.6	70	18	12	5	90	7.5	1.8	0	0

Setting: In bogs in southeastern Wisconsin; the original herbaceous vegetation is undisturbed in most places; mean ann. soil temp. 44°F, mean ann. ppt. 30 in. Place of proposal, unknown; not suitable for oats or pine; catena no. 56.

Shiffer sl-l (C7); Aquollic Hapludalf, fine-loamy over sandy or sandy-skeletal, mixed, mesic (Gray-Brown Podzolic); somewhat poorly drained; solum is 20 to 40 in. thick, developed in loamy deposits of that thickness over acid sand and gravel glacial outwash. Horizons: O1, O2, A1, A2, B1, B2t, IIB3, IIC.

Some properties of Shiffer loam

Horizon	Depth in.	OM %	Sand %	Silt %	Clay %	CEC me/100g	BS %	pH	BD g/cc
A2	4-13	1.5	40	40	20	15	80	5.7	1.2
B2t	18-24	0.5	37	38	25	12	60	5.3	1.4
IIB3	24-28	0.3	55	30	15	6	65	5.2	1.6
IIC	28-54	0.1	88	8	4	1	70	5.6	1.9

Other: The solum is mottled. Setting: On glacial outwash plains with slope gradients of 0 to 2%; the northern mesic forest is largely replaced by fields of hay, small grains, and corn; irrigation and drainage are both practiced in some areas; mean ann. soil temp. 44°F, mean ann. ppt. 30 in. Proposed, Eau Claire Co., Wis., 1950; 55 bu. oats; 400-450 bd. ft. white pine; catena no. 28.

Shiocton sl-sil (E5, E8, E11, I16) (Figs. 11-5, 11-7); Aquic Eutrochrept, coarse-loamy, mixed, mesic (Brown Podzolic);

somewhat poorly drained; solum is 20 to 30 in. thick. Horizons: O1, O2, A1, A2, A3, B, C.

Some properties of Shiocton sandy loam

Horizon	Depth in.	OM %	Sand %	Silt %	Clay %	CEC me/ 100g	BS %	pH	BD g/cc
A2	5-10	1.3	58	30	12	10	95	6.8	1.3
B	14-22	0.7	16	70	14	8	98	7.5	1.5
C	22-40	0.1	60	30	10	4	10	7.8	1.7

Other: Calcareous at about 40 in. Setting: On glacial lake plains on slopes of 0 to 3% gradient; the southern mesic forest was logged off, and is replaced by new-growth forest, pasture, and, where drained, by fields of hay, small grains, corn, and cabbage; mean ann. soil temp. 44°F, mean ann. ppt. 29 in. Proposed, Winnebago Co., Wis., 1948; 65 bu. oats; 450-500 bd. ft. white pine; catena no. 88.

Shullsburg sil (A4); Aquic Argiudoll, fine, mixed, mesic (Brunizem); somewhat poorly drained; solum is 24 to 30 in. thick, developed in 15 to 30 in. of loess over neutral to calcareous shale. Horizons: O1, O2, A1, A3, B21tg, IIB22t, IIB23t, R.

Some properties of Shullsburg silt loam

Horizon	Depth in.	OM %	Sand %	Silt %	Clay %	CEC me/ 100g	BS %	pH	BD g/cc
A1	0-7	8.0	4	70	26	23	95	6.5	1.2
B21tg	12-16	1.0	2	65	33	17	98	7.2	1.5
IIB22t	16-23	0.5	5	45	50	20	100	8.1	1.6

Other: The profile is mottled; concretions of oxides of iron and manganese are present in the solum. Setting: On rock-controlled uplands on slope gradients of 1 to 3%; the original prairie vegetation is replaced by pasture, hay, small grains, and corn; mean ann. soil temp. 47°F, mean ann. ppt. 32 in. Estab., Stephenson Co., Ill., 1969; 60 bu. oats; not suitable for pine; catena no. 12.

Sisson sl-sil (B16); Typic Hapludalf, fine-loamy, mixed, mesic (Gray-Brown Podzolic); well drained; solum is 24 to 42 in. thick, developed in calcareous fine sand and silt. Horizons: O1, O2, A1, A2, B1, B2t, B3, IIC.

Some properties of Sisson sandy loam

Horizon	Depth in.	OM %	Sand %	Silt %	Clay %	CEC me/ 100g	BS %	pH	BD g/cc
A2	8-11	1.5	56	30	14	12	90	6.5	1.2
B2t	15-27	0.4	14	60	26	12	85	6.5	1.5
IIC	30-60	0.1	5	85	10	4	100	8.5*	1.6

*calcareous

Other: The IIC horizon is calcareous at about 40 in.; fine sand lenses are numerous in the IIC horizon. Setting: On glacial lake plains with slope gradients of 2 to 6%; the northern mesic forest is largely replaced by fields of hay, small grains, corn, soybeans, and by pasture; mean ann. soil temp. 45°F, mean ann. ppt. 29 in. Estab., Lapeer Co., Mich., 1966; 70 bu. oats; 400-450 bd. ft. red pine; 450-500 bd. ft. white pine; catena no. 87.

Skillet sil (A10); Aquic Hapludalf, fine-silty, mixed, mesic (Gray-Brown Podzolic); somewhat poorly drained; solum is 24 to 40 in. thick, developed in the same thickness of loess over quartzite. Horizons: O1, O2, A1, A2, B1, B2t, R.

Some properties of Skillet silt loam

Horizon	Depth in.	OM %	Sand %	Silt %	Clay %	CEC me/ 100g	BS %	pH	BD g/cc
A2	4-13	1.5	17	70	13	11	75	5.5	1.3
B2t	18-36	0.6	10	60	30	14	50	5.2	1.6

Other: A gritty silty clay lower B subhorizon may be present; solum is mottled. Setting: On rock-controlled uplands with slope gradients of 0 to 2% on flats and 2 to 20% at seepage sites on hillsides; oak savanna and southern mesic forest cover has been replaced over considerable areas by pasture and, where drained, by hay, small grains, and corn; mean ann. soil temp. 44°F, mean ann. ppt. 30 in. Proposed, Sauk Co., Wis., 1955; where artificially drained, 65 bu. oats; catena no. 11.

Sogn l-c (A2, A3) (Fig. 7-5); Lithic Haplustoll, loamy, mixed, mesic (Brunizem); excessively drained; solum ranges from 3 to 12 in. in depth, developed in that thickness of silt and clayey residuum over dolomite. Horizons: O1, O2, A1, R.

Some properties of Sogn clay

Horizon	Depth in.	OM %	Sand %	Silt %	Clay %	CEC me/ 100g	BS %	pH	BD g/cc
A1	0-11	12.0	10	30	60	42	100	8.5*	1.6

*calcareous

Other: Outcrops may occupy 25% of the surface area. Setting: On rock-controlled uplands on slope gradients of 2 to 35%; the original oak savanna and prairie have largely been altered by grazing; mean ann. soil temp. 47°F, mean ann. ppt. 32 in. Estab., Goodhue Co., Minn., 1913; Robinson and Klingelhoets, 1961; Watson, 1966; not suitable for oats or pine; catena no. 9.

Solona sl-l (E3, E4, E5); Aquic Eutroboralf, fine-loamy, mixed, frigid (Gray Wooded); somewhat poorly drained; solum is 24 to 30 in. thick, developed in 15 to 30 in. of loamy material over calcareous reddish-brown (5YR 5/4) sandy loam glacial till. Horizons: O1, O2, A1, A2, B1, B2t, C.

Some properties of Solona sandy loam

Horizon	Depth in.	OM %	Sand %	Silt %	Clay %	CEC me/ 100g	BS %	pH	BD g/cc
A2	6-15	1.7	57	30	13	10	90	7.3	1.2
B2t	17-26	1.0	37	28	35	15	95	7.5	1.6
C	26-30	0.1	73	15	12	8	100	8.5*	1.8

*calcareous

Other: Coarse fragments of dolomite occupy less than 15% by volume of the solum. Setting: On moraines on rock-controlled uplands on slope gradients of 0 to 2%; the northern mesic forest has in many areas been cleared and, after some drainage, the soil has been used to produce hay, small grains, and silage corn, or for pasture; mean ann. soil temp. 44°F, mean ann. ppt. 28 in. Proposed, Shawano Co., Wis., 1947; 65 bu. oats; 300-375 bd. ft. white pine; catena no. 136.

Spalding mucky peat (J13); Typic Borohemist, dysic, frigid (Bog); very poorly drained; solum is 50 in. to 15 ft. thick, developed in mixed herbaceous and woody material. Horizons: Oe1, Oe2, Oe3, Oe4, IIC.

Some properties of Spalding mucky peat

Hori-zon	Depth in.	OM %	Sand %	Silt %	Clay %	CEC me/ 100g	BS %	pH	BD g/cc	Fibers unrub. %	Fibers rub. %
Oe1	0-15	73.0	0	16	11	190	50	4.4	0.1	50	30
Oe2	15-35	90.0	0	6	4	150	30	4.4	0.2	55	35
IIC	60-70	0.5	84	10	6	10	75	5.8	1.8	0	0

Other: More than 35% of the fibers in the solum are woody. Setting: In bogs of northern Wisconsin; the original swamp conifer forest is largely undisturbed; mean ann. soil temp. 40°F, mean ann. ppt. 28 in. Estab., Chippewa Co., Mich., 1927; Hole et al., 1962; Milfred, Olson, and Hole, 1967; not suitable for oats or pine; catena no. 96.

Sparta s-lfs (C5, C6, C7, C8) (Figs. 7-4, 16-4); Entic Hapludoll, sandy, mixed, mesic (Brunizem); excessively drained; solum is 24 to 40 in. thick, developed in acid glacial outwash sand. Horizons: O1, O2, A1, A3, B, C.

Some properties of Sparta sand

Hori-zon	Depth in.	OM %	Sand %	Silt %	Clay %	CEC me/ 100g	BS %	pH	BD g/cc
A1	0-11	1.7	91	5	4	3	57	5.5	1.3
C	34-60	0.1	98	1	1	1	10	6.2	1.8

Other: The A1 horizon ranges in thickness from 10 to 24 in. Setting: On glacial outwash plains with slope gradients of 0 to 3% common, and ranging on stablilized dunes to 8%; the original prairie cover has been largely replaced by irrigated fields of hay, small grains, corn, and soybeans, and by severely wind-eroded pastures and oak and jack pine savanna; mean ann. soil temp. 47°F, mean ann. ppt. 32 in. Estab., Monroe Co., Wis., 1923; Hole, 1956a; Robinson and Klingelhoets, 1959; Beatty, 1960; Soil Conservation Service, 1967a; 40 bu. oats; 300-375 bd. ft. white pine; catena no. 33.

Spencer sil (F12, F22); Typic Glossoboralf, fine-silty, mixed, frigid (Gray-Brown Podzolic); moderately well drained; solum is 30 to 40 in. thick, developed in 36 to 50 in. of loess over reddish-brown (5YR 4/4) acid loam glacial till. Horizons: O1, O2, A1, A2, A&B, B2t, B3t, C1, IIC2.

Some properties of Spencer silt loam

Hori-zon	Depth in.	OM %	Sand %	Silt %	Clay %	CEC me/ 100g	BS %	pH	BD g/cc
A1	0-4	12.0	8	76	16	41	48	5.1	1.2
A2	3-13	2.0	8	82	10	10	15	4.8	1.4
B2t	17-28	0.2	19	60	21	17	44	4.9	1.5
IIC2	40-48	0.1	63	21	16	10	62	5.4	1.7

Other: Some stones have worked up from the till into the silty solum. Setting: On glacial moraines with slope gradients of 3 to 7%; the original northern mesic forest is largely replaced by fields of hay, small grains, and silage corn, and by pasture; mean ann. soil temp. 43°F, mean ann. ppt. 30 in. Proposed, Barron Co., Wis., about 1940; Robinson et al., 1958; Soil Conservation Service, 1967a; 80 bu. oats; 400-450 bd. ft. white pine; catena no. 94.

Spinks s-sl (B15) (Fig. 11-5); Psammentic Hapludalf, sandy, mixed, mesic (Gray-Brown Podzolic); excessively drained; solum is 36 to 85 in. thick, developed in calcareous outwash and dune sand. Horizons: O1, O2, A1, A2, B, A2&B2t, C1, C2.

Some properties of Spinks sand

Hori-zon	Depth in.	OM %	Sand %	Silt %	Clay %	CEC me/ 100g	BS %	pH	BD g/cc
A2	4-6	1.0	90	5	5	7	75	6.5	1.5
B2t	22-85*	0.3	93	4	3	5	55	5.8	1.7
C1	85-90	0.1	83	8	9	4	85	7.2	1.8
C2	108-112	0.1	95	3	2	1	100	8.5†	1.9

†calcareous at 9 ft. *The B2t is repetitive, occurring as bands.

Other: The B2 horizon occurs as bands 0.5 to 5 in. thick that are convoluted and branching, separated by A2 horizon of sand. Setting: On outwash plains and dunes, with slope gradients of 0 to 40%; the original oak savanna has been disturbed in places by grazing; mean ann. soil temp. 47°F, mean ann. ppt. 32 in. Proposed, Lapeer Co., Mich., 1954; Milfred and Hole, 1970; 40 bu. oats; 400-450 bd. ft. white pine; catena no. 74.

Stambaugh sil (F5, F16, F24, G2, G16, G17, G25, G27) (Figs. 2-54, 12-9); Alfic Fragiorthod, coarse-loamy, mixed, frigid (Podzol); well-drained; solum is 20 to 40 in. thick, developed in the same depth of silty sediment over acid glacial outwash sand and gravel. Horizons: O1, O2, A1, A2, Bhir, Bir, A'2x, B'2x, IIC.

Some properties of Stambaugh silt loam

Hori-zon	Depth in.	OM %	Sand %	Silt %	Clay %	CEC me/ 100g	BS %	pH	BD g/cc
A2	0-3	2.5	15	71	14	6	40	5.5	1.4
Bhir	3-7	2.1	17	67	16	18	20	5.0	1.1
A'2x	9-16	1.0	27	60	13	6	20	5.0	1.9
B'2x	16-25	0.3	20	62	18	7	50	5.5	1.8
IIC	25-55	0.1	87	10	3	2	55	6.2	1.8

Setting: On glacial outwash plains with slope gradients of 0 to 20%; the northern mesic forest was logged off and is replaced by new-growth forest, and fields of hay, small grains, and silage corn; mean ann. soil temp. 40°F, mean ann. ppt. 32 in. Estab., Iron Co., Mich., 1930; Hole et al., 1962; 75 bu. oats; 500-550 bd. ft. red or white pine; catena no. 141.

Stronghurst sil (A3, A6, A9); Aeric Ochraqualf, fine-silty, mixed, mesic (Gray-Brown Podzolic); somewhat poorly drained; solum is 3.5 to 5 ft. thick, developed in deep loess over residuum on dolomite. Horizons: O1, O2, A1, A2, B1, B2t, B3, C.

Some properties of Stronghurst silt loam

Hori-zon	Depth in.	OM %	Sand %	Silt %	Clay %	CEC me/ 100g	BS %	pH	BD g/cc
A2	4-11	1.7	6	80	14	13	70	5.6	1.3
B2t	15-35	0.8	3	66	31	25	60	4.7	1.6
C	47-60	0.1	10	75	15	20	68	5.7*	1.7

*calcareous at 9 ft. and pH = 8.5

Other: The profile is mottled. Setting: On rock-controlled upland ridges, on slope gradients of 0 to 5%; the original oak savanna forest has been replaced by fields of hay, small grains, corn, and soybeans, and by pasture; mean ann. soil temp. 47°F, mean ann. ppt. 32 in. Estab., Henderson Co., Ill., 1947; Hole, 1956a; Robinson and Klingelhoets, 1961; 70 bu. oats; not suitable for pine; catena no. 3.

Suamico muck (E6); Terric Borosaprist, clayey, euic, frigid (Bog); very poorly drained; solum is 16 to 50 in. thick, developed in herbaceous material over clayey substratum. Horizons: Oa1, Oa2, Oa3, IIC.

Some properties of Suamico muck

Horizon	Depth in.	OM %	Sand %	Silt %	Clay %	CEC me/100g	BS %	pH	BD g/cc	Fibers unrub. %	Fibers rub. %
Oa1	0-10	53.0	0	27	20	130	15	4.0	0.1	15	5
Oa2	10-25	68.0	0	17	15	125	20	4.5	0.2	20	8
IIC	36-42	0.5	78	14	8	5	60	5.8	1.8	0	0

Setting: In bogs in northern Wisconsin; the original sphagnum and leatherleaf vegetative cover is undisturbed in most areas; mean ann. soil temp. 39°F, mean ann. ppt. 30 in. Place of proposal unknown; not suitable for oats or pine; catena no. 140.

Summerville sl-sil (E6, I3); Entic, Lithic Haplorthod, loamy, mixed, frigid (Podzol); well to moderately well drained; solum is 10 to 20 in. thick, developed in the same depth of glacial drift, and loess and some residuum over dolomite. Horizons: O1, O2, A1, Bhir, Bir, R.

Some properties of Summerville loam

Horizon	Depth in.	OM %	Sand %	Silt %	Clay %	CEC me/100g	BS %	pH	BD g/cc
A1	0-2	12.0	45	40	15	12	95	8.0	1.3
Bhir	5-10	2.0	40	43	17	8	98	8.0	1.4

Other: A dark reddish-brown silty clay layer may be present immediately over the dolomite. Setting: On nearly level rock-controlled upland; the northern mesic forest cover was largely logged off and has been replaced by regrowth forest and by pasture; mean ann. soil temp. 44°F, mean ann. ppt. 28 in. Estab., Alpena Co., Mich., 1924; not suitable for oats; 400-450 bd. ft. white pine; catena no. 140.

Superior sl-l (I1, I9, I18) (Fig. 14-5); Alfic Haplorthod, coarse-loamy over clayey, mixed, frigid (Gray Wooded); moderately well drained; solum is 18 to 30 in. thick, developed in 10 to 20 in. of loamy deposit over calcareous reddish-brown clay. Horizons: O1, O2, A1, A2, Bhir, Bir, A′2, IIB′2t, B3, IIC.

Some properties of Superior silt loam

Horizon	Depth in.	OM %	Sand %	Silt %	Clay %	CEC me/100g	BS %	pH	BD g/cc
A1	0-2	15.0	18	70	12	19	65	5.3	1.2
A2	2-5	5.0	18	72	10	13	60	5.2	1.4
Bhir	5-10	4.0	17	65	18	11	60	5.3	1.3
A′2	15-16	0.5	30	60	10	5	48	5.3	1.5
IIB′2t	16-22	0.6	6	52	42	20	75	6.5	1.6
IIC	28-35	0.1	10	35	55	18	100	8.5*	1.7

*calcareous

Setting: On glacial lake plains with slope gradients of 0 to 12%; the boreal and northern mesic forests were logged off and are replaced by new-growth forest, by fields of hay, oats, and potatoes, and by pasture; mean ann. soil temp. 40°F, mean ann. ppt. 32 in. Estab., Alger Co., Mich., 1904; Ableiter and Hole, 1961; Hole et al., 1962; 60 bu. oats; 400-450 bd. ft. red pine; 450-500 bd. ft. white pine; catena no. 151.

Tama sil (A1) (Figs. 1-5, 2-53, 2-54, 4-6, 7-5, 7-6, 7-7, 7-8); Typic Argiudoll, fine-silty, mixed, mesic (Brunizem); well drained; solum is 36 to 60 in. thick, developed in deep loess over residuum on dolomite. Horizons: O1, O2, A1, B1, B2t, B3t, C.

Some properties of Tama silt loam

Horizon	Depth in.	OM %	Sand %	Silt %	Clay %	CEC me/100g	BS %	pH	BD g/cc
A1	0-14	6.5	12	60	28	26	60	5.7	1.0
B2t	18-32	1.0	11	57	32	25	89	5.0	1.3
C	45-60	0.1	12	73	15	20	100	5.4*	1.6

*at 6 to 10 ft., 8.5 and calcareous

Other: Bulk density is higher in cultivated areas. Setting: On bedrock-controlled upland ridges in the "Driftless Area"; the original prairie vegetation has been replaced by hay, small grains, corn, soybeans, and pasture; mean ann. soil temp. 47°F, mean ann. ppt. 32 in. Estab., Black Hawk Co., Ia., 1917; Hole, 1956a; Robinson and Klingelhoets, 1961; Baxter and Hole, 1967; Slota, 1969; Bouma and Hole, 1971; 75 bu. oats; not suitable for pine unless forest soil has been used as an inoculant; catena no. 1.

Tawas muck (E6); Terric Borosaprist, sandy, euic, frigid (Bog); very poorly drained; solum is 16 to 50 in. thick, developed in mixed herbaceous and woody material of the same depth, overlying sandy glacial drift. Horizons: Oa1, Oa2, Oa3, IIC.

Some properties of Tawas muck

Horizon	Depth in.	OM %	Sand %	Silt %	Clay %	CEC me/100g	BS %	pH	BD %	Fibers unrub. %	Fibers rub.
Oa1	0-7	85.0	0	8	7	178	15	6.5	0.1	15	5
Oa2	7-20	97.0	0	2	1	132	24	6.8	0.2	20	10
IIC	30-42	0.2	74	16	10	4	65	7.1	1.8	0	0

Other: At least 35% of the fibers of the solum are of wood. Setting: In bogs of northern Wisconsin; the original swamp conifer and hardwood forest is undisturbed except for a small area used for pasture; mean ann. soil temp. 42°F, mean ann. ppt. 31 in. Estab., Sanilac Co., Mich., 1955; not suitable for oats or pine; catena no. 145.

Tedrow s-ls (B4); Aquic Udipsamment, sandy, mixed, mesic (Regosol); somewhat poorly drained; solum is 24 to 54 in. thick, developed in deep outwash and dune sands. Horizons: O1, O2, A1, B2, B3, C.

Some properties of Tedrow sand

Horizon	Depth in.	OM %	Sand %	Silt %	Clay %	CEC me/100g	BS %	pH	BD g/cc
A1	0-8	3.0	91	5	4	5	90	6.4	1.3
B2	8-31	0.4	89	6	5	3	85	6.2	1.6
C	33-50	0.1	94	3	3	1	98	8.5*	1.8

*calcareous

Other: Depth to carbonate is the same as thickness of solum; the profile is mottled. Setting: On glacial lake beaches, swales in dunes, outwash plains on slope gradients of 0 to 4%; the original oak savanna is still present in about a quarter of the area, the rest being devoted to corn, soybeans, small grains, and hay; mean ann. soil temp. 46°F, mean ann. ppt. 31 in. Estab., Paulding Co., O., 1957; 45 bu. oats; 250-325 bd. ft. white pine; catena no. 81.

Tell sil (A11, A12); Typic Hapludalf, fine-silty over sandy or sandy-skeletal, mixed, mesic (Gray-Brown Podzolic); well drained; solum is 24 to 40 in. thick, developed in silty material of the same thickness, over acid glacial outwash sand and gravel. Horizons: O1, O2, A1, A2, B1, B2t, IIB3, IIC.

Some properties of Tell silt loam

Horizon	Depth in.	OM %	Sand %	Silt %	Clay %	CEC me/100g	BS %	pH	BD g/cc
A2	4-14	1.5	10	75	15	12	78	5.7	1.3
B2t	18-28	0.6	5	70	25	14	62	5.4	1.5
IIC	32-40	0.1	92	5	3	3	80	5.0	1.8

Setting: On glacial outwash terraces with slope gradients of 0 to 8%; the original oak savanna has been largely replaced by hay, small grains, and corn; mean ann. soil temp. 47°F, mean ann. ppt. 30 in. Estab., Richland Co., Wis., 1940; Robinson and Klingelhoets, 1959; Beatty, 1960; 65 bu. oats; 400-450 bd. ft. red pine; 450-500 bd. ft. white pine; catena no. 26.

Terril fsl-l (J1); Cumulic Hapludoll, fine-loamy, mixed, mesic (Alluvial soil); well drained; solum is 3 to 4.5 ft. thick, developed in loam alluvium. Horizons: O1, A1, B, C.

Some properties of Terril loam

Horizon	Depth in.	OM %	Sand %	Silt %	Clay %	CEC me/100g	BS %	pH	BD g/cc
A1	0-24	8.0	40	40	20	22	90	6.4	1.1
B	31-60	4.0	34	36	30	18	95	7.2	1.5
C	60-70	0.1	33	42	25	12	98	7.5	1.7

Setting: In drainageways in the "Driftless Area"; slope gradients are 0 to 9%; the original prairie vegetation is largely replaced by hay, small grains, corn, and pasture; mean ann. soil temp. 47°F, mean ann. ppt. 32 in. Estab., Dickinson Co., Ia., 1944; Haszel, 1968; 70 bu. oats; not suitable for pine; catena no. 40.

Theresa sil (B12, B17, B24) (Figs. 8-4, 8-7, 8-8); Typic Hapludalf, fine-loamy, mixed, mesic (Gray-Brown Podzolic); well drained; solum is 24 to 40 in. thick, developed in 20 to 30 in. of loess over highly calcareous (about 48% $CaCO_3$ equiv.) gravelly loam glacial till. Horizons: O1, O2, A1, A2, B1, B21t, IIB22t, IIB3t, IIC.

Some properties of Theresa silt loam

Horizon	Depth in.	OM %	Sand %	Silt %	Clay %	CEC me/100g	BS %	pH	BD g/cc
A2	3-9	1.5	26	60	14	8	90	6.8	1.3
IIB3t	29-35	0.6	39	23	38	20	85	6.5	1.6
IIC	35-40	0.1	55	32	13	5	100	8.5*	1.9

*calcareous; $CaCO_3$ equiv. is 55%.

Other: Very dark brown (10YR 2/2) clay films and organic stains occur on ped surfaces of the IIB2 horizon; till appears "fluffy" and yellowish. Setting: On glacial moraines and drumlins with slope gradients of 0 to 12%; the original southern mesic forest is largely replaced by fields of hay, small grains, and corn, and by pasture; mean ann. soil temp. 47°F, mean ann. ppt. 32 in. Proposed, Dodge Co., Wis., 1955; Milfred and Hole, 1970; Link and Demo, 1970; Parker, Kurer, and Steingraeber, 1970; Schmude, 1971; Steingraeber and Reynolds, 1971; 65 bu. oats; 300-350 bd. ft. red pine; 350-400 bd. ft. white pine; catena no. 62.

Toddville sil (A11); Typic Argiudoll, fine-silty, mixed, mesic (Brunizem); moderately well drained; solum is 40 to 50 in. thick, developed in 36 to 50 in. of silty material over acid glacial sand and gravel outwash. Horizons: O1, O2, A1, A3, B1, B2t, C1, IIC2.

Some properties of Toddville silt loam

Horizon	Depth in.	OM %	Sand %	Silt %	Clay %	CEC me/100g	BS %	pH	BD g/cc
A1	0-10	14.0	4	70	26	23	88	6.8	1.2
B2t	20-34	1.0	4	65	31	18	70	6.4	1.5
C1	34-45	0.2	14	68	18	8	90	5.6	1.7
IIC2	45-50	0.1	92	5	3	1	100	8.5*	1.8

*calcareous

Setting: On glacial outwash terraces in the "Driftless Area"; slope gradients are 0 to 3%; the original prairie vegetation is largely replaced by hay, small grains, and corn; mean ann. soil temp. 47°F, mean ann. ppt. 32 in. Estab., La Crosse Co., Wis., 1957; Slota and Garvey, 1961; Thomas, Carroll, and Wing, 1962; 70 bu. oats; not suitable for pine; catena no. 24

Trempe s-ls (C5, C8); Entic Hapludoll, sandy, mixed, mesic (Brunizem); well drained; solum is 18 to 24 in. thick, developed in acid reddish-brown (5YR 4/6) alluvium and outwash. Horizons: O1, O2, A1, A3, C.

Some properties of Trempe loamy sand

Horizon	Depth in.	OM %	Sand %	Silt %	Clay %	CEC me/100g	BS %	pH	BD g/cc
A1	0-14	3.0	82	10	8	11	80	5.4	1.4
C	20-30	0.1	91	5	4	2	88	5.3	1.8

Setting: On high bottoms and outwash terraces in the "Driftless Area"; the original prairie vegetation is replaced by corn, small grains, hay, and pasture; mean ann. soil temp. 47°F, mean ann. ppt. 32 in. Estab., La Crosse Co., Wis., 1957; Beatty, 1960; Thomas, Carroll, and Wing, 1962; 40 bu. oats; 400-450 bd. ft. red pine; 450-500 bd. ft. white pine; catena no. 30.

Trempealeau sl-sil (C5, C8); Typic Argiudoll, fine-loamy over sandy or sandy-skeletal, mixed, mesic (Brunizem); well drained; solum is 20 to 40 in. thick, developed in the same thickness of loamy material over reddish (5YR 4/8) acid glacial outwash sand. Horizons: O1, O2, A1, A3, B2t, IIB3, IIC.

Some properties of Trempealeau loam

Horizon	Depth in.	OM %	Sand %	Silt %	Clay %	CEC me/100g	BS %	pH	BD g/cc
A1	0-7	11.0	40	40	20	23	85	5.7	1.3
B2t	11-20	1.0	39	37	24	16	50	5.4	1.5
IIB3	20-26	0.6	45	25	30	14	60	5.5	1.7
IIC	26-60	0.1	92	5	3	1	65	5.6	1.8

Setting: On sandy outwash terraces in the "Driftless Area"; the original prairie vegetation has been replaced by hay, small grains, and corn; mean ann. soil temp. 45°F, mean ann. ppt. 30 in. Estab., Trempealeau Co., Wis., 1927; Beatty, 1960; 70 bu. oats; 450-500 bd. ft. red pine; 475-550 bd. ft. white pine; catena no. 29.

Troxel sil (F13); Typic Argiudoll, fine-silty, mixed, mesic (Brunizem); well to moderately well drained; solum is 4 to 9 ft. thick, developed in silty colluvium. Horizons: O1, O2, A1, A3, B2t, B3, IIC.

Some properties of Troxel silt loam

Horizon	Depth in.	OM %	Sand %	Silt %	Clay %	CEC me/100g	BS %	pH	BD g/cc
A1	0-33	15.0	15	65	20	30	85	5.7	1.2
IIB22t	53-74	3.0	7	60	33	20	90	5.6	1.5
IIIC1	100-104	0.1	87	8	5	1	100	8.5*	1.9

*calcareous

Setting: On outwash deposits in lowlands in glacial drift landscapes of southeastern Wisconsin on slope gradients of 0 to 2%; the original prairie vegetation has been replaced by hay, small grains, corn, and soybeans; mean ann. soil temp. 45°F, mean ann. ppt. 31 in. Estab., Kendall Co., Ill., 1941; Haszel, 1971; 75 bu. oats; not suitable for pine unless forest soil is introduced as an inoculant; catena no. 92.

Tustin ls-sl (E8, E11, I13, I14, I17, I21, J7) (Fig. 11-5); Arenic Hapludalf, fine, mixed, mesic (Gray-Brown Podzolic); well drained; solum is 30 to 50 in. thick, formed in 20 to 40 in. of sandy sediment over calcareous clayey glacial lake deposits. Horizons: O1, O2, A1, A2, B1, IIB2t, IIC.

Some properties of Tustin loamy sand

Horizon	Depth in.	OM %	Sand %	Silt %	Clay %	CEC me/100g	BS %	pH	BD g/cc
A2	3-12	1.0	82	10	8	5	88	5.7	1.3
IIB2t	30-40	0.3	8	50	42	22	70	6.3	1.6
IIC	40-60	0.1	7	55	38	19	100	8.5*	1.9

*calcareous

Setting: On glacial lake plains on slope gradients of 1 to 8%; the original northern mesic forest was logged off and has been replaced by fields of hay, small grains, and silage corn, and by pasture and woodlots; mean ann. soil temp. 43°F, mean ann. ppt. 30 in. Proposed, Waushara Co., Wis., 1942; Peck and Lee, 1961; 45 bu. oats; 450-500 bd. ft. red or white pine; catena no. 91.

Underhill sl-sil (E4, E9); Typic Eutroboralf, fine-loamy, mixed, frigid (Gray Wooded); well drained; solum is 45 to 55 in. thick, developed in more than 30 in. of loamy material over reddish-brown (2.5YR 3/4) calcareous sandy loam to loam glacial till. Horizons: O1, O2, A1, A2, Bhir, Bir, A′2&IIBt, IIB′2t, IIC.

Some properties of Underhill sandy loam

Horizon	Depth in.	OM %	Sand %	Silt %	Clay %	CEC me/100g	BS %	pH	BD g/cc
A1	0-1	4.5	61	32	7	10	40	5.1	1.2
A2	1-6	2.0	54	38	8	4	28	5.3	1.3
Bhir	6-12	1.5	54	37	9	5	17	5.4	1.4
A′2	16-21	0.2	68	24	8	3	25	5.7	1.7
IIB′2t	39-48	0.2	66	15	19	7	98	7.0	1.8
IIC	48-66	0.1	63	27	10	5	100	8.0*	1.8

*calcareous

Other: The A′2 and IIB′2 are fragipans; the upper one tongues down into the lower one. Setting: On glacial moraines with slope gradients of 1 to 20%; the northern mesic forest has been logged off in most places and replaced by new-growth forest, by fields of hay and small grains, and by pasture; mean ann. soil temp. 41°F, mean ann. ppt. 29 in. Proposed, Oconto Co., Wis., 1958; Milfred, Olson, and Hole, 1967; 70 bu. oats; 400-450 bd. ft. red or white pine; catena no. 135.

Urne fsl-sil (D1, D2); Dystric Eutrochrept, coarse-loamy, mixed, mesic (Lithosol); excessively drained; solum is 24 to 30 in. thick, the thickness of loose material over indurated glauconitic sandstone. Horizons: O1, O2, A1, B, C, R.

Some properties of Urne sandy loam

Horizon	Depth in.	OM %	Sand %	Silt %	Clay %	CEC me/100g	BS %	pH	BD g/cc
A1	0-2	5.0	70	20	10	8	88	6.6	1.3
B	2-24	0.5	66	22	12	7	92	6.8	1.5
C	24-30	0.1	69	20	11	6	98	7.6	1.7

Other: The C horizon is olive brown (2.5Y 4/4). Setting: On rock-controlled ridges and valley slopes, with slope gradients of 12 to 40%; the original oak savanna has been replaced by mixed deciduous forest and some fields of hay and small grains; mean ann. soil temp. 45°F, mean ann. ppt. 30 in. Estab., Buffalo Co., Wis., 1955; Thomas, Carroll, and Wing, 1962; Thomas, 1964; 55 bu. oats; 400-450 bd. ft. red or white pine; catena no. 19.

Varna l-sil (B9, B20) (Figs. 2-53, 4-5, 8-10); Typic Argiudoll, fine, illitic, mesic (Brunizem); moderately well drained; solum is 20 to 42 in. thick, formed in less than 20 in. of silty material over calcareous silty clay loam to clay loam till. Horizons: O1, O2, A1, A3, B21t, IIB22t, IIB3t, IIC.

Some properties of Varna silty clay loam

Horizon	Depth in.	OM %	Sand %	Silt %	Clay %	CEC me/100g	BS %	pH	BD g/cc
A1	0-8	7.0	11	52	37	27	73	6.3	1.3
B21t	12-15	2.1	13	47	40	20	76	6.7	1.6
IIB22t	15-30	1.5	9	38	53	21	80	7.4	1.7
IIC	40-48	0.1	9	49	42	10	100	8.0*	1.8

*calcareous

Other: The profile is mottled below 24 in. Setting: On glacial moraines with slope gradients of 3 to 12% in southeastern Wisconsin; the original prairie vegetation has been replaced by hay, small grains, corn, and soybeans; mean ann. soil temp. 47°F, mean ann. ppt. 32 in. Estab., Marshall Co., Ill., 1934;

Soil Conservation Service, 1967a; Link and Demo, 1970; 70 bu. oats; not suitable for pine; catena no. 53.

Veedum sil (D13); Typic Humaquept, fine-loamy, mixed, acid, frigid (Low Humic Gley); poorly drained; solum is 30 to 45 in. thick, developed in 20 to 36 in. of silty material over acid, shaly sandstone. Horizons: O1, O2, A1, A2g, IIB2g, IIIC.

Some properties of Veedum silt loam

Hori-zon	Depth in.	OM %	Sand %	Silt %	Clay %	CEC me/100g	BS %	pH	BD g/cc
A1	0-10	12.0	28	60	12	9	50	5.0	1.0
IIB2g	16-35	0.5	10	55	35	15	30	4.7	1.5
IIIC	35-60	0.1	23	45	32	12	25	4.5	1.7

Other: The A1 is a mucky silt loam; below the C horizon is stratified fine-grained sandstone and silty shale. Setting: On rock-controlled lowlands with slope gradients of 0 to 2%; the original vegetation was swamp hardwoods and sedge meadows; where drainage has been introduced, hay and small grains and pasture are present; mean ann. soil temp. 42°F, mean ann. ppt. 30 in. Proposed, Wood Co., Wis., 1942; 60 bu. oats; not suitable for pine; catena no. 14.

Vesper sil (D13) (Fig. 10-8); Humic Haplaquept, fine-loamy, mixed, acid, frigid (Low Humic Gley); poorly drained; solum is 30 to 40 in. thick, developed in 10 to 30 in. of loess over acid shaly sandstone. Horizons: O1, O2, A1, A2, IIB21g, IIIB22g, IVCg.

Some properties of Vesper silt loam

Hori-zon	Depth in.	OM %	Sand %	Silt %	Clay %	CEC me/100g	BS %	pH	BD g/cc
A2	6-16	1.5	24	65	11	8	60	5.2	1.3
IIB21g	16-25	0.7	46	40	14	6	35	5.3	1.5
IIIB22g	25-34	0.2	13	45	42	18	30	4.5	1.6
IVCg	34-65	0.1	89	7	4	1	70	4.2	1.8

Setting: On rock-controlled lowlands with slope gradients of 0 to 2% in north-central Wisconsin; the original vegetation was swamp conifers and hardwoods; this is replaced in some areas by pasture, and, where drained, by fields of hay and small grains; mean ann. soil temp. 43°F, mean ann. ppt. 30 in. Estab., Wood Co., Wis., 1915; Whitson et al., 1918; 65 bu. oats; not suitable for pine; catena no. 14.

Vilas s-ls (G3, G4, G5, G7, G11, G18, G20, G25, G27, H1, H2, H3, H4, H5, H6) (Figs. 2-53, 2-54, 14-5); Entic Haplorthod, sandy, mixed, frigid (Podzol); excessively drained; solum is 18 to 36 in. thick, developed in deep, acid glacial drift. Horizons: O1, O2, A1, A2, Bhir, Bir, B3, C.

Some properties of Vilas loamy sand

Hori-zon	Depth in.	OM %	Sand %	Silt %	Clay %	CEC me/100g	BS %	pH	BD g/cc
A1	0-1	3.4	83	12	5	15	44	5.0	1.3
A2	1.4	0.8	87	10	3	8	23	4.5	1.4
Bhir	4-11	1.2	86	9	5	9	20	5.0	1.4
C	25-60	0.1	96	2	2	1	30	6.0	1.8

Other: A somewhat coherent layer (incipient fragipan) may occur between the B and C horizons. Considerable gravel is present in parts of Douglas Co. Setting: On glacial drift uplands with slope gradients of 1 to 20%; the original northern mesic forest was logged off and is replaced by new-growth forest and by pasture; mean ann. soil temp. 40°F, mean ann. ppt. 29 in. Estab., Iron Co., Mich., 1930; Hole and Schmude, 1959; Ableiter and Hole, 1961; Hole et al., 1962; Milfred, Olson, and Hole, 1967; 35 bu. oats; 400-450 bd. ft. red or white pine; catena no. 146.

Virgil sil (J8); Mollic Ochraqualf, fine-silty, mixed, mesic (Gray-Brown Podzolic intergrading to Brunizem); somewhat poorly drained; solum is 40 to 60 in. thick, developed in 36 to 50 in. of loess over calcareous loam glacial till. Horizons: O1, O2, A1, A2, B1, B2t, B31t, IIB32t, IIC.

Some properties of Virgil silt loam

Hori-zon	Depth in.	OM %	Sand %	Silt %	Clay %	CEC me/100g	BS %	pH	BD g/cc
A2	5-13	1.8	4	70	26	21	70	5.4	1.3
B2t	17-44	1.0	13	55	32	18	80	5.2	1.5
IIB32t	49-58	0.5	40	30	30	15	92	6.8	1.7
IIC	58-60	0.1	70	18	12	5	100	8.1*	1.8

*calcareous below 45 or 70 in.

Other: The profile is mottled; iron-manganese concretions are present in the B2 horizon. Setting: On glacial moraines with slope gradients of 0 to 7%; the original savanna vegetation is replaced by hay, small grains, corn, soybeans, and pasture; mean ann. soil temp. 30°F, mean ann. ppt. 31 in. Estab., Kendall Co., Ill., 1941; Schmude, 1971; Steingraeber and Reynolds, 1971; 65 bu. oats; not suitable for pine; catena no. 48.

Vlasaty l-sil (A4); Glossaquic Hapludalf, fine-loamy, mixed, mesic (Gray-Brown Podzolic); moderately well drained; solum is 40 to 60 in. deep, developed in 12 to 24 in. of loess over calcareous loam or clay loam glacial till. Horizons: O1, O2, A1, A2, B1, IIB2t, IIB3, IIC.

Some properties of Vlasaty silt loam

Hori-zon	Depth in.	OM %	Sand %	Silt %	Clay %	CEC me/100g	BS %	pH	BD g/cc
A2	4-9	1.5	70	16	11	75	5.7	1.2	
IIB2t	17-35	0.5	27	40	33	14	80	5.6	1.5
IIC	42-50	0.1	38	38	24	10	100	8.1*	1.8

*calcareous

Other: A stone line may be present at the top of the IIB2 horizon. Setting: On glacial moraines with slope gradients of 1 to 4%; the oak savanna vegetation has been largely replaced by hay, small grains, corn, and soybeans; mean ann. soil temp. 44°F, mean ann. ppt. 29 in. Estab., Dodge Co., Minn., 1956; Haszel, 1968; 70 bu. oats; 400-450 bd. ft. red pine; 450-500 bd. ft. white pine; catena no. 106.

Wallace ls-s (H1, H2, H3, H4, H5, H6, H7) (Fig. 1-7); Typic Haplorthod, sandy, mixed, frigid (Podzol); well drained; solum is 40 to 50 in. thick, developed in acid outwash sand. Horizons: O1, O2, A1, A2, Bhirm, Bhir, B3, C.

Some properties of Wallace loamy sand

Hori-zon	Depth in.	OM %	Sand %	Silt %	Clay %	CEC me/100g	BS %	pH	BD g/cc
A2	1-9	2.5	79	15	6	6	17	4.1	1.5
Bhirm	9-21	0.5	77	11	12	12	29	5.3	1.2
C	46-86	0.1	96	2	2	2	37	5.4	1.6

Other: An ortstein is present just below the A2 horizon; this soil may have started as an Aquod, but with dewatering of the landscape by natural and artificial drainage improvement, has become well drained. Setting: On low-lying benches adjacent to bogs; largely in forests, but in places cleared for use as pastureland and cropland (corn for silage, small grains, hay); mean ann. soil temp. 43F, mean ann. ppt. 30 in. Estab., Menominee Co., Mich., 1925; Lambert, 1970; Lambert and Hole, 1971; formerly called AuTrain, as described by Gaikawad and Hole in Hole et al., 1962; catena no. 167.

Wakefield l-sil (G23); Alfic Fragiorthod, fine-loamy, mixed, frigid (Podzol); well drained; solum is 24 to 36 in. thick, developed in the same thickness of loamy material over acid sandy clay loam to loam glacial till. Horizons: O1, O2, A2, Bhir, Bir, A′2&B′t, B′2tx, C.

Some properties of Wakefield loam

Horizon	Depth in.	OM %	Sand %	Silt %	Clay %	CEC me/ 100g	BS %	pH	BD g/cc
A2	1-3	2.3	44	40	16	24	70	4.9	1.3
Bhir	3-7	3.2	42	40	18	16	50	5.0	1.6
B'2tx	28-34	0.5	47	32	21	10	40	4.9	1.8
C	34-40	0.1	48	25	27	12	80	6.2	1.8

Other: A fragipan may be present below the Bhir horizon. Setting: On glacial moraines with slope gradients of 2 to 30%; the original northern mesic forest was logged off, and is replaced by new-growth forest, by fields of hay and small grains, and by pasture; mean ann. soil temp. 40°F, mean ann. ppt. 28 in. Estab., Gogebic Co., Mich., 1952; Hole et al., 1962; 70 bu. oats; 400-450 bd. ft. red pine; 450-500 bd. ft. white pine; catena no. 130.

Wallkill sil (J1) (Fig. 8-7); Thapto-Histic Fluvaquent, fine-loamy, mixed, mesic (Alluvial soil); very poorly drained; solum is 15 to 40 in. thick, the depth of local alluvial mineral material overlying a buried peat or muck. Horizons: O1, A1, IIOab.

Some properties of Wallkill silt loam

Horizon	Depth in.	OM %	Sand %	Silt %	Clay %	CEC me/ 100g	BS %	pH	BD g/cc
A1	0-30	5.0	10	70	20	17	85	6.5	1.3

Setting: In depressions at the edges of peat bogs; vegetation is swamp hardwoods and sedge meadows except where drained for production of hay, small grains, corn, and potatoes; mean ann. soil temp. 45°F, mean ann. ppt. 30 in. Estab., Orange Co., N.Y., 1912; Thomas, Carroll, and Wing, 1962; Link and Demo, 1970; Schmude, 1971; Haszel, 1971; Steingraeber and Reynolds, 1971; 60 bu. oats; not suitable for pine; catena no. 93.

Warsaw sl-sil (B32);Typic Argiudoll, fine-loamy over sandy or sandy-skeletal, mixed, mesic (Brunizem); well drained; solum is 20 to 42 in. thick, developed in less than 20 in. of silty material over calcareous sand and gravel glacial outwash. Horizons: O1, O2, A1, B2t, IIC.

Some properties of Warsaw loam

Horizon	Depth in.	OM %	Sand %	Silt %	Clay %	CEC me/ 100g	BS %	pH	BD g/cc
A1	0-12	12.0	40	40	20	25	80	6.5	1.3
B2t	12-36	1.0	38	38	24	15	70	6.2	1.6
IIC	36-42	0.1	87	10	3	1	100	8.5*	1.9

*calcareous

Other: The B2 tongues down into the IIC horizon; a dark beta B overlies the IIC Horizon. Setting: On glacial outwash deposits with slope gradients of 0 to 6%; the original prairie vegetation is replaced by hay, small grains, corn, soybeans, and pasture; mean ann. soil temp. 45°F, mean ann. ppt. 31 in. Estab., Kosciusko Co., Ind., 1922; Link and Demo, 1970; Milfred and Hole, 1970; Haszel, 1971; Steingraeber and Reynolds, 1971; 60 bu. oats; not suitable for pine without inoculant from forest soil; catena no. 75.

Washtenaw sil (J1); Typic Haplaquent, fine-loamy, mixed, nonacid, mesic (Alluvial soil); very poorly and poorly drained; solum is 20 to 40 in. thick, the thickness of overwash deposits on a buried Argiaquoll silt loam. Horizons: O1, O2, A1, C, IIA1b, IIB2gtb, IICb.

Some properties of Washtenaw silt loam

Horizon	Depth in.	OM %	Sand %	Silt %	Clay %	CEC me/ 100g	BS %	pH	BD g/cc
A1	0-8	4.0	5	75	20	17	90	6.5	1.3
C	8-24	0.5	8	70	22	12	95	6.4	1.6

Setting: On slope gradients of 0 to 6% at the edges of low-lying bodies of very poorly drained mineral soils; vegetation is swamp hardwoods and sedge meadows except where drained for production of corn, soybeans, small grains, and hay, and for pasture; mean ann. soil temp. 44°F, mean ann. ppt. 30 in. Estab., Washtenaw Co., Mich., 1930; 65 bu. oats; not suitable for pine; catena no. 92.

Waukegan sil (A14, F8); Typic Hapludoll, fine-silty over sandy or sandy-skeletal, mixed, mesic (Brunizem); well drained; solum is 20 to 42 in. thick, developed in 20 to 40 in. of silty sediment over calcareous sand and gravel glacial outwash. Horizons: O1, O2, A1, A3, B2, B31, IIB32, IIC.

Some properties of Waukegan silt loam

Horizon	Depth in.	OM %	Sand %	Silt %	Clay %	CEC me/ 100g	BS %	pH	BD g/cc
A1	0-12	12.0	6	70	24	28	85	6.5	1.2
B2	15-30	1.0	9	65	26	16	90	6.3	1.6
IIC	38-45	0.1	86	10	4	1	100	8.5*	1.8

*calcareous

Setting: On glacial outwash plains and terraces with slope gradients of 0 to 6%; the original prairie vegetation is replaced by corn, soybeans, small grains, and hay; mean ann. soil temp. 44°F, mean ann. ppt. 30 in. Estab., Washington Co., Minn., 1940; Beatty, 1960; Slota and Garvey, 1961; Thomas, Carroll, and Wing, 1962; Haszel, 1968; 65 bu. oats; not suitable for pine unless forest soil is introduced as an inoculant; catena no. 111.

Wauseon sl-l (E9, J7); Typic Haplaquoll, coarse-loamy over clayey, mixed, mesic (Humic Gley); poorly drained; solum is 18 to 40 in. thick, the thickness of sandy deposits over calcareous, clayey glacial drift. Horizons: O1, O2, A1, B1g, IIB2tg, IICg.

Some properties of Wauseon sandy loam

Horizon	Depth in.	OM %	Sand %	Silt %	Clay %	CEC me/ 100g	BS %	pH	BD g/cc
A1	0-11	10.0	78	12	10	15	90	6.5	1.3
IIB2tg	26-30	0.5	2	46	52	23	98	7.2	1.5
IICg	30-60	0.1	20	30	50	21	100	8.5*	1.8

*calcareous

Setting: In depressions in glacial lake plains on slopes of 0 to 2% gradient; the original cover of swamp hardwoods is largely replaced by drained fields of hay, small grains, corn, and soybeans; mean ann. soil temp. 45°F, mean ann. ppt. 30 in. Estab., Fulton Co., O., 1922; Peck and Lee, 1961; 60 bu. oats; not suitable for pine; catena no. 91.

Whalan sl-sil (B6, B7); Typic Hapludalf, fine-loamy, mixed, mesic (Gray-Brown Podzolic); well drained; solum is 20 to 42 in. thick, developed in less than 20 in. of silts over glacial till on limestone. Horizons: O1, O2, A1, A2, B1, B21t, IIB22t, IIC, IIIR.

Some properties of Whalan silt loam

Horizon	Depth in.	OM %	Sand %	Silt %	Clay %	CEC me/ 100g	BS %	pH	BD g/cc
A2	4-9	1.5	17	65	18	15	80	6.2	1.3
B21t	12-17	0.5	23	45	32	18	78	6.5	1.5
IIB22t	17-22	0.3	42	20	38	19	90	7.0	1.7
IIC	22-24	0.1	70	18	12	5	100	8.5*	1.8

*calcareous

Other: Pockets of dark IIB22 interpenetrate into the till and limestone. Setting: On rock-controlled morainic uplands with slope gradients of 2 to 15%; the southern mesic forest has been in many places replaced by cropped fields and pasture. Proposed, Fillmore Co., Minn., 1940; Haszel, 1968; Milfred and Hole, 1970; 60 bu. oats; 300-350 bd. ft. red or white pine; catena no. 69.

Wien sil (F21); Mollic Ochraqualf, fine-loamy, mixed, frigid (Low Humic Gley); poorly drained; solum is 30 to 36 in. deep, developed in 15 to 30 in. of silty material over acid clay loam glacial till. Horizons: O1, O2, A1, A2g, B2tg, B3tg, Cg.

Some properties of Wien silt loam

Horizon	Depth in.	OM %	Sand %	Silt %	Clay %	CEC me/ 100g	BS %	pH	BD g/cc
A2g	7-15	6.0	10	70	20	15	80	5.6	1.3
B2tg	15-27	0.6	22	40	38	20	60	5.4	1.6
Cg	34-40	0.1	27	35	38	21	55	5.5	1.8

Setting: Depressions in Precambrian bedrock-controlled morainic upland; swamp hardwoods are present except where drainage has permitted production of hay and small grains, and use as pasture; mean ann. soil temp. 42°F, mean ann. ppt. 31 in. Proposed, Marathon Co., Wis., 1942; 60 bu. oats; not suitable for pine; catena no. 98.

Will sl-sil (B31, J9); Typic Haplaquoll, fine-loamy over sandy or sandy-skeletal, mixed, mesic (Humic Gley); poorly drained; solum is 20 to 42 in. thick, the thickness of silty deposits over calcareous sand and gravel glacial outwash. Horizons: O1, O2, A1, B1g, B2g, IIC.

Some properties of Will silt loam

Horizon	Depth in.	OM %	Sand %	Silt %	Clay %	CEC me/ 100g	BS %	pH	BD g/cc
A1	0-10	20.0	34	40	26	30	88	6.4	1.3
B2g	18-35	1.0	33	40	27	15	92	7.0	1.6
IIC	35-45	0.1	88	8	4	1	100	8.5*	1.9

*calcareous

Setting: In depressions on glacial outwash plains; the natural cover of swamp hardwoods and sedge meadows is replaced, where drainage has been done, by hay, small grains, and corn, and by pasture; mean ann. soil temp. 45°F, mean ann. ppt. 31 in. Estab., Will Co., Ill., 1951; Milfred and Hole, 1970; 65 bu. oats; not suitable for pine; catena no. 76.

Withee sil (F11, F21) (Figs. 2-53, 2-54, 10-8, 12-8); Aeric Glossaqualf, fine-loamy, mixed, frigid (Gray-Brown Podzolic); somewhat poorly drained; solum is 24 to 48 in. thick, developed in 15 to 30 in. of silty material over acid heavy loam glacial till. Horizons: O1, O2, A1, A2g, IIB&A, IIB2t, IIB3t, IIC.

Some properties of Withee silt loam

Horizon	Depth in.	OM %	Sand %	Silt %	Clay %	CEC me/ 100g	BS %	pH	BD g/cc
A1	0-4	5.0	12	72	16	14	29	5.0	1.2
A2g	4-11	1.0	14	74	12	9	22	4.9	1.5
IIB2t	24-36	0.4	49	28	23	16	46	4.0	1.8
IIC	40-60	0.1	55	27	18	14	60	4.5	1.9

Other: The profile is mottled; some stones have been moved up from the IIC material into the silty upper solum. Setting: On glacial moraines with slope gradients of 1 to 3%; the original northern mesic forest was logged off, and is replaced by fields of hay, small grains, and corn, by pasture, and by new-growth forest; mean ann. soil temp. 43°F, mean ann. ppt. 30 in. Proposed, Clark Co., Wis., 1957; Soil Conservation Service, 1967a; 70 bu. oats; 450-500 bd. ft. white pine; catena no. 97.

Worthen sil (J1); Cumulic Hapludoll, fine-silty, mixed, mesic (Alluvial soil); well drained; solum is 30 to 42 in. thick, developed in alluvium. Horizons: O1, A1, A3, B2, B3, C.

Some properties of Worthen silt loam

Horizon	Depth in.	OM %	Sand %	Silt %	Clay %	CEC me/ 100g	BS %	pH	BD g/cc
A1	0-17	6.0	5	75	20	22	90	6.3	1.2
B2	20-29	1.2	7	72	21	16	95	6.4	1.6
C	37-45	0.1	15	70	15	8	98	6.7	1.7

Setting: In floodplains of small valleys in the "Driftless Area"; the original prairie vegetation has been replaced by hay, small grains, corn, and pasture; mean ann. soil temp. 47°F, mean ann. ppt. 32 in. Estab., Jackson Co., Ill., 1929; Watson, 1966; Haszel, 1968; Link and Demo, 1970; 70 bu. oats; not suitable for pine; catena no. 36.

Wyeville ls-l (C18, E10) (Fig. 11-6); Aquic Arenic Hapludalf, coarse-loamy over clayey, mixed, mesic (Gray-Brown Podzolic); somewhat poorly drained; solum is 20 to 40 in. thick, developed in sandy deposits of nearly the same thickness over acid clayey glacio-lacustrine deposits. Horizons: O1, O2, A1, A2, IIB2t, IIC.

Some properties of Wyeville loamy sand

Horizon	Depth in.	OM %	Sand %	Silt %	Clay %	CEC me/100g	BS %	pH	BD g/cc
A2	5-24	1.2	83	10	7	6	65	5.8	1.3
IIB2t	24-30	0.7	12	50	38	17	50	5.5	1.6
IIC	30-35	0.1	6	52	42	20	60	5.2	1.7

Setting: In depressions in old lake plains; swamp hardwoods are present except where artificial drainage has permitted production of hay, small grains, and corn; mean ann. soil temp. 43°F, mean ann. ppt. 30 in. Place of proposal unknown; 55 bu. oats; not suitable for pine; catena no. 123.

Wykoff sl-sil (G8); Typic Hapludalf, fine-loamy over sandy-skeletal, mixed, mesic (Gray-Brown Podzolic); excessively drained; solum is 15 to 30 in. thick, developed in sandy glacial drift. Horizons: O1, O2, A1, A2, B1, B2t, B3t, C.

Some properties of Wykoff loam

Horizon	Depth in.	OM %	Sand %	Silt %	Clay %	CEC me/100g	BS %	pH	BD g/cc
A2	5-12	1.4	42	40	18	21	60	5.5	1.3
B2t	19-30	0.6	51	26	23	15	50	5.2	1.6
C	35-40	0.1	82	10	8	2	55	5.4	1.8

Setting: On glacial moraines with slope gradients of 0 to 20%; the original oak savanna cover has been replaced by hay, small grains, and corn; mean ann. soil temp. 45°F, mean ann. ppt. 30 in. Proposed, Fillmore Co., Minn., about 1950; Haszel, 1968; 60 bu. oats; 450-500 bd. ft. red or white pine; catena no. 109.

Wyocena ls-sl (C2, C3, G6, G15, G24) (Fig. 9-4); Typic Hapludalf, coarse-loamy, mixed, mesic (Gray-Brown Podzolic); well drained; solum is 24 to 40 in. thick, developed in weakly calcareous loamy sand glacial till. Horizons: O1, O2, A1, A2, B1, B2t, B3t, C.

Some properties of Wyocena loamy sand

Horizon	Depth in.	OM %	Sand %	Silt %	Clay %	CEC me/100g	BS %	pH	BD g/cc
A2	4-10	1.0	83	10	7	8	85	6.4	1.4
B2t	15-31	0.4	68	20	12	7	90	6.2	1.6
C	36-54	0.1	87	8	5	1	100	8.5*	1.8

*calcareous

Setting: On glacial moraines with slope gradients of 2 to 25%; oak savanna has been replaced by new-growth hardwood forest, pasture, and fields of hay, small grains, and corn; mean ann. soil temp. 43°F, mean ann. ppt. 30 in. Proposed, Columbia Co., Wis., 1942; Peck and Lee, 1961; 55 bu. oats; 450-500 bd. ft. red pine; 475-500 bd. ft. white pine; catena no. 73.

Zittau l-sicl (J11); Aquollic Hapludalf, fine over sandy or sandy-skeletal, mixed, mesic (Gray-Brown Podzolic); somewhat poorly drained; solum is 24 to 36 in. thick, the depth of silty and clayey material over calcareous sand and gravel outwash. Horizons: O1, O2, A1, A2, A3, B2t, B3t, IIC.

Some properties of Zittau silty clay loam

Horizon	Depth in.	OM %	Sand %	Silt %	Clay %	CEC me/100g	BS %	pH	BD g/cc
A2	5-7	1.6	10	60	30	21	70	5.8	1.3
B2t	13-22	0.6	7	50	43	20	80	5.6	1.6
IIC	30-40	0.1	88	7	5	2	100	8.5*	1.9

*calcareous

Setting: Nearly level glacial lake plains; the original southern mesic forest is largely replaced by fields of hay, small grains, and corn, and by pasture; mean ann. soil temp. 43°F, mean ann. ppt. 30 in. Proposed, Winnebago Co., Wis., about 1955; 65 bu. oats; not suitable for pine; catena no. 89.

Zwingle sil-sicl (A11, A12); Typic Albaqualf, fine, mixed, mesic (Gray-Brown Podzolic); somewhat poorly drained; solum is 24 to 36 in. thick, developed in 15 to 30 in. of silty sediment over calcareous lacustrine clays. Horizons: O1, O2, A1, A2, B1, IIB2t, IIC.

Some properties of Zwingle silt loam

Horizon	Depth in.	OM %	Sand %	Silt %	Clay %	CEC me/100g	BS %	pH	BD g/cc
A2	4-12	1.4	14	60	26	22	80	5.4	1.3
IIB2t	16-30	0.4	10	55	35	16	75	6.3	1.6
IIC	30-35	0.1	8	50	42	18	100	8.5*	1.7

*calcareous

Setting: In slight depressions on stream terraces in the Wisconsin and Mississippi River valleys; the original hardwood forest has been largely replaced by fields of hay, small grains, and corn; mean ann. soil temp. 47°F, mean ann. ppt. 32 in. Place of proposal unknown; 65 bu. oats; not suitable for pine; catena no. 35.

CHAPTER **18**

Major Soil Toposequences of Wisconsin

Soil bodies lie on Wisconsin landscapes in repeating patterns. Table 18-1 shows relationships of soils in terms of topographic sequences, called soil toposequences or soil catenas (soil chains). Toposequence number 52 in the table is an example of a complete catena. This means that at the top of a hill one might expect to find a well drained Miami soil body, and at progressively lower positions down-slope, the moderately well drained Celina, the somewhat poorly drained Conover, the poorly drained Brookston, the very poorly drained Kokomo (all silt loams to silty clay loams), and the very poorly drained Carlisle muck (an organic soil). Except for the Carlisle, all of these soils formed from a loess covering and an underlying calcareous glacial till, that may also underlie the Carlisle muck. This sequence of soils may repeat itself up one hill and down the next.

Not all toposequences are "complete." Toposequences numbers 1, 2, and 3 have three, two, and four members, respectively. In southwestern Wisconsin, where wetlands are relatively few and far apart, many toposequences consist of only one soil series, as is the case with the Norden series (number 18 in Table 18-1). This is hardly a sequence at all, but by convention is placed in the table, nevertheless. It is a record of the fact that soil bodies have developed from this geologic material only in naturally well drained landscape positions. The grouping of soils into toposequences has helped to make landscapes much more understandable in terms of ecosystems and land use.

As soil series names are changed (see Chapter 20), the nomenclature of toposequences will change.

Table 18-1. Major soil toposequences (catenas of soil series) of Wisconsin (after Lee et al., 1968)

(Symbols after soil series names have the following meaning in terms of natural drainage conditions: 1 = well drained; 2 = moderately well drained; 3 = somewhat poorly drained; 4 = poorly drained; 5 = very poorly drained. Absence of symbol signifies excessive drainage.)

Catena no.	Members
1	Tama,^{1} Muscatine,^{3} Garwin^{4}
2	Downs,^{1} Atterberry^{3}
3	Fayette,^{1} Rozetta,^{2} Stronghurst,^{3} Sable^{4}
4	Port Byron,^{1} Joy^{3}
5	Seaton,^{1} Decorra^{2}
6	Lamont
7	Ashdale^{1}
8	Palsgrove^{1}
9	Sogn, Dubuque,^{1} Norwalk^{2}
10	Dodgeville^{1}
11	Baraboo,^{1} Skillet^{3}
12	Schapville,^{1} Shullsburg,^{3} Calamine^{5}
13	Derinda^{1}
14	Hiles,^{1} Kert,^{3} Veedum,^{4} Vesper^{5}
15	Humbird,^{1} Merrillan,^{3} Elm Lake^{4}
16	Gale^{1}
17	Northfield^{1}
18	Norden^{1}
19	Urne
20	Hesch^{1}
21	Hixton^{1}
22	Boone
23	Chelsea
24	Richwood,^{1} Toddville,^{2} Rowley,^{3} Marshan^{4}
25	Bertrand,^{1} Jackson,^{2} Curran^{3}
26	Tell^{1}
27	Dakota^{1}
28	Meridian,^{1} Shiffer^{2}
29	Trempealeau,^{1} Seckler^{3}
30	Trempe^{1}
31	Rockbridge^{1}
32	Hubbard
33	Sparta
34	Plainfield, Nekoosa,^{2} Morocco,^{3} Newton,^{4} Dawson^{5} (frigid equivalents: Sartell, Friendship,^{2} Lino,^{3} Isanti^{4})
35	Medary,^{1} Zwingle,^{3} Denrock,^{3} Perrot^{4}
36	Worthen,^{1} Littleton,^{3} Millington^{4}
37	Chaseburg,^{1} Osseo,^{3} Sawmill^{4}
38	Huntsville,^{1} Lawson^{3}
39	Arenzville,^{1} Orion^{3}
40	Terril,^{1} Kickapoo^{1}
41	Akan,^{3} Ettrick^{4}
42	Boaz^{3}
43	Ogle^{1}
44	Flagg^{1}
45	Durand^{1}
46	Pecatonica,^{1} Mingo^{3}
47	Plano,^{1} Elburn,^{3} Pella^{4}
48	Batavia,^{1} Virgil^{3}
49	St. Charles,^{1} Kendall,^{3} Lena^{5}
50	Saybrook,^{1} Lisbon,^{3} Palms^{5}
51	Dodge,^{1} Mayville,^{2} Lamartine,^{3} Houghton,^{5} Adrian^{5}
52	Miami,^{1} Celina,^{2} Conover,^{3} Brookston,^{4} Kokomo,^{5} Carlisle^{5}
53	Varna,^{1} Elliott,^{3} Ashkum^{4}
54	Morley,^{1} Blount^{3}
55	Ozaukee,^{1} Mequon^{3}
56	Kewaunee,^{1} Manawa,^{3} Poygan,^{4} Sheboygan^{5}
57	Hortonville^{1}
58	Ringwood^{1}
59	McHenry^{1}
60	Pardeeville^{1}
61	Lapeer^{1}
62	Theresa^{1}
63	Hochheim,^{1} Nenno^{3}
64	Lomira^{1}
65	LeRoy^{1}
66	Ripon^{1}
67	Knowles^{1}
68	Rockton^{1}
69	Whalan^{1}
70	Puchyan^{1}
71	Metea^{1}
72	Mecan^{1}
73	Wyocena^{1}
74	Spinks
75	Warsaw^{1}
76	Fox,^{1} Matherton,^{3} Will^{4}
77	Casco,^{1} Fabius^{3}
78	Rodman
79	Oshtemo^{1}
80	Boyer^{1}
81	Oakville, Tedrow,^{3} Granby,^{4} Maumee^{5}
82	Leeman, Shawano^{1}
83	Oshkosh^{1}
84	Briggsville^{1}

Catena no.	Members
85	Saylesville,[1] Del Rey,[3] Montgomery[4]
86	Zurich[1]
87	Sisson,[1] Kibbie,[3] Colwood[4]
88	Salter,[1] Shiocton,[3] Keowns[4]
89	Borth,[1] Zittau,[3] Poy[4]
90	Hebron,[1] Mosel,[3] Aztalan,[3] Navan[5]
91	Tustin,[1] Berrien,[2] Rimer,[3] Wauseon[4]
92	Troxel,[1] Radford,[3] Washtenaw[4]
93	Juneau,[1] Pistakee,[3] Wallkill[4]
94	Otterholt,[1] Spencer,[2] Almena,[3] Auburndale,[4] Adolph,[5] Cable[5]
95	Jewett[1]
96	Santiago,[1] Freeon,[2] Freer,[3] Spalding[5]
97	Loyal,[1] Withee,[3] Marshfield[4]
98	Cassel,[3] Wien[4]
99	Norrie[1]
100	Kennan,[1] Greenwood[5]
101	Marathon[1]
102	Fenwood,[1] Rietbrock[3]
103	Rozellville,[1] Mead (Meadland)[3]
104	Rudolph,[1] Dolph,[3] Altdorf[4]
105	Ostrander,[1] Floyd[3]
106	Renova,[1] Vlasaty,[2] Skyberg,[3] Sargeant[4]
107	Cushing,[1] Alstad[3]
108	Arland[1]
109	Wykoff[1]
110	Scandia
111	Waukegan,[1] Rib[4]
112	Antigo,[1] Brill,[2] Poskin[3]
113	Onamia[1]
114	Burkhardt[1]
115	Chetek,[1] Crown[3]
116	Richford[1]
117	Kellner[1]
118	"Dalbo,"[1] Dalbo,[3] Bluffton (Brickton)[4]
119	Campia,[1] Crystal Lake,[2] Comstock,[3] Barronett[4]
120	Alban[1]
121	Bevent[1]
122	Braham,[1] Blomford[4]
123	Delton,[1] Wyeville,[3] Wautoma[4]
124	Goodman[1]
125	Iron River,[1] Monico[3]
126	Ahmeek[1]
127	Lafont,[1] Clifford[3]
128	Gogebic[1]
129	Munising[1]
130	Wakefield[1]
131	Marenisco[1]
132	Cloquet
133	Hibbing,[1] Selkirk,[3] Pickford[4]
134	Leonidas[1]
135	Underhill,[1] Angelica,[4] Carbondale[5]
136	Onaway,[1] Solona,[3] Lupton[5]
137	Emmet[1]
138	Kolberg[1]
139	Longrie,[1] Bonduel[3]
140	Summerville,[1] Detour,[4] Ruse,[5] Suamico[5]
141	Stambaugh[1]
142	Padus[1]
143	Pence[1]
144	Kiva
145	Omega, Saugatuck,[3] Kinross,[4] Tawas[5]
146	Vilas, Au Gres[3]
147	Hiawatha,[1] Roscommon[4]
148	Alpena
149	Guenther,[1] Dancy[4]
150	Ontonagon,[1] Rudyard,[3] Pickford,[4] Bergland,[5] Rifle[5]
151	Superior[1]
152	Manistee,[1] Ogemaw (Allendale)[3]
153	Bibon,[1] Orienta[3]
154	Fence[1]
155	Rousseau[1]
156	Brule[1]
157	Caryville[1]
158	DePere,[1] Stinson[3]
159	Edith
160	Elkmound
161	Ewen[1]
162	Gotham
163	Milaca[1]
164	Neda,[2] Ashippun[3]
165	Ankeny
166	Keltner[1]
167	Wallace

CHAPTER **19**

Acreages of Wisconsin Soil Associations

Table 19-1. Acreages of Wisconsin soil associations, as listed in the legend of the soil map (Plate 1)

Soil Region A

Map symbol	Acres	Square miles	Percent of land in state
A1	282,300	441	0.8
A2	197,300	308	0.6
A3	263,400	412	0.7
A4	32,000	50	0.1
A1-A4	775,000	1,211	2.2
A5	657,500	1,027	1.8
A6	1,194,500	1,866	3.4
A7	214,000	335	0.6
A8	6,600	10	0.2
A9	447,000	698	1.3
A10	34,500	55	0.1
A5-A10	2,554,100	3,991	7.4
A11	102,900	161	0.3
A12	150,050	234	0.4
A13	21,500	34	0.1
A14	177,200	277	0.5
A11-A14	451,650	706	1.3
A1-A14	**3,780,750**	**5,908**	**10.9**

Soil Region B

Map symbol	Acres	Square miles	Percent of land in state
B1	10,100	15	0.1
B2	69,075	110	0.2
B3	24,075	38	0.1
B4	143,800	225	0.4
B1-B4	247,050	386	0.8
B5	139,600	218	0.4
B6	31,580	49	0.1
B7	151,550	237	0.4
B8	89,300	140	0.3
B9	31,280	49	0.1
B10	55,750	87	0.2
B11	128,710	201	0.4
B12	225,600	353	0.6
B13	303,640	474	0.7
B14	71,400	112	0.2
B15	170,840	267	0.5
B16	10,200	16	0.1
B17	98,800	154	0.3
B18	24,900	39	0.1
B5-B18	1,533,150	2,396	4.4

Soil Region B, *continued*

Map symbol	Acres	Square miles	Percent of land in state
B19	313,945	491	0.9
B20	127,050	199	0.4
B21	24,497	38	0.1
B22	353,050	552	1.0
B23	138,290	215	0.4
B24	489,040	764	1.3
B25	198,615	310	0.6
B26	74,960	117	0.2
B27	113,500	177	0.3
B28	67,825	106	0.2
B29	49,105	77	0.1
B30	284,735	445	0.8
B31	81,770	128	0.3
B19-B31	2,316,382	3,619	6.6
B32	197,250	308	0.5
B33	35,415	55	0.1
B34	55,320	86	0.2
B32-B34	287,985	449	0.8
B1-B34	**4,384,567**	**6,850**	**12.6**

Soil Region C

Map symbol	Acres	Square miles	Percent of land in state
C1	49,700	78	0.1
C2	216,150	338	0.6
C3	186,375	291	0.5
C4	237,864	372	0.7
C5	261,653	409	0.8
C6	187,380	292	0.5
C7	64,650	101	0.2
C8	64,650	101	0.2
C9	66,310	104	0.2
C1-C9	1,334,732	2,086	3.8
C10	212,500	332	0.6
C11	49,630	78	0.1
C12	39,884	62	0.1
C13	68,360	107	0.2
C14	54,270	85	0.2
C15	102,065	159	0.3
C16	35,025	55	0.1
C17	521,207	814	1.5
C18	55,000	86	0.2
C10-C18	1,137,941	1,778	3.3
C1-C18	**2,472,673**	**3,864**	**7.1**

Soil Region D

Map symbol	Acres	Square miles	Percent of land in state
D1	1,333,700	2,084	3.9
D2	360,595	563	1.0
D3	69,510	109	0.2
D4	377,400	590	1.1
D5	195,710	306	0.6
D6	109,080	170	0.3
D7	149,645	234	0.4
D1-D7	2,595,640	4,056	7.5

Soil Region D, *continued*

Map symbol	Acres	Square miles	Percent of land in state
D8	113,815	178	0.3
D9	74,175	116	0.2
D10	90,515	141	0.3
D8-D10	278,505	435	0.8
D11	55,051	86	0.2
D12	116,920	183	0.3
D13	223,500	349	0.6
D11-D13	395,471	618	1.1
D1-D13	**3,269,616**	**5,109**	**9.4**

Soil Region E

Map symbol	Acres	Square miles	Percent of land in state
E1	65,260	102	0.2
E2	235,275	367	0.7
E1-E2	300,535	469	0.9
E3	92,570	145	0.3
E4	439,777	686	1.1
E5	134,380	210	0.4
E6	199,570	312	0.6
E7	14,068	22	0.1
E3-E7	880,365	1,375	2.5
E8	85,117	133	0.3
E9	42,375	66	0.1
E10	28,920	45	0.1
E11	72,368	113	0.2
E12	64,113	100	0.2
E13	111,150	174	0.3
E8-E13	404,043	631	1.2
E1-E13	**1,584,943**	**2,475**	**4.6**

Soil Region F

Map symbol	Acres	Square miles	Percent of land in state
F1	139,070	217	0.4
F2	51,363	80	0.2
F3	131,520	206	0.4
F4	142,805	223	0.4
F5	420,000	656	1.2
F6	105,800	165	0.3
F7	68,360	107	0.2
F8	47,140	74	0.1
F1-F8	1,106,058	1,728	3.2
F9	175,706	275	0.5
F10	547,925	855	1.6
F11	295,040	461	0.8
F12	284,980	445	0.8
F13	174,680	273	0.5
F14	316,900	495	0.9
F15	60,080	94	0.2
F16	15,700	25	0.1
F17	73,240	114	0.2
F9-F17	1,944,251	3,037	5.6

Soil Region F, *continued*

Map symbol	Acres	Square miles	Percent of land in state
F18	393,100	614	1.1
F19	112,960	177	0.3
F20	426,660	667	1.3
F21	595,000	930	1.7
F22	351,338	549	1.0
F23	80,550	126	0.2
F24	69,266	108	0.2
F25	338,345	529	1.0
F26	38,270	60	0.1
F18-F26	2,405,489	3,760	6.9
F1-F26	**5,455,798**	**8,525**	**15.7**

Soil Region G

Map symbol	Acres	Square miles	Percent of land in state
G1	111,105	174	0.3
G2	239,475	374	0.7
G3	87,340	136	0.3
G4	206,725	323	0.7
G5	116,210	182	0.3
G6	49,215	77	0.1
G7	81,600	128	0.2
G8	50,510	79	0.1
G1-G8	942,180	1,473	2.7
G9	83,140	130	0.2
G10	111,275	174	0.3
G11	828,564	1,294	2.4
G12	113,742	178	0.3
G13	413,423	645	1.2
G14	329,970	516	0.9
G15	76,618	120	0.2
G16	135,602	212	0.4
G17	63,575	99	0.2
G18	157,510	246	0.5
G19	91,200	143	0.3
G9-G19	2,404,619	3,757	6.9
G20	472,990	739	1.4
G21	388,926	608	1.1
G22	278,800	436	0.8
G23	332,335	519	1.0
G24	70,195	110	0.2
G25	298,410	466	0.8
G26	253,250	396	0.7
G20-G26	2,094,906	3,274	6.0
G27	196,465	307	0.6
G28	173,115	270	0.5
G27-G28	369,580	577	1.1
G1-G28	**5,811,285**	**9,081**	**16.7**

Soil Region H

Map symbol	Acres	Square miles	Percent of land in state
H1	367,310	574	1.1
H2	732,400	1,144	2.1
H3	220,695	345	0.6
H4	574,035	897	1.6
H1-H4	1,894,440	2,960	5.4
H5	181,785	284	0.5
H6	340,370	532	1.0
H7	172,945	270	0.5
H5-H7	695,100	1,086	2.0
H1-H7	**2,589,540**	**4,046**	**7.4**

Soil Region I

Map symbol	Acres	Square miles	Percent of land in state
I1	58,985	92	0.2
I2	53,245	83	0.2
I3	19,270	30	0.1
I1-I3	131,500	205	0.5
I4	526,090	822	1.5
I5	77,155	121	0.2
I6	36,538	57	0.1
I4-I6	639,783	1,000	1.8
I7	75,177	117	0.2
I8	211,804	331	0.6
I9	58,780	92	0.1
I10	111,155	174	0.3
I11	99,355	155	0.2
I12	54,565	85	0.2
I13	176,560	277	0.5
I14	33,558	52	0.1
I15	40,810	64	0.1
I16	129,495	202	0.4
I17	33,950	53	0.1
I7-I17	1,025,209	1,602	2.8
I18	117,014	183	0.3
I19	249,005	389	0.7
I20	206,355	322	0.6
I21	174,410	273	0.5
I22	50,850	79	0.2
I18-I22	797,634	1,246	2.3
I1-I22	**2,594,126**	**4,053**	**7.4**

Soil Region J

Map symbol	Acres	Square miles	Percent of land in state
J1	61,890	97	0.2
J2	199,985	312	0.6
J1-J2	261,875	409	0.8
J3	73,075	114	0.2
J4	91,615	143	0.3
J5	266,415	416	0.7
J6	50,120	78	0.1
J7	102,730	161	0.3
J8	157,680	246	0.5
J9	122,145	191	0.3
J10	63,650	99	0.2
J11	91,687	143	0.3
J3-J11	1,019,117	1,591	2.9
J12	307,400	481	0.9
J13	295,993	462	0.8
J14	448,447	702	1.3
J15	523,870	819	1.5
J12-J15	1,575,710	2,464	4.5
J1-J15	**2,856,702**	**4,464**	**8.2**
A1-J15 (soil and rock)	34,800,000	54,375	100.0
Water[a]	1,138,560	1,779	—
Total	**35,938,560**	**56,154**	—

[a]Waters of Lakes Superior and Michigan within the state boundaries cover an additional 6,439,680 acres (10,062 square miles).

CHAPTER 20

Soil Series Name Changes

The discipline of soil classification is as subject as any other to the phenomenon of changes in nomenclature. Dr. Charles E. Kellogg, head of the USDA Soil Survey for many years, once said: "When we stop learning about soils, we will stop changing soil names!"

A soil series name, like that of a tree species, for example, stands for a whole group of characteristics. A person familiar with these can judge which soil is good for a specific crop, or stable enough to build a house or road on, or capable of absorbing sewage effluent safely. The purpose of soil survey reports and maps is to make this kind of information available to the public, to land-use planners and to other responsible agents. Changes in soil nomenclature from bulletin to bulletin is confusing, unless a key to soil names, old and new, is provided, as in Table 20-1.

Two examples will help to clarify the matter. Soil nomenclature has been changed whenever soil scientists noted that the same name was being used for two different soils in different regions. At least one of the soils had to be given a new name. Thus the Colby silt loam of Wood, Clark, and Marathon counties, Wisconsin, was changed in the 1940s to Spencer silt loam, because the name Colby was in use in Kansas for a very different soil. Soon thereafter, the Spencer soil was divided into two series, a moderately well drained one, labeled Spencer, and a somewhat poorly drained one, called Almena. Subdivision of a soil series into two or more series is likely when field men conclude that a particular soil has such a wide range of properties that land-use interpretations are not specific enough for practical crop, silvicultural, and engineering interpretations and recommendations. This is why the forest soil, Knox silt loam, of the early soil survey of Dane County (Whitson et al., 1917) was subdivided into the Seaton, Fayette, and Dubuque silt loams. The Dubuque silt loam was later subdivided on the basis of depth to cherty red clay and thickness of the clay over limestone bedrock into several series (see Lee et al., 1968). The name Knox is no longer used in Wisconsin. New classifications of geologic materials and scientific discoveries, such as those concerning shrink-swell potential of soil clays, may bring about recognition of new soil series and groupings of them.

Table 20-1 is intended to help the reader to relate soil names of earlier published soil maps and reports to names in modern publications.

Table 20-1. Listing of old soil series names with modern equivalents

(Note that this list contains some familiar names such as Miami because the original series name included several modern soil series, including the named one (such as Miami) in its modern restricted sense. Please consult the Soil Conservation Service to bring this list up-to-date.)

Old soil series name	Modern equivalent
Ackley	Spencer, Kennan
Acton	Fox, Warsaw
Addison	Theresa, Hochheim
Alden	Hubbard
Allen	Lows
Amherst	Kennan, Coloma
Arlington	Plano
Astico	Sisson
Attica	Chaseburg
Atwater	Markesan
Auburn	Boone
Augusta	Vesper
Aurorahville	Salter, Shiocton
Au Train	Wallace
Baldwin	Renova
Bancroft	Antigo, Plainfield
Bark	Manistee
Bates	Hesch, Sylvester
Baxter	Dubuque
Bay Port	Waukegan, Dakota
Bellchester	Boone
Bellefontaine	Casco, Fox, Lapeer
Belmore	Fox
Beloit	Westville
Berrien	Tustin
Bluffton	Brickton
Boardman	Milaca
Bogus	Alluvial land (sandy)
Bono	Pella

Old soil series name	Modern equivalent
Boone	Boone, Gale, Hixton
Bradford	Cushing
Bridgman	Shawano (dune phase)
Brillion	Hebron
Bristol	Elburn
Brokaw	Santiago
Brownsville	LeRoy
Burnett	Markesan
Cady	Renova, Sargeant
Calamus	St. Charles
Canton	Calamine
Carnot	Solona
Carrington	Plano, Mendota, Sun Prairie, Saybrook
Cary	Vesper
Cashton	Rozetta
Casimer	Cassel
Cassoday	Onamia
Catawba	Spirit
Celina	Mayville
Centerville	Ettrick
Chelsea	Kenman, Emmert, Elderon (modern Chelsea is in S.W. Wisconsin)
Clarno	Unsettled—see Cadiz

Old soil series name	Modern equivalent
Clayton	Barronett
Clinton	Fayette
Clyde	Pella, Harpster
Clyman	Kendall, Lamartine, Virgil
Cochrane	Burkhardt
Cogan	Renova
Colburn	Altoona
Colby	Almena, Spencer
Coloma	Plainfield, Coloma, Wyocena
Columbus	Puchyan
Cooley	Chetek, Padus
Cornucopia	Hiawatha
Crane	Elburn
Crawfish	Aztalan
Crawford	Sogn, Dunbarton
Crosby	Conover, Lamartine
Cylon	Santiago
Dane	St. Charles
Dell	Dells
Delphi	St. Charles
Derinda, deep phase	Eleroy
Detour	Ruse
Dodgeville, deep phase	Ashdale
Dodgeville, shallow phase	Edmund
Dorchester	Lawson, Arenzville

Old soil series name	Modern equivalent
Dryden	Lapeer
Dubuque, deep phase	Palsgrove
Dubuque, shallow phase	Dunbarton
Dunfries	Antigo
Dunkirk	Au Gres
Dunning	Newton, Granby
Durand	Norden
Duroy	Gaastra
Eagle	Warsaw
Edgerton	Miami, Parr
Eel	Arenzville, Orion
Ehler	Pella
Elba	Pella
El Paso	Derinda
Emerald	Santiago
Emily	Pella
Endeavor	Darroch
Erling	Delton
Esdaile	Renova
Estella	Rozetta
Exile	Otterholt
Fall River	Colwood, Keowns
Foresman	Jasper
Galesville	Ettrick
Garwin	Sable
Genesee	Arenzville
Gilbert	Anson, Dakota
Glandon	Freer
Glarus	Brill
Gloucester	Kennan, Iron River, Pence, Vilas
Goetz	Onamia
Granton	Loyal
Hackett	Spinks, Chelsea
Hahns	Del Rey
Hanover	Miami
Harrison	Kennan, Vilas
Hartland	Fayette
Hatley	Alban
Hegg	Wallkill
Henrietta	Stronghurst, Rozetta
Herbert	Octagon
Hersey	Sargeant
Hessel	Alstad
Hewitt	Hiles
Hines	Renova
Homer	Matherton, Brady
Ino	Manistee, Bibon
Janesville	Plano, Parr
Joel	Poskin
Joslin	Dakota
Judson	Worthen
Junction	Guenther
Kaiser	Goodman
Kasson	Racine
Keyser	Plano
Knife Lake	Campia, Medary
Knowlton	Guenther
Knox	Fayette, Seaton, Dubuque
Kokomo	Pella, Harpster
La Crosse	Burkhardt
Langdon	Dakota, Pillot
Langlois	Batavia
Lannon	Knowles, Whalan
Leipsig	Hochheim, LeRoy
Lima	Norden, Pepin
Lindina	Dillon, Wautoma
Lindley	Pecatonica, Renova
Lintonia	Bertrand
Lohn	Northfield
Lost Lake	Kibbie, Shiocton
Lowell	Elburn
Lucas	Fox
Luggarville	Gaastra
Lynn	Lafont
Mason	Munising, Saugatuck
Mead	Meadland
Melita	Bibon
Mellen	Gogebic, Marenisco, Iron River, Wakefield
Mercer	Pence
Meridean	Onamia, Plainfield, Omega
Merrimac	Onamia, Antigo
Miami	Miami, Lapeer, Dodge, St. Charles, and others
Milaca	Amery
Milltown	Antigo
Mingo	Kendall
Myron	Marathon
Nekoosa	Brems
Nye	Knowles
Oakdale	Scandia
Oakfield	Elliott, Ashippun
Oak Grove	Plano
Ockley	St. Charles
Ogemaw	Manistee
O'Neill	Burkhardt
Oquawka	Sparta
Osman	Kewaunee
Ostrander	Prairie analogue of Renova
Ottawa	Tustin
Parr	Saybrook, Mendota, Sun Prairie, Plano, Parr
Parr (shallow phase)	Ripon
Pearl	Richford
Peshtigo	Au Gres, Tedrow
Pigeon	Plainfield, Sparta
Pipe	Matherton
Posen	Longrie
Pray	Alluvial soils
Prentice	Goodman
Primrose	Norden
Quandahl	Stony Chaseburg
Ray	Arenzville
Rice Lake	Onamia
Rio	Puchyan
Rock Elm	Spencer
Rockton, deep phase	Hitt
Rome	Symerton
Roselawn	Vilas, Hiawatha
Rozell	Rozellville
Rubicon	Vilas, Hiawatha
Rudo	Rudolph
Saukville	Casco-Hochheim-Sisson Complex
Schapville, deep phase	Keltner
Sharon	Arenzville
Sheboygan	Sisson
Shoals	Orion
Sinissippi	Ashippun, poorly drained variant
Sleeth	Kendall
Stettin	Fenwood
Stanley	Brill, Poskin
Sterling	Plainfield
Superior	Oshkosh, Ontonagon, Superior, Briggsville, Bibon
Suring	Au Gres, Tedrow
Ten Mile	Newton
Thackery	St. Charles
Thornapple	Omega, Vilas
Thorp	Marshan
Thurman	Dark (mollic) Chelsea
Thurston	Ringwood
Tipler	Gaastra
Tippecanoe	Plano
Trade River	Alban
Trappe	Freeon
Trow	Arland
Truax	Anson
Tuscola	Sisson
Ubet	Hubbard
Union	Fayette, Seaton, Dubuque
Unity	Ettrick
Vlasity	Renova, Sargeant
Wabash	Lawson
Waukegan	Pillot
Waukesha	Richwood, Warsaw, Waukegan
Waupun	Plano
Wea	Plano
Webster	Antigo
Westerfield	Boaz
Westville	Pecatonica, Renova
Whalan, deep phase	Woodbine, Knowles
Whitman	Alluvial soil associated with Almena, Freer, Vesper
Wilbur	Lowe
Willow	Borth
Withrow	Onamia, Chetek
York	Wallkill
Zell	Mead (Meadland)

CHAPTER **21**

Acreages of Major Soil Series in Wisconsin

Table 21-1. Estimated acreages (in thousands of acres) of some major soil series and types and land types in Wisconsin

(The following abbreviations are used: cl = clay loam; fsl = fine sandy loam; gr. = gravelly; l = loam; lfs = loamy fine sand; ls = loamy sand; s = sand; sil = silt loam; sicl = silty clay loam; sl = sandy loam; st. = stony; sw. = shallow. Capital letter symbols refer to the soil region in which the soil or land type is most common. The first entry in the table is interpreted as follows:

Iron River .... Soil series
G .... common in Soil Region G
800: .... occupies an estimated 800,000 acres in Wisconsin
l 630 .... Iron River loam occupies about 630,000 acres
sl 170 .... Iron River sandy loam occupies about 170,000 acres

The many soils with acreages less than 50,000 have been omitted from this table. Note that slope phases are not reported here.

For classification of soil series not defined in this publication, consult state soil keys of the Soil Conservation Service and the Geological and Natural History Survey.)

Soil series and land types	Soil region	Acreage and texture
		800
Iron River	G	800: l 630; sl 170
		600-800
Fayette	A	660: sil 530; st. sil 1; sil, valley phase, 129
Pence	G, H, F	636: sl 636
Plainfield	C	658: ls 478; s 180
Rocky and st. land	A, D	616
		500-600
Alluvial lands	J	560: wet 335; not wet 225
Dubuque	A, B	552: sil 523; st. sil 29
Hixton	D	556: l 240; st. l 12; sl 304
		400-500
Kewaunee	I	435: cl 19; st. l 5; sil 367; sl 44
Miami	B	475: l 405; st. l 61; sl 9
Milaca (Amery)	G, F	437: l 335; st. l 2; l in complex with peat 100
Omega	H, G	481: ls 416; s 65
Palsgrove	A	425: sil 425
		300-400
Boone	D	379: ls 146; s 233
Gale	D	328: sil 328
Onamia	G, F	335: l 259; sl 76
Santiago	F	337: sil 333; st. sil 4
Vilas	H, G	347: ls 343; s 4
		200-300
Almena	F	202: sil 198; st. sil 4
Antigo	F	287: sil 205; deep sil 12; sw. sil 70
Carbondale	J	215: muck 215
Casco	B	285: l 138; sil 107; in complex with Rodman 25; in complex with Sisson 15
		200-300, *continued*
Crivitz	G	235: ls 235
Dunbarton	A	227: sil 227
Elderon	G	222: l 222
Fox	B	243: sil 201; sl 42
Freeon	F	214: sil 214
Gogebic	G	291: l 291
Kennan	G	243: sil 186; st. sil 31; sl 26
Norden	D	280: sil 280
Onaway	B	203: l 203
Ontonagon	I	255: sicl 255
Oshkosh	I	250: sil 198; scl 40; sl 12
St. Charles	B	246: sil 246
Stambaugh	F	230: sil 230
Withee	F	284: sil 284
		100-200
Adolph	F, J	176: sil 113; st. sil 63
Adrian	J	136: muck 136
Arland	D	100: l 30; sl 70
Auburndale	F	133: sil 133
Au Gres	H, J	187: ls 187
Bergland	I, J	105: sil 105
Billett	A	111: sl 111
Brems (Nekoosa)	C	197: ls 197
Carlisle	J	121: muck 121
Cathro	J	194: peat 194
Chaseburg	A, J	127: sil 127
Chetek	G, D	192: l 192
Cloquet	G	131: ls 13; sl 118
Dodge	B	106: sil 106
Dodgeville	A	140: sil 140
Fenwood	F	117: sil 117
Freer	F	136: sil 124; st. sil 12
Goodman	F	138: sil 138

Soil series and land types	Soil region	Acreage and texture
	100-200, *continued*	
Greenwood-Spalding	J	180: peat 180
Houghton	J	122: muck 122
Kokomo	B, J	119: sil 110; st. sil 9
Lapeer	B	107: l 63; sl 44
Manawa	I	145: sil 145
Markey	J	141: muck 141
Marsh	A	161
Merrillan	D	134: sl 134
Otterhold	F	165: sil 165
Padus	G	154: l 154
Parr	B	159: sil 159
Plano	B	196: sil 196
Rifle	J	107: peat 107
Solona	E	138: l 138
Sparta	C	109: ls 102; s 7
Tama	A	193:sil 193
Terrace escarpments	A, C, H, G	151
Wyocena	C	174: ls 93; sl 72; st. sl 9
	50-100	
Ahmeek	G	86: sil 86
Alcona	I	61: fsl 61
Ashdale	A	82: sil 82
Ashkum	B	51: sil 48; sicl 3
Bertrand	A	62: sil 62
Brill	F	51: sil 51
Burkhardt	A	64: sl 64
Coloma	C	95: ls 43; s 52
Comstock	F	62: sil 62
Conover	B	76: sil 76
Dakota	A	60: l 33; sl 27
Dawson	J	79: peat 79
Downs	A	56: sil 56
Elkmound	D	55: sl 55
Emmert	G	84: sl 60; gr. l 24
Emmet	E	76: fsl 57; ls 19
Ettrick	A, J	75: sil 75
Gaastra	F	53: sil 47; st. sil 6
Granby	E	83: ls 36; sl 47
Hiawatha	H	92: ls 92
Hochheim	B	68: sil 68
Keowns	B	76: l 76
Kert	F	85: sil 85

Soil series and land types	Soil region	Acreage and texture
	50-100, *continued*	
Lafont	F	72: sil 72
Lamartine	B	87: sil 87
Longrie	E	92: l 11; sw. l 81
Loyal	F	67: sil 67
Marathon	F	88: l 80; st. l 8
Maumee	C	53: sl 53
Mead (Meadland)	F	86: sil 86
Monico	F	55: sil 55
Morley	B	88: sil 88
Newton	C	94: ls 59; sl 35
Norrie	F	74: sil 74
Northfield	D	66: sil 66
Ogden	J	67: muck 67
Pecatonica	B	60: sil 60
Pella	B, J	94: sil 94
Poskin	F	69: sil 69
Rietbrock	F	67: sil 65; st. sil 3
Rock land	A	55
Rough broken land	A	84
Rozellville	F	66: sil 59; sl 7
Sargeant	F	53: sil 53
Seaton	A	82: sil 82
Shawano	E	89: lfs 89
Shiocton	B	80: l 80
Sisson	B	72: l 50; in complex with Casco 22
Sogn	A	81: sil 81
Spencer	F	98: sil 98
Tawas	J	56: muck 56
Tilleda	E	60: sil 60
Trenary	E	53: l 53
Tustin	B	85: ls 53; sl 32
Underhill	E	75: sil 57; sl 18
Urne	D	68: l 68
Vesper	D	76: sil 76
Warman	F, J	62: l 62
Warsaw	B	61: l 55; sl 6
Wauseon	B	67: sl 67
Willette	J	65: muck 65
Worthen	A	83: sil 83

Source: Compiled from the Conservation Needs Inventory data of the Soil Conservation Service, USDA, February, 1971.

Appendices

APPENDIX 1
Glossary

Aeolian—Processes and deposits brought about by wind action.

Aggregate, soil—A cluster of soil particles. Synonym for *ped.*

A horizon—The surface horizon of an undisturbed mineral soil. It is usually subdivided into several subhorizons. The A1 is dark colored and high in content of organic matter; the A2 is usually light colored and leached; the A3 is transitional to the B horizon; the Ap is the plow layer (0 to 7 inches in depth) of surface soil in a field that has been cultivated. If this layer contains a notable accumulation of lime it is termed the Apca horizon. Some soils have all of these subdivisions; others do not.

Albaqualfs—Aqualfs (see Chapter 5, Table 5-4, Fig. 5-1) with impeded drainage (somewhat poor to poor).

Albic horizon—A leached surface or lower soil horizon (A2 or A'2) that is less brown than the underlying B horizon and less dark than the overlying A1 horizon (if present).

Alfisol—See Chapter 5, Table 5-4, Fig. 5-1.

Allophane—A noncrystalline (amorphous) gel-like clay present in small amounts in most Wisconsin soils.

Alluvium—Soil material deposited by streams.

Apatite—A common primary mineral containing phosphorus and calcium.

Argillans—Clay skins; coatings (cutans), composed mostly of clay, on the surfaces of blocky peds and of stones in the subsoil.

Association, soil—See *Soil association.*

Bar—A term used as an international unit of pressure equal to .987 atmospheres. Negative pressure or tension of soil water is measured in bars (b) and millibars (mb).

B horizon—Master horizon or layer in a soil profile usually found below the A horizon. It is generally characterized by stronger colors (usually brown) than those in horizons above or below, by an accumulation of iron, clay, or organic matter, and by a blocky structure. It is usually subdivided into several subhorizons.

B1—Horizon that is transitional from A3 to B2.

B2—The part of the B horizon in which diagnostic properties dominate. If clay accumulation is diagnostic, the horizon may be labeled Bt or B2t (B21t and B22t are subdivisions thereof). If accumulation of organic matter and iron oxide is diagnostic, the horizon is designated Bhir.

B3—Horizon that is transitional from B2 to C.

Beta B—A particular kind of B2t horizon that has formed at the interface between finer textured (above) and coarser textured (below) materials. In the Fox soil series, for example, the beta B overlies and penetrates into the calcareous C horizon. This is the most clayey part of the B2t horizon (see Fig. 1-6).

Biocycling—The process by which vegetation (and some associated fauna) brings nutrients such as calcium from the soil to aboveground plant parts, returns the nutrients to the soil by leaf fall, leaf decomposition, and leaching, and continues to move the same materials through this cycle year after year.

Biosequence—A sequence of soils that are similar except for the plant and animal associations influencing them. For example, there are places in southwestern Wisconsin where one can walk from Fayette, across Downs, to Tama. This is a biosequence despite the fact that the soils may now all be in corn.

Biota—The ensemble of plants and animals in a particular ecosystem.

Bisequum—A double sequence of soil horizons, commonly including 0, A1, A2, B, A'2, B'2, C (see Fig. 12-10).

Bog (peat)—A peat deposit, commonly consisting of moss peat, upon which plants are growing. Bogs are generally found in enclosed depressions.

Bog (Soil)—An organic soil.

Boralfs—Alfisols (see Chapter 5, Table 5-4, Fig. 5-1) of cool humid temperate forest regions, with A2 (albic) horizons that tongue down into the Bt horizons.

Brunizem—The kind of soil which usually developed under prairie vegetation in southern and western Wisconsin. These soils have a thick, dark surface horizon over C or over B and C horizons.

Calcareous (soil)—Soil containing free lime (carbonates) which effervesces when diluted (1:10) HCl is applied.

Cambic horizon—An altered subsurface diagnostic soil horizon. In Wisconsin, a B horizon, of finer texture than loamy sand, that is obviously different from the underlying C in color, or structure, but does not have accumulations of clay or organic matter and iron as the Bt and Bhir horizons do.

Carbonate—A salt or ester of carbonic acid; a compound containing the radical CO_3^{2-}. Calcite ($CaCO_3$) and dolomite ($CaMg(CO_3)_2$) are the most common carbonates in Wisconsin soils and their initial materials.

Catena—A group of soils developed from similar initial materials but differing in morphology because of differences in natural drainage conditions. A kind of toposequence (see Fig. 2-52).

Channery—Term for stony soil, with fragments of rock that are thin and flat and measure up to 6 inches long.

Chlorite—A crystalline mica-like clay mineral common in Wisconsin soils. It has a cation exchange capacity of about 5 me/100g. In the laboratory, Mg-saturated and glycerol-solvated chlorite has a 14 to 14.3 Å spacing that does not collapse with K saturation and heating to 550°C. This is a 2:2 layer lattice silicate mineral.

C horizon—A layer of relatively unweathered material similar to the material from which at least a part of the soil above it was formed. Soil initial material.

Chronosequence—An array of soils formed under the influence of the same factors of soil formation except for the time intervals involved; a time or developmental sequence of soils.

Clay—The smallest mineral grains, less than 0.002 mm in diameter.

Clay (texture)—Soil that contains 40% or more clay, less than 45% sand, and less than 40% silt.

Clay loam—Soil consisting of 27 to 40% clay and 20 to 45% sand.

Clay skins—See *Argillans.*

Colluvium—Deposit of soil accumulated at the base of a slope under the influence of gravity. Slope wash.

Color of soil—Soil color designations given in parentheses in this bulletin are from the scientific Munsell Color Chart (Pendleton and Nickerson, Munsel Co., Baltimore, Md., 1951) in terms of (1) hue, such as 10YR; (2) value (white to black); and (3) chroma (intensity of color). 10YR 4/4 is a designation for dark yellowish-brown (Soil Survey Staff, 1951). In the field, workers sometimes generalize color terminology, using such adjectives as "pink," "gray," "yellow" (see Fig. 8-11).

Complex, soil—Several soils, so closely intermingled that they cannot be shown separately on a map at the scale being used.

Consistence, soil—The resistance of soil to separation or deformation. Soil consistence varies with moisture content. It is described in such terms as loose, friable, firm, hard, sticky.

Cradle-knoll—A shallow pit, 6 to 30 inches deep and several feet across, and adjacent mound, 12 to 36 inches high and several feet across, created by the tipping of a tree in a windstorm. The root mass of the falling tree pulled soil up from the pit site (cradle) and dumped it in the form of the mound (knoll).

Cuesta—A Spanish word denoting an unsymmetrical ridge with one slope (the dip slope of a resistant formation) long and gentle, and the other slope steep, cutting across the resistant ledge.

Cycle—A regularly recurring series of events or phenomena.

Dolomite—Magnesian limestone. A limestone consisting mostly of the mineral dolomite: $CaMg(CO_3)_2$. See *Carbonate.*

Dolomitic—Containing dolomite (as opposed to calcite, for example; see *Carbonate*).

Drainage, soil—Natural soil drainage refers to the speed with which water is removed from the soil surface and through the soil itself. Seven classes have been recognized: excessive, somewhat excessive, well, moderately well, somewhat poor (imperfect), poor, and very poor. Artificial drainage refers to removal of water by ditching, tiling, and construction of surface waterways and terraces.

Drift—Glacial deposits, both ice-laid and water-laid.

Drumlin—An oval or fish-shaped hill of glacial drift (usually till), ordinarily with its long axis parallel to the movement of ice that formed it.

Duricrust—The hardened crust formed in soil and porous rock by cementation, particularly with siliceous, ferruginous, or aluminous precipitates.

Dysic—A family category of peats and mucks (Histosols) including the most acid soils, having a pH below 4.5. See *Euic.*

Earth—See *Fine earth.*

Eluvial (*horizon*)—A horizon that has lost bases, iron, clay, etc., by processes of soil formation. A2 horizons are eluvial.

Entisol—See Chapter 5, Table 5-4, Fig. 5-1.

Epipedon—Soil horizon that forms at the surface of the soil. The mollic epipedon is an example.

Esker—Serpentine ridge of rudely stratified glacial outwash sand and gravel deposited in tunnels under stagnant glacial ice.

Euic—A family category of peats and mucks (Histosols) including soils with a pH of 4.5 and above. See *Dysic.*

Eutrophied—A condition of a body of water which has been enriched with nutrients and depleted of oxygen.

Exchangeable cations—Postively charged ions (including those of hydrogen, and the alkali and alkaline earth metals, calcium, magnesium, potassium, and sodium, all of which are important plant nutrients) attached in "available" forms to clay and organic constituents of soils. These plant nutrients can be exchanged with each other and with other positively charged ions in soil solutions.

Fabric—The relative sizes, shapes, and arrangements of constituents of a soil or initial material.

Feature, soil—A very general term for any property of a soil, including soil horizons, krotovinas, clay skins, and all the attendant patterns of color, distribution of moisture and temperature and biota, and slope of the surfaces of land and water table at a given site.

Field grading of soil texture—For more than 70 years soil surveyors have routinely done field grading of soil texture by rubbing soil between the fingers. With experience, a person can judge by the feel of the soil how much sand, silt, and clay are present. According to the proportions of material of these three sizes, textural class names are given to soil from different horizons and different profiles. These textural terms and their definitions are summarized best in a chart known as the textural triangle (Soil Survey Staff, 1951). Some people find it helpful to have a copy of this triangle with them in the field. However, an attempt is made here to describe briefly how the different textural classes of soil feel. These classes are defined elsewhere in the glossary. Some phrases are taken from C. F. Shaw (Soil Survey Staff, 1951). It is a good idea to accompany a professional soil surveyor from time to time in order to compare field grading judgments. Textures intermediate between those listed below can be recognized by the relative amounts of gritty, soft, and sticky material in them.

1. Stones, cobbles, and gravel. These coarse fragments, all with diameters greater than 2 mm (0.079 inch), can be recognized by eye and measured with a rule and are not rubbed between the fingers as is finer soil. Fragments of gravel size are less than 3 inches in diameter, cobbles are between 3 and 10 inches in diameter, and stones (or boulders) are larger than that.
2. Sand. Feels gritty and harsh. Individual grains can be seen and felt. Squeezed when moist, the soil forms a fragile cast.
3. Sandy loam. Feels quite gritty but also somewhat loamy. Individual sand grains can be seen and felt. There is enough silt and clay to soften the feel of this soil. Squeezed when dry, the soil forms a somewhat stable cast.
4. Loam. Feels somewhat gritty, somewhat smooth, and possibly a little sticky and plastic. Sometimes the observer decides to call the soil a loam chiefly because it is not sandy enough to be a sandy loam, silty enough to be a silt loam, nor clayey enough to be a clay loam. Squeezed when dry, it forms a fragile cast. Squeezed when moist, it forms a stable cast.
5. Silt loam. In a dry state, lumps and clods prove to be very fragile. When rubbed, this soil feels soft like flour and forms a fairly stable cast when squeezed. In a moist state this soil feels smooth and mellow. The moist cast is stable. Moist soil will not form a polished ribbon when rubbed between the thumb and finger, but will appear as a somewhat rough and noncoherent coating on the thumb.
6. Clay loam. The dry soil is hard and lumpy. Moist soil is plastic; forms a very stable cast when squeezed; and, when rubbed between the thumb and finger, forms a thin, somewhat fragile ribbon, with a somewhat polished surface. The moist soil can be kneaded in the hand into a compact mass which does not readily crumble.
7. Clay. The soil is very hard and lumpy when dry. Moist soil is very plastic and sticky; it forms a cast which is stable; elongated casts may sag under their own weight; when rubbed between the thumb and finger, the soil forms a long flexible ribbon which has a good polish on the surface.

Fine earth—Soil particles less than 2 mm (0.079 inch) in diameter. Most soil analyses are made of fine earth (sand, silt, and clay), excluding coarser fragments (gravel, stones, cobbles).

Fragic—Having properties of a fragipan, or nearly so.

Fragiochrepts—Ochrepts (see Chapter 5, Table 5-4, Fig. 5-1) that have a fragipan.

Fragipan—A loamy subsurface horizon that is brittle (and not plastic) when moist and seemingly cemented when dry. Hand specimens rupture abruptly under increasing pressure.

Frigid—A soil temperature class. The soil has a mean annual temperature at a depth of 50 cm (20 in.) of less than 8°C (47°F). There is a difference of more than 5°C (9°F) between mean summer and mean winter temperatures.

g—Symbol for a soil horizon that is gleyed.

Glacial drift—See *Drift.*

Glacial till—Unsorted glacial drift transported and deposited by ice.

Glacio-fluvial deposits—Sediments deposited by glacial streams. These deposits are usually sandy or gravelly and are typically stratified.

Glacio-lacustrine deposits—Sediments deposited in glacial lakes. These include fine sands, silts, and clays. They may be stratified or varved.

Gleyed (*soil*)—Soil material which is olive gray or bluish gray in color. Gleyed horizons are usually found below a dark-colored surface layer in poorly drained soils.

Gray-Brown Podzolic—The kind of soil that usually developed under forest vegetation in southern Wisconsin. These soils have light-colored subsurface horizons, brown illuvial (clayey) subsoils, and are generally acid.

Gray Wooded—Northern equivalent of Gray-Brown Podzolic. The A2 tongues down into the Bt horizon.

Histosol—See Chapter 5, Table 5-4, Fig. 5-1.

Horizon, diagnostic—The new soil taxonomy (Soil Conservation Service, 1970) designates certain soil horizons as diagnostic in classifying soils. The reader is referred to the taxonomy for complete definitions. General ones are given below for horizons found in Wisconsin.

1. Diagnostic suface horizons (epipedons)
   a) Mollic epipedon. Dark, thick, soft or friable, fertile A1 horizon. This horizon is diagnostic of Mollisols (Brunizems; Humic Gleys).
   b) Ochric epipedon. Thin or only moderately dark A1 horizon. This horizon is common in Alfisols, Inceptisols, and Entisols.
   c) Histic epipedon. Peat or muck less than about a foot thick.
2. Diagnostic subsurface horizons
   a) Argillic horizon. Clay-enriched B2t horizon. Clay skins are present. This horizon is diagnostic of Alfisols (Gray-Brown Podzolics). It may be present in Mollisols.
   b) Spodic horizon. Bhir horizons enriched in humus (h), iron (ir), and aluminum. This horizon is diagnostic of Spodosols (Podzols).
   c) Cambic horizon. "Weak" B horizon in which no accumulation has taken place of clay, humus, iron, or aluminum, but which is different in color and structure and leached state from the C horizon. The horizon is diagnostic of most Inceptisols in Wisconsin.

Note that the above list of diagnostic horizons does not include these important horizons:

1. Albic horizon. The A2 horizon, which is paler than the overlying A1 or O horizon and has lost some materials to the underlying B horizon.
2. The O1 and O2 horizons. The subhorizons, O1 (litter) and O2 (humus), are of vital importance in biocycling and in formation of underlying horizons.
3. The Oi, Oe, and Oa horizons. These peat and muck horizons constitute Histosols, all of which are deeper than about a foot. These horizons are fibrous and peaty (Oi), or mucky and pasty (Oa), or a blend of the two (Oe).

Horizon, soil—A layer of soil more or less parallel to the land surface and having characteristics produced by processes of soil formation.

Humic Gley—A naturally poorly drained soil having a thick, dark-colored surface horizon and a gray (gleyed) subsoil.

Humus—The organic layer of a forest soil consisting of well-decomposed organic matter, the origin of which cannot be determined by observation of the material with the naked eye.

Illite (*mica*)—A mica clay mineral that is very common in Wisconsin soils and initial materials (particularly shales and limestones and glacial drift derived from them). The cation exchange capacity is about 30 me/100g when the illite is pure, but it commonly is intermixed with vermiculite enough to raise this somewhat. In the laboratory, the Mg-saturated, glycerol-solvated 2:1 clay has a 10.1 Å spacing. Ions of K hold clay units together tightly, thereby reducing the cation exchange capacity from a potential of 250 to an actual 30 me/100g.

Illuvial (*horizon*)—A horizon that has received material (bases, clay, etc.) from an eluvial horizon. B horizons of Gray-Brown Podzolic and Podzol soils are illuvial.

Inceptisol—See Chapter 5, Table 5-4, Fig. 5-1.

Initial material—The material from which a soil formed, such as sandy loam glacial till, deep sand, or woody peat. This term is preferred to "parent material."

Intergrade—A soil that does not clearly belong to any major soil category but has some characteristics of several categories.

Ion—A positively or negatively charged atom or group of atoms.

Kame—A roughly conical hill of rudely stratified glacial outwash sand and gravel deposited in a cavern under a glacier.

Kaolinite—A crystalline silicate clay mineral (1:1) common in Wisconsin soils, with a cation exchange capacity of 10 me/100g. In the laboratory the Mg-saturated, glycerol-solvated clay has a spacing of 7.3 Å. It loses its crystallinity on heating to 550°C.

Kettle—An enclosed depression in glacial drift produced by the melting of a buried glacial ice block and accompanying collapse of the overlying and surrounding drift.

Krotovina—A filling of an animal burrow in soil.

Lacustrine—Pertaining to a lake, as in the phrase "lacustrine deposits."

Landscape—The portion of the land or territory or region that the eye can comprehend in a single view, including all that is visible to the zenith. See *Soilscape.*

Leaching—Removal of material from soil in solution by percolating water. For example, the removal of lime from the upper part of a soil is a leaching process.

Lithosequence—A sequence of soils that have formed under similar conditions except for initial materials. For example, where a prairie once extended for many centuries across contiguous bodies of sand, silt, and clay, a lithosequence of deep sandy, moderately deep silty, and shallow clayey Mollisols (Brunizems) is found today.

Lithosol—A shallow soil consisting of a dark-colored surface soil underlain by bedrock.

Litter layer—The mat of dead, decomposing organic material above the mineral soil. See *Biocycling.*

Loam (*texture*)—Soil that contains 7 to 27% clay, 28 to 59% silt, and less than 52% sand.

Loamy sand—Soil that contains at the upper limit 85 to 90% sand, and the percentage of silt plus 1.5 times the percentage of clay is less than 15. At the lower limit it contains not less than 70 to 85% sand, and the percentage of silt plus twice the percentage of clay does not exceed 30.

Loess—A wind-blown deposit of silt-size particles of a wide variety of minerals, including quartz, feldspar, and carbonates. Loess and loess-derived soils are generally fertile.

Marl—An earthy deposit consisting of calcium carbonate (lime), silt, and clay. It is found in lake bottoms or below peat.

Marsh—A wet area supporting sedge, grass, and reed vegetation.

Mesic—A soil temperature class. The soil has a mean annual temperature at a depth of 50 cm (20 in.) of 8°C (47°F) or more. There is a difference of more than 5°C (9°F) between mean summer and mean winter temperatures.

Mica—See *Illite.*

Milliequivalent—One milligram of hydrogen or the amount of any other ion that will combine with it or displace it. Thus, a clay with an cation exchange capacity of 10 milliequivalents (10 me/100g) is capable of adsorbing and holding 10 milligrams of hydrogen or its equivalent for every 100 grams of dry clay.

Mineral soil—Soil containing less than 20% (for sands) to 30% (for clays) organic matter by weight. See Table 5-4.

Mollisol—See Chapter 5, Table 5-4, Fig. 5-1.

Montmorillonite—A crystalline clay mineral (2:1) common in Wisconsin soils, with a cation exchange capacity of 100 me/100g, and with a remarkable capacity to shrink on drying and to swell on wetting.

In the laboratory, Mg-saturated, glycerol-solvated clay has a spacing of 17.7 Å, that collapses with K-saturation to 14 Å, shifting to 10 Å.

Morphology, soil—A term referring to the physical constitution of the soil, including such characteristics as the color, texture, structure, and consistency of the various horizons, their thickness, and arrangement in the soil profile.

Mottled—Somewhat spotted appearance, as in the case of soil that shows splotches of rust and gray colors. Mottling in most of the soils in Wisconsin indicates that natural drainage is restricted, or that the water table rises to or near the surface periodically.

Muck—Organic soil material that is partially decomposed. Muck is usually dark in color.

Mull—A mixture of humus and mineral soil that is intimate. The A1 soil horizon.

O horizon—An organic soil horizon, that is, a soil horizon containing more than about 25% by weight of organic matter, ovendry. Sandy soil horizons are termed organic if the organic matter content exceeds 20%; clayey horizons are termed organic if the organic matter content exceeds 30%.

Organic soil—Soil formed from organic materials. Peat and muck are organic soils and are classified in the Bog great soil group. (Histosol.)

Ortstein—An indurated layer in the B horizon of a Podzol soil, cemented by illuviated sesquioxides, mostly iron oxides, as well as organic matter.

Outwash—Sorted sand and gravel deposited by glacial melt waters flowing out from the glacier.

Paleophreatic—Pertaining to an earlier condition of high stand of the water table, at a presently well-drained site.

Parent material—Synonym for *initial material,* which see.

Particle size distribution (*of soil*)—A term synonymous with *texture,* referring to the per cent by weight of clay and silt (commonly determined by a hydrometer method) and sands (determined with sieves) in dry material.

Peat—Organic soil material that is relatively undecomposed. This material may be broken (disintegrated), but plant parts can still be recognized. When peat undergoes decomposition it becomes muck.

Ped (*soil*)—A soil aggregate. A ped may be blocky, platy, prismatic, or granular in shape.

Pedon—A basic unit of a polypedon that is a meter in diameter (or one half of a small-scale complex of two units, i.e., <3.5 meters in diameter), and is as deep as the common rooting of native perennial plants.

Pedosphere—The entire soil continuum on lands of the earth.

Pedota—The total list of soil species in a given region (see Chapter 6, page 49).

pH—A notation used to designate the acidity or alkalinity of a soil. A pH of 7.0 indicates neutrality. Lower values indicate acidity and higher values, alkalinity.

Phase, soil—A subdivision of a soil unit based on features significant to man's use of a soil. For example: sloping phase, stony phase.

Phreatic zone—Goundwater zone: the saturated zone below the water table.

Piping—A process of soil erosion by which running water does its work in subsurface channels.

Planosols—Hydromorphic soils (see Table 5-1), poorly to somewhat poorly drained.

Pleistocene—A geologic interval characterized by glaciation. See Table 2-10.

Polypedon—An assemblage of contiguous like pedons on a landscape. See *Soil body.*

Profile, soil—A vertical section through a soil, exposing all of its horizons, including the initial material.

Quartzite—A metamorphic rock consisting mostly of silica; a former sandstone cemented with quartz.

Reductant-soluble Fe—A term referring to free iron in soil. Free iron is determined by reducing and complexing the iron in a neutral system.

Regosol—A shallow soil consisting of an A horizon over unweathered, unconsolidated initial material.

Sand—Mineral grains having diameters ranging between 2 and 0.05 mm.

Sand (*texture*)—Soil consisting of 85% or more sand. The percentage of silt plus 1.5 times the percentage of clay does not exceed 15. Coarse sand, sand, fine sand, and very fine sandy subclasses are recognized.

Sandy clay loam—Soil that consists of 20 to 35% clay, less than 28% silt, and 45% or more sand.

Sandy loam—Soil consisting of either (1) 20% clay or less, and the percentage of silt plus twice the percentage of clay exceeds 30, and 52% or more of sand; or (2) less than 7% clay, less than 50% silt, and between 43 and 52% sand.

Sequence, soil—An arrangement of soils along a continuum. See *Biosequence, Chronosequence, Lithosequence, Toposequence.*

Sequum—A sequence of an eluvial horizon and its related illuvial horizon.

Silt—Mineral grains ranging in size from 0.05 to 0.002 mm in diameter. Soil material containing more than 80% silt and less than 12% clay is included in the silt class.

Silt loam—Soil consisting of (1) 50% or more silt and 12 to 27% clay; or (2) 50 to 80% silt and less than 12% clay.

Silty clay—Soil that contains 40% or more clay and 40% or more silt.

Silty clay loam—Soil consisting of 27 to 40% clay and less than 20% sand.

Skeletal—A family category of mineral soils including soils with more than 35% by volume, of material composed of particles coarser than 2 mm in diameter.

Skeletan—A coating (cutan) consisting of skeleton grains (sand and silt particles) on the surfaces of voids in the soil.

Slope, of soil—The slope of soil is reported as gradient, i.e., the number of feet of rise (or fall) per 100 feet of horizontal distance. If the soil surface rises 10 feet vertically in a horizontal distance of 100 feet, the slope gradient is 10%.

Soil association—A natural geographic grouping of soils. The legend of the soil map (Plate 1) accompanying this volume lists some 200 soil associations. See *Soilscape.*

Soil body—A single soil individual on the landscape. A unit of the soilscape. See *Polypedon.*

Soilscape—The soil portion of a landscape, bounded by the surface of leaf litter above and the lower boundary of the rooting zone of native perennial plants (trees, prairie plants). In the text this term is used interchangeably with *soil association.*

Soil series—The lowest category in the Soil Taxonomy of the U.S. National Cooperative Soil Survey. A soil series is named after a place. The Antigo soil series, for example, was named after the city of Antigo, Wisconsin, near which the soil was first observed and described in detail.

Solum, soil—The soil solum contains the A and B horizons taken together. The C horizon is not included.

Spodosol—See Chapter 5, Table 5, Fig. 5-1.

Structure, soil—A term referring to the aggregation of primary soil particles into compound particles such as granules, blocks, prisms, or plates.

Swamp—A wet area supporting woody vegetation, usually tamarack.

Texture, soil—A term referring to the relative proportions of the various size groups of individual soil grains. See *Field grading of soil texture.*

Till—See *Glacial till.*

Toposequence—A sequence of soils that have formed under similar conditions except for topographic position. See *Catena.*

Vadose zone—The unsaturated subsurface zone above the water table.

Variant, soil—A soil of limited or unknown extent but having characteristics unique enough to set it apart from related series. The term *variant* is usually used temporarily until the soil can be studied further. *Phase* is sometimes used in the same sense.

Vermiculite—A crystalline clay mineral common in Wisconsin soils. It has a cation exchange capacity of about 150 me/100g. In the laboratory, Mg-saturated, glycerol-solvated clay has a spacing of 14.4 Å which collapses to 10 Å with K saturation and heating to 300°C.

Water table—The upper surface of a zone of saturation except where that surface is formed by an impermeable body; the locus of all points in soil water at which the pressure is equal to atmospheric pressure.

APPENDIX 2

Common and Scientific Names of Plants Mentioned in the Text

Alfalfa. *Medicago sativa*
Arrowhead. *Sagittaria* sp.
Ash. *Fraxinus* sp.
Aspen. *Populus* sp.
Aster. *Aster lateriflorus*
Basswood. *Tilia americana*
Birch, white. *Betula papyrifera*
Birch, yellow. *Betula lutea*
Bluegrass. *Poa pratensis*
Blue joint. *Calamagrostis canadensis*
Bluestem, big. *Andropogon gerardi*
Bluestem, little. *Andropogon scoparius*
Bulrush. *Scirpus americanus*
Cactus, prickly. *Opuntia compressa*
Cattail. *Typha angustifolia*
Cedar, red. *Juniperus virginiana*
Cedar, white. *Thuja occidentalis*
Cherry, black. *Prunus serotina*
Corn. *Zea mays*
Cranberry. *Vaccinium angustifolium*
Elm. *Ulmus* sp.
Fern, bracken. *Pteridium aquilinum*
Fern, sweet. *Myrica asplenifolia*
Fir, balsam. *Abies balsamea*
Gentian, bottle. *Gentiana andrewsii*
Hawkweed. *Hieracium aurantiacum*
Heath family. *Ericaceae* sp.
Hemlock. *Tsuga canadensis*
Kentucky coffee tree. *Gymnocladus dioica*
Labrador tea. *Ledum groenlandicum*
Leatherleaf (shrubby). *Chamaedaphne calyculata*
Lichens. *Cladonia* sp.
Lilac. *Syringa vulgaris*
Lupine. *Lupinus perennis*
Maple, red. *Acer rubrum*
Maple, silver (soft). *Acer saccharinum*
Maple, sugar. *Acer saccharum*
Moss, sphagnum. *Sphagnum* sp.
Moss, spike. *Selaginella rupestris*
Pine, jack. *Pinus banksiana*
Pine, red. *Pinus resinosa*
Pine, white. *Pinus strobus*
Pitcher plant. *Sarracenia purpurea*
Oak, black. *Quercus velutina*
Oak, bur. *Quercus macrocarpa*
Oak, Hill's. *Quercus ellipsoidalis*
Oak, red. *Quercus borealis*
Oak, white. *Quercus alba*
Quackgrass. *Agropyron repens*
Ragweed. *Ambrosia* sp.
Rose. *Rosa* sp.
Sagebrush. *Artemisia caudata*
Sedge. *Carex* sp.
Spruce, black. *Picea mariana*
Spruce, white. *Picea glauca*
Squirrel corn. *Dicentra canadensis*
Sundew. *Drosera rotundifolia*
Tamarack. *Larix laricina*
Trillium. *Trillium grandiflorum*
Willow, black. *Salix nigra*

APPENDIX 3

Legend of the Soil Map (Plate 1) in Terms of Classification Categories above the Series Level

A. Soils of the Southwestern Ridges and Valleys: Alfisols, Mollisols, and Entisols.

Undulating, rolling, and hilly soils on limestone ridges with shale in places.

A1 Typic Argiudolls, with Mollic Hapludalfs and Aquic Argiudolls, fine-silty, mixed, mesic.

A2 Typic Argiudolls, fine-silty, mixed, mesic; with Lithic Haplustolls, loamy, mixed, mesic.

A3 Typic Hapludalfs, with Typic Argiudolls, fine-silty, mixed, mesic; and Lithic Haplustolls, loamy, mixed, mesic.

A4 Typic Argiudolls and Hapludalfs, fine, mixed, mesic; with Typic Argiaquolls, fine, illitic, mesic; and Glossaquic Hapludalfs, fine-loamy, mixed, mesic.

Gently rolling to very steep soils on limestone ridges or on quartzite uplands.

A5, A6, A7, A8, A9 Typic Hapludalfs, fine-silty, mixed, mesic; with steep rocky land.

A10 Typic and Aquic Hapludalfs, fine-silty, mixed, mesic.

Nearly level to sloping soils on stream terraces.

A11 Typic Argiudolls and Hapludalfs, fine-silty, mixed, mesic.

A12 Typic Hapludalfs and Udollic Ochraqualfs, fine-silty, mixed, mesic; with Typic Argiudolls and Mollic Hapludalfs, fine-loamy over sandy or sandy-skeletal, mixed, mesic; and Typic Udifluvents, coarse-silty, mixed, nonacid, mesic.

A13 Typic Hapludalfs, fine-silty over sandy or sandy-skeletal, mixed, mesic; and Udollic Ochraqualfs, fine-silty, mixed, mesic; with Typic Argiaquolls, fine-silty, mixed, mesic.

A14 Typic Argiudolls, with Typic Glossoboralfs and Eutroboralfs, fine-loamy and fine-silty over sandy, mixed, mesic and frigid.

B. Soils of the Southeastern Upland: Alfisols, Mollisols, Entisols, Inceptisols, Spodosols, and Histosols.

Hilly to rolling soils of drumlins, moraines, and bedrock escarpments.

B1 Typic Hapludalfs, fine-silty and fine, mixed and illitic, mesic; with rocky land.

B2 Typic Argiudolls and Hapludalfs, with Typic Argiaquolls, fine-loamy, mixed, mesic.

B3 Typic Hapludalfs, fine-silty and loamy, mixed, mesic; some soils stony.

Hilly to rolling soils of moraines and terrace escarpments.

B4 Typic Hapludalfs, fine-loamy over sandy or sandy-skeletal, and coarse-loamy, mixed, mesic; with Typic Hapludolls, sandy-skeletal, mixed, mesic.

Rolling to undulating soils of glaciated uplands.

B5 Typic Argiudolls, fine-loamy, mixed, mesic.

B6 Typic Hapludalfs, fine-silty and fine-loamy, mixed, mesic.

B7 Typic Hapludalfs, fine-loamy and fine-silty, mixed, mesic.

B8 Typic Hapludalfs, coarse-loamy and fine-loamy, mixed, mesic; with rock outcrops.

B9 Typic and Aeric Hapludalfs, with Typic Argiudolls, fine, illitic, mesic.

B10 Typic Hapludalfs, fine-silty and loamy, mixed, mesic; with Aeric Ochraqualfs, fine-silty, mixed, mesic.

B11 Typic Hapludalfs, with Aeric Ochraqualfs and Typic Argiudolls, fine-loamy, mixed, mesic.

B12 Typic Hapludalfs, with Typic and Aquic Argiudolls, fine-loamy, mixed, mesic.

B13 Typic Hapludalfs, fine-silty and fine-loamy, mixed, mesic; with Typic Haplaquolls, fine-silty, mixed, mesic.

B14 Typic Hapludalfs, fine-loamy and coarse-loamy, mixed, mesic; with Typic Argiudolls, fine-loamy, mixed, mesic.

B15 Typic and Mollic Hapludalfs, coarse-loamy and fine-loamy, mixed, mesic.

B16 Typic Hapludalfs and Argiudolls, fine-silty and fine-loamy and fine-loamy over sandy or sandy-skeletal, mixed, mesic.

B17 Typic Hapludalfs, fine-loamy and fine-loamy over sandy or sandy-skeletal, mixed, mesic; and Alfic Haplorthods, fine-loamy, mixed, frigid; and Typic Eutrochrepts, coarse-loamy, mixed, mesic.

B18 Typic Hapludalfs, fine-loamy over sandy or sandy-skeletal and fine-silty over sandy, mixed, mesic.

Gently undulating to rolling soils of glaciated uplands.

B19 Typic Hapludalfs, fine, illitic and mixed, mesic; with Aeric Ochraqualfs, fine, illitic, mesic; and Typic Haplaquolls, fine, mixed, mesic.

B20 Typic Hapludalfs, fine, illitic and mixed, mesic; with Typic Haplaquolls, fine, mixed, mesic.

B21 Typic Argiudolls, fine-silty and fine-loamy, mixed, mesic; with Typic Haplaquolls, fine-silty, mixed, mesic.

B22 Typic Argiudolls, fine-silty and fine-loamy, mixed, mesic; with Aquic Argiudolls and Typic Haplaquolls, fine-silty, mixed, mesic.

B23 Typic Hapludalfs, with Typic Argiaquolls, fine-loamy, mixed, mesic; and Histosols.

B24 Typic Hapludalfs, with Typic and Aquic Argiudolls, fine-loamy, mixed, mesic.

B25 Typic Hapludalfs, fine-silty and fine-loamy, mixed, mesic; with Typic Haplaquolls, fine-silty, mixed, mesic.

B26 Arenic Hapludalfs, with Typic Hapludalfs, fine-loamy and coarse-loamy, mixed, mesic.

B27 Typic Hapludalfs, fine-silty and fine-loamy, mixed, mesic.

B28 Typic and Mollic Hapludalfs, coarse-loamy and fine-loamy, mixed, mesic.

B29 Typic Hapludalfs, fine-silty, mixed, mesic.

B30 Typic Hapludalfs, fine-loamy over sandy or sandy-skeletal, and coarse-loamy, mixed, mesic.

B31 Typic Hapludalfs, with Typic Haplaquolls, fine-loamy over sandy or sandy-skeletal, mixed, mesic.

Nearly level and gently undulating soils of outwash plains and uplands.

B32 Typic Argiudolls and Hapludalfs, fine-silty and fine-loamy over sandy or sandy-skeletal, mixed, mesic.

B33 Typic Hapludalfs, fine-loamy and fine-loamy over sandy or sandy-skeletal, mixed, mesic; with Aeric Ochraqualfs, fine, illitic, mesic.

B34 Typic Hapludalfs, fine-silty and fine-loamy, commonly over sandy or sandy-skeletal, mixed, mesic.

C. Soils of the Central Sandy Uplands and Plains: Alfisols, Entisols, Mollisols, Spodosols, Inceptisols, and Histosols.

Rolling, hilly, undulating, and nearly level soils of uplands and plains.

C1 Typic Hapludalfs, coarse-loamy, mixed, mesic; with Psammentic Hapludalfs and Typic Udipsamments, sandy, mixed, mesic.

C2 Typic Hapludalfs, coarse-loamy, mixed, mesic; with Typic Udipsamments and Psammentic Hapludalfs, sandy, mixed, mesic.

C3 Typic Udipsamments and Psammentic Hapludalfs, sandy, mixed, mesic; with Typic Hapludalfs, coarse-loamy, mixed, mesic.

Nearly level and undulating soils of plains with included hilly and steep outliers of sandstone.

C4 Typic Quartzipsamments, mesic, uncoated; with Typic Udipsamments, sandy, mixed, mesic; and Entic Haplaquods, sandy, mixed, frigid; and Histosols.

Nearly level and gently undulating soils of stream terraces and long narrow outwash plains.

C5 Entic Hapludolls and Typic Udipsamments, with Psammentic Hapludalfs, sandy, mixed, mesic.

C6 Typic Udipsamments and Entic Hapludolls, with Psammentic Hapludalfs, sandy, mixed, mesic.

C7 Mollic and Aquollic Hapludalfs, fine-loamy over sandy or sandy-skeletal, mixed, mesic; with Typic Udipsamments and Entic Hapludolls, sandy, mixed, mesic.

C8 Entic Hapludolls, sandy, mixed, mesic; with Typic Argiudolls and Mollic Hapludalfs, fine-loamy over sandy or sandy-skeletal, mixed, mesic.

C9 Typic Argiudolls and Eutroboralfs and Mollic Hapludalfs, fine-loamy over sandy or sandy-skeletal, mixed, mesic and frigid; with Typic Hapludolls, sandy, mixed, mesic.

Nearly level and undulating soils of broad outwash plains.

C10 Typic Udipsamments, with Typic Humaquepts, sandy, mixed, mesic; and Histosols.

C11 Typic Udipsamments and Psammentic Hapludalfs, sandy, mixed, mesic.

C12 Typic and Aquic Udipsamments, with Typic Humaquepts, sandy, mixed, mesic.

C13 Aquic Udipsamments, sandy, mixed, mesic; with Entic Haplaquods and Typic Udipsamments, sandy, mixed, frigid.

C14 Typic and Aquic Udipsamments, with Typic Humaquepts, sandy, mixed, mesic.

C15 Typic Udipsamments and Arenic Hapludalfs, sandy, mixed, mesic; with Typic Argiudolls, fine-loamy over sandy or sandy-skeletal, mixed, mesic.

C16 Typic Argiudolls and Eutroboralfs, fine-loamy over sandy or sandy-skeletal, mixed, mesic and frigid.

C17 Alfic Haplorthods, sandy over loamy, mixed, frigid; and Aeric Glossaqualfs, fine-loamy, mixed, frigid; with Typic Udipsamments and Humaquepts, sandy, mixed, mesic.

C18 Arenic Hapludalfs (some Aquic), coarse-loamy over clayey, mixed, mesic; with Typic Glossoboralfs, coarse-loamy, mixed, mesic.

D. Soils of Western Sandstone Uplands, Valley Slopes, and Plains: Alfisols, Entisols, Inceptisols, Mollisols, Spodosols, and Histosols.

Steep to rolling soils on partially dissected uplands underlain by sandstone and some limestone.

D1 Steep rocky land; with Typic Hapludalfs, fine-loamy, commonly over sandy or sandy-skeletal, and fine-silty, mixed, mesic.

Hilly, rolling, and steep soils on dissected sandstone uplands.

D2 Typic Hapludalfs, fine-silty, commonly over sandy or sandy-skeletal, mixed, mesic.

D3 Typic Hapludalfs, fine-loamy, fine-silty over sandy or sandy-skeletal, and fine-silty, mixed, mesic.

D4 Typic Hapludalfs, fine-loamy and fine-loamy over sandy or sandy-skeletal; with Lithic Hapludalfs, loamy, mixed, mesic; and Typic Quartzipsamments, mesic, uncoated.

D5 Typic Hapludalfs, fine-loamy and fine-loamy or fine-silty over sandy or sandy-skeletal, mixed, mesic; with Typic Quartzipsamments, mesic, uncoated.

D6 Typic Quartzipsamments, mesic, uncoated; and Lithic Hapludalfs, loamy, mixed, mesic.

D7 Typic Hapludalfs and Eutric Glossoboralfs, fine-loamy over sandy or sandy-skeletal, and fine-loamy, mixed, mesic and frigid.

Gently rolling and rolling soils on sandstone uplands.

D8 Aqualfic Haplorthods, coarse-loamy over clayey, mixed, frigid; with Typic Quartzipsamments, mesic, uncoated; and Lithic Hapludalfs, loamy, mixed, mesic; and Typic Haplaquents, sandy over loamy, mixed, acid, frigid; and Eutric Glossoboralfs, fine-loamy over sandy or sandy-skeletal, mixed, frigid.

D9 Typic Hapludalfs, fine-loamy and fine-silty over sandy or sandy-skeletal, mixed, mesic; with Lithic Hapludalfs, loamy, mixed, mesic; with Eutric Glossoboralfs and Glossic Eutroboralfs, fine-loamy and coarse-loamy over sandy or sandy-skeletal, mixed, frigid.

D10 Typic Hapludalfs, fine-loamy over sandy or sandy-skeletal, mixed, mesic; and Typic Eutroboralfs, fine-loamy over sandy or sandy-skeletal, and coarse-loamy, mixed, frigid.

Nearly level and undulating soils on sandstone plains.

D11 Typic Haplaquents, sandy over loamy, mixed, acid, frigid; with Aqualfic and Alfic Haplorthods, coarse-loamy over clayey, mixed, frigid; and Typic Quartzipsamments, mesic, uncoated; and Lithic Hapludalfs, loamy, mixed, mesic.

D12 Aqualfic and Alfic Haplorthods, coarse-loamy over clayey, mixed, frigid; with Typic Haplaquents, sandy over loamy, mixed, acid, frigid.

D13 Aquic Glossoboralfs, fine-loamy, mixed, frigid; with Humic Haplaquepts and Typic Humaquepts, fine-loamy, mixed, acid, frigid.

E. Soils of the Northern and Eastern Sandy and Loamy Reddish Drift Uplands and Plains: Alfisols, Entisols, Inceptisols, Mollisols, Spodosols, and Histosols.

Soils of rolling to undulating uplands.

E1 Alfic Haplorthods, coarse-loamy and fine-loamy, mixed, frigid; with Entic Haplorthods, sandy, mixed, frigid.

E2 Alfic Haplorthods and Aquic Eutroboralfs, fine-loamy and coarse-loamy, mixed, frigid; with Typic Eutroboralfs, fine-loamy, mixed, frigid; and Aeric Haplaquepts, fine-loamy, mixed, nonacid, frigid.

Soils of undulating uplands.

E3 Alfic Haplorthods, coarse-loamy and fine-loamy, mixed, frigid; with Aquic Eutroboralfs and Aeric Haplaquepts, fine-loamy, mixed, nonacid, frigid; and Entic Haplorthods, sandy, mixed, frigid.

E4 Alfic Haplorthods, fine-loamy and coarse-loamy, mixed, frigid; and Typic Eutroboralfs, fine-loamy, mixed, frigid; with Typic Glossoboralfs, coarse-loamy, mixed, frigid.

E5 Aquic Eutroboralfs and Alfic Haplorthods, fine-loamy, mixed, frigid; with Glossoboric Hapludalfs, fine-loamy, mixed, mesic; with Aquic Eutrochrepts, coarse-loamy, mixed, mesic; and Aeric Haplaquepts, fine-loamy, mixed, nonacid, frigid.

E6 Entic, Alfic, and Lithic Haplorthods, coarse-loamy, loamy, and fine-loamy, mixed, frigid; with Aquic Eutroboralfs, fine-loamy, mixed, frigid; and limestone rock outcrops.

E7 Alfic and Entic Haplorthods, fine-loamy and coarse-loamy, mixed, frigid; with Aquic and Lithic Eutrochrepts, fine-loamy, mixed, frigid.

Soils of nearly level plains.

E8 Aquic Eutrochrepts, coarse-loamy, mixed, mesic; and Arenic Hapludalfs, clayey, mixed, mesic; with Typic Udipsamments, sandy, mixed, frigid; with Typic Hapludalfs and Haplaquolls, fine and very fine, mixed, mesic; and Histosols.

E9 Alfic Haplorthods and Typic Eutrochrepts, fine-loamy, mixed, frigid; with Aeric Haplaquepts, fine-loamy, mixed, nonacid, frigid; and Typic Haplaquolls, coarse-loamy over clayey, mixed, mesic.

E10 Arenic Hapludalfs and Aquic Udorthents, coarse-loamy over clayey, mixed, mesic; with Typic Haplaquolls, fine, mixed, mesic; and Histosols.

E11 Arenic Hapludalfs, clayey, mixed, mesic; and Aquic Eutrochrepts, coarse-loamy, mixed, mesic; with Aquollic Hapludalfs, fine-loamy, mixed, mesic.

E12 Typic Udipsamments, with Typic Haplaquolls, sandy, mixed, mesic; and Mollic Haplaquepts, coarse-loamy, mixed, nonacid, mesic; and Entic Haplaquods, sandy, mixed, frigid.

E13 Typic Udipsamments, sandy, mixed, frigid; and Typic Haplaquolls, sandy, mixed, mesic; with Histosols.

F. Soils of the Northern Silty Uplands and Plains: Spodosols, Alfisols, Mollisols, Inceptisols, and Histosols.

Soils of rolling to undulating uplands.

F1 Alfic Haplorthods and Haplaquods, coarse-loamy, mixed, frigid; with Typic Glossaqualfs, fine-silty, mixed, frigid.

F2 Typic Glossoboralfs and Aeric Ochraqualfs, fine-loamy, mixed, frigid; with Glossic Eutroboralfs, coarse-loamy, mixed, frigid; and Typic Haplaquepts, coarse-loamy, mixed, nonacid, frigid.

F3 Typic Glossoboralfs and Aeric Ochraqualfs and Glossaqualfs, fine-loamy and fine-silty, mixed, mesic; with Typic Haplaquolls, coarse-loamy, mixed, frigid.

F4 Typic Glossoboralfs and Hapludalfs, fine-loamy, mixed, frigid; with Typic Eutroboralfs, fine-loamy over sandy or sandy-skeletal, mixed, frigid; and Histosols.

F5 Alfic Fragiorthods, coarse-loamy, mixed, frigid; and Alfic Haplorthods, coarse-silty and coarse-loamy, mixed, frigid.

F6 Typic Glossoboralfs and Eutroboralfs, fine-silty and fine-loamy over sandy or sandy-skeletal, mixed, frigid; with Typic Glossoboralfs and Hapludalfs, fine-loamy, mixed, frigid and mesic.

F7 Glossic and Glossaquic Eutroboralfs, fine-loamy, mixed, frigid; with Mollic Ochraqualfs, fine, mixed, frigid.

F8 Eutric Glossoboralfs, fine-loamy, mixed, frigid; and Typic Argiudolls, fine-silty and fine-loamy over sandy or sandy-skeletal, mixed, mesic.

Soils of undulating uplands.

F9 Alfic Haplaquods, coarse-loamy, mixed, frigid; and Typic Glossaqualfs, fine-silty, mixed, frigid; with Histosols.

F10 Typic and Aeric Glossoboralfs, with Glossic Eutroboralfs, fine-loamy, mixed, frigid; and Typic Haplaquepts, coarse-loamy, mixed, acid, frigid.

F11 Typic Glossoboralfs and Aeric Glossaqualfs, with Eutric Glossoboralfs, fine-loamy over sandy or sandy-skeletal, mixed, frigid; and Typic Ochraqualfs, fine-loamy, mixed, frigid.

F12 Typic and Aeric Glossoboralfs, with Typic Glossaqualfs, fine-silty, mixed, frigid; and Typic Haplaquolls, coarse-loamy, mixed, frigid.

F13 Typic Hapludalfs and Argiudolls, fine-loamy, mixed, mesic; with Aquic Hapludolls and Typic Glossaqualfs, fine-loamy, mixed, mesic.

F14 Typic Glossoboralfs, coarse-loamy, mixed, frigid; with Typic Haplaquepts, coarse-loamy, mixed, nonacid, frigid; some stony areas.

F15 Typic and Aquic Glossoboralfs, coarse-loamy, mixed, frigid.

F16 Alfic Fragiorthods, coarse-loamy, mixed, frigid; with Histosols.

F17 Typic and Aquic Glossoboralfs and Typic Eutroboralfs, fine-silty over sandy or sandy-skeletal, mixed, frigid.

Soils of nearly level plains.

F18 Alfic Haplaquods, coarse-loamy, mixed, frigid; and Typic Glossaqualfs, fine-silty, mixed, frigid; with Histosols; some stony areas.

F19 Alfic Haplaquods and Haplorthods, coarse-loamy, mixed, frigid; with Typic Haplaquolls, fine-silty, mixed, frigid.

F20 Aeric and Typic Ochraqualfs and Glossaqualfs, fine-loamy and fine-silty, mixed, frigid.

F21 Aeric and Typic Glossaqualfs, fine-loamy, mixed, frigid; with Typic Haplaquolls, coarse-loamy, mixed, frigid; and Histosols.

F22 Aeric and Typic Glossoboralfs, with Typic Glossaqualfs, fine-silty, mixed, frigid; and Histosols.

F23 Aquic and Aeric Glossoboralfs, coarse-loamy, mixed, frigid; some stony areas.

F24 Alfic Fragiorthods and Haplorthods and Typic Haplorthods, coarse-loamy, mixed, frigid.

F25 Typic Glossoboralfs and Eutroboralfs, fine-silty and fine-loamy over sandy or sandy-skeletal, mixed, frigid.

F26 Aquic and Typic Glossoboralfs and Typic Eutroboralfs, fine-silty and fine-loamy over sandy or sandy-skeletal, mixed, frigid.

G. Soils of the Northern Loamy Uplands and Plains: Spodosols, Alfisols, Entisols, Inceptisols, and Histosols.

Soils of hilly, rolling to undulating uplands.

G1 Alfic Fragiorthods, coarse-loamy, mixed, frigid; with bedrock outcrops.

G2 Alfic Fragiorthods and Typic and Alfic Haplorthods, coarse-loamy and coarse-silty, mixed, frigid; with Aquic Dystrochrepts, coarse-loamy, mixed, frigid; and Histosols; some stony areas.

G3 Alfic Fragiorthods and Typic Haplorthods, coarse-loamy, mixed, frigid; with Entic Haplorthods, sandy, mixed, frigid; and Histosols.

G4 Glossic Eutroboralfs, with Typic Haplorthods, coarse-loamy, mixed, frigid; with Dystric Eutrochrepts, sandy, mixed, frigid; and Histosols.

G5 Typic Hapludalfs, fine-loamy, mixed, mesic; and Alfic Haplorthods, coarse-loamy, mixed frigid; with Typic Haplorthods, coarse-loamy, mixed, frigid; and Entic Haplorthods, sandy, mixed, frigid; and Histosols.

G6 Typic Hapludalfs, fine-loamy and coarse-loamy, mixed, mesic; with Typic Eutroboralfs, fine-loamy over sandy or sandy-skeletal, mixed, frigid.

G7 Alfic and Typic Haplorthods, coarse-loamy, mixed, frigid; with Entic Haplorthods, sandy, mixed, frigid; and Histosols.

G8 Eutric Glossoboralfs, coarse-loamy, mixed, frigid; and Typic Hapludalfs, coarse-loamy, mixed, mesic; with Entic Haplorthods, sandy, mixed, mesic; and Histosols.

Soils of rolling to undulating uplands, with some broad valleys.

G9 Alfic Fragiorthods, with Typic Fragiorthods, coarse-loamy, mixed, frigid; and Typic Haplorthods, sandy, mixed, frigid; and bedrock outcrops.

G10 Alfic and Typic Fragiorthods, coarse-loamy, mixed, frigid; and Typic Haplorthods, sandy, mixed, frigid; with Typic Haplaquepts, coarse-loamy, mixed, nonacid, frigid; and Histosols.

G11 Alfic Fragiorthods and Typic Haplorthods, coarse-loamy, mixed, frigid; with Entic Haplorthods, sandy, mixed, frigid; and Histosols.

G12 Dystric Eutrochrepts, sandy, mixed, frigid; with Alfic Fragiorthods and Typic Haplorthods, coarse-loamy, mixed, frigid; and Histosols.

G13 Glossic Eutroboralfs, with Alfic Fragiorthods, coarse-loamy, mixed, frigid; and Dystric Eutrochrepts, sandy, mixed, frigid; with Typic Haplaquepts, coarse-loamy, mixed, nonacid, frigid; and Histosols.

G14 Typic Hapludalfs, fine-loamy, mixed, mesic; with Alfic Fragiorthods and Typic Haplorthods, coarse-silty and sandy, mixed, frigid; with Histosols.

G15 Typic Hapludalfs, fine-loamy and coarse-loamy, mixed, mesic; with Typic Eutroboralfs, fine-loamy over sandy or sandy-skeletal, mixed, frigid; with Alfic Haplorthods, sandy, mixed, frigid; and Histosols.

G16 Alfic and Typic Haplorthods, coarse-loamy and sandy, mixed, frigid; with Entic Haplorthods, sandy, mixed, frigid; and Histosols.

G17 Typic Haplorthods, sandy, mixed, frigid; and Typic Dystrochrepts, coarse-loamy over sandy or sandy-skeletal, mixed, frigid; with Alfic Haplorthods, coarse-loamy, mixed, frigid; and Histosols.

G18 Typic and Entic Haplorthods, sandy, mixed, frigid; and Histosols.

G19 Typic Eutroboralfs, fine-loamy, mixed, frigid; and Eutric Glossoboralfs, coarse-loamy over sandy or sandy-skeletal, mixed, frigid; and Typic Argiudolls, fine-loamy over sandy or sandy-skeletal, mixed, mesic.

Soils of undulating uplands.

G20 Alfic Fragiorthods, coarse-loamy, mixed, frigid; with Typic Dystrochrepts, coarse-loamy over sandy or sandy-skeletal, mixed, frigid; with Entic Haplorthods, sandy, mixed, frigid; and Typic Haplaquepts, coarse-loamy, mixed, nonacid, frigid.

G21 Alfic Fragiorthods, coarse-loamy, mixed, frigid; and Typic Haplaquepts, coarse-loamy, mixed, nonacid, frigid; and Histosols.

G22 Glossic Eutroboralfs, coarse-loamy, mixed, frigid; and Typic Dystrochrepts, coarse-loamy over sandy or sandy-skeletal, mixed, frigid; with Typic Glossoboralfs and Aeric Ochraqualfs, fine-loamy, mixed, frigid; with Typic Haplaquepts, coarse-loamy, mixed, nonacid, frigid; and Histosols.

G23 Alfic Fragiorthods and Aquic Dystrochrepts, coarse-loamy, mixed, frigid; with Typic Haplorthods, sandy, mixed, frigid; and Histosols.

G24 Typic Hapludalfs, fine-loamy and coarse-loamy, mixed, mesic; with Typic Eutroboralfs, fine-loamy over sandy or sandy-skeletal, mixed, frigid.

G25 Typic and Alfic Haplorthods, coarse-loamy and sandy, mixed, mesic; with Entic Haplorthods, sandy, mixed, frigid.

G26 Typic Eutroboralfs and Glossoboralfs, coarse-loamy, and fine-loamy over sandy or sandy-skeletal, mixed, frigid; with Histosols.

Soils of nearly level plains, pitted and unpitted.

G27 Typic and Entic Haplorthods, sandy, mixed, frigid; with Alfic Fragiorthods, coarse-loamy, mixed, frigid; and Histosols.

G28 Typic Eutroboralfs and Eutric Glossoboralfs, coarse-loamy and fine-loamy over sandy or sandy-skeletal, mixed, frigid; with Typic Glossoboralfs, fine-silty over sandy or sandy-skeletal, mixed, frigid; and Histosols.

H. Soils of the Northern Sandy Uplands and Plains: Spodosols, Entisols, Alfisols, and Histosols.

Soils of hilly to rolling uplands.

H1 Entic and Typic Haplorthods, sandy, mixed, frigid; with Histosols.

Soils of rolling uplands.

H2 Entic and Typic Haplorthods, sandy, mixed, frigid; with Histosols.

Soils of undulating uplands.

H3, H4 Entic and Typic Haplorthods, sandy, mixed, frigid; with Histosols.

Soils of nearly level plains.

H5 Entic and Typic Haplorthods, sandy, mixed, frigid; with Histosols.

H6 Entic and Typic Haplorthods, sandy, mixed, frigid; and Eutric Glossoboralfs, coarse-loamy, mixed, frigid; with Histosols.

H7 Typic and Entic Haplorthods, with Entic Haplaquods, sandy, mixed, frigid.

I. Soils of the Northern and Eastern Clayey and Loamy Reddish Drift Uplands and Plains: Alfisols, Mollisols, Spodosols, Inceptisols, and Histosols.

Soils of rolling to hilly uplands.

I1 Typic Eutroboralfs, fine, mixed, frigid; and Dystric Eutrochrepts, coarse-loamy over clayey, mixed, frigid; with Alfic Haplorthods, coarse-loamy over clayey, mixed, frigid; and Aeric Haplaquods, sandy over loamy, mixed, ortstein, frigid.

I2 Typic and Glossic Eutroboralfs, fine and very fine, mixed, frigid; and Aeric Haplaquepts, fine, mixed, nonacid, frigid; with Typic Haplorthods, sandy, mixed, frigid.

I3 Glossic Eutroboralfs, fine, mixed, frigid; and Entic and Lithic Haplorthods, loamy, mixed, frigid; with Typic Hapludalfs, fine, mixed, mesic; with limestone and shale rockland.

Soils of rolling to undulating uplands.

I4 Typic, Glossoboric, and Aquollic Hapludalfs, fine and fine-loamy, mixed, mesic; with Typic Haplaquolls, fine, mixed, mesic.

I5 Glossoboric, Typic, and Aquollic Hapludalfs, fine-loamy and fine, mixed, mesic; with Typic Haplaquolls, fine, mixed, mesic.

I6 Alfic Haplorthods, fine-loamy, mixed, frigid; and Typic Hapludalfs, fine-loamy and fine-loamy over sandy or sandy-skeletal, mixed, mesic; with Glossoboric Hapludalfs, fine-loamy, mixed, mesic.

Soils of undulating plains.

I7 Typic Eutroboralfs, fine, mixed, frigid; and Dystric Eutrochrepts, coarse-loamy over clayey, mixed, frigid; with Alfic Fragiorthods, coarse-loamy, mixed, frigid; and Typic Haplorthods, sandy, mixed, frigid.

I8 Typic and Aquic Eutroboralfs, fine and very fine, mixed, frigid; with Aeric Haplaquepts, fine, mixed, nonacid, frigid; and Typic Haplorthods, sandy, mixed, frigid.

I9 Alfic Haplorthods, coarse-loamy over clayey and sandy over clayey, mixed, frigid; and Aquic and Typic Haplorthods, sandy, mixed, frigid.

I10, I11 Typic and Aquollic Hapludalfs, with Typic Haplaquolls, fine, mixed, mesic.

I12 Typic and Aquollic Hapludalfs, fine, mixed, mesic.

I13 Typic and Aquollic Hapludalfs, with Typic Haplaquolls, fine, mixed, mesic; and Glossoboric Hapludalfs, fine-loamy, mixed, mesic; and Arenic Hapludalfs, clayey, mixed, mesic.

I14 Aquollic Hapludalfs and Typic Haplaquolls, fine, mixed, mesic; with Arenic Ochraqualfs, coarse-loamy, mixed, mesic; and Arenic Hapludalfs, clayey, mixed, mesic.

I15 Typic Hapludalfs, with Aquollic Hapludalfs, fine, mixed, mesic; and Glossic Eutroboralfs, fine, mixed, frigid.

I16 Glossoboric Hapludalfs, fine-loamy, mixed, mesic; and Aquollic Hapludalfs, fine, mixed, mesic; with Typic Haplaquolls, fine, mixed, mesic; and Aquic Eutrochrepts, coarse-loamy, mixed, mesic.

I17 Typic Hapludalfs, fine and coarse-loamy, mixed, mesic; and Typic Haplaquolls, fine, mixed, mesic; with Arenic Hapludalfs, clayey, mixed, mesic.

Soils of nearly level plains.

I18 Typic and Aquic Eutroboralfs, fine and very fine, mixed, frigid; with Aeric Haplaquepts, fine, mixed, nonacid, frigid; and Alfic Haplorthods, coarse-loamy over clayey, mixed, frigid.

I19 Typic and Aquic Eutroboralfs, fine and very fine, mixed, frigid.

I20 Typic and Aquollic Hapludalfs, fine and very fine, mixed, mesic; with Typic Haplaquolls, fine, mixed, mesic.

I21 Typic Hapludalfs, very fine, mixed, mesic; with Aquollic Hapludalfs, fine, mixed, mesic; and Typic Haplaquolls, fine, mixed, mesic.

I22 Arenic Eutrochrepts and Mollic Haplaquepts, sandy over loamy, mixed, frigid.

J. Soils of the Stream Bottoms and Major Wetlands: Entisols, Histosols, Mollisols, Spodosols, Inceptisols, and Alfisols.

Soils of nearly level bottoms, with local cutbanks.

J1 Typic and Aquic Udifluvents, coarse-silty, mixed, nonacid, mesic; and Tyic Argiaquolls, fine-silty, mixed, mesic.

J2 Haplaquolls, Haplaquents, and Fluvaquents, undifferentiated.

Wet mineral soils of nearly level plains.

J3 Typic Haplaquolls, sandy, mixed, mesic; with Typic Udipsamments, sandy, mixed, frigid; and Alfic Haplorthods, coarse-loamy, mixed, frigid; with shallow Histosols.

J4 Typic Humaquepts, with Typic and Aquic Udipsamments, sandy, mixed, mesic; and shallow Histosols.

J5 Typic Humaquepts, with Aquic Udipsamments and Typic Haplaquolls, sandy, mixed, mesic; and shallow Histosols.

J6 Typic Haplaquepts and Aeric Ochraqualfs, fine-loamy, mixed, frigid; and Aquic Dystrochrepts, coarse-loamy, mixed, frigid; with Typic Glossaqualfs, fine-silty, mixed, frigid; and Histosols.

J7 Typic Haplaquolls, coarse-loamy over clayey, mixed, mesic; and Mollic Haplaquepts, coarse-loamy, mixed, nonacid, mesic; with Arenic Ochraqualfs, coarse-loamy, mixed, mesic.

J8 Typic Haplaquolls and Argiaquolls and Mollic Ochraqualfs, fine-silty, mixed, mesic.

J9 Udollic Ochraqualfs and Typic Haplaquolls, fine-loamy over sandy or sandy-skeletal, and fine-silty, mixed, mesic.

J10 Typic Argiaquolls, fine-loamy, mixed, mesic; with Typic Hapludalfs and Aquic Argiudolls, fine-loamy, mixed, mesic; and Typic Haplaquolls, fine-silty, mixed, mesic.

J11 Aquollic and Typic Hapludalfs, clayey over sandy or sandy-skeletal, mixed, mesic; and Typic Haplaquolls, fine, and clayey over sandy or sandy-skeletal, mixed, mesic.

Wet organic soils of nearly level plains.

J12 Fibrists and Hemists; with Entic Haplaquods, sandy, mixed, frigid; and Typic Haplaquepts, coarse-loamy, mixed, nonacid, frigid.

J13 Fibrists and Hemists; with Typic Haplaquepts, coarse-loamy, mixed, nonacid, frigid; and Aeric Ochraqualfs, fine-silty, mixed, frigid.

J14 Fibrists, Hemists, and Saprists; with Entic Haplaquods, sandy, mixed, frigid; with Typic Humaquepts and Aquic Udipsamments, sandy, mixed, mesic.

J15 Saprists, Hemists, and Fibrists; with Typic Haplaquolls, fine and fine-silty, mixed, mesic; and Typic Argiaquolls, fine-loamy, mixed, mesic.

APPENDIX 4

Errata in the Soil Map of Wisconsin (Plate 1)

For A1, read A12, in Trempealeau County, T.20N., R.7W.

For B, read B28, in Green Lake County, T.16N., R.11E.

For a blank strip, in Waukesha County, T.6N., R.20E., southward the soil boundaries interrupted.

For a missing label in a soil body in Menominee County, T.29N., R.14E., insert G14.

For 4, read I4, in Ozaukee County, T.12N., R.22E.

For I0 read I10, in Brown County, T.21N., R.19E.

For I1 read I11, in Calumet County, T.19N., R.19E.

In the legend:

F15, for Meade read Mead.

G10, for Marensico read Marenisco.

Note: For clarification of soil boundaries interrupted by map lettering or crossed by streams, see the overlay soil map referred to in the lower left-hand corner of the map sheet.

APPENDIX 5

List of Soil Survey Publications for Wisconsin

County	Publisher (date)[1]
Adams	GNHS (1924)
Barron	USDA (1958); GNHS (1948)
Bayfield	USDA (1961); GNHS (1929)
Brown	GNHS (1929-OP)
Buffalo	USDA (1962); GNHS (1917)
Calumet	GNHS (1925)
Columbia	GNHS (1916-RD)
Crawford	USDA (1961); GNHS (1930)
Dane	GNHS (1917)
Dodge	GNHS (Part, 1953[2])
Door	GNHS (1919)
Florence	GNHS (1962)
Fond du Lac	USDA (1973); GNHS (1914-OP)
Grant	USDA (1961); GNHS (1952;[2] 1956)
Green	GNHS (1930)
Green Lake	GNHS (1919)
Iowa	USDA (1962); GNHS (1914-OP)
Jackson	GNHS (1923)
Jefferson	GNHS (1970; 1916-OP)
Juneau	GNHS (1914-RD)
Kenosha	USDA (1970); GNHS (1923)
Kewaunee	GNHS (1914-OP)
La Crosse	USDA (1966)
Lafayette	USDA (1966)
Langlade	GNHS (1947)
Manitowoc	GNHS (1926)
Marquette	GNHS (1961)
Menominee	GNHS (1967)
Milwaukee	USDA (1971); GNHS (1919-OP)
Monroe	GNHS (1931)
Oneida	GNHS (1959)
Outagamie	GNHS (1922-OP)
Ozaukee	USDA (1970); GNHS (1926)
Pepin	USDA (1964)
Pierce	USDA (1968); GNHS (1930)
Portage	GNHS (1918)
Racine	USDA (1970); GNHS (1923)
Richland	USDA (1959); GNHS (1950)
Rock	GNHS (1922)
Sauk	GNHS (1925)
Sheboygan	GNHS (1929)
Trempealeau	GNHS (1927)
Vernon	USDA (1969); GNHS (1928)
Vilas	GNHS (1915)
Walworth	USDA (1971); GNHS (1924-OP)
Washington	GNHS (1926)
Waukesha	USDA (1971); GNHS (1965; 1914-OP)
Waupaca	GNHS (1921)
Waushara	GNHS (1913-OP)
Winnebago	GNHS (1927)
Wood	GNHS (1918)

1. Publishing agencies are: GNHS = Geological and Natural History Survey, University of Wisconsin Extension, Madison, Wisconsin 53706. USDA = U.S. Department of Agriculture, Soil Conservation Service. The abbreviations "OP" and "RD," respectively, signify "out of print" and "restricted distribution."

2. Folder detailed soil maps are available for several towns in this county.

State Soil Maps (in color)

Leaflet map GNHS (1968)
Wall map GNHS (1968)
Overlay map (to be used on USGS 1:250,000 quadrangles) GNHS (1968)

Miscellaneous Publications

Bouma, J., et al. Soil potential for disposal of septic tank effluent: a field study of major soils of Wisconsin GHNS (1972-OP)
Hole, F. D., et al. What's in that soil map? GNHS (1953)
Hole, F. D., and G. B. Lee. Introduction to Soils of Wisconsin GNHS (1955)

Bibliography
Index

Bibliography

Ableiter, J. K., and F. D. Hole. 1961. Soil survey of Bayfield County, Wisconsin. U.S. Dept. Agr. Soil Conserv. Service Ser. 1939, No. 30. 77 p.

Aebischer, D. C., M. T. Beatty, I. O. Hembre, and F. D. Hole. 1969. How good is your land? A handbook for agricultural workers on land judging in Wisconsin. University of Wisconsin Extension, Madison. 31 p.

Ahlgren, H. L., M. L. Wall, R. J. Muckenhirn, and J. M. Sund. 1946a. Yields of forage from woodland pastures on sloping land in southern Wisconsin. J. Forest. 44:709-711.

Ahlgren, H. L., M. L. Wall, R. J. Muckenhirn, and J. M. Sund. 1946b. Yields of renovated and unimproved permanent pastures on sloping land in southern Wisconsin. J. Amer. Soc. Agron. 38:914-922.

Akers, R. H. 1964. Unusual surficial deposits in the Driftless Area of Wisconsin. Ph.D. Thesis. Univ. Wisconsin. 119 p.

Albert, A. R. 1945. Status of organic soil use in Wisconsin. Soil Sci. Soc. Amer. Proc. 10:275-278.

Albert, A. R. 1951. Better crops and incomes from sandy soils. Univ. Wis. Ext. Coll. Agr. Circ. 402. 32 p.

Albert, A. R., and O. R. Zeasman. 1953. Farming muck and peat in Wisconsin. Univ. Wis. Ext. Coll. Agr. Circ. 456. 35 p.

Alden, W. C. 1904. The Delavan Lobe of the Lake Michigan glacier of the Wisconsin stage of glaciation and related phenomena. U.S. Geol. Surv. Prof. Paper 34. 106 p.

Alden, W. C. 1905. The drumlins of southeastern Wisconsin. U.S. Geol. Surv. Bull. 273. 43 p.

Alden, W. C. 1918a. Map of southeastern Wisconsin, bedrock configuration, 1:250,000. *In* U.S. Geol. Surv. Prof. Paper 106.

Alden, W. C. 1918b. The Quaternary geology of southeastern Wisconsin. U.S. Geol. Surv. Prof. Paper 106. 356 p.

Al-Rawi, A. H., F. D. Hole, M. L. Jackson, and J. K. Syers. 1969. Genesis of layer silicates in representative soils in a glacial landscape of northeastern Wisconsin, p. 105-141. *In* A. H. Al-Rawi, 1969, Quantitative mineralogical analysis of some soils in Iraq, and Antigo silt loam catena in Menominee County, Wisconsin. Ph.D. Thesis. Univ. Wisconsin. 141 p.

Anderson, H. O., and I. O. Hembre. 1955. The Coon valley watershed project: a pioneer venture in good land use. J. Soil Water Conserv. 10:180-184.

Anderson, J. R. 1967. Major land uses (map of U.S.A.). *In* A. C. Gerlach, 1970. The national atlas of the United States of America. U.S. Geological Survey, Washington, D.C. 417 p.

Arnold, R. W., and F. F. Riecken. 1964. Grainy gray coatings in some Brunizem soils. Ia. Acad. Sci. Proc. 71:350-360.

Baldwin, M., C. E. Kellogg, and J. Thorp. 1938. Soil classification, p. 979-1001. *In* Soils and Men, U.S. Dept. Agr. Yearbook of Agriculture. U.S. Government Printing Office, Washington, D.C.

Ballagh, T. M., and E. C. A. Runge. 1970. Clay-rich horizons over limestone—illuvial or residual? Soil Sci. Soc. Amer. Proc. 34:534-536.

Bartelli, L. J., and R. T. Odell. 1960a. Field studies of a clay-enriched horizon in the lowest part of the solum of some Brunizem and Gray-Brown Podzolic soils in Illinois. Soil Sci. Soc. Amer. Proc. 24:388-390.

Bartelli, L. J., and R. T. Odell. 1960b. Laboratory studies and genesis of a clay-enriched horizon in the lowest part of the solum of some Brunizem and Gray-Brown Podzolic soils in Illinois. Soil Sci. Soc. Amer. Proc. 24:390-395.

Batson, D. M. 1940. Minerals of the clay and other textural separates of the various horizons of Miami silt loam and their relations to soil forming processes. Ph.D. Thesis. Univ. Wisconsin. 77 p.

Bauer, K. W. 1964. Regional planning in southeastern Wisconsin. Wis. Acad. Rev. 11(3 & 4):2-6.

Baxter, F. P. 1968. Genesis of upland soils, p. 91-101. *In* C. J. Milfred, G. W. Olson, and F. D. Hole, Soil resources and forest ecology of Menominee County, Wisconsin. Wis. Geol. Nat. Hist. Surv. Bull. 85, Soil Ser. No. 60.

Baxter, F. P., and F. D. Hole. 1967. Ant (*Formica cinerea*) pedoturbation in a prairie soil. Soil Sci. Soc. Amer. Proc. 31:425-428.

Beaton, J. D. 1971. Market potential for fertilizer sulphur, p. 16-29. *In* a symposium, Marketing fertilizer sulphur. T.V.A. and Sulphur Institute, Washington, D.C.

Beatty, M. T. 1960. Soil survey of La Crosse County, Wisconsin. U.S. Dept. Agr. Soil Conserv. Service Ser. 1956, No. 7. 93 p., 30 map sheets.

Beatty, M. T., and R. B. Corey. 1961. General subsoil fertility groups. Map in L. M. Walsh and E. E. Schulte [eds.], 1970, Computer-programmed soil test recommendations for field and vegetable crops. Soils Department, College of Agriculture, University of Wisconsin, Madison. 47 p.

Beatty, M. T., and R. B. Corey. 1962. Subsoil fertility in Wisconsin soils. Wis. Acad. Rev. 9:22-23.

Beatty, M. T., and A. E. Peterson. 1963. Principles of soil conservation and soil improvement as they apply to certain groups of soils in southeastern Wisconsin. Trans. Wis. Acad. Sci. Arts Letters 52: 205-212.

Beatty, M. T., A. J. Klingelhoets, D. A. Rohweder, and E. A. Brickbauer. 1966. What yields from Wisconsin soils? Univ. Wis. Ext. Coll. Agr. Spec. Circ. 65. 19 p.

Beatty, M. T., J. A. Schoenemann, R. B. Schuster, G. C. Kingbiel, and J. H. von Elbe. 1964. Vegetable and fruit potential for production and processing in central Wisconsin. Extension Service, College of Agriculture, University of Wisconsin, Madison. 32 p., colored map.

Beaver, A. J. 1963. Characteristics of some bisequal soils in the Valderan drift region of eastern Wisconsin. M.S. Thesis. Univ. Wisconsin. 117 p.

Beaver, A. J. 1966. Characteristics and genesis of some bisequal soils in eastern Wisconsin. Ph.D. Thesis. Univ. Wisconsin. 220 p.

Becking, L. G., I. R. Kaplan, and D. Moore. 1960. Limits of the natural environment in terms of pH and oxidation-reduction potentials. J. Geol. 68:243-284.

Berger, K. C., F. D. Hole, and J. M. Beardsley. 1952. A soil productivity score card. Soil Sci. Soc. Amer. Proc. 16:307-311.

Bidwell, O. W., and F. D. Hole. 1963. Numerical taxonomy and soil classification. Soil Sci. 97:58-62.

Bidwell, O. W., and F. D. Hole. 1965. Man as a factor of soil formation. Soil Sci. 99:65-72.

Black, R. F. 1964. The physical geography of Wisconsin, p. 171-177. *In* The Wisconsin Blue Book. Legislative Reference Bureau, State of Wisconsin, Madison.

Black, R. F. 1965a. Pothole and associated gravels of Devil's Lake State Park. Trans. Wis. Acad. Sci. Arts Letters 54:165-175.

Black, R. F. 1965b. Ice-wedge casts of Wisconsin. Trans. Wis. Acad. Sci. Arts Letters 54:187-222.

Black, R. F. 1968. Geomorphology of Devil's Lake area, Wisconsin. Trans. Wis. Acad. Sci. Arts Letters 56:117-148.

Black, R. F. 1969a. Slopes in southeastern Wisconsin, U.S.A.—periglacial or temperate? Biul. Peryglacjalny 18:69-81.

Black, R. F. 1969b. Glacial geology of northern Kettle Moraine State Forest, Wisconsin. Trans. Wis. Acad. Sci. Arts Letters 57:99-119.

Black, R. F. 1970a. Residuum and ancient soils of the Driftless Area of southwestern Wisconsin, p. I-1 through I-12. *In* R. F. Black, N. K. Bleuer, F. D. Hole, N. P. Lasca, and L. J. Maher, Jr., Pleistocene geology of southern Wisconsin. Wis. Geol. Nat. Hist. Surv. Info. Circ. No. 15.

Black, R. F. 1970b. Glacial geology of Two Creeks Forest Bed, Valderan type locality, and northern Kettle Moraine State Forest. Wis. Geol. Nat. Hist. Surv. Info. Circ. No. 13. 40 p.

Black, R. F., and T. D. Hamilton. 1973. Mass-movement studies near Madison, Wisconsin. Manuscript on file at University of Connecticut. 179 p.

Black, R. F., and M. Rubin. 1968. Radiocarbon dates of Wisconsin. Trans. Wis. Acad. Sci. Arts Letters 56:99-116.

Black, R. F., N. K. Bleuer, F. D. Hole, N. P. Lasca, and L. J. Maher, Jr. 1970. Pleistocene geology of southern Wisconsin. Wis. Geol. Nat. Hist. Surv. Info. Circ. No. 15. 173 p.

Bleuer, N. K. 1970. Glacial stratigraphy of southcentral Wisconsin, p. J-1 through J-35. *In* R. F. Black, N. K. Bleuer, F. D. Hole, N. P. Lasca, and L. J. Maher, Jr., Pleistocene geology of southern Wisconsin. Wis. Geol. Nat. Hist. Surv. Info. Circ. No. 15.

Bleuer, N. K. 1971. Glacial stratigraphy of south central Wisconsin. Ph.D. Thesis, Univ. Wisconsin. 173 pp.

Borchardt, G. A. 1966. Mineralogy and clay genesis of some soils of north-central U.S.A. M.S. Thesis. Univ. Wisconsin. 100 p.

Borchardt, G. A., F. D. Hole, and M. L. Jackson. 1968. Genesis of layer silicate in representative soils in a glacial landscape of southeastern Wisconsin. Soil Sci. Soc. Amer. Proc. 32:399-403.

Borchardt, G. A., M. L. Jackson, and F. D. Hole. 1966. Expansible layer silicate genesis in soil depicted in mica pseudomorphs. Proc. Int. Clay Conf. 1:3-13. Jerusalem.

Borchert, J. R. 1950. The climate of the central North American grassland. Ann. Ass. Amer. Geog. 40:1-29.

Bouma, J. 1973. Use of physical methods to better define the soil survey concept of soil drainage. Soil Sci. Soc. Amer. Proc. 37:413-421.

Bouma, J., and F. D. Hole. 1965. Soil peels and a method for estimating biopore size distribution in soils. Soil Sci. Soc. Amer. Proc. 29:483-485.

Bouma, J., and F. D. Hole. 1971. Soil structure and hydraulic conductivity of adjacent virgin and cultivated pedons at two sites. A Typic Argiudoll (silt loam) and a Typic Eutrochrept (clay). Soil Sci. Soc. Amer. Proc. 35:316-319.

Bouma, J., D. I. Hillel, F. D. Hole, and C. R. Amerman. 1971. Field measurement of unsaturated hydraulic conductivity by infiltration through artificial crusts. Soil Sci. Soc. Amer. Proc. 35:362-364.

Bouma, J., W. J. Ziebell, W. G. Walker, P. Olcott, E. McCoy, and F. D. Hole. 1972. Soil absorption of septic tank effluent: a field study of some major soils of Wisconsin. Wis. Geol. Nat. Hist. Surv. Info. Circ. No. 20. 200 p.

Brewer, R. 1964. Fabric and mineral analysis of soils. John Wiley and Sons, Inc., New York. 470 p.

Bryson, R. A. 1957. Mean annual precipitation minus estimated potential evapotranspiration. Map of the eastern U.S.A. Department of Meteorology, University of Wisconsin, Madison.

Bryson, R. A., and W. M. Wendland. 1966. Tentative climatic patterns for some late glacial and post-glacial episodes in central North America, p. 271-298. *In* W. J. Mayer-Oakes, Life, land and water, Lake Agassiz region. Univ. Manitoba Dept. Anthropol. Occ. Papers No. 1.

Bryson, R. A., and W. M. Wendland. 1967. Radiocarbon isochrones of the retreat of the Laurentide ice sheet. Univ. Wis. Dept. Meteorol. Tech. Rep. No. 35. 27 p.

Buffalo County Board of Supervisors. 1965. Buffalo County zoning ordinance. Courthouse, Alma, Wisconsin. 28 p.

Buol, S. A., and F. D. Hole. 1959. Some characteristics of clay skins on peds in the B horizon of a Gray-Brown Podzolic soil. Soil Sci. Soc. Amer. Proc. 23:239-241.

Buol, S. A., F. D. Hole, and R. J. McCracken. 1973. Soil genesis and classification. Iowa State University Press, Ames. 360 p.

Burley, M. W. 1964. The climate of Wisconsin, p. 143-148. *In* The Wisconsin Blue Book. Legislative Reference Bureau, State of Wisconsin, Madison.

Carroll, P. C. 1959. Genetic and morphological characteristics of Gray Wooded soils in Taylor County, Wisconsin. Unpublished report. U.S. Department of Agriculture, Soil Conservation Service, Madison, Wisconsin.

Carroll, P. C. 1964. Santiago soils. U.S. Department of Agriculture, Soil Conservation Service, Madison, Wisconsin. 68 p.

Catenhusen, J. 1950. Secondary successions in the peat lands of glacial Lake Wisconsin. Trans. Wis. Acad. Sci. Arts Letters 40:29-48.

Chamberlin, T. C. 1883. Geology of Wisconsin. The Commissioners of Public Printing, Madison, Wisconsin. 4 vol.

Chiang, S. L., and G. W. Petersen. 1970. Soil catena for hydrologic interpretations. J. Soil Water Conserv. 25:225-227.

Ciolkosz, E. J. 1964. Mineralogy of the very fine sand and silt separates of three loess-derived soils of Wisconsin. M.S. Thesis. Univ. Wisconsin. 124 p.

Ciolkosz, E. J. 1967. I. The mineralogy and genesis of the Dodge catena of southeastern Wisconsin. II. Rhizosphere weathering and synthesis of soil minerals. Ph.D. Thesis. Univ. Wisconsin. 133 p.

Collins, C. W. 1971. The influence of drumlin topography on field patterns in Dodge County, Wisconsin. Trans. Wis. Acad. Sci. Arts Letters 59:55-66.

Curtis, J. T. 1959. The vegetation of Wisconsin: an ordination of plant communities. University of Wisconsin Press, Madison. 657 p.

Dallman, J. 1968. Mastodons in Dane County. Wis. Acad. Rev. 15(2): 9-13.

Dalziel, I. W. D., and R. H. Dott, Jr. 1970. Geology of the Baraboo District, Wisconsin. Wis. Geol. Nat. Hist. Surv. Info. Circ. No. 14. 164 p., 7 pl.

Dana, M. H., J. Zimmerman, and K. Lettau. 1963. Lilac blossoming and Wisconsin weather. Wis. Acad. Rev. 10(1):33-36.

Denning, J. L., F. D. Hole, and J. Bouma. 1973. Ant (*Formica cinerea*) pedoturbation in a Typic Haplaquoll in Wisconsin, p. 69-105. *In* J. L. Denning, 1973. I. Measurement and calculation of hydraulic conductivity using physical and morphometric techniques. II. Ant pedoturbation in a poorly drained Calamine silty clay loam (Typic Haplaquoll). M.S. Thesis. Univ. Wisconsin. 105 p.

Dixon, J. B., and M. L. Jackson. 1960. Mineralogical analysis of soil clays involving vermiculite-chlorite-kaolinite differentiation. Clays Clay Minerals 8:274-286.

Dixon, R. M., and A. E. Peterson. 1971. Water infiltration control: a channel system concept. Soil Sci. Soc. Amer. Proc. 35:668-673.

Dokuchaev, V. V. 1879. Cartography of Russian soils, St. Petersburg. [All writings of V.V.D. are included in "DoKuchaev, Sochineniia," USSR Acad. Sci., Moscow, 1951.]

Dolan, E. F., Jr. 1959. Green universe, the story of Alexander von Humboldt. Dodd, Mead & Co., New York. 244 p.

Dosen, R. C., S. F. Peterson, and D. T. Pronin. 1950. Effect of ground water on the growth of red pine and white pine in central Wisconsin. Trans. Wis. Acad. Sci. Arts Letters 40:79-82.

Dott, R. H., Jr., and R. L. Batten. 1971. Evolution of the earth. McGraw-Hill Book Company, New York. 649 p.

Durand, L. 1962. The retreat of agriculture in Milwaukee County, Wisconsin. Trans. Wis. Acad. Sci. Arts Letters 51:197-218.

Dury, G. R., and J. C. Knox. 1971. Duricrusts and deep weathering profiles in southwestern Wisconsin. Science 174:291.

Dutton, C. E., and R. E. Bradley. 1970. Lithologic, geophysical and mineral commodity maps of precambrian rocks in Wisconsin. Misc. Geol. Invest. Map I-631. U.S. Geological Survey, Washington, D.C.

Ekern, P. C. 1950. Raindrop impact as the force initiating soil erosion. Ph.D. Thesis. Univ. Wisconsin. 81 p.

Emiliani, C. 1972. Quaternary paleotemperatures and the duration of the high-temperature intervals. Science 178:398-401.

Engel, M. S. and A. W. Hopkins. 1956. The prairie and its people. Wis. Agr. Exp. Sta. Bull. 520. 37 p.

Fanning, D. S., and M. L. Jackson. 1966a. Zirconium content of coarse silt from some Wisconsin soils and sediments. Soil Sci. 103: 253-260.

Fanning, D. S., and M. L. Jackson. 1966b. Clay mineral weathering in southern Wisconsin soils developed in loess and in shale-derived till. Clays Clay Minerals 13:175-191.

Farb, P. 1968. Man's rise to civilization as shown by the Indians of North America from primeval times to the coming of the industrial state. E. P. Dutton & Co., Inc., New York. 332 p.

Finch, V. C., G. T. Trewartha, A. H. Robinson, and E. H. Hammond. 1957. Elements of geography. McGraw-Hill Book Company, New York. 678 p.

Fitzpatrick, E. A. 1956. An indurated horizon formed by permafrost. J. Soil Sci. 7:248-254.

Flint, R. F. 1945. Glacial map of North America. 1st ed. 1:4,555,000. Geological Society of America, New York.

Flint, R. F. 1971. Glacial and quaternary geology. John Wiley & Sons, Inc., New York. 892 p.

Foss, J. E., and R. H. Rust. 1968. Soil genesis study of a lithologic discontinuity in glacial drift in western Wisconsin. Soil Sci. Soc. Amer. Proc. 32:393-398.

Fox, G. R. 1959. The prehistoric garden beds of Wisconsin and Michigan and the Fox Indians. Wis. Archeol. 40:1-19.

Franzmeier, D. P., and E. P. Whiteside. 1963. A chronosequence of Podzols in northern Michigan. Mich. State Univ. Agr. Exp. Sta. Quart. Bull. 46:2-20.

Frazier, B. E., and G. B. Lee. 1971. Characteristics and classification of three Wisconsin Histosols. Soil Sci. Soc. Amer. Proc. 35:776-780.

Frye, J. C., H. B. Willman, and R. F. Black. 1965. Outline of glacial geology of Illinois and Wisconsin, p. 43-61. *In* H. E. Wright, Jr., and D. G. Frey [eds.], The Quaternary of the United States. Princeton University Press, Princeton, N.J.

Gaikawad, S. T., and F. D. Hole. 1961. Characteristics and genesis of a Podzol soil in Florence County, Wisconsin. Trans. Wis. Acad. Sci. Arts Letters 50:183-190.

Gaikawad, S. T., and F. D. Hole. 1965. Characteristics and genesis of a gravelly Brunizemic regosol. Soil Sci. Soc. Amer. Proc. 29:483-485.

Gardner, W. R. 1971. Laboratory measurement of available soil water. Soil Sci. Soc. Amer. Proc. 35:852.

Geib, W. J., M. J. Edwards, E. H. Bailey, A. C. Anderson, T. J. Dunnewald, J. F. Fudge, O. L. Stockstad, and H. Chapman. 1925. Soil survey of Sauk County, Wisconsin. Wis. Geol. Nat. Hist. Surv. Bull. No. 60C, Soil Ser. No. 36. 32 p.

Gleason, H. A., and A. Cronquist. 1963. Manual of vascular plants of northeastern United States and adjacent Canada. D. Van Nostrand Co., Princeton, N.J. 810 p.

Glenn, R. C. 1959. Phosphate and silicate weathering during soil formation. Ph.D. Thesis. Univ. Wisconsin. 127 p.

Glenn, R. C., M. L. Jackson, and F. D. Hole. 1960. Chemical weathering of layer silicate clay in loess-derived Tama silt loam of southwestern Wisconsin. Clays Clay Minerals 8:63-83.

Glocker, C. L. 1966. Soils of the University Experimental Farm at Lancaster, Wisconsin. M.S. Thesis. Univ. Wisconsin. 214 p.

Glover, W. H. 1952. Farm and college. The College of Agriculture of the University of Wisconsin. A History. University of Wisconsin Press, Madison. 462 p.

Graven, E. H., O. J. Attoe, and D. Smith. 1965. Effect of liming and flooding on manganese toxicity in alfalfa. Soil Sci. Soc. Amer. Proc. 29:702-706.

Green, P. A. 1950. Ecological composition of high prairie relics in Rock County, Wisconsin. Trans. Wis. Acad. Sci. Arts Letters 40: 159-172.

Grittinger, T. F. 1971. Vegetational patterns and ordination in Cedarburg Bog, Wisconsin. Trans. Wis. Acad. Sci. Arts Letters 59:79-106.

Hamilton, T. D. 1963. Quantitative study of mass movements, southwestern Wisconsin. M.S. Thesis. Univ. Wisconsin. 104 p.

Harper, K. T. 1963. Structure and dynamics of the maple-basswood forest of southern Wisconsin. Ph.D. Thesis. Univ. Wisconsin. 131 p.

Harris, R. F., G. Chesters, and O. N. Allen. 1966. Dynamics of soil aggregation. Advances Agron. 18:107-169.

Haszel, O. 1968. Soil survey of Pierce County, Wisconsin. U.S. Department of Agriculture, Soil Conservation Service, Washington, D.C. 171 p., 102 map sheets.

Haszel, O. 1971. Soil survey of Walworth County, Wisconsin. U.S. Department of Agriculture, Soil Conservation Service, Washington, D.C. 107 p., 96 map sheets.

Hayes, O. E., A. G. McCall, and F. G. Bell. 1948. Investigations in erosion control and the reclamation of eroded land at the Upper Mississippi Valley Conservation Experiment Station near La Crosse, Wisconsin, 1933-1943. U.S. Dept. Agr. Tech. Bull. 973. 87 p.

Hilgard, E. W. 1892. A report on the relation of soils to climate. U.S. Dept. Agr. Meteorol. Bur. Bull. 2. 59 p.

Hobbs, H. W. 1963. Hydrologic data for experimental agricultural watersheds in the United States, 1956-59. U.S. Dept. Agr. Misc. Pub. 945. [Includes Soil Conservation Service map entitled, Soil resource regions and major land resources areas of the United States, coterminous 48 states.] 760 p. with map.

Hogan, J. D. 1961. Age and properties of buried paleosols and overlying loess deposit in southwestern Wisconsin. M.S. Thesis. Univ. Wisconsin. 100 p.

Hogan, J. D., and M. T. Beatty. 1963. Age and properties of a buried paleosol and overlying loess deposit in southwestern Wisconsin. Soil Sci. Soc. Amer. Proc. 27:345-350.

Hole, F. D. 1943. Correlation of the glacial border drift of north-central Wisconsin. Amer. J. Sci. 241:498-516.

Hole, F. D. 1953. Suggested terminology for describing soils as three-dimensional bodies. Soil Sci. Soc. Amer. Proc. 17:131-135.

Hole, F. D. 1955. Soil productivity ratings by towns in Wisconsin. Unpublished manuscript. Geological and Natural History Survey, University of Wisconsin Extension, Madison. 80 p.

Hole, F. D. 1956a. Soil survey of Grant County, Wisconsin. Wis. Geol. Nat. Hist. Surv. Bull. 80, Soil Ser. No. 55. 54 p.

Hole, F. D. 1956b. Soil survey of Waukesha County, Wisconsin. Wis. Geol. Nat. Hist. Surv. Bull. 81, Soil Ser. No. 56. 140 p.

Hole, F. D. 1961. A classification of pedoturbations and some other processes and factors of soil formation in relation to istropism and anisotropism. Soil Sci. 91:375-377.

Hole, F. D. 1962. History and progress in soil survey in Wisconsin. Wis. Acad. Rev. 9(4):167-169.

Hole, F. D. 1968. The Mindoro Cut. Soil Surv. Horizons 9(3):15-17.

Hole, F. D. 1971. Antipedogenic factors and human survival. Soil Surv. Horizons 12(2):11-15.

Hole, F. D. 1973. Soil association suitabilities analysis (with three maps at scale 1:500,000). Department of Administration, State of Wisconsin, Madison, 102 p.

Hole, F. D. 1974a. Wild soils of the Pine-Popple River Basin. Trans. Wis. Acad. Sci. Arts Letters 62:37-50.

Hole, F. D. 1974b. Soil association productivity analysis (with two maps at scale 1:500,000). Department of Administration, State of Wisconsin, Madison. 136 p.

Hole, F. D. 1975. Some relationships between forest vegetation and Podzol B horizons in soils of Menominee tribal lands, Wisconsin, U.S.A. Pochvovedeniye 12:115-124. [In Russian.]

Hole, F. D., and G. A. Nielsen. 1970. Soil genesis under prairie, p. 28-34. *In* Proceedings of a symposium on prairie and prairie restoration, 1968. Knox College, Galesburg, Illinois.

Hole, F. D., and K. O. Schmude. 1959. Soil survey of Oneida County, Wisconsin. Wis. Geol. Nat. Hist. Surv. Bull. 82, Soil Ser. No. 57. 140 p., colored map.

Hole, F. D., N. P. Dahlstrand, and R. J. Muckenhirn. 1947. Soils of Langlade County, Wisconsin. Wis. Geol. Nat. Hist. Surv. Bull. 74, Soil Ser. No. 52. 61 p., colored map.

Hole, F. D., F. D. Peterson, and G. H. Robinson. 1952. The distribution of soils and slopes on the major terraces of southern Richland County, Wisconsin. Trans. Wis. Acad. Sci. Arts Letters 41:73-81.

Hole, F. D., G. B. Lee, I. O Hembre, and D. C. Aebischer. 1953. What's in that soil map? Geological and Natural History Survey, University of Wisconsin Extension, Madison. 16 p.

Hole, F. D., G. W. Olson, K. O. Schmude, and C. J. Milfred. 1962. Soil survey of Florence County, Wisconsin. Wis. Geol. Nat. Hist. Surv. Bull. 84, Soil Ser. No. 59. 140 p., soil map.

Holstrom, B. K. 1972. Drainage area data for Wisconsin streams. File report. U.S. Geological Survey, Madison, Wisconsin. 74 p.

Holt, C. L. R., Jr., K. B. Young, and W. H. Cartwright. 1964. The water resources of Wisconsin, p. 178-198. *In* The Wisconsin Blue Book. Legislative Reference Bureau, State of Wisconsin, Madison.

Holt, C. L. R., Jr., R. D. Cotter, J. H. Green, and P. G. Olcott. 1970. Hydrogeology of the Rock-Fox River Basin of southeastern Wisconsin. Wis. Geol. Nat. Hist. Surv. Info. Circ. No. 17. 47 p.

Horn, M. E. 1957. Some characteristics of glacial drift soil parent material in Dodge County, Wisconsin. M.S. Thesis. Univ. Wisconsin. 93 p.

Horn, M. E. 1959. A pedological study of red clay soils and their parent materials in eastern Wisconsin. Ph.D. Thesis. Univ. Wisconsin. 238 p.

Isirimah, N. O. 1969. Chemical differentiation and nitrogen transformations of some Wisconsin organic soils. M.S. Thesis. Univ. Wisconsin. 127 p.

Jackson, M. L. 1964. Chemical composition of soils, p. 71-141. *In* F. E. Bear [ed.], Chemistry of the soil. 2d ed. Van Nostrand Reinhold Company, New York.

Jacobson, A. E. 1969. Characteristics of some Spodosols and associated soils of Washburn County, Wisconsin, with emphasis on the spodic horizon. M.S. Thesis. Univ. Wisconsin. 217 p.

Jakobson, L., W. E. Rennebohm, W. F. Weege, G. Roubal, D. E. Rosenbrook, and W. M. Larsen. 1963. Land use in Wisconsin. Department of Resource Development, State of Wisconsin, Madison. 58 p.

Jaehnig, M. E. W. 1968. A buried soil at Aztalan, Wisconsin. Unpublished manuscript. Geological and Natural History Survey, University of Wisconsin Extension, Madison. 17 p.

Janke, W. E. 1962. Characteristics and distribution of soil parent materials in the Valderan drift region of eastern Wisconsin. Ph.D. Thesis. Univ. Wisconsin. 152 p.

Jenny, H. 1961. E. W. Hilgard and the birth of modern soil science. Agrochimica, Pisa, Italy, and Farallon Publications, Berkeley.

Johnson, W. K., and D. Carley. 1963. A plan for Wisconsin. Department of Resource Development, State of Wisconsin, Madison. 103 p.

Jones, R. L., B. W. Ray, J. B. Fehrenbacher, and A. H. Beavers. 1967. Mineralogical and chemical characteristics of soils in loess overlying shale in northwestern Illinois. Soil Sci. Soc. Amer. Proc. 31:800-804.

Kaddou, N. S. 1960. Clay mineralogy of some alluvial soils of Iraq and a Dubuque silt loam and underlying dolomitic limestone of Wisconsin. Ph.D. Thesis. Univ. Wisconsin. 116 p.

Keller, T., and K. G. Watterston. 1962. Forest cutting and spread of sphagnum in northern Wisconsin. Trans. Wis. Acad. Sci. Arts Letters 51:79-82.

King, F. H. 1895. The soil. Macmillan, London. 303 p.

King, F. H. 1911. Farming of forty centuries. Democrat Printing Co., Madison, Wisconsin. 441 p.

Klingelhoets, A. J. 1959. Fayette soils, a summary of pertinent data for use in classification and interpretation. U.S. Department of Agriculture, Soil Conservation Service, Madison, Wisconsin. 39 p.

Klingelhoets, A. J. 1962. Soil survey of Iowa County, Wisconsin. U.S. Dept. Agr. Soil Conserv. Service Ser. 1958, No. 22. 101 p., 44 map sheets.

Knox, J. C. 1972. Valley alluviation in southwestern Wisconsin. Ann. Ass. Amer. Geog. 62:401-410.

Kowalke, O. L. 1952. Location of drumlins in the Town of Liberty Grove, Door County, Wisconsin. Trans. Wis. Acad. Sci. Arts Letters 41:15-16.

Kubiena, W. L. 1953. Bestimmungsbuch und Systematik der Böden Europas. Ferdinand Enke, Stuttgart. 392 p.

Lambert, J. L. 1970. The hydrologic and growth responses of a young pine plantation to weed removal. Ph.D. Thesis. Univ. Wisconsin. 45 p.

Lambert, J. L., and F. D. Hole. 1971. Hydraulic properties of an ortstein horizon. Soil Sci. Soc. Amer. Proc. 35:785-787.

Langton, J. E., and G. B. Lee. 1964. Characteristics and genesis of some organic soil horizons as determined by morphological studies and chemical analysis. Trans. Wis. Acad. Sci. Arts Letters 53:149-158.

Lapham, I. A. 1873. The great fires of 1871 in the northwest. Reprinted, 1963, in Wis. Acad. Rev. 12:6-9.

Lee, G. B. 1949. Soils and sites in the University Arboretum suited to the maple-basswood associates. M.S. Thesis. Univ. Wisconsin. 56 p.

Lee, G. B., and M. E. Horn. 1972. Pedology of the Two Creek section, Manitowoc County, Wisconsin. Trans. Wis. Acad. Sci. Arts Letters 60:183-199.

Lee, G. B., and B. Manoch. 1974. Macromorphology and micromorphology of a Wisconsin Saprist, p. 47-62. *In* A. R. Aandahl, S. W. Buol, D. E. Hill, and H. H. Bailey [eds.], Histosols, their characteristics, classification and use. Soil Sci. Soc. Amer. Spec. Pub. No. 6.

Lee, G. B., W. E. Janke, and A. J. Beaver. 1962. Particle size analysis of Valders drift in eastern Wisconsin. Science 138:154-155.

Lee, G. B., D. E. Parker, and D. A. Yanggen. 1972. Development of new techniques for delineation of flood plain hazard zones. Part 1: By means of detailed soil surveys. Final Tech. Rep., OWRR B-002-Wis. University of Wisconsin Water Resources Center, Madison.

Lee, G. B., A. J. Klingelhoets, P. H. Carroll, R. E. Fox, F. D. Hole, and C. J. Milfred. 1968. Classification of Wisconsin soils. Univ. Wis. Coll. Agr. Life Sci. Spec. Bull. 12. 41 p.

Leopold, A. 1943. Deer irruptions. Trans. Wis. Acad. Sci. Arts Letters 35:351-366.

Leopold, A. 1949. A sand county almanac. Oxford University Press, New York. 295 p.

Lewis, P. H., Jr. 1964. The landscape resources of Wisconsin, p. 130-142. *In* The Wisconsin Blue Book. Legislative Reference Bureau, State of Wisconsin, Madison.

Link, E. G., and O. R. Demo. 1970. Soil survey of Kenosha and Racine Counties, Wisconsin. U.S. Department of Agriculture, Soil Conservation Service, Washington, D.C. 113 p.

Love, J. R., A. E. Peterson, and L. E. Engelbert. 1960. Lime and fertilizer incorporation for alfalfa production. Trans. Wis. Acad. Sci. Arts Letters 49:161-169.

McColley, P. D. 1971. Interim soil survey report for Columbia County, Wisconsin. U.S. Department of Agriculture, Soil Conservation Service, Portage, Wisconsin. 187 p.

Madison, F. W. 1963. Mineralogic studies of some sandy soils in Wisconsin. M.S. Thesis. Univ. Wisconsin. 118 p.

Madison, F. W. 1972. Genesis and mineralogy of selected soils and soil parent materials in the Valderan drift region of eastcentral Wisconsin. Ph.D. Thesis. Univ. Wisconsin. 140 p.

Madison, F. W., and G. B. Lee. 1965. Some mineralogic characteristics of sandy soils in Wisconsin. Trans. Wis. Acad. Sci. Arts Letters 54:223-230.

Marbut, C. F. 1935. Soils of the United States, pt. 3. 98 p. *In* O. E. Baker [ed.], 1936, Atlas of American agriculture. U.S. Department of Agriculture, Washington, D.C.

Martin, L. 1932. The physical geography of Wisconsin. Wis. Geol. Nat. Hist. Surv. Bull. 36, Educ. Ser. No. 4. 608 p.

Miedema, R., and S. Slager. 1972. Micromorphological quantification of clay eluviation. J. Soil Sci. 23:309-314.

Milfred, C. J. 1966. Pedography of three soil profiles of Wisconsin representing Fayette, Tama and Underhill series. Ph.D. Thesis. Univ. Wisconsin. 227 p.

Milfred, C. J., and F. D. Hole. 1970. Soils of Jefferson County, Wisconsin. Wis. Geol. Nat. Hist. Surv. Bull. 86, Soil Ser. No. 61. 173 p.

Milfred, C. J., C. W. Olson, and F. D. Hole. 1967. Soil resources and forest ecology of Menominee County, Wisconsin. Wis. Geol. Nat. Hist. Surv. Bull. 85, Soil Ser. No. 60. 203 p., 3 maps.

Muckenhirn, R. J., L. T. Alexander, R. S. Smith, W. D. Shrader, F. F. Riecken, P. R. McMiller, and H. H. Krussekopf. 1955. Field descriptions and analytical data of certain loess-derived Gray-Brown Podzolic soils in the Upper Mississippi Valley. N. Cent. Region Pub. 46, Ill. Agr. Exp. Sta. Bull. 587. 80 p.

Muckenhirn, R. J., and NCR-3 Technical Committee on Soil Survey. 1960. Soils of the North Central Region of the United States. Univ. Wis. Agr. Exp. Sta. Bull. 544. 192 p., classification charts, colored map.

Murdock, J. T., and L. E. Engelbert. 1958. The importance of subsoil phosphorus to corn. Soil Sci. Soc. Amer. Proc. 22:53-57.

Musolf, G. E., and F. D. Hole. 1972. Lower Wisconsin River valley soil resources and use potentials. Trans. Wis. Acad. Sci. Arts Letters 60:201-209.

Nelson, A. R. 1973. Age relationships of the deposits of the Wisconsin Valley and Langlade glacial lobes of north-central Wisconsin. M.S. Thesis. Univ. Wisconsin. 105 p.

Nelson, L. B. 1942. The nature of two associated Wisconsin soils as influenced by postglacial erosion, topography and substratum. Trans. Wis. Acad. Sci. Arts Letters 34:63-72.

Nielsen, G. A., and F. D. Hole. 1963. A study of natural processes of incorporation of organic matter into soil in the University of Wisconsin Arboretum. Trans. Wis. Acad. Sci. Arts Letters 52:213-227.

Nielsen, G. A., and F. D. Hole. 1964. Earthworms and the development of coprogenous A1 horizons in forest soils of Wisconsin. Soil Sci. Soc. Amer. Proc. 28:426-430.

Noel, W. A., and F. D. Hole. 1958. Soil color as an indication of nitrogen content in some Wisconsin soils. Trans. Wis. Acad. Sci. Arts Letters 47:11-16.

Nygard, I. J., and F. D. Hole. 1949. Soil classification and soil maps: units of mapping. Soil Sci. 67:163-168.

Nygard, I. J., P. R. McMiller, and F. D. Hole. 1952. Characteristics of some Podzolic, Brown Forest and Chernozem soils of the northern portion of the Lake States. Soil Sci. Soc. Amer. Proc. 16:123-219.

Olcott, P. G. 1968. Water resources of Wisconsin Fox-Wolf River Basin. Hydrologic investigations atlas HA-321. U.S. Geological Survey, Washington, D.C.

Olson, G. W., and F. D. Hole. 1967. The fragipan in soils of northeastern Wisconsin. Trans. Wis. Acad. Sci. Arts Letters 56:174-184.

Orvedal, A. C. 1970. World great soil groups [map scale 1:75,000,000], p. 22-23. *In* E. B. Espenshade, Jr. [ed.], Goode's world atlas. 13th ed. Rand McNally & Company, Chicago.

Orvedal, A. C., and M. J. Edwards. 1941. General principles of technical grouping of soils. Soil Sci. Soc. Amer. Proc. 6:386-391.

Owens, D. W. 1968. Some characteristics of glacial till soil parent materials and their influences on soils in the Cary drift region of southern Wisconsin. M.S. Thesis. Univ. Wisconsin. 113 p.

Packer, B. G., and E. J. Delwiche. 1918. Farm making in upper Wisconsin; hints for the settler. Wis. Agr. Exp. Sta. Bull. 290. 71 p.

Palmquist, R. 1965. Geomorphic development of part of the Driftless Area, southwest Wisconsin. Ph.D. Thesis. Univ. Wisconsin. 182 p.

Parker, D. E., D. C. Kurer, and J. A. Steingraeber. 1970. Soil survey of Ozaukee County, Wisconsin. U.S. Department of Agriculture, Soil Conservation Service, Washington, D.C. 92 p., 41 photo maps.

Parker, M. E. 1969. Corrosion by soils, chapter 6. *In* National Association of Corrosion Engineers basic course. National Association of Corrosion Engineers, Houston, Texas.

Parsons, R. B., W. H. Scholtes, and F. F. Riecken. 1962. Soils of Indian mounds in northeastern Iowa as benchmarks for studies of soil genesis. Soil Sci. Soc. Amer. Proc. 26:491-496.

Peck, T. R., and G. B. Lee. 1961. Soil survey of Marquette County, Wisconsin. Wis. Geol. Nat. Hist. Surv. Bull. 83, Soil Ser. No. 58. 95 p.

Pedersen, E. J. 1954. Correlation and genesis of the Elliot series on Cary and Tazewell tills in southeastern Wisconsin and northeastern Illinois. M.S. Thesis. Univ. Wisconsin. 84 p.

Petersen, G. W. 1965. I. Comparison of the red glacio-lacustrine clay sediments of northern and eastern Wisconsin. II. A new procedure for the determination of calcite and dolomite. Ph.D. Thesis. Univ. Wisconsin. 106 p.

Petersen, G. W., G. B. Lee, and G. Chesters. 1968. A comparison of red clay glacio-lacustrine sediments in northern and eastern Wisconsin. Trans. Wis. Acad. Sci. Arts Letters 56:185-196.

Peterson, A. E. 1964. Factors affecting infiltration rates of Wisconsin soils, p. 12-17. *In* Papers presented at the Soil and Water Conservation and Use section, Farm and Home Conference. University of Wisconsin, Madison.

Peterson, A. E., and L. E. Engelbert. 1959. Growing corn in Wisconsin without plowing. Trans. Wis. Acad. Sci. Arts Letters 48: 135-140.

Phillip, J. 1970. Wetland soils: a review. Wis. Dept. Nat. Resour., Resour. Rep. 57. 27 p.

Pierce, R. S. 1951. Prairie-like mull humus, its physico-chemical and micro-biological properties. Soil Sci. Soc. Amer. Proc. 15:362-364.

Ranney, R. W. 1966. Soil forming processes in Glossoboralfs of westcentral Wisconsin. Ph.D. Thesis. Univ. Wisconsin. 114 p.

Ranney, R. W., and M. T. Beatty. 1969. Clay translocation and albic tongue formation in two Glossoboralfs of westcentral Wisconsin. Soil Sci. Soc. Amer. Proc. 33:768-775.

Riecken, F. F. 1965. Present soil-forming factors and processes in temperate regions. Soil Sci. 99:58-64.

Riecken, F. F., and E. Poetsch. 1967. Genesis and classification considerations of some prairie-formed soil profiles from local alluvium in Adair County, Iowa. Ia. Acad. Sci. Proc. 67:268-276.

Rieger, S. 1947. Loessial soil materials in southern Wisconsin, their characteristics and distribution. M.S. Thesis. Univ. Wisconsin. 25 p.

Ritzenthaler, R. E. 1953. Prehistoric Indians of Wisconsin. Milw. Public Mus. Pop. Sci. Handbook No. 4. 43 p.

Robinson, G. H. 1950. Soil carbonate and clay contents as criteria of rate of soil genesis. Ph.D. Thesis. Univ. Wisconsin. 114 p.

Robinson, G. H., and A. J. Klingelhoets. 1959. Soil survey of Richland County, Wisconsin. U.S. Dept. Agr. Soil Conserv. Service Ser. 1949, No. 9. 42 p., 51 map sheets.

Robinson, G. H., and A. J. Klingelhoets. 1961. Soil survey of Grant County, Wisconsin. U.S. Dept. Agr. Soil Conserv. Service Ser. 1951, No. 10. 98 p., 72 map sheets.

Robinson, G. H., and C. I. Rich. 1960. Characteristics of the multiple yellowish-red bands common to certain soils in the southeastern U.S. Soil Sci. Soc. Amer. Proc. 24:226-230.

Robinson, G. H., A. J. Vessel, R. A. Erickson, and F. D. Hole. 1958. Soil survey of Barron County, Wisconsin. U.S. Dept. Agr. Soil Conserv. Service, Ser. 1948, No. 1. 103 p., 4 map sheets.

Rolands, W. A. 1963. The history of rural zoning in Wisconsin. Wis. Acad. Rev. 10(2):55-57.

Ruhe, R. V., and W. H. Scholtes. 1956. Ages and development of soil landscapes in relation to climatic and vegetational change in Iowa. Soil Sci. Soc. Amer. Proc. 20:264-273.

Russell, E. J. 1961. Soil conditions and plant growth. Longmans, Green & Co., Ltd., London. 688 p.

Salem, M. Z., and F. D. Hole. 1968. Ant (*Formica exsectoides*) pedoturbation in a forest soil. Soil Sci. Soc. Amer. Proc. 32:563-567.

Schmude, K. O. 1971. Soil survey of Washington County, Wisconsin. U.S. Department of Agriculture, Soil Conservation Service, Washington, D.C. 105 p., 72 photo maps.

Scholz, H. F., and F. B. Trenk. 1959. Timber yields, wood increment, and composition of regeneration in a managed hardwood forest on morainal soils. Trans. Wis. Acad. Sci. Arts Letters 48:11-29.

Schorger, A. W. 1937. The range of the bison in Wisconsin. Trans. Wis. Acad. Sci. Arts Letters 30:117-130.

Schorger, A. W. 1954. The elk in early Wisconsin. Trans. Wis. Acad. Sci. Arts Letters 43:5-23.

Schorger, A. W. 1965. The beaver in early Wisconsin. Trans. Wis. Acad. Sci. Arts Letters 54:147-180.

Simonson, R. W. 1959. Outline of a generalized theory of soil genesis. Soil Sci. Soc. Amer. Proc. 23:152-156.

Slota, R. W. 1969. Soil survey of Vernon County, Wisconsin. U.S. Department of Agriculture, Soil Conservation Service, Washington, D.C. 86 p., 140 map sheets.

Slota, R. W., and G. D. Garvey. 1961. Soil survey of Crawford County, Wisconsin. U.S. Dept. Agr. Soil Conserv. Service Ser. 1958, No. 18. 85 p., 39 map sheets.

Small, T. W. 1973. The spatial distribution of hillslope form and process in selected Driftless Area watersheds, southwestern Wisconsin. Ph.D. Thesis. Univ. Wisconsin. 185 p.

Smith, H. T. U. 1949. Periglacial features in the Driftless Area of southern Wisconsin. J. Geol. 57:196-214.

Soil Conservation Service. 1952. Field and laboratory data on some Podzol, Brown Podzolic, Brown Forest and Gray Wooded soils in northern United States and southern Canada. U.S. Dept. Agr. Soil Surv. Lab. Mem. No. 1. 191 p.

Soil Conservation Service. 1960. Soil classification—a comprehensive system: 7th approximation. U.S. Department of Agriculture, Washington, D.C. 265 p.

Soil Conservation Service. 1963. Land resource regions and major land resource areas of the United States, map. U.S. Department of Agriculture, Washington, D.C.

Soil Conservation Service. 1967a. Soil survey laboratory data and descriptions for some soils of Wisconsin. U.S. Dept. Agr. Soil Conserv. Service Soil Surv. Invest. Rep. No. 17 [in cooperation with Wis. Agr. Exper. Sta.]. 225 p.

Soil Conservation Service. 1967b. Soils: distribution of principal kinds of soils; orders, suborders and great groups, sheet no. 86. *In* National atlas. U.S. Geological Survey, Washington, D.C.

Soil Conservation Service. 1967c. Supplement to the soil classification section (7th approximation). U.S. Department of Agriculture, Washington, D.C. 207 p.

Soil Conservation Service. 1969. Wisconsin soils, their properties and uses. U.S. Department of Agriculture, Madison, Wisconsin. 216 p.

Soil Conservation Service. 1970. Selected chapters from the unedited text of the soil taxonomy of the National Cooperative Soil Survey. Washington, D.C.

Soil Science Department. 1958. Mimeographed report of samples collected for the cooperative subsoil fertility project, October, 1955—December, 1957. College of Agriculture, University of Wisconsin, Madison.

Soil Survey Staff. 1951. Soil survey manual. U.S. Dept. Agr. Soil Conserv. Service Handbook No. 18. 503 p.

Soil Survey Staff. 1962. Supplement to agricultural handbook no. 18, soil survey manual: identification and nomenclature of soil horizons. U.S. Department of Agriculture, Soil Conservation Service, Washington, D.C. 16 p.

Stauber, J. L. 1956. Conservation—a banker views it, p. 63-69. *In* Resource training for business, industry, government. Natural Resources Study Committee, The Conservation Foundation, New York.

Steingraeber, J. A., and C. A. Reynolds. 1971. Soil survey of Milwaukee and Waukesha counties, Wisconsin. U.S. Department of Agriculture, Soil Conservation Service, Washington, D.C. 177 p., 138 photo maps.

Strahler, A. N. 1957. Quantitative analysis of watershed geomorphology. Amer. Geophys. Union Trans. 38:913-920.

Tamm, C. O., and H. G. Östlund. 1960. Radiocarbon dating of soil humus. Nature 185:207-212.

Tanner, C. B. 1964. Wisconsin water budget, p. 7-10. *In* Papers presented at the Soil and Water Conservation and Use section, Farm and Home Conference, University of Wisconsin, Madison.

Tanner, C. B., and E. R. Lemon. 1962. Radiant energy utilized in evapotranspiration. Agron. J. 54:207-212.

Thomas, D. D. 1964. Soil survey of Pepin County, Wisconsin. U.S. Dept. Agr. Soil Conserv. Service Ser. 1958, No. 29. 141 p., 16 map sheets.

Thomas, D. D., P. H. Carroll, and G. N. Wing. 1962. Soil survey of Buffalo County, Wisconsin. U.S. Dept. Agr. Soil Conserv. Service Ser. 1957, No. 13. 103 p., 45 map sheets.

Thornbury, W. D. 1965. Regional geomorphology of the United States. John Wiley & Sons, Inc. New York. 60 p.

Thorp, J. 1965. The nature of the pedological record in the Quaternary. Soil Sci. 99:1-8.

Thorp, J., and E. E. Gamble. 1972. Annual fluctuations of water levels in soils of the Miami catena, Wayne County, Indiana. Earlham Coll. Sci. Bull. No. 5. 26 p.

Thorp, J., and G. D. Smith. 1949. Higher categories of soil classification: order, suborder and great soils groups. Soil Sci. 67:77-191.

Threinen, C. W., and R. Poff. 1963. The geography of Wisconsin trout streams. Trans. Wis. Acad. Sci. Arts Letters 52:57-75.

Thwaites, F. T. 1929. Glacial geology of part of Vilas County, Wisconsin. Trans. Wis. Acad. Sci. Arts Letters 24:109-125.

Thwaites, F. T. 1943. Pleistocene of part of northeastern Wisconsin. Bull. Geol. Soc. Amer. 54:87-144.

Thwaites, F. T. 1947. Outline of glacial geology. Edwards Bros., Ann Arbor, Michigan. 129 p.

Thwaites, F. T. 1958. Land forms of the Baraboo District, Wisconsin. Trans. Wis. Acad. Sci. Arts Letters 47:137-159.

Thwaites, F. T. 1960. Evidence of dissected erosion surfaces in the Driftless Area. Trans. Wis. Acad. Sci. Arts Letters 49:17-50.

Thwaites, F. T., and K. Bertrand. 1957. Pleistocene geology of the Door Peninsula. Bull. Geol. Soc. Amer. 69:831-880.

Twenhofel, W. H. 1936. The greensands of Wisconsin. Econ. Geol. 31: 472-487.

Van Rooyen, D. J. 1972. I. Nature and significance of hydraulic properties of the Withee, Marshfield and Mann Soils in the drainage basin of the University of Wisconsin Agricultural Experiment Station, Marshfield, Wisconsin. II. Soils of the Lake Wingra Basin (a progress report for the International Biological Program). M.S. Thesis. Univ. Wisconsin. 130 p.

Van Rooyen, D. J. 1973. I. Organic carbon and nitrogen status in two Hapludalfs under prairie and deciduous forest, as related to moisture regime, some morphological features, and response to manipulation. II. Comparison of the hydrologic regimes of adjacent virgin and cultivated pedons at two sites. Ph.D. Thesis. Univ. Wisconsin. 176 p.

Vogl, R. J. 1964. The effects of fire on the vegetation composition of bracken grasslands. Trans. Wis. Acad. Sci. Arts Letters 51:67-82.

Walsh, L. M., and R. G. Hoeft. 1971. Survey of sulphur needs on alfalfa, p. 11-16. *In* Proceedings of the 1971 Wisconsin Fertilizer and Aglime Conference. Department of Soil Science, College of Agriculture, University of Wisconsin, Madison.

Wascher, H. L., B. W. Ray, J. D. Alexander, J. B. Fehrenbacher, A. H. Beavers, and L. R. Jones. 1971. Loess soils of northwest Illinois. Univ. Ill. Coll. Agr. Bull. 739. 112 p.

Watson, B. G. 1961. Characterization and classification of Morley and associated soils in southeastern Wisconsin. M.S. Thesis. Univ. Wisconsin. 113 p.

Watson, B. G. 1963. Characteristics and classification of glacial tills and associated soils in Columbia County, Wisconsin. U.S. Department of Agriculture, Soil Conservation Service, Madison, Wisconsin. 46 p., with glacial till and soil maps.

Watson, B. G. 1966. Soil survey of Lafayette County, Wisconsin. U.S. Dept. Agr. Soil Conserv. Service Ser. 1960, No. 27. 137 p., 100 photo maps.

Weaver, J. E., V. H. Hougen, and M. D. Weldon. 1935. Relation of root distribution to organic matter in prairie soils. Bot. Gaz. 96: 389-420.

Weaver, R. M. 1970. Montmorillonite genesis in soils as influenced by the activities of monosilicic acid and various cations in the matrix solution. Ph.D. Thesis. Univ. Wisconsin. 243 p.

Weidman, S. 1903. Preliminary report on the soils agricultural conditions of north central Wisconsin. Wis. Geol. Nat. Hist. Surv. Bull. 11, Econ. Ser. No. 7. 72 p.

Weidman, S. 1907. The geology of north central Wisconsin. Wis. Geol. Nat. Hist. Surv. Bull. 16, Scientific Ser. No. 4. 697 p., 2 maps.

Weidman, S. 1914. Soil survey of the south part of northwestern Wisconsin. Wis. Geol. Nat. Hist. Surv. Bull. 23, Econ. Ser. No. 14. 102 p.

Whitson, A. R. 1927. Soils of Wisconsin. Wis. Geol. Nat. Hist. Surv. Bull. 68, Soil Ser. No. 49. 270 p.

Whitson, A. R., and O. E. Baker. 1912. The climate of Wisconsin and its relation to agriculture. Univ. Wis. Agr. Exp. Sta. Bull. 223. 65 p.

Whitson, A. R., and H. W. Ullsperger. 1919. Marsh soils. Univ. Wis. Agr. Exp. Sta. Bull. 309. 32 p.

Whitson, A. R., T. J. Dunnewald, and C. Thompson. 1919. The soils of northern Wisconsin. Univ. Wis. Agr. Exp. Sta. Bull. 306. 45 p.

Whitson, A. R., W. J. Geib, G. W. Conrey, and A. E. Taylor. 1917. Soil survey of Dane County, Wisconsin. Wis. Geol. Nat. Hist. Surv. Bull. 53A, Soil Ser. No. 20. 86 p.

Whitson, A. R., W. J. Geib, G. W. Conrey, W. C. Boardman, and C. B. Post. 1918. Soil survey of Wood County, Wisconsin. Wis. Geol. Nat. Hist. Surv. Bull. 52B, Soil Ser. No. 17. 85 p.

Whitson, A. R., W. J. Geib, A. H. Meyer, H. G. Frost, and J. A. Chucka. 1928. Soil survey of Vernon County, Wisconsin. Wis. Geol. Nat. Hist. Surv. Bull. 60E, Soil Ser. No. 38. 43 p.

Whitson, A. R., W. J. Geib, T. J. Dunnewald, C. B. Post, W. C. Boardman, A. R. Albert, A. E. Taylor, L. R. Schoenmann, and Carl Thompson. 1916. Soil survey of north part of north central Wisconsin. Wis. Geol. Nat. Hist. Surv. Bull. 50, Soil Ser. No. 15. 78 p.

Whitson, A. R., W. J. Geib, T. J. Dunnewald, H. D. Chapman, Kenneth Whitson, F. J. O'Connell, A. C. Anderson, Ernest H. Bailey, and Max J. Edwards. 1929. Soil survey of Green Lake County, Wisconsin. Wis. Geol. Nat. Hist. Surv. Bull. 61C, Soil Ser. No. 41. 79 p.

Wilde, S. A. 1940. The subdivision of the Gray-Brown Podzolic soils from ecological and silvicultural viewpoints. Soil Sci. Soc. Amer. Proc. 5:336-340.

Wilde, S. A. 1954. Forest humus: its genetic classification. Trans. Wis. Acad. Sci. Arts Letters 43:137-163.

Wilde, S. A. 1965. Growth of Wisconsin coniferous plantations in relation to soils. Univ. Wis. Agr. Exp. Sta. Res. Bull. 262. 81 p.

Wilde, S. A., F. G. Wilson, and D. P. White. 1949. Soils of Wisconsin in relation to silviculture. Wis. Conserv. Dept. Pub. No. 525-49. 171 p.

Willman, H. B., H. D. Glass, and J. C. Frye. 1963. Mineralogy of glacial tills and their weathering profiles in Illinois. Part I. Glacial tills. Ill. Geol. Surv. Circ. 347. 55 p.

Willman, H. B., and J. C. Frye. 1970. Pleistocene stratigraphy of Illinois. Ill. Geol. Survey Bull. 94. 204 p.

Wilson, L. R., and R. M. Webster. 1942. Microfossil studies of three northcentral Wisconsin maps. Trans. Wis. Acad. Sci. Arts Letters 34:177-193.

Wisconsin Conservation Needs Committee. 1970. Wisconsin soil and water needs inventory. U.S. Department of Agriculture, Lincoln, Nebraska. 122 p.

Wojta, A. J., F. V. Burcalow, R. F. Johannes, and A. E. Peterson. 1960. Land forming. The Wojta system of surface drainage. Univ. Wis. Ext. Coll. Agr. Circ. 587. 8 p.

Wright, H. E., Jr., and R. V. Ruhe. 1965. Glaciation of Minnesota and Iowa, p. 29-41. *In* H. E. Wright, Jr., and D. G. Frey [eds.], The Quaternary of the United States. Princeton University Press, Princeton, N.J.

Wurman, E. 1961. The geologic material. Its impact on soil profile characteristics in westcentral Wisconsin. Trans. Wis. Acad. Sci. Arts Letters 50:191-202.

Yanggen, D. A., M. T. Beatty, and A. J. Brovold. 1966. Use of detailed soil maps for zoning. J. Soil Water Conserv. 21:123-126.

Yehle, L. A. 1954. Soil tongues and their confusion with certain indicators of periglacial climate. Amer. J. Sci. 252:532-546.

Young, H. E., P. N. Carpenter, and R. A. Altenberger. 1965. Preliminary tables of some chemical elements in seven tree species in Maine. University of Maine Agricultural Experiment Station, Orono. 89 p.

Zakrzewska, B. 1970. Paleo-geographic implications of clay ball deposits under Valderan till in eastern Wisconsin. Trans. Wis. Acad. Sci. Arts Letters 58:153-162.

Zeasman, O. R., and I. O. Hembre. 1963. A brief history of soil erosion control in Wisconsin. State Soil and Water Conservation Committee and the College of Agriculture, Madison, Wisconsin. 48 p.

Zicker, W. A. 1955. An analysis of Jefferson County vegetation using surveyors records and present day data. M.S. Thesis. Univ. Wisconsin. 56 p.

Index

DESIGNED BY IRVING PERKINS
COMPOSED BY FOX VALLEY TYPESETTING, MENASHA, WISCONSIN
MANUFACTURED BY KINGSPORT PRESS, KINGSPORT, TENNESSEE
PLATES 1-8 & FIGURE 5-1 (COLOR) SUPPLIED BY
THE GEOLOGICAL AND NATURAL HISTORY SURVEY, WISCONSIN
TEXT IS SET IN TIMES ROMAN, DISPLAY LINES IN NEWS GOTHIC

Library of Congress Cataloging in Publication Data
Hole, Francis Doan, 1913-
Soils of Wisconsin.
(Bulletin—Geological and Natural History
Survey ; 87) (Soil series ; 62)
Bibliography: pp. 207-13.
Includes index.
1. Soils—Wisconsin. I. Beatty, Marvin T.,
joint author. II. Lee, Gerhard Bjarne, 1917-
joint author. III. Title. IV. Series: Wisconsin.
Geological and Natural History Survey. Bulletin ; 87.
S599.W5H58 631.4'9775 75-12209
ISBN 0-299-06830-7